Antibacterial Agents

Antibacterial Agents

Chemistry, Mode of Action, Mechanisms of Resistance and Clinical Applications

ROSALEEN J. ANDERSON
Sunderland Pharmacy School, University of Sunderland, UK

PAUL W. GROUNDWATER
Faculty of Pharmacy, University of Sydney, Australia

ADAM TODD
Sunderland Pharmacy School, University of Sunderland, UK

ALAN J. WORSLEY
Department of Pharmacology and Pharmacy, The University of Hong Kong, Hong Kong SAR

A John Wiley & Sons, Ltd., Publication

This edition first published 2012
© 2012 John Wiley & Sons, Ltd

Registered office
John Wiley & Sons Ltd, The Atrium, Southern Gate, Chichester, West Sussex, PO19 8SQ, United Kingdom

For details of our global editorial offices, for customer services and for information about how to apply for permission to reuse the copyright material in this book please see our website at www.wiley.com.

Library of Congress Cataloging-in-Publication Data

Antibacterial agents : chemistry, mode of action, mechanisms of resistance,
and clinical applications / Rosaleen Anderson . . . [et al.].
     p. ; cm.
  Includes bibliographical references and index.
  ISBN 978-0-470-97244-1 (cloth) – ISBN 978-0-470-97245-8 (pbk.)
  I. Anderson, Rosaleen J.
  [DNLM: 1. Anti-Bacterial Agents. QV 350]

  615.7′922–dc23

                        2012006648

A catalogue record for this book is available from the British Library.

HB ISBN: 9780470972441
PB ISBN: 9780470972458

Set in 10/12pt Times by Thomson Digital, Noida, India.
Printed and bound in Singapore by Markono Print Media Pte Ltd.

FOR OUR FAMILIES

Contents

Preface

Since the introduction of benzylpenicillin (penicillin G) in the 1940s, it is estimated that over 150 antibacterials have been developed for use in humans, and many more for veterinary use. It is the use of antibacterials in the treatment of infections caused by pathogenic bacteria that led to them being labelled as 'miracle drugs', and, considering their often simple pharmacology, the effect that they have had upon infectious diseases and population health is remarkable. We are lucky enough to have been one of the generations for whom antibiotics have been commonly available to treat a wide variety of infections. In comparison, our grandparents were from an era where bacterial infection was often fatal and where chemotherapeutic agents were limited to the sulfonamides and antiseptic agents. This golden age of antibacterial agents may, however, soon come to an end as more and more bacteria develop resistance to the classes of antibacterial agents available to the clinician.

The timescale of antibacterial development occupies the latter half of the 20th century, with the introduction of the sulfonamides into clinical use in the 1930s, shortly followed by the more successful penicillin group of antibiotics. The discovery of penicillin by Sir Alexander Fleming in 1928, for which he received the Nobel Prize jointly with Howard Florey and Ernst Boris Chain, represents one of the major events in drug discovery and medicine. The subsequent development and wartime production of penicillin was a feat of monumental proportions and established antibiotic production as a viable process. This discovery prompted research which was aimed at discovering other antibiotic agents, and streptomycin (the first aminoglycoside identified) was the next to be isolated, by Albert Schatz and Selman Waksman in 1943, and produced on a large scale. Streptomycin became the first antibiotic to be used to successfully treat tuberculosis, for which every city in the developed world had had to have its own specialised sanatorium for the isolation and rudimentary treatment of the 'consumptive' infected patients. It was estimated at the time that over 50% of the patients with tuberculosis entering a sanatorium would be dead within 5 years, so the introduction of streptomycin again proved a significant step in the treatment of infectious disease.

The development of antibacterials continued throughout the latter part of the 20th century, with the introduction into the clinic of the cephalosporins, chloramphenicol, tetracyclines, macrolides, rifamycins, quinolones, and others. All of these agents have contributed to the arsenal of antibacterial chemotherapy and all have a specific action on the bacterial cell and thus selective activity against specific bacteria. We hope that this book will serve to highlight the development of the major antibacterial agents and the synthesis (where plausible) of these drugs. In addition, as health care professionals, we hope that students of medicinal chemistry, pharmacy, pharmaceutical sciences, medicine, and other allied sciences will find this textbook invaluable in explaining the known mechanisms of action of these drugs. We believe that knowledge of the mode of action and pharmacology of antibacterial agents is essential to our understanding of the multidrug approach to the treatment of bacterial infections. Several administered antibiotics acting upon different bacterial cell functions, organelles, or structures simultaneously can potentiate the successful eradication of infection. In addition, by understanding the action of the antibacterial agents at a cellular level, we are able to envisage those mechanisms involved in drug toxicity and drug interactions. As is demonstrated with the majority of the available therapeutic agents, antibacterial toxicities are observed with increased doses, as well as idiosyncratically in some patients and in combination with other therapeutic agents in the form of a drug interaction.

We have endeavoured to provide the major clinical uses of each class of antibiotic at the time of writing. As bacterial resistance may develop towards these therapeutic agents, and as other antibiotics are developed, the prescribed indications of these agents may change. Antibacterial prescribing worldwide is a dynamic process due to the emergence of resistance, and consequently some drugs have remained in clinical use, while others have 'limited' use.

In most developed countries in the world, the use of antibiotics is second to analgesic use, and with such extensive use, antibacterial resistance has inevitably become a major global concern. Rational prescribing of antibiotics is a key target for the World Health Organization, which endeavours to limit the use of antibiotics in an attempt to reduce the incidence of drug resistance. Despite these attempts, it is the nature of bacteria that resistance will inevitably occur to some agents, and this should prompt the further development of new antibiotics by the pharmaceutical industry. If we revisited the topic of antibiotic use, development, and mechanisms of action in 10–20 years (this is not to be taken as a hint as to when we might revise this book), we would hopefully find that several new drugs had been developed, while some of the classes with which we are familiar would have disappeared. Perhaps the clinical picture would appear to be similar, but drug treatments would probably have changed.

1

Introduction to Microorganisms and Antibacterial Chemotherapy

1.1

Microorganisms

Key Points

- Bacteria can be classified according to their staining by the Gram stain (Gram positive, Gram negative, and mycobacteria) and their shape.
- Most bacterial (prokaryotic) cells differ from mammalian (eukaryotic) cells in that they have a cell wall *and* cell membrane, have no nucleus or organelles, and have different biochemistry.
- Bacteria can be identified by microscopy, or by using chromogenic (or fluorogenic) media or molecular diagnostic methods (e.g. real-time polymerase chain reaction (PCR)).
- Bacterial resistance to an antibacterial agent can occur as the result of alterations to a target enzyme or protein, alterations to the drug structure, and alterations to an efflux pump or porin.
- Antibiotic stewardship programmes are designed to optimise antimicrobial prescribing in order to improve individual patient care and slow the spread of antimicrobial resistance.

1.1.1 Classification

There are two basic cell types: prokaryotes and eukaryotes, with prokaryotes predating the more complex eukaryotes on earth by billions of years. Bacteria are prokaryotes, while plants, animals, and fungi (including yeasts) are eukaryotes. For our purposes in the remainder of this book, we will further subdivide bacteria into Gram positive, Gram negative, and mycobacteria (we will discuss prokaryotic cell shapes a little later).

As you are probably already aware, we can use the Gram stain to distinguish between groups of bacteria, with Gram positive being stained dark purple or violet when treated with Gentian violet then iodine/potassium iodide (Figures 1.1.1 and 1.1.2). Gram negative bacteria do not retain the dark purple stain, but can be visualised by a counterstain (usually eosin or fuschin, both of which are red), which does not affect the Gram positive cells. Mycobacteria do not retain either the Gram stain or the counterstain and so must be visualised using other staining methods. Hans Christian Joachim Gram developed this staining technique in 1884, while trying to develop a new method for the visualisation of bacteria in the sputum of patients with pneumonia, but

Antibacterial Agents: Chemistry, Mode of Action, Mechanisms of Resistance and Clinical Applications, First Edition.
Rosaleen J. Anderson, Paul W. Groundwater, Adam Todd and Alan J. Worsley.
© 2012 John Wiley & Sons, Ltd. Published 2012 by John Wiley & Sons, Ltd.

Gentian violet

Eosin

Fuschin

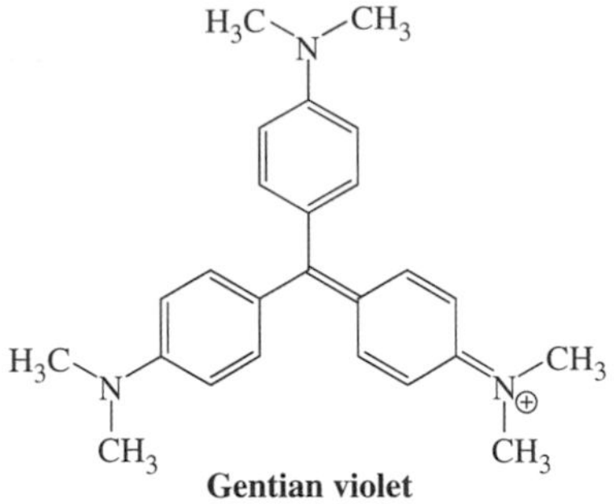

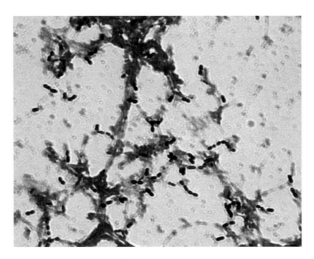

Figure 1.1.1 *Dyes used in the Gram stain*

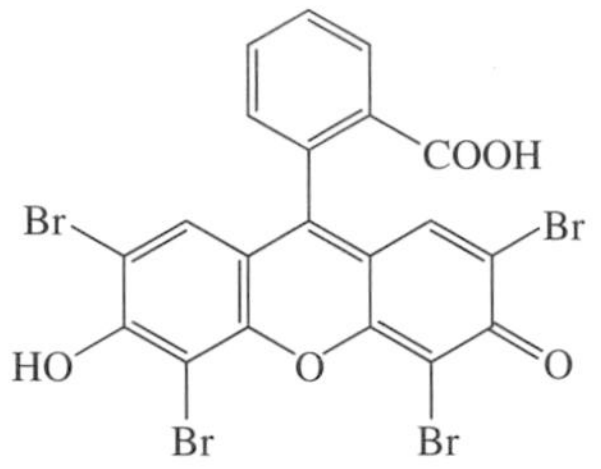

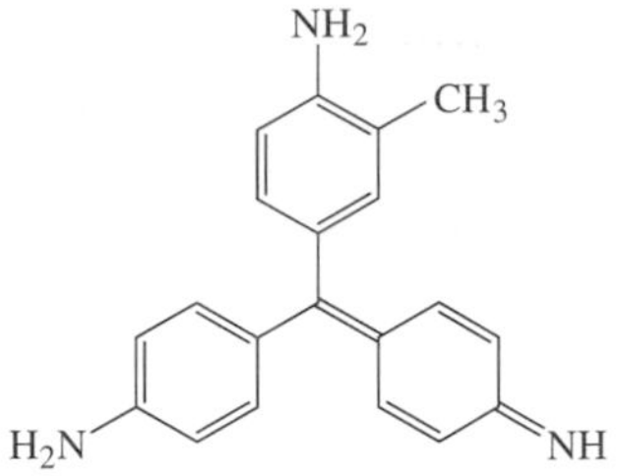

Figure 1.1.2 *Example of a Gram stain showing* ***Gram positive*** *(*Streptococcus pneumoniae*) and* ***Gram negative*** *bacteria (Image courtesy of Public Health Image Library, Image ID 2896, Online, [http://phil.cdc.gov/phil/home.asp, last accessed 26th March 2012].)*

the mechanism of staining, and how it is related to the nature of the cell envelopes in these different classes of bacteria, is still unclear.

Some of the Gram positive and Gram negative bacteria, as well as some mycobacteria, which we shall encounter throughout this book, are listed in Table 1.1.1.

1.1.2 Structure

The ultimate aim of all antibacterial drugs is selective toxicity – the killing of pathogenic[1] bacteria (bactericidal agents) or the inhibition of their growth and multiplication (bacteriostatic agents), without affecting the cells of the host. In order to understand how antibacterial agents can achieve this desired selectivity, we must first understand the differences between bacterial (prokaryote) and mammalian (eukaryote) cells.

[1] ‘Pathogenic’ means ‘disease-causing’.

Table 1.1.1 *Examples of Gram positive and Gram negative bacteria, and mycobacteria*

Gram positive	Gram negative	Mycobacteria
Bacillus subtilis	*Burkholderia cenocepacia*	*Mycobacterium africanum*
Enterococcus faecalis	*Citrobacter freundii*	*Mycobacterium avium* complex (MAC)
Enterococcus faecium	*Enterobacter cloacae*	*Mycobacterium bovis*
Staphylococcus epidermis	*Escherichia coli*	*Mycobacterium leprae*
Staphylococcus aureus	*Morganella morganii*	*Mycobacterium tuberculosis*
Meticillin-resistant *Staphylococcus aureus* (MRSA)	*Pseudomonas aeruginosa*	
Streptococcus pyogenes	*Salmonella typhimurium*	
Listeria monocytogenes	*Yersinia enterocolitica*	

The name 'prokaryote' means 'pre-nucleus', while eukaryote cells possess a true nucleus, so one of the major differences between bacterial (prokaryotic) and mammalian (eukaryotic) cells is the presence of a defined nucleus (containing the genetic information) in mammalian cells, and the absence of such a nucleus in bacterial cells. Except for ribosomes, prokaryotic cells also lack the other cytoplasmic organelles which are present in eukaryotic cells, with the function of these organelles usually being performed at the bacterial cell membrane.

A schematic diagram of a bacterial cell is given in Figure 1.1.3, showing the main features of the cells and the main targets for antibacterial agents. As eukaryotic cells are much more complex, we will not include a schematic diagram for them here, and will simply list the major differences between the two basic cell types:

- Bacteria have a cell wall and plasma membrane (the cell wall protects the bacteria from differences in osmotic pressure and prevents swelling and bursting due to the flow of water into the cell, which would occur as a result of the high intracellular salt concentration). The plasma membrane surrounds the cytoplasm and between it and the cell wall is the periplasmic space. Surrounding the cell wall, there is often a capsule (there is also an outer membrane layer in Gram negative bacteria). Mammalian eukaryotic cells only have a cell membrane, whereas the eukaryotic cells of plants and fungi also have cell walls.
- Bacterial cells do not have defined nuclei (in bacteria the DNA is present as a circular double-stranded coil in a region called the 'nucleoid', as well as in circular DNA plasmids), are relatively simple, and do not contain

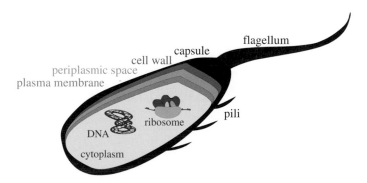

Figure 1.1.3 *Simplified representation of a prokaryotic cell, showing a cross-section through the layers surrounding the cytoplasm and some of the potential targets for antibacterial agents*

organelles, whereas eukaryotic cells have nuclei containing the genetic information, are complex, and contain organelles,[2] such as lysosomes.

- The biochemistry of bacterial cells is very different to that of eukaryotic cells. For example, bacteria synthesise their own folic acid (vitamin B9), which is used in the generation of the enzyme co-factors required in the biosynthesis of the DNA bases, while mammalian cells are incapable of folic acid synthesis and mammals must acquire this vitamin from their diet.

Whenever we discuss the mode of action of a drug, we will be focussing on the basis of any selectivity. As you will see from the section headings, we have classified antibacterial agents into those which target DNA (Section 2), metabolic processes (Section 3), protein synthesis (Section 4), and cell-wall synthesis (Section 5). In some cases, the reasons for antibacterial selectivity are obvious, for example mammalian eukaryotic cells do not have a peptidoglycan-based cell wall, so the agents we will discuss in Section 5 (which target bacterial cell-wall synthesis) should have no effect on mammalian cells. In other cases, however, the basis for selectivity is not as obvious, for example agents targeting protein synthesis act upon a process which is common to both prokaryotic and eukaryotic cells, so that in these cases selective toxicity towards the bacterial cells must be the result of a more subtle difference between the ribosomal processes in these cells.

We will now look at these antibacterial targets in detail, in preparation for our in-depth study of the modes of action of antibacterial agents and bacterial resistance in the remaining sections.

1.1.3 Antibacterial targets

1.1.3.1 DNA replication

DNA replication is a complex process, during which the two strands of the double helix separate and each strand acts as a template for the synthesis of complementary DNA strands. This process occurs at multiple, specific locations (origins) along the DNA strand, with each region of new DNA synthesis involving many proteins (shown in italics below), which catalyse the individual steps involved in this process (Figure 1.1.4):

- The separation of the two strands at the origin to give a replication fork (*DNA helicase*).
- The synthesis and binding of a short **primer** DNA strand (*DNA primase*).

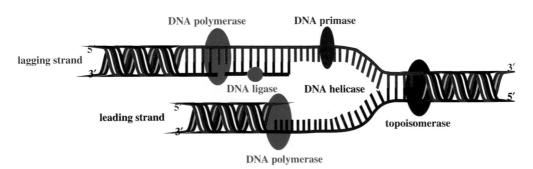

Figure 1.1.4 *DNA replication fork (adapted from http://commons.wikimedia.org/wiki/File:DNA_replication_en. svg, last accessed 7 March 2012.)*

---

[2] Specialised cellular subunits with a specific function.

- DNA synthesis, in which the base (A, T, C, or G) that is complementary to that in the primer sequence is added to the growing chain, as its triphosphate; this process is continued along the template strand, with the new base always being added to the $3'$-end of the growing chain (*DNA polymerase*) in the leading strand.
- The meeting and termination of replication forks.
- The proofreading and error-checking process to ensure the new DNA strand's fidelity; that is, that this strand (red) is exactly complementary to the template (black) strand (*DNA polymerase and endonucleases*).

Due to the antiparallel nature of DNA, synthesis of the strand that is complementary (black) to the lagging strand (red) must occur in the opposite direction, and this is more complex than the process which takes place in the leading strand.

DNA helicase is the enzyme which separates the DNA strands and in so doing, as a result of the right-handed helical nature of DNA, produces positive supercoils (knots) ahead of the replication site. In order for DNA replication to proceed, these supercoils must be removed by enzymes (known as topoisomerases) relaxing the chain. By catalysing the formation of negative supercoils, through the cutting of the DNA chain(s) and the passing of one strand through the other, these enzymes remove the positive supercoils and give a tension-free DNA double helix so that the replication process can continue. Type I topoisomerases relax DNA by cutting one of the DNA strands, while, you've guessed it, type II cut both strands (Champoux, 2001). In Section 2.1 we will look at a class of drugs which target the topoisomerases: the quinolone antibacterials, which, as DNA replication is obviously common to both prokaryotes and eukaryotes, must act on some difference in the DNA relaxation process between these cells.

1.1.3.2 Metabolic processes (folic acid synthesis)

As mentioned above, metabolic processes represent a key difference between prokaryotic and eukaryotic cells and an example of this is illustrated by the fact that bacteria require *para*-aminobenzoic acid (PABA), an essential metabolite, for the synthesis of folic acid. Bacteria lack the protein required for folate uptake from their environment, whereas folic acid is an essential metabolite for mammals (as it cannot be synthesised by mammalian cells and must therefore be obtained from the mammalian diet). Folic acid is indirectly involved in DNA synthesis, as the enzyme co-factors which are required for the synthesis of the purine and pyrimidine bases of DNA are derivatives of folic acid. If the synthesis of folic acid is inhibited, the cellular supply of these co-factors will be diminished and DNA synthesis will be prevented.

Bacterial synthesis of folic acid (actually dihydrofolic acid[3]) involves a number of steps, with the key steps shown in Schemes 1.1.1 and 1.1.2. A nucleophilic substitution is initially involved, in which the free amino group of PABA substitutes for the pyrophosphate group (OPP) introduced on to 6-hydroxymethylpterin by the enzyme 6-hydroxymethylpterinpyrophosphokinase (PPPK). In the next step, amide formation takes place between the free amino group of L-glutamic acid and the carboxylic acid group derived from PABA (Achari *et al.*, 1997).

Dihydrofolic acid (FH_2) is further reduced to tetrahydrofolic acid (FH_4), a step which is catalysed by the enzyme dihydrofolate reductase (DHFR), and FH_4 is then converted into the enzyme co-factors N^5,N^{10}-methylenetetrahydrofolic acid (N^5,N^{10}-CH_2-FH_4) and N^{10}-formyltetrahydrofolic acid (N^{10}-CHO-FH_4) (Scheme 1.1.2).

The tetrahydrofolate enzyme co-factors are the donors of one-carbon fragments in the biosynthesis of the DNA bases. Crucially, each time these co-factors donate a C-1 fragment, they are converted back to dihydrofolic acid, which, in an efficient cell cycle, is reduced to FH_4, from which the co-factors are regenerated. For example, in the biosynthesis of deoxythymidine monophosphate (from deoxyuridine monophosphate), the enzyme thymidylate synthetase utilises N^5,N^{10}-CH_2-FH_4 as the source of the methyl group introduced on to the pyrimidine ring (Scheme 1.1.3).

---

[3] The two hydrogens added to folic acid to give dihydrofolic acid are highlighted in purple in Scheme 1.1.1.

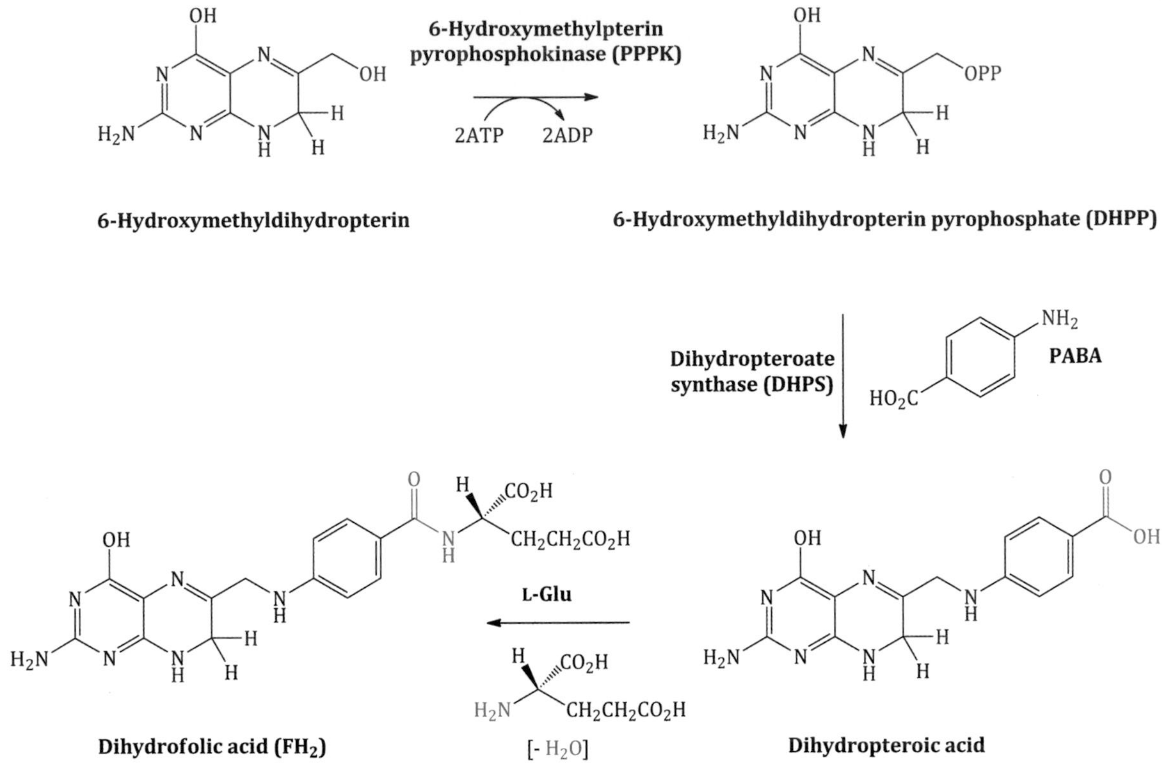

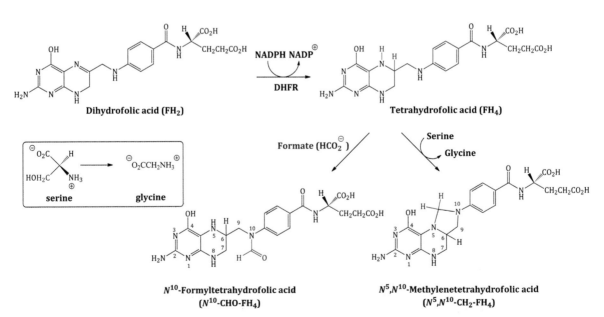

Scheme 1.1.1   *Bacterial synthesis of dihydrofolic acid*

Scheme 1.1.2   *Formation of the tetrahydrofolate enzyme co-factors*

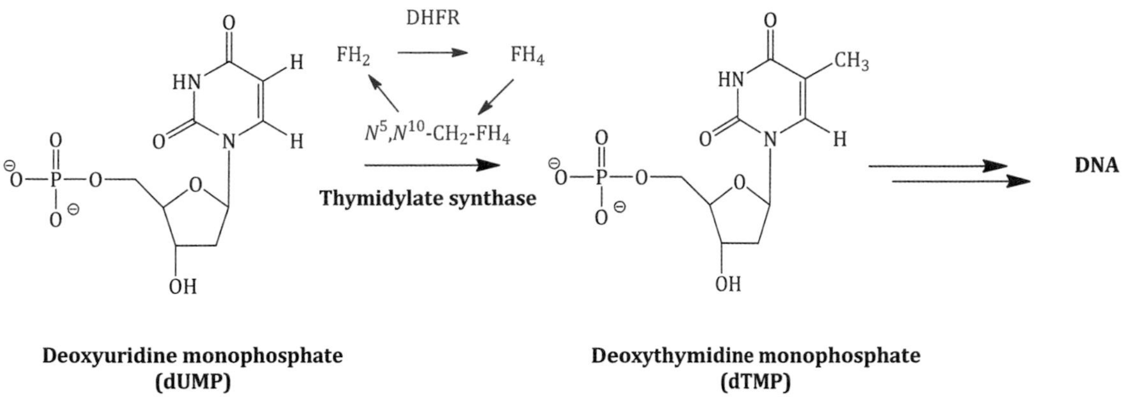

Scheme 1.1.3 *Biosynthesis of deoxythymidine monophosphate (dTMP)*

Similarly, N^{10}-CHO-FH$_4$ serves as the source of a formyl group in the biosynthesis of the purines and, once again, is converted to dihydrofolic acid (which must be converted to tetrahydrofolic acid and then N^{10}-CHO-FH$_4$, again in a cyclic process).

Cells which are proliferating thus need to continually regenerate these enzyme co-factors due to their increased requirement for the DNA bases. If a drug interferes with any step in the formation of these co-factors then their cellular levels will be depleted and DNA replication, and so cell proliferation, will be halted. In Section 3 we will look more closely at drugs which target these processes: the sulfonamides (which interfere with dihydrofolic acid synthesis) and trimethoprim (a DHFR inhibitor).

1.1.3.3 Protein synthesis

Protein synthesis, like DNA replication, is a truly awe-inspiring process, involving:

- Transcription – the transfer of the genetic information from DNA to messenger RNA (mRNA).
- Translation – mRNA carries the genetic code to the cytoplasm, where it acts as the template for protein synthesis on a **ribosome**, with the bases complementary to those on the mRNA being carried by transfer RNA (tRNA).
- Post-translational modification – chemical modification of amino acid residues.
- Protein folding – formation of the functional 3D structure.

Throughout this process, any error in transcription or translation may result in the inclusion of an incorrect amino acid in the protein (and thus a possible loss of activity), so it is essential that all of the enzymes involved in this process carry out their roles accurately. (For further information on protein synthesis, see Laursen *et al.*, 2005; Steitz, 2008.)

During **transcription**, DNA acts as a template for the synthesis of mRNA (Figure 1.1.5), a process which is catalysed by DNA-dependent RNA polymerase (RNAP), a nucleotidyl transferase enzyme (Floss and Yu, 2005; Mariani and Maffioli, 2009). In bacteria, the transcription process can be divided into a number of distinct steps in which the RNAP holoenzyme[4] binds to duplex promoter DNA to form the RNAP-promoter complex, then a series of conformational changes leads to local unwinding of DNA to expose the transcription start site. RNAP

---

[4] An apoenzyme is an enzyme which requires a co-factor but does not have it bound. A holoenzyme is the active form of an enzyme, consisting of the co-factor bound to the apoenzyme.

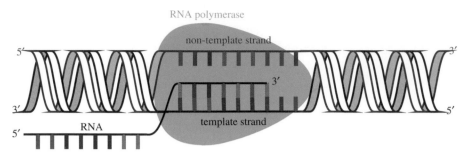

Figure 1.1.5   *DNA transcription*

can then initiate transcription, directing the synthesis of short RNA products, with synthesis of the RNA taking place in the $5' \rightarrow 3'$ direction (with the DNA template strand being read in the $3' \rightarrow 5'$ direction).

RNAP is a complex system, comprising five subunits ($\alpha_2\beta\beta'\omega$), each of which has a different function. The α subunits assemble the enzyme and bind regulatory factors, the β subunit contains the polymerase, the β' subunit binds non-specifically to DNA, and the ω subunit promotes the assembly of the subunits and constrains the β' unit. The core structure of RNAP is thought to resemble a crab's claw, with the active centre on the floor of the cleft between the two 'pincers', the β and β' subunits, and also contains a secondary channel, by which the nucleotide triphosphates access the active centre, and an RNA-exit channel (for a really good interactive tutorial showing the structure of RNAP, see http://www.pingrysmartteam.com/RPo/RPo.htm, last accessed 26 March 2012). Bacterial RNAP contains only these conserved subunits, while eukaryotic RNAP contains these and seven to nine other units (Ebright, 2000).

In bacteria, the transcription of a particular gene requires the binding of a further subunit, a σ factor (a transcription initiation factor), which increases the specificity of RNAP binding to a particular promoter region and is involved in promoter melting, and so results in the transcription of a particular DNA sequence. Once the assembly process is complete, the holoenzyme (the active form containing all the subunits: $\alpha_2\beta\beta'\omega\sigma$) catalyses the synthesis of RNA, which is complementary to the DNA sequence characterised by the σ factor (Figure 1.1.5) (eukaryotic RNAP also requires the binding of transcription factors, as do some bacterial RNAP). Proofreading of the transcription process is less effective than that involved in the copying of DNA, so this is the point in the transfer of genetic information which is most susceptible to errors. As we will see in Subsection 2.2.4, DNA-dependent RNA polymerase is the target of the rifamycin antibiotics.

Ochoa and Kornberg were awarded the Nobel Prize for Physiology or Medicine in 1959 'for their discovery of the mechanisms in the biological synthesis of ribonucleic acid and deoxyribonucleic acid' (http://nobelprize.org/nobel_prizes/medicine/laureates/1959/#, last accessed 26 March 2012).

Once the mRNA has been synthesised, it moves to the cytoplasm, where it binds to the ribosome, a giant ribonucleoprotein which catalyses protein synthesis from an mRNA template (**translation**). In 2009, Ramakrishnan, Steitz, and Yonath were awarded the Nobel Prize in Chemistry for their 'studies of the structure and function of the ribosome' (http://nobelprize.org/nobel_prizes/chemistry/laureates/2009/press.html, last accessed 26 March 2012).

The ribosome (Steitz, 2008), a large assembly consisting of RNA and proteins (ribonucleoproteins), has two subunits (30S and 50S in bacteria (complete ribosome 70S), 40S and 60S in eukaryotic cells (complete ribosome 80S)), and the large ribosome subunit has three binding sites, peptidyl-tRNA (P), aminoacyl-tRNA (A), and the exit (E) site, in the peptidyl transferase centre (PTC). Protein synthesis is initiated by the binding of a tRNA charged with methionine[5] to its AUG codon on the mRNA. tRNAs (or charged

---

[5] In prokaryotes and mitochondria, this methionine is formylated (NH-CHO); in eukaryotic cytoplasm, it is free methionine.

tRNAs) then carry amino acids to the ribosome site where mRNA binds. tRNA has three nucleotides which code for a specific amino acid (a triplet) and bind to the complementary sequence on the mRNA. The ribosome moves along the mRNA from the 5′- to the 3′-end and, once the peptide bond has formed, the non-acylated tRNA leaves the P site and the peptide-tRNA moves from the A to the P site. A new tRNA-amino acid (as specified by the mRNA codon) then enters the A site and the peptide chain grows as amino acids are added, until a stop codon is reached, when it leaves the ribosome through the nascent protein exit tunnel (Figure 1.1.6). One thing which has probably already occurred to you is that every protein does not have a methionine residue at its amino terminus; this is a result of modifications once the protein has been synthesised. In bacteria, the formyl group is removed by peptide deformylase and the methionine is then removed by a methionine aminopeptidase. Although you might not agree, this is actually a simplification of protein synthesis, which also involves other processes and species, including initiation factors, elongation factors, and release factors.

In Section 4 we will look at several drug classes which target protein synthesis by interfering with different aspects of the ribosomal translation process highlighted above. As with DNA replication, these antibiotics target processes common to both prokaryotes and eukaryotes and so any selectivity will be based on subtle differences in the structures of the ribosomes in the different cell types.

1.1.3.4   Bacterial cell-wall synthesis

As mentioned earlier, bacteria have a cell wall and a cell (plasma) membrane, while mammalian eukaryotic cells only have a cell membrane. The prokaryotic cell wall is composed of peptidoglycan (a polymer consisting of sugar and peptide units) and other components, depending upon the type of bacterium.

Gram positive bacteria (which are stained dark purple/violet by Gentian violet-iodine complex) are surrounded by a plasma membrane and cell wall containing peptidoglycan (Figure 1.1.7) linked to lipoteichoic acids (which consist of an **acylglycerol** linked via a **carbohydrate (sugar)** to a poly(**glycerophosphate**) backbone, Figure 1.1.8).

The cell wall of **Gram negative bacteria** is more complex. They have a plasma membrane and a thinner cell wall (peptidoglycan and associated proteins) surrounded by an outer membrane of phospholipid and lipopolysaccharide and proteins called porins (Figure 1.1.9). The outer membrane is thus the feature of the Gram negative cell wall which represents the greatest difference to that of Gram positive bacteria. The **lipopolysaccharide** (LPS) consists of: a phospholipid containing glucosamine rather than glycerol (**lipid A**[6]), a **core polysaccharide** (often containing some rather unusual sugars), and an *O*-antigen polysaccharide side chain (Figure 1.1.10). As this outer membrane poses a significant barrier for the uptake of any non-hydrophobic molecules, the outer membrane contains porins: protein pores which allow hydrophilic molecules to diffuse through the membrane. As a result of their more complex cell wall and membranes, Gram negative bacteria are not stained dark blue/violet by the Gram stain, but can be visualised with a counterstain (usually the pink dye fuschin).

Finally, **mycobacteria** have a structure which includes a cell wall (Figure 1.1.11), composed of peptidoglycan and arabinogalactan, to which are anchored mycolic acids (long-chain α-alkyl-substituted β-hydroxyacids which can contain cyclopropyl or alkenyl groups, as well as a range of oxygenated functional groups); see Figure 1.1.12. Mycobacteria are resistant to antibacterial agents that target cell-wall synthesis (such as the β-lactams).

---

[6] LPS is toxic and produces a strong immune response in the host. If Gram negative cell walls are broken by the immune system, the release of components of the cell wall containing the toxic lipid A results in fever and possibly septic shock.

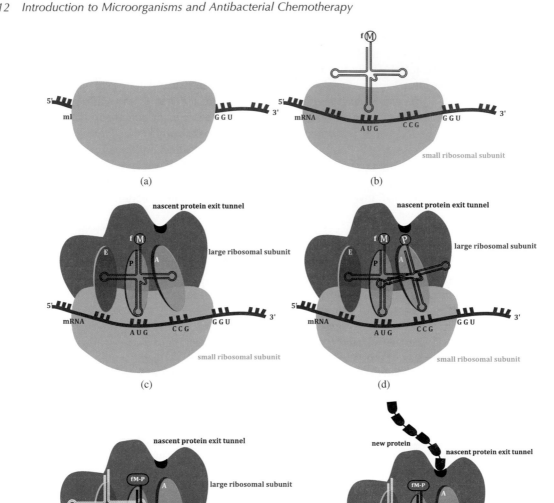

Figure 1.1.6   *The sequence of events leading to protein synthesis on the ribosome: (a) the small ribosomal subunit binds to the **mRNA** product of transcription; (b) the initiation complex is formed as the **initiator tRNA(formyl-methionine)** binds; (c) the **large ribosomal subunit** binds – tRNA(formyl-methionine) bound in the P (peptidyl-tRNA) site of the peptidyl transfer centre (the small subunit is transparent to allow a view of molecular events within the ribosome); (d) the mRNA codon (CCG) dictates that **tRNA(proline)**, with an anticodon of GGC, binds to the A (aminoacyl-tRNA) site of the peptidyl transfer centre; (e) a peptide bond forms between methionine (M) and proline (P), the ribosome moves along mRNA in the 5′ → 3′ direction, **tRNA bearing M-P** binds to the P site, leaving the A site free to bind the tRNA encoded by the next three bases of the mRNA. The exit (E site) binds the free tRNA before it exits the ribosome; (f) as the amino acids are added, the **new protein** exits the ribosome into the cytoplasm via the nascent protein exit tunnel*

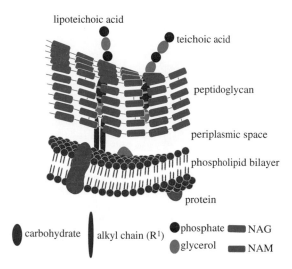

Figure 1.1.7 *Schematic representation of the plasma membrane and cell wall of Gram positive bacteria*

The common components of the bacterial cell wall and plasma membrane are thus a phospholipid bilayer and a peptidoglycan layer. You will probably already be familiar with the phospholipid bilayer, in which a membrane is formed by the association of the hydrophobic (nonpolar) lipid tails of the phospholipids with the external part of the bilayer consisting of the hydrophilic polar head groups (Figure 1.1.13).

We will concentrate here on the biosynthesis of peptidoglycan (the target for the antibacterial agents discussed in Section 5) and leave further discussion of the mycobacterial cell wall to Section 5.4 (Isoniazid).

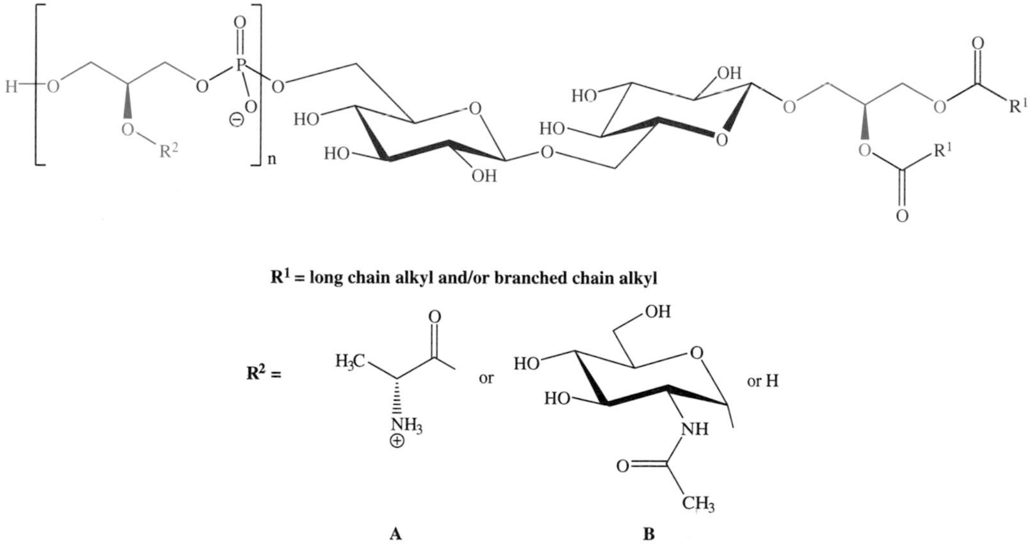

R¹ = long chain alkyl and/or branched chain alkyl

$R^2 =$ ... A ... or ... B ... or H

Figure 1.1.8 *General structure of the lipoteichoic acid from* Staphylococcus aureus *(n = 40–50; ratio of R^2 side-chains is D-Ala **A** (≈ 70%): N-acetylglucosamine **B** (≈ 15%): H (≈ 15%) (Reprinted from A. Stadelmaier, S. Morath, T. Hartung, and R. R. Schmidt, Angew. Chem. Int. Ed., 42, 916–920, 2003, with permission of John Wiley & Sons.)*

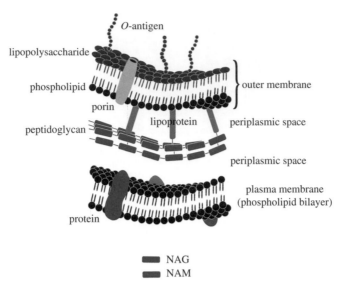

Figure 1.1.9 *Schematic representation of the plasma membrane, cell wall, and outer membrane of Gram negative bacteria*

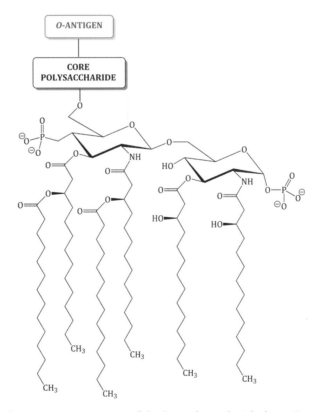

Figure 1.1.10 *Schematic representation of the lipopolysaccharide from Gram negative bacteria*

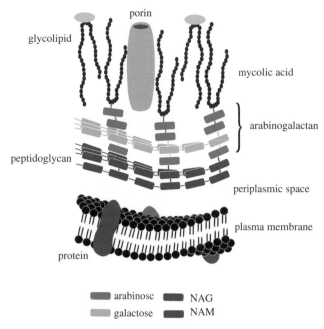

Figure 1.1.11 *Schematic representation of the plasma membrane, cell wall, and mycomembrane (mycolic acid layer) of mycobacteria*

Peptidoglycan (or murein) consists of parallel sugar backbones composed of alternating *N*-acetylglucosamine (NAG) and *N*-acetylmuramic acid (NAM) (Figure 1.1.14). As with cellulose fibres, these chains have strength in only one direction and, in order to form the peptidoglycan structure which will give the cell wall its rigidity, they must be crosslinked. This crosslinking takes place via peptide chains attached to the *N*-acetylmuramic acid residue through the carboxylic acid group. These chains are then linked together in a series of steps catalysed by the penicillin binding proteins (PBPs), enzymes which are located at the outer portion of the plasma membrane and have a range of activities, including: D-alanine carboxypeptidase (removal of D-ala from the peptidoglycan precursor), peptidoglycan transpeptidase, and peptidoglycan endopeptidase.

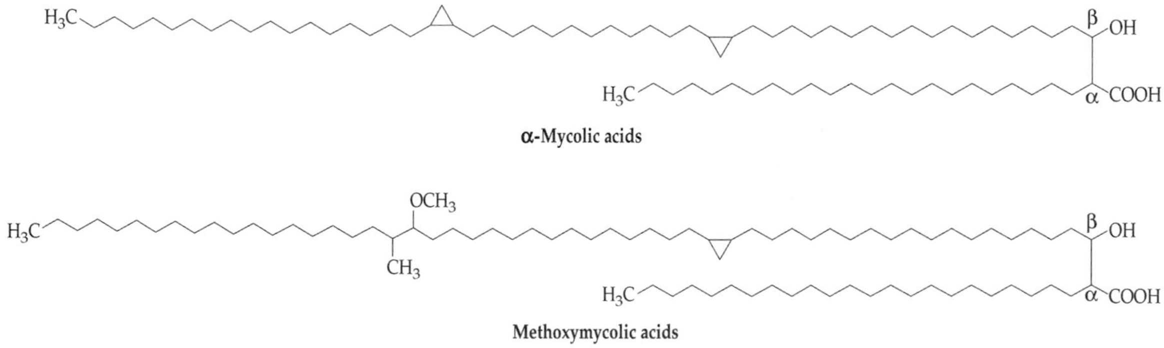

α-Mycolic acids

Methoxymycolic acids

Figure 1.1.12 *Examples of the structures of mycolic acids from mycobacterial cell walls (Langford et al., 2011)*

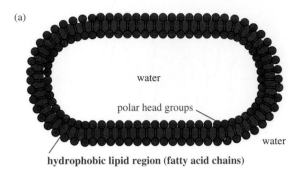

(i) **Phosphatidyl glycerol (monoanionic)**   (ii) **Phosphatidyl ethanolamine (zwitterionic)**

e. g. $R^1 = R^2 = C_{15}H_{31}$ (palmitoyl)

Figure 1.1.13   *(a) Phospholipid bilayer formation, with the solvated polar head groups extending into the aqueous layer and the fatty acid chains forming the hydrophobic region; and (b) examples of (i) an anionic **phospholipid** derived from **fatty acids** and glycerol and (ii) a zwitterionic (doubly-charged) **phospholipid** derived from **fatty acids**, glycerol, and **ethanolamine***

As can be seen from Scheme 1.1.4, crosslinking of the peptide side chains involves the PBP acting as a serine-acyl transpeptidase (a serine residue at the active site attacks the terminal D-Ala-D-Ala sequence to generate an acyl-enzyme intermediate, with the loss of the terminal D-Ala[7]). Being an ester, this intermediate is more reactive than the amide it replaced and is attacked by the amino group of a glycine residue to give the amide crosslink.

As this crosslinking process is occurring at many places along the peptide-NAG-NAM chains, the net result is a rigid scaffold which gives the cell wall its strength.

As we've just seen, the formation of crosslinks in Gram positive bacteria involves the attack of the *N*-terminal amino of a glycine (Gly) residue on an acyl-enzyme intermediate to give a new Gly-D-Ala bond. The Gly residue is the last in a run of five Gly residues on a side chain of the pentapeptide attached to NAM. This sequence of Gly residues is attached to the pentapeptide through a dibasic amino acid (lysine, Lys). In

---

[7] The loss of this D-Ala from the carboxyl end of the peptide chain is catalysed by D-alanine carboxypeptidase activity of the PBP, a carboxypeptidase being the enzyme which removes the amino acid group from the carboxyl terminus (the end with the free COOH) of a peptide.

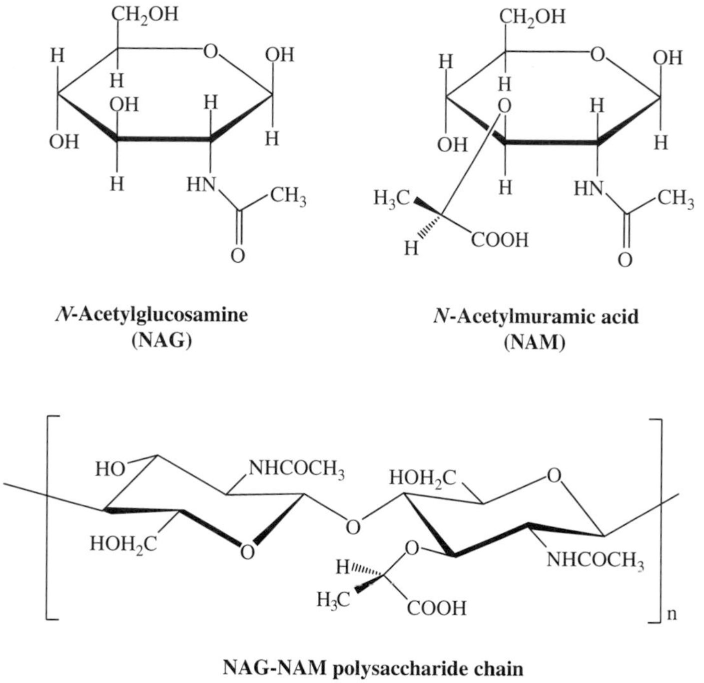

N-Acetylglucosamine
(NAG)

N-Acetylmuramic acid
(NAM)

NAG-NAM polysaccharide chain

Figure 1.1.14   *Structures of N-acetylglucosamine (NAG) and N-acetylmuramic acid (NAM) and a NAG-NAM polysaccharide chain*

Gram negative bacteria, this dibasic acid is mostly diaminopimelic acid (A2pm) and the crosslink formation involves the direct attack of the ε-amino group on the acyl-enzyme intermediate, as shown in Scheme 1.1.5.

The biosynthesis of the cell-wall precursors takes place within the cytoplasm and these are then transported across the plasma membrane to the periplasmic space (van Heijenoort, 2001) and ultimately to the growing cell wall where they are required (Figure 1.1.15). The lipid which carries the 'monomer' units across the plasma membrane is derived from **pyrophosphoryl undecaprenol** and is shown in Figure 1.1.16. Once this lipid has delivered the monomeric unit to the growing cell wall, it returns to the cytoplasm to recruit another monomeric unit.

In Section 5, we will look at antibiotics which target cell-wall synthesis. As cell walls are unique to prokaryotic cells, these agents have the potential for selective toxicity: killing the bacterial cells but not affecting the eukaryotic (mammalian) cells. You are probably already aware that the β-lactams target cell-wall synthesis – the fact that the penicillin binding proteins (a rather misleading name as the function of these proteins is to catalyse peptidoglycan synthesis) are involved in cell-wall synthesis is a bit of a giveaway. Other agents which target the different steps involved in peptidoglycan synthesis are D-cycloserine and the glycopeptides, vancomycin and teicoplanin.

1.1.4   Bacterial detection and identification

The detection and identification of bacteria is important in a variety of settings, including food hygiene, but we will concentrate here on the detection of pathogenic bacteria since, as we shall see, this is an increasingly important aspect of global health care systems.

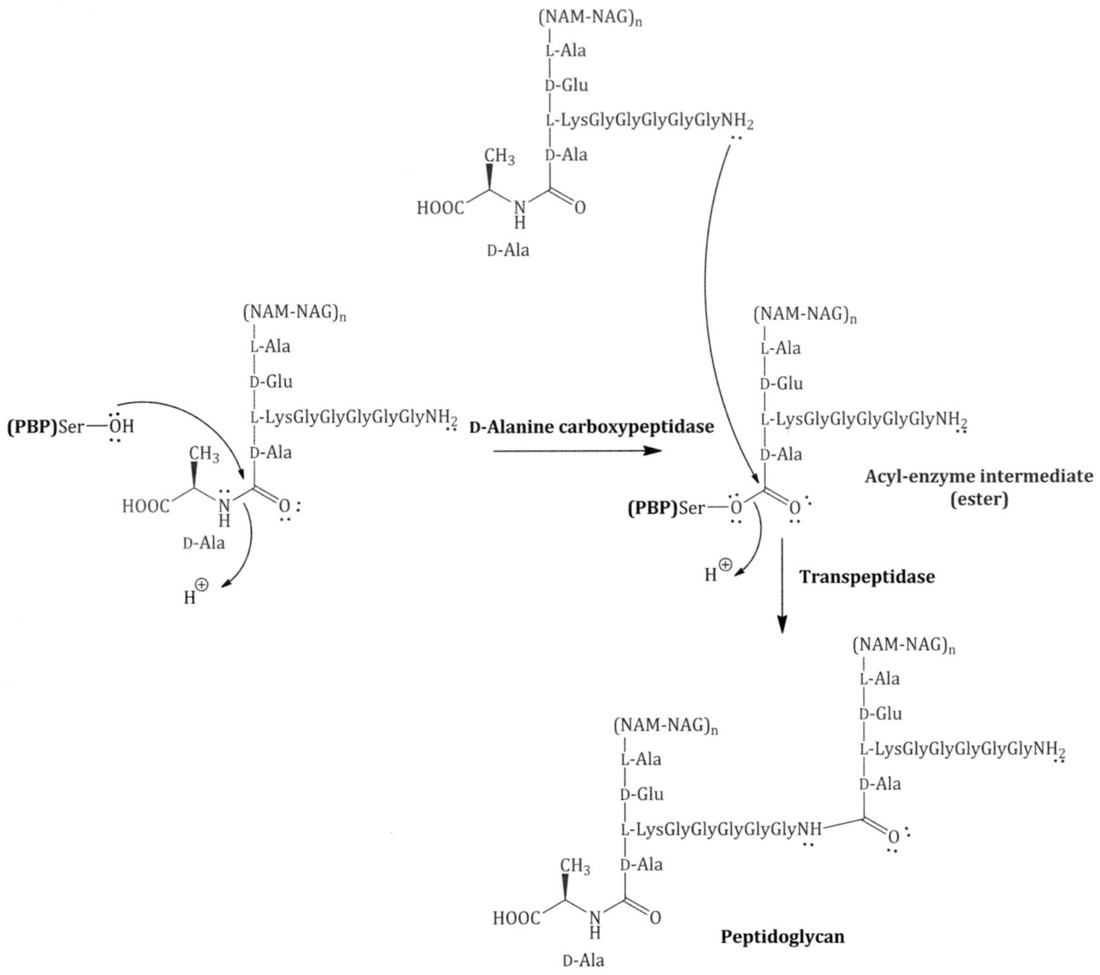

Scheme 1.1.4   *Penicillin binding protein (PBP)-catalysed peptidoglycan cross-coupling in Gram positive bacteria*

In the UK, health care-associated infections (HAIs), including multiresistant organisms (MROs), were estimated to cost the National Health Service (NHS) in excess of £1 billion in 2004 (National Audit Office Report, 2004) and it has been estimated that the overall direct cost of HAI in the United States in 2007 was between $36 and $45 billion. If 20% of these infections are preventable then an annual health care saving of between $5.7 and $6.8 billion could be achieved in the United States alone (Scott, 2009). The development of new surveillance methods, which are key components of effective infection prevention and control, is, therefore, essential. Rapid identification of bacterial pathogens would also inform a more effective directed clinical treatment of the infection.

MROs are an increasing clinical problem, with particular concerns being cross-infection of patients and the transmission of resistance between these bacteria, which could ultimately lead to strains with limited, or no, susceptibility to current antibacterial agents. For example, although the incidence of glycopeptide-resistant enterococci (GRE) is currently much lower in the rest of the world than in the USA (where more than 20% of enterococcal isolates are vancomycin resistant), the report in 2003 of the *in vivo* transmission of vancomycin resistance from GRE to meticillin-resistant *Staphylococcus aureus* (MRSA) highlights the significant risk

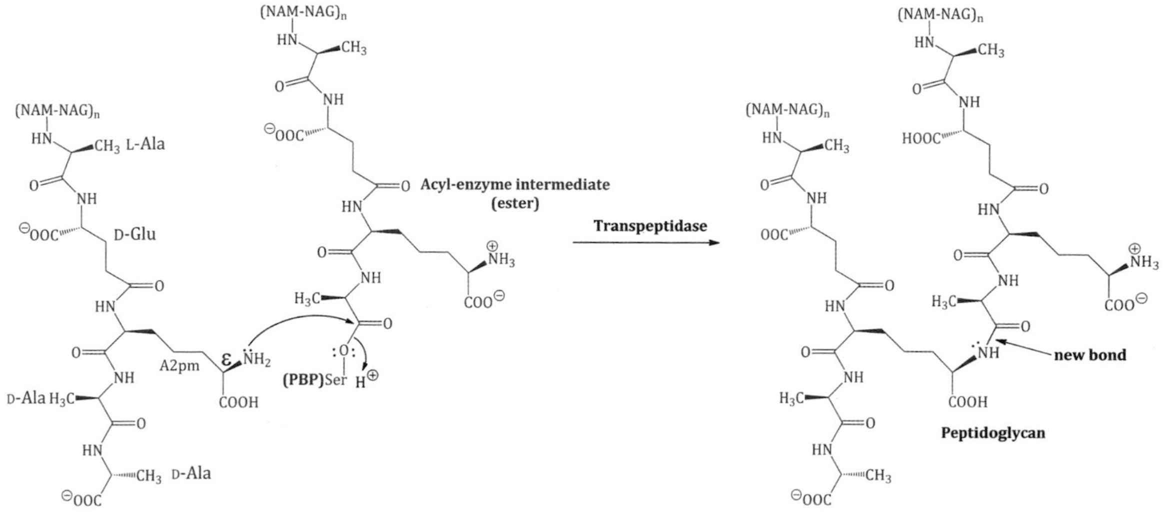

Scheme 1.1.5 *Synthesis of peptidoglycan crosslink in Gram negative bacteria*

associated with having co-existing, non-isolated infections due to these pathogens (Chang *et al.*, 2003). It should also be noted that MRSA is susceptible to very few agents, including the glycopeptides (vancomycin and teicoplanin), quinupristin-dalfopristin, and linezolid, and that cases of meticillin- and quinuprustin-dalfopristin-resistant *Staphylococcus aureus* have already been reported in Europe (Werner *et al.*, 2001).

Although there is no global consensus as to the most appropriate means of screening for MROs, timely active screening to identify colonised/infected patients should form the basis of an organism-specific approach to transmission-based precautions (NHMRC, 2010). Effective infection prevention relies upon rapid and reliable analysis of patient specimens and the introduction of contact precautions (such as patient isolation in a single-patient room or cohorting patients with the same strain of MRO in designated patient-care areas). For example, the use of a universal surveillance strategy was followed by a significant reduction in the rates of colonisation and infection of patients with MRSA (i.e. a change in the prevalence density from 8.91 to 3.88 per 10 000 patient days, compared to the case where no surveillance was undertaken) (Robicsek *et al.*, 2008).

From April 2009, the Care Quality Commission (www.cqc.org.uk) took over responsibility for health and social care regulation in the UK from the Healthcare Commission and 'The Health and Social Care Act 2008, Code of Practice for the NHS on the prevention and control of healthcare associated infections and related guidance', published in January 2009, describes how the CQC will assess compliance with the requirements regarding health care-associated infections, as set out in the Regulations made under Section 20(5) of this Act. Relevant NHS bodies must have, and adhere to, policies for the control of outbreaks and infections associated with both MRSA and *Clostridium difficile*, while acute NHS trusts must have similar policies for other specific alert organisms. With specific regard to MRSA, this policy should make provision for the screening of all patients on admission (including the screening of all elective admissions since March 2009 and the provision for screening of emergency admissions at presentation as soon as practical). This screening should then be used to inform the need for decontamination and/or isolation of colonised patients. Acute NHS trusts[8] must also

---

[8] A hospital is an acute NHS (or secondary care) trust.

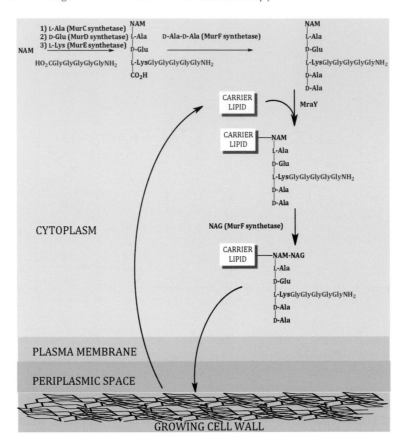

Figure 1.1.15   *Assembly of the peptidoglycan precursors in Gram positive bacteria and transport to the growing cell wall. The enzymes which catalyse each step are shown in brackets (MraY = phospho-N-acetylmuramoyl-pentapeptide transferase). The GlyGlyGlyGlyGly sequence is shown in red only, to indicate that it is not entirely clear when in the cytoplasmic processes this pentapeptide is added to the lysine residue)*

have policies for other specific alert organisms (for example, glycopeptide-resistant enterococci (GRE), Acinetobacter and other antibiotic-resistant bacteria, and tuberculosis (TB), including multidrug-resistant TB (MDR-TB)) (Health and Social Care Act, 2008; Groundwater *et al.*, 2009).

The Health Protection Agency publishes data derived from the mandatory surveillance of MRSA, *C. diff.*, and VRE bacteraemia, and the data show that January–March 2009 saw a 2.1% increase in MRSA bacteraemia compared to the previous quarter (but a reduction of 29% compared to the corresponding quarter in 2008),

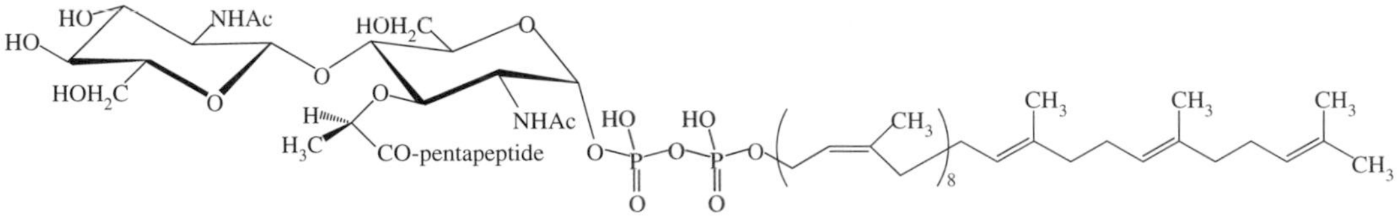

Figure 1.1.16   *Lipid II, the immediate peptidoglycan precursor (Reprinted from J. van Heijenoort, Nat. Prod. Rep., 18, 503–520, 2001, with permission of the RSC.)*

while there was a 34% decrease in the number of reported MRSA bacteraemias in the financial year 2008/2009 (HPA Mandatory Surveillance Report, 2008).

The need for rapid and simple methods for the detection of pathogenic bacteria, such as MRSA, GRE, NDM-1 metallo-β-lactamase producing organisms, *Pseudomonas aeruginosa*, Group B *streptococci* (*Streptococcus agalactiae*), and *Acinetobacter baumannii*, is hopefully self-evident.

Traditionally, pathogenic bacteria are detected by Gram staining and microscopy and/or on the basis of their colonial appearance, after inoculation of a culture medium, which facilitates the growth of a wide range of organisms.

Prokaryotes have various shapes (Figure 1.1.17), and these, together with their appearance after the Gram stain, are used for their initial identification. The four basic shapes are:

- Cocci (spherical);
- Bacilli (rod-shaped);
- Spirochaetes (spirals);
- Vibrio (comma-shaped).

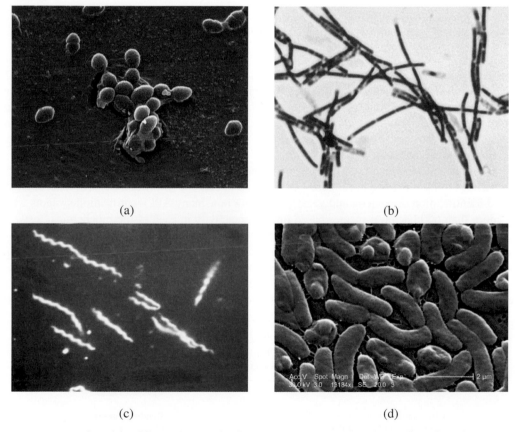

(a)　　　　　　　　　　　　　　　　(b)

(c)　　　　　　　　　　　　　　　　(d)

Figure 1.1.17 *Examples of the different bacterial cell shapes: (a) Cocci (*Enterococcus faecalis *(photo ID12803)); (b) Bacilli (*Bacillus anthracis *(photo ID1064)); (c) Spirochaetes (*Borrelia Burgdorferi *(ID6631)); (d) Vibrio (*Vibrio vulnificus *(ID7815)) (Image courtesy of Public Health Image Library, Images a) ID12803, b) ID1064, c) 6631 d) 7815, Online, [http://phil.cdc.gov/phil/home.asp, last accessed 29th March 2012].)*

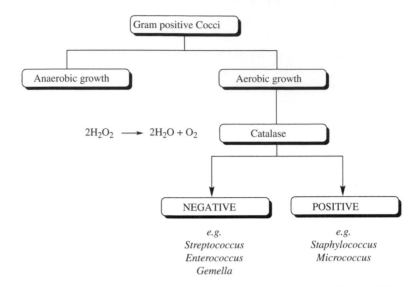

$$2H_2O_2 \longrightarrow 2H_2O + O_2$$

Figure 1.1.18   *BSOP ID 1 Identification Flowchart for Gram positive Cocci* (**catalase activity is detected via the production of oxygen upon addition of hydrogen peroxide**)

Bacterial identification requires the skills of an experienced clinical microbiologist and often requires further testing of commensal bacteria, which may have similar morphological characteristics to pathogenic bacteria. In the UK, bacteria are identified according to the National Standards (*Introduction to the Preliminary Identification of Medically Important Bacteria, BSOP ID 1*; http://www.hpa-standardmethods.org.uk/documents/bsopid/pdf/bsopid1.pdf, last accessed 30 April 2012) and a typical flowchart for bacterial identification is shown in Figure 1.1.18.

In the 1970s, the introduction of API strips, which consist of a series of miniaturized biochemical tests (such as the catalase activity mentioned in Figure 1.1.18), used in conjunction with extensive databases, allowed more rapid identification of bacteria and yeasts. There are now many API identification systems which can identify more than 600 bacterial species based on their reactivity in each of the biochemical tests.

Specific chromogenic media, in which a non-coloured enzyme substrate (a targeting molecule linked to a chromogenic compound) is added to the culture medium, have been employed for over 20 years in the detection of pathogenic bacteria (Figure 1.1.19) (Perry and Freydière, 2007). Ideally, this is a substrate for an enzyme which is unique to a particular bacterium, and cleavage of a key bond liberates a chromogen, which can be detected against a background of other, colourless colonies (as these do not contain the requisite enzyme for

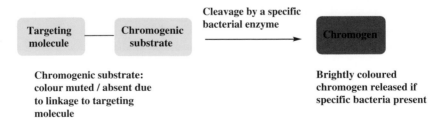

Figure 1.1.19   *The principle behind the chromogenic detection of bacteria*

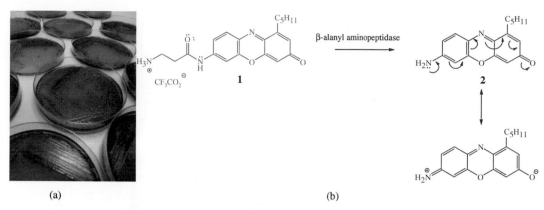

(a)   (b)

Figure 1.1.20   *(a) Detection of* Pseudomonas aeruginosa *colonies using chromID (picture courtesy of Larissa Laine, Freeman Hospital, Newcastle upon Tyne, UK). (b) P.* aeruginosa *and the origin of the purple colour (Reprinted from A. V. Zaytsev, R. J. Anderson, A. Bedernjak, et al., Org. Biomol. Chem., 8, 682–692, 2008, with permission of the RSC.)*

cleavage of the chromogenic substrate). Often more than one chromogenic substrate can be employed in a single culture plate to help in the differentiation of commensal and pathogenic bacteria.

Among the benefits of the use of chromogenic media are that they can be sufficiently specific that no further testing is required, and that they can give indicative colours for bacterial colonies, although the time required (usually 24–48 hours) for the growth of the colonies (and so the development of colour) is a limiting factor in the development of a rapid test that could be applied to all patients on admission to hospital. Fluorogenic media (in which a fluorescent compound is released upon enzymatic cleavage) offer the possibility of more rapid bacterial detection and the simultaneous detection of more than one bacterium, if fluorogens with different emission wavelengths are used to target different enzymatic activities.

Chromogenic media have been employed for the detection of MRSA (Perry *et al.*, 2004), VRE (Randall *et al.*, 2009), ESBL-producing organisms (Ledeboer *et al.*, 2007), and *C. diff.* (Perry *et al.*, 2010). A recent example of a medium that is selective for the detection of *P. aeruginosa*, the most common respiratory pathogen in patients with cystic fibrosis, employs the pale yellow-coloured β-alanyl-1-pentylresorufamine **1**, which is selectively cleaved by β-alanyl aminopeptidase (an enzyme specific to *P. aeruginosa*, *Burkholderia cenocepacia*, and *Serratia marcescens*) to give 1-pentylresorufamine **2**, which is retained within the bacteria and gives rise to purple colonies with a metallic sheen, which are easily detected by the naked eye (Figure 1.1.20) (Zaytsev *et al.*, 2008).

Molecular diagnostic methods (Tenover, 2007) offer the advantage that they are more rapid (results can typically be obtained within a few hours), can be highly specific, and, like the automated culture-based methods, can be performed in closed systems and have the capacity for automation. For example, one real-time multiplex PCR method[9] for the detection of MRSA in a mixture of staphylococci employs DNA primers that are specific for the *mecA*- and *S. aureus*-specific *orfX* genes (to allow for variations in the staphylococcal cassette chromosome (SCC*mec*)), in addition to a series of molecular beacon probes for the detection of the single-stranded PCR product (Figure 1.1.21) (Huletsky *et al.*, 2004). Molecular beacons are single-stranded probes which have a specific DNA recognition sequence with a fluorescent dye at one end and a quencher at the other. If the recognition sequence matches the DNA sequence, the probe opens up and binds to the DNA, and

---

[9] For an excellent description of PCR, see the Dolan DNA Learning Center's Web site (http://www.youtube.com/DNALearningCenter#p/f/9/2KoLnIwoZKU, last accessed 30 March 2012).

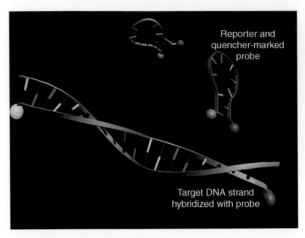

Figure 1.1.21   *How a molecular beacon works (Reproduced from 'Molecular Beacons', Wikimedia, Online, [http://en.wikipedia.org/wiki/File:Molecular_Beacons.jpg].)*

the resulting separation between the fluorescent dye and quencher means that the fluorescence is no longer quenched.

The SCC*mec* is a mobile genetic element that carries the *mecA*[10] gene (which encodes the β-lactam-resistant penicillin binding protein PBP2′ (see Section 5)). In this study, of the 1657 MRSA isolates, 98.7% were detected in under 1 hour using this technique, with only 26 of the 569 meticillin-susceptible *S. aureus* (MSSA) strains being mistakenly identified as MRSA. Real-time PCR assays are available for the detection of MRSA (Cepheid Xpert MRSA (CA, USA) and BD GeneOhm MRSA (CA, USA)), based on the *mecA* gene, and of vancomycin-resistant enterococci (VRE) (BD GeneOhm VanR assay (CA, USA)), based upon the *VanA* gene. As we shall see later (in Section 5), the *VanA* phenotype produces a D-Ala-D-Lactate ligase which synthesises an ester (D-Ala-D-Lactate) rather than an amide (D-Ala-D-Ala) – this ester can still act as a substrate in peptidoglycan (cell wall) synthesis but has 1000-fold reduced binding affinity for vancomycin, which exerts its antibiotic effect by binding to the terminal D-Ala-D-sequence.

Another apparently appealing technique is the Matrix-Assisted Laser Desorption Ionisation-Time-of-Flight (MALDI-TOF) Mass Spectrometry (MS) detection of microorganisms (Hsieh *et al.*, 2008), but this has so far not been able to identify mixed organisms in the same blood culture specimen, requires a number of sample handing procedures, and, like PCR, requires expensive instrumentation and expert operators/analysts. In this technique, unlike most MS applications, the compounds giving rise to the individual MS peaks are not identified, but spectral fingerprints are obtained (which vary between microorganisms) (Figure 1.1.22). Among the compounds detected in the spectrum, some peaks (molecular masses) are specific to genus or species (and sometimes to subspecies). The mass spectra obtained are reproducible as long as the bacteria are grown and harvested under the same conditions, and protein extraction and analysis are also performed under standard conditions.

For example, a number of studies have attempted to differentiate between MRSA and MSSA based upon their MS profiles, which are different (with MRSA containing more peaks); however, no specific profile for MRSA has been identified (but individual strains do have very similar profiles) (Hsieh *et al.*, 2008).

---

[10] Don't worry about these genes and their functions just now. We'll discuss them more fully as they become important when discussing bacterial resistance to the different antibiotic drug classes.

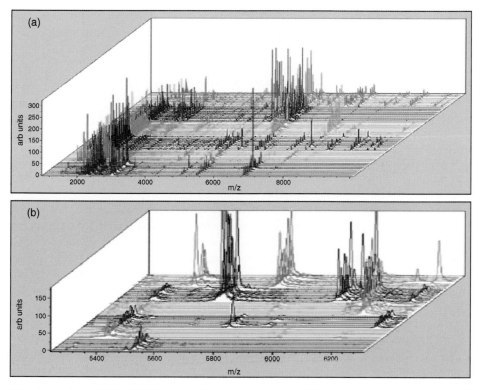

Figure 1.1.22 *Examples of MALDI-TOF spectral fingerprints obtained from a number of micro-organisms: (a) full spectra obtained from strains of* **S. aureus***,* Streptococcus group B*,* **E. coli***,* Klebsiella pneumoniae*,* **Salmonella serotype B***, and* Pseudomonas aeruginosa*; (b) expansion of the 5000–8500 Dalton/z region (Reprinted from S.-Y. Hsieh, C.-L. Tseng, Y. S. Lee, et al., Mol. Cell. Proteomics,* **7***, 448–456, 2008, with permission of The American Society for Biochemistry and Molecular Biology.)*

1.1.5   Other than its mode of action, what factors determine the antibacterial activity of a drug?

In Subsection 1.1.2 we saw that the different structures of the bacterial cell walls in mycobacteria and Gram negative and Gram positive bacteria have an effect on the staining of these bacteria, and we might imagine that these structural differences would also have an effect on the uptake of antibacterial agents by the cells. This is undoubtedly the case, and infections due to Gram negative bacteria are often more difficult to treat than those caused by Gram positive bacteria. The antibacterial activity of a drug is not, however, solely governed by its mode of action and its ability to cross the cell membrane. There are a number of other factors which are important and which we will consider now. (For further information, see Adembri and Novelli, 2009; Barbour *et al.*, 2010.)

1.1.5.1   Bacteriostatic or bactericidal?

Agents which kill bacteria are referred to as **bactericidal**, while those which prevent them from multiplying (and thus rely on the host's immune system to kill and remove them) are **bacteriostatic**. Seems simple, doesn't it? In fact, some agents are bacteriostatic against some bacteria and bactericidal against others and whether an agent is bacteriostatic or bactericidal can even depend upon its concentration (see the aminoglycosides in

Section 4.1). **Bactericidal** agents are actually defined as those which produce a **'99.9% reduction in viable bacterial density in an 18–24 h period'** (Pankey and Sabath, 2004), so that agents which produce a 99.9% reduction after 48 hours, and are essentially bactericidal, are not classified as such. For example, the penicillins, vancomycin, and fluoroquinolones (which are generally referred to as bactericidal agents) kill *S. aureus* and *S. pneumoniae*, but are only considered to be bacteriostatic against enterococci *in vitro* as they do not produce a 99.9% reduction within 24 hours (Chambers, 2003).

1.1.5.2   Minimum inhibitory concentration (MIC) and minimum bactericidal concentration (MBC)

The efficacy of an antibiotic agent is usually indicated by the minimum inhibitory concentration (MIC), the lowest concentration which inhibits the visible growth of a particular organism after overnight incubation. The MIC is important for the following reasons:

- It gives an indication of the susceptibility of an organism to a particular antimicrobial agent: the lower the MIC, the more susceptible is the microorganism (this is particularly useful in the determination of bacterial resistance, as resistant strains will have higher MICs than those which are susceptible to an agent).
- It is used to inform the treatment of a bacterial infection (the antibacterial agent chosen would normally be that with the lowest MIC against the organism) and the dosing regimen for the antibacterial used.

MICs are determined by a standardised method (Andrews, 2001) and there are databases available which list the MICs for microorganisms (see for example the Antimicrobial Index Knowledgebase, http://antibiotics. toku-e.com/, last accessed 30 March 2012). For example, one of the methods of determining the MIC (which is used in the measurement of bacterial resistance) is the E-test, in which a strip containing an increasing concentration gradient of an antibacterial agent is placed on an agar plate inoculated with the bacterium of interest. After overnight incubation of the bacteria, a zone of inhibition appears as the bacteria grow around the strip and the MIC is determined as the position where this zone intersects with the strip.

Related to the MIC is the minimum bactericidal concentration (MBC), which is defined as the lowest concentration of an antimicrobial which will prevent the growth of an organism when subcultured on to antibiotic-free media.

1.1.5.3   Time- versus concentration-dependence

For bactericidal agents, the killing of bacteria is dependent upon both the concentration of the drug relative to the MIC and the length of time for which the bacteria are exposed to the drug. Bactericidal agents can exhibit either time- or concentration-dependent killing of bacteria, depending upon which effect predominates:

- If the bactericidal agent has time-dependent bactericidal action then, providing the drug concentration is above the MIC, the most important factor is the time the drug is in contact with the bacteria (the exact concentration doesn't matter) and there will be a constant rate of bacterial kill if the drug concentration is greater than the MIC.
- If the bactericidal agent has concentration-dependent bactericidal action then the absolute concentration of the drug is the most important factor. In this case, the rate of kill increases with increased drug concentration (which must be greater than the MIC) and one single large dose may be sufficient to eradicate all the pathogenic bacteria.

Some examples of concentration- and time-dependent antibacterial agents are given in Table 1.1.2 (Filho *et al.*, 2007).

Table 1.1.2   *Concentration- and time-dependent antibacterial agents (Filho et al., 2007)*

Antibacterial agents	Bacterial killing
Aminoglycosides (see Section 4.1, e.g. gentamicin)	Concentration
β-Lactams (see Section 5.1, e.g. ceftazidime)	Time
Fluoroquinolones (see Section 2.1, e.g. ciprofloxacin)	Concentration
Macrolides (see Section 4.2, e.g. azithromycin)	Time
Oxazolidinones (see Section 4.5, e.g. linezolid)	Time

1.1.6   Bacterial resistance

It was Sir Alexander Fleming who first warned that the inappropriate use of penicillin could lead to the selection of resistant forms of *S. aureus*. He was proved correct as, in less than a year of the widespread introduction of penicillin, resistant *S. aureus* strains were discovered and there were epidemics of 'hospital *Staphylococcus*', a strain of *S. aureus* resistant to penicillin, chloramphenicol, erythromycin, and tetracyclines (Levy, 2002). Since the initial use of antibacterial agents, more and more bacteria have developed resistance, and some, such as *P. aeruginosa*, *Acinetobacter baumannii*, and *Klebsiella pneumoniae*, have now developed such multidrug resistance that clinical isolates have emerged which are susceptible to only one class of antibacterial agent! (Falagas and Bliziotis, 2007).

Antibiotic resistance can be described as microbiological or clinical, with the latter being the failure to achieve an *in vivo* antibacterial concentration which inhibits the growth of the bacteria (Wickens and Wade, 2005).

Microbiological resistance can be classified as intrinsic or acquired. We will not concentrate much on intrinsic (or innate) resistance here, save to say that, if we think about it, it is obvious why some bacteria will have natural resistance to antibiotics. Streptomyces species, for example, produce a range of antibiotics, which we will study in the following sections, such as streptomycin, chloramphenicol, macrolides, tetracyclines. The Streptomyces species produce these agents to inhibit the growth of competing microorganisms and so must have natural resistance to these antibiotics themselves (otherwise they would be committing suicide by producing agents which kill themselves).

As we have said, resistance was observed soon after the introduction of antibiotics – the ability of resistant bacteria to transfer resistance to antibiotic-susceptible strains was linked to an 'R factor', later identified as a plasmid, which carries the genetic information required to confer the resistance, through a number of possible mechanisms:

- By alterations to a target enzyme or a group of bacterial enzymes linked to a biosynthetic pathway.
- By alterations to a protein, such as the PBP or the ribosome (a ribonucleoprotein).
- By alterations to the drug structure (rendering it inactive).
- By alterations to an efflux pump or porin, or other changes to the cell wall that confer impermeability.

We will come across many examples of each of these types of resistance in the sections on the individual classes of antibacterial agent.

Resistance to antibiotics is driven by three main conditions: pressure of selection (either continuously or periodically), the emergence of stable resistance genes, and the ability of these genes to be transferred via resistance vectors, such as plamids, transposons, and epidemic strains (Amyes and Towner, 1990). Selective pressure is a result of *most* of the susceptible bacteria in the host being eliminated by the action of an antibacterial agent. Those bacteria which have resistance will be unaffected and continue to multiply, and the use of the antibacterial agent will thus lead to the selection of these resistant microorganisms.

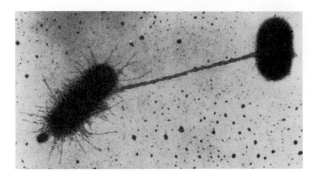

Figure 1.1.23 *Bacterial conjugation (Reprinted from G. Karp, Cell and Molecular Biology, 2010, with permission of John Wiley & Sons.)*

Resistance due to a chance mutation in the chromosomal DNA of the bacteria, leading to a beneficial change in a bacterial protein, can result in survival of that microorganism in the presence of an antibacterial agent. Although such chance mutations happen only once in every 10^7 cells produced, since some bacteria can divide every 30 minutes (or less), it may take as little as 6 hours for a mutant daughter cell to be produced. Once this altered (mutated) bacterial DNA has been produced, it can also be transferred to other bacteria by conjugation, transformation, or transduction (Alanis, 2005; Alekshun and Levy, 2007).

Conjugation is the most common mechanism for the transmission of resistance and is mediated by plasmids (circular fragments of DNA), which are transmitted between bacteria via a pilus – a hollow tube which is responsible for bringing the cells into intimate contact, thereby allowing the plasmids to transfer from one to another (Figure 1.1.23).

The most successful resistance plasmids are transferable between bacteria and carry the genes for resistance to a number of unrelated antibiotics; they can involve transposons[11] that can move from one plasmid to another, creating a mobile and effective communication of resistance information from antibiotic-resistant to antibiotic-susceptible bacteria, across strains, species, and even genera (Amyes and Towner, 1990). Although Gram positive and Gram negative bacterial resistance can have similar mechanisms, they are sufficiently different to be studied and reviewed independently. Gram positive bacterial resistance is most commonly plasmid-borne (Berger-Bächi, 2002; Woodford, 2005; Mlynarczyk *et al.*, 2010). Among pathogenic Gram negative bacteria, integrons[12] play a major role in the spread of antibiotic resistance genes. Integrons are predominantly found in plasmids, located in transposons, and are linked to the insertion, excision, and expression of mobile gene cassettes (White *et al.*, 2001).

Transformation involves the transfer of free DNA (in this case that containing the mutated sequence which confers some advantage on the recipient microorganism) between bacteria.

Transduction usually involves a virus known as a bacteriophage. Viruses require other cells to multiply and need to insert their genetic information into that of a host in order for it to be copied. In this case, the virus contains the mutated bacterial DNA and this becomes incorporated into that of the bacterium infected, so that every time this cell multiplies it will also produce the viral DNA (and so produce copies of the bacteriophage which can infect other bacterial cells).

---

[11] Transposons are DNA sequences that can move to new positions within the genome, using one of two processes with which you may, in your university life, already be familiar: 'copy and paste' and 'cut and paste'.

[12] An integron is a genetic element that possesses a site at which further DNA, in the form of gene cassettes (usually linear sequences of a larger DNA molecule, such as a bacterial chromosome), can be integrated by site-specific recombination. It also encodes an enzyme – integrase – which mediates this site-specific recombination.

1.1.7   The 'post-antibiotic age'?

As mentioned in the previous subsection, and as we will see throughout this textbook, bacteria have remarkable ability to develop resistance to the antibacterial agents with which they are treated. In addition, humans have been very complacent with the use of antibiotics, believing that there was a never-ending supply of antibacterial agents. We even continue to use them in animal feedstuffs as growth promoters (e.g. between 1992 and 1997, 56% of all antibiotics employed annually in Australia were used in feedstock). Avoparcin, a vancomycin analogue, was used as a growth promoter in pig feedstock, but was withdrawn from the market when a link was discovered between its use and the emergence of *Enterococcus faecium* expressing the *vanA* gene. The mass exposure of farm animals to antibiotics has probably been a key factor in the rapid emergence of resistance in some drug classes and the transfer of resistant bacteria to humans through the food chain (Turnidge, 2001).

As an example of the speed with which resistance can be transmitted around the world, we need only consider the metallo-β-lactamases (MBLs), a class of carbapenemases. As we shall see in Section 5.1, there are over 300 different β-lactamases, enzymes which cleave the β-lactam ring of these antibiotics, rendering them inactive. Different β-lactamases have different substrates and some, such as extended-spectrum β-lactamases (ESBLs), hydrolyse even third-generation cephalosporins (but not carbapenems) (Tumbarello *et al.*, 2004). MBLs, enzymes which hydrolyse the β-lactam ring of carbapenems through a mechanism involving zinc ion catalysis, have recently emerged in *Enterobacteriaceae* and represent the greatest threat to the use of these antibiotics. For example, the gene for the newest MBL, NDM-1 (bla_{NDM-1}), was only characterised in 2008 (it originated in India and Pakistan), but NDM-1 positive isolates have now been detected all across the globe, including the USA, UK, Europe, and Australia. These isolates have been detected in patients who have visited India (in an age where air travel can take us *and our infections* anywhere in the globe within 24 hours, this is not surprising), often for surgery (Walsh, 2010).

More worryingly, the plasmid which carries the NDM-1 gene also confers resistance to the macrolides, aminoglycosides, rifampicin, sulfamethoxazole, and aztreonam, leaving very few treatment options for infections caused by organisms such as *K. pneumoniae* or *A. baumannii* which possess it (Yong *et al.*, 2009).

The plasmid that carries the bla_{NDM-1} gene is just one example that confers resistance to a whole range of antibacterial agents, and as you will already be aware, there has been an alarming increase in the last decade in the number of bacteria which have resistance to more than one class of antibiotic, such as meticillin-resistant *Staphylococcus aureus* (MRSA). There are no prizes available for guessing one of the drugs which MRSA is resistant to, but the name doesn't indicate that MRSA strains are frequently also resistant to other clinically useful β-lactams, erythromycin (a macrolide) and ciprofloxacin (a quinolone), leaving very few agents available for treatment of MRSA infections (with vancomycin, a glycopeptide, being the agent of choice).

Other multiresistant bacterial infections include multidrug-resistant tuberculosis (MDR-TB, in which the tuberculosis is resistant to at least the two main first-line TB drugs – isoniazid and rifampicin) and extensively drug-resistant tuberculosis (XDR-TB, which is MDR-TB that is also resistant to three or more of the six classes of second-line drugs). Given that more than one third of the world's population has been exposed to TB, and that 90% of the exposed population has latent TB (asymptomatic), these are particularly worrying developments (World Health Organization, 2010).

We all have a vested interest in the availability of antibacterial agents for the treatment of infections (the authors have an even greater vested interest as, if there were no effective antibacterial agents remaining, there would be no need for students to study them, and hence no need for this book). These trends in increased levels of resistance and pan-resistant bacteria have been taken by some researchers to mean that we are entering (or have entered) a post-antibiotic era (Alanis, 2005; Walsh *et al.*, 2011).

Some of the questions which have to be asked in relation to the future availability of antibacterial treatments are:

- Will we soon witness the re-emergence of relatively simple life-threatening infections, which we believed were a thing of the past when clinicians had a range of potential antibacterial agents at their disposal?
- How soon will it be before some of the multidrug-resistant strains are no longer sensitive to any antibacterial agent (e.g. when vancomycin-resistant *Staphylococcus aureus* acquires the gene conferring resistance to all the other major classes of antibacterial agent)?
- What can we do to stop the further emergence of pan-resistant bacteria?
- What new classes of antibiotic are in development?

These are not easy questions to answer. Serious infections caused by multiresistant bacteria are already a global health care problem and antibiotic stewardship programmes are in operation throughout the world, through which hospitals seek to optimise antimicrobial prescribing in order to improve individual patient care and slow the spread of antimicrobial resistance (as well as reduce hospital costs) (MacDougall and Polk, 2005). Such programmes have a number of elements, including: restricting the dispensing of specific antibacterial agents to approved indications; the scheduled rotation of antibacterials used in a particular unit (if appropriate); further education of all members of the health care team to ensure the appropriate use of antibacterial agents (in particular the prescription of antibiotics with a narrower spectrum of activity as part of a directed clinical treatment regimen, rather than agents with a broad spectrum of activity in an empirical approach); and the education of the general public as to their expectations for the prescription of antibacterial agents and their use (e.g. advising patients to complete the course of antibiotic therapy, even though they may feel better after only a few days). The probability is that bacterial infections will soon emerge which have resistance to all the major classes of antibacterial agent, but, by adhering to such antibiotic stewardship principles, we may be able to delay this for a while yet.

You may well be thinking that all of the last couple of paragraphs are alarmist and that there will always be new members of a class of antibacterial agents (or even new classes) available to clinicians. Unfortunately, this is not the case. The 'golden age' of antibiotic discovery is now well gone and between 1983 and 2001 only 47 new antibiotics were approved in North America (by the US Federal Drug Administration (FDA) or Canadian Health Ministry). Of the nine new antibiotics approved since 1998, only two had a new mode of action and they are the only truly novel antibacterial agents launched in the last 30 years: linezolid (Pharmacia and Pfizer) and daptomycin (Cubist) (Overbye and Barrett, 2005).

And what of antibacterial agents in development? Surely every pharmaceutical company recognises the need for novel antibacterial agents and has research programmes aimed at discovering and developing them? Once again, the answers are not comforting. Pharmaceutical companies must make a profit to continue to be viable and so target chronic diseases, aiming for blockbuster drugs which will be given to patients for lengthy periods, in order to maximise their return on the massive costs associated with drug discovery and the short period during patent protection in which they, and they alone, can make a profit from the sale of a particular drug. Narrow-spectrum antibacterial agents, which will be used for the treatment of acute illnesses, and which are at risk of quickly becoming obsolete as a result of the emergence of bacterial resistance, are not commercially attractive to pharmaceutical companies. Lastly, as many antibacterial discovery programmes in industry were disbanded many years ago, much of the research expertise has been lost and would have to be re-introduced if companies were given incentives (such as increased patent lifetimes to allow them to maximise profits) to develop new discovery programmes.

All is not doom and gloom, however, and we will conclude the discussion of each class of antibacterial agent in the remainder of this book with a section on any new agents which are currently under development. New antibacterial targets obviously represent the ideal scenario for development, since resistance should develop more slowly to an agent with a novel mode of action, but new agents in the already established classes, with different antibacterial spectra of activity, are welcome additions to the dwindling arsenal of agents which are available to clinicians.

References

A. Achari, D. O. Somers, J. N. Champness, P. K. Bryant, J. Rosenmond, and D. K. Stammers, *Nature Struct. Biol.*, 1997, **4**, 490–497.

C. Adembri and A. Novelli, *Clin. Pharmacokin.*, 2009, **48**, 517–528.

A. J. Alanis, *Arch. Med. Res.*, 2005, **36**, 697–705.

M. N. Alekshun and S. B. Levy, *Cell*, 2007, **128**, 137–150.

S. G. B. Amyes and K. J. Towner (Eds.), *J. Med. Microbiol.*, 1990, **31**, 1–19.

J. M. Andrews, *J. Antimicrob. Chemother.*, 2001, *48* (S1), 5–16.

A. Barbour, F. Scaglione, and H. Derendorf, *Intn. J. Antimicrob. Agents*, 2010, **35**, 431–438.

B. Berger-Bächi, *Int. J. Med. Microbiol.*, 2002, **292**, 27–35.

H. F. Chambers, *Clin. Updates Infect. Dis.*, 2003, **VI**, 1–6.

J. J. Champoux, *Ann. Rev. Biochem.*, 2001, **70**, 369–413.

S. Chang, D. M. Sievert, J. C. Hageman, M. L. Boulton, F. C. Tenover, F. P. Downes, S. Shah, J. T. Rudrik, G. R. Pupp, W. J. Brown, D. Cardo, and S. K. Fridkin, *New Engl. J. Med.*, 2003, **348**, 1342–1347.

R. H. Ebright, *J. Mol. Biol.*, 2000, **304**, 687–698.

M. E. Falagas and I. A. Bliziotis, *Intn. J. Antimicrob. Agents*, 2007, **29**, 630–636.

L. S. Filho, J. L. Kuti, and D. P. Nicolau, *Braz. J. Microbiol.*, 2007, **38**, 183–193.

H. G. Floss and T. W. Yu, *Chem. Rev.*, 2005, **105**, 621–632.

P. W. Groundwater, A. Todd, A. J. Worsley, and R. J. Anderson, *Pharm. J.*, 2009, **283**, 281–282.

Health and Social Care Act, 2008 (http://www.dh.gov.uk/prod_consum_dh/groups/dh_digitalassets/documents/digitalasset/dh_110435.pdf, last accessed 8 March 2012).

HPA Mandatory Surveillance Report, 2008 (http://www.hpa.org.uk/web/HPAwebFile/HPAweb_C/1229502459877, last accessed 8 March 2012).

S.-Y. Hsieh, C.-L. Tseng, Y. S. Lee, A. J. Kuo, C. F. Sun, Y. H. Lin, and J. K. Chen, *Mol. Cell. Proteomics*, 2008, **7**, 448–456.

A. Huletsky, R. Giroux, V. Rossbach, M. Gagnon, M. Vaillancourt, M. Bernier, F. Gagnon, K. Truchon, M. Bastein, F. J. Pickard, A. van Belkum, M. Ouellette, P. H. Roy, and M. G. Bergeron, *J. Clin. Microbiol.*, 2004, **42**, 1875–1884.

K. W. Langford, B. Penkov, I. M. Derrington, and J. H. Gundlach, *J. Lipid Res.*, 2011, **52**, 272–277.

B. S. Laursen, H. P. Srensen, K. K. Mortensen, and H. U. Sperling-Petersen, *Microbiol. and Mol. Biol. Rev.*, 2005, **69**, 101–123.

N. A. Ledeboer, K. Das, M. Eveland, C. Roger-Dalbert, S. Mailler, S. Chatellier, and W.M. Dunne, *J. Clin. Microbiol.*, 2007, **45**, 1556.

S. B. Levy, 'From tragedy the antibiotic era is born' in S. B. Levy (Ed.) *The Antibiotic Paradox: How the Misuse of Antibiotics Destroys Their Curative Powers*, 2nd ed., Perseus Publishing, Cambridge, MA, 2002, pp. 1–14.

C. MacDougall and R. E. Polk, *Clin. Microbiol. Rev.*, 2005, **18**, 638–656.

R. Mariani and S. I. Maffioli, *Curr. Med. Chem.*, 2009, **16**, 430–454.

B. Mlynarczyk, A. Mlynarczyk, M. Kmera-Muszynska, S. Majewski, and G. Mlynarczyk, *Mini Rev. Med. Chem.*, 2010, **10**, 928–937.

National Audit Office Report, Improving patient care by reducing the risk of hospital acquired infection: a progress report, 2004 (http://image.guardian.co.uk/sys-files/Society/documents/2004/07/14/0304876.pdf, last accessed 8 March 2012).

NHMRC, Australian Guidelines for the Prevention and Control of Infection in Healthcare. Commonwealth of Australia, 2010 (http://www.nhmrc.gov.au/_files_nhmrc/publications/attachments/cd33_complete.pdf, last accessed 8 March 2012).

K. M. Overbye and J. F. Barrett, *Drug Disc. Today*, 2005, **10**, 45–52.

G. A. Pankey and L. D. Sabath, *Clin. Infect. Dis.*, 2004, **38**, 864–870.

J. D. Perry and A. M. Freydière, *J. Appl. Microbiol.*, 2007, **103**, 2046–2055.

J.D. Perry, A. Davies, L. A. Butterworth, A. L. J. Hopley, A. Nicholson, and F. K. Gould, *J. Clin. Microbiol.*, 2004, **42**, 4519.

J. D. Perry, K. Asir, D. Halimi, S. Orenga, J. Dale, M. Payne, R. Carlton, J. Evans, and F. K. Gould, *J. Clin. Microbiol.*, 2010, **48**, 3852.

L. P. Randall, M. Kirchner, C. J. Teale, N. G. Coldham, E. Liebana, and F. Clifton-Hanley, *J. Antimicrob. Chemother.*, 2009, **63**, 302.

A. Robicsek, J. L. Beaumont, S. M. Paule, D. M. Hacek, R. B. Thomson, K. L. Kaul, P. King, and L. R. Peterson, *Ann. Intern. Med.*, 2008, **148**, 409–418.

R.D. Scott, The direct medical costs of healthcare-associated infections in U.S. hospitals and the benefits of prevention (Publication No. CS200891-A). Centers for Disease Control and Prevention, 2009 (http://www.cdc.gov/HAI/pdfs/hai/Scott_CostPaper.pdf, last accesssed 8 March 2012).

A. Stadelmaier, S. Morath, T. Hartung, and R. R. Schmidt, *Angew. Chem. Int. Ed.*, 2003, **42**, 916–920.

T. A. Steitz, *Nature Rev., Mol. Cell. Biol.*, 2008, **9**, 242–253.

F. C. Tenover, *Clin. Infect. Dis.*, 2007, **44**, 418–423.

M. Tumbarello, R. Citton, T. Spanu, M. Sanguinetti, L. Romano, G. Fadda, and R. Cauda, *J. Antimicrob. Chemother.*, 2004, **53**, 277.

J. Turnidge, *Aust. Prescriber*, 2001, **24**, 26–27.

J. van Heijenoort, *Nat. Prod. Rep.*, 2001, **18**, 503–519.

T. R. Walsh, *Intn. J. Antimicrob. Agents*, 2010, **36**, S8–S14.

T. R. Walsh, J. Weeks, D. M. Livermore, and M. A. Toleman, *Lancet Inf. Dis.*, 2011, **11**, 355–362.

G. Werner, C. Cuny, F.-J. Schmitz, and W. Witte, *J. Clin. Microbiol.*, 2001, **39**, 3586–3590.

P. A. White, C. J. McIver, and W. D. Rawlinson, *Antimicrob. Agents Chemother.*, 2001, **45**, 2658–2661.

H. Wickens and P. Wade, *Pharmaceutical J.*, 2005, **274**, 501–504.

N. Woodford, *Clin. Microbiol. Infect.*, 2005, **11** (Suppl. 3), 2–21.

World Health Organization, Tuberculosis Fact Sheet No. 104, 2010 (http://www.who.int/mediacentre/factsheets/fs104/en/, last accessed 8 March 2012).

D. Yong, M. A. Toleman, C. G. Giske, H. S. Cho, K. Sundman, K. Lee, and T. R. Walsh, *Antimicrob. Agents Chemother.*, 2009, **53**, 5046–5054.

A. V. Zaytsev, R. J. Anderson, A. Bedernjak, P. W. Groundwater, Y. Huang, J. D. Perry, S. Orenga, C. Roger-Dalbert, and A. James, *Org. Biomol. Chem.*, 2008, **8**, 682–692.

Questions

(1) In Figure 1.1.5, if the pink colour represents guanine, deduce the colours of the other bases.

(2) In Figure 1.1.6, what will be the next amino acid in the protein after proline?

(3) What is the codon (mRNA) and anticodon (tRNA) for each of the following:

- tryptophan;
- glycine;
- STOP.

(4) The discovery of new classes of selective antibacterial agents is partly dependent upon the identification of specific new bacterial targets. Within this section there is a reference to an enzyme which could be such a target for which there are no inhibitors currently in clinical use. Did you spot the enzyme?

(5) Would *Staphylococcus aureus*:

- be Gram positive or negative?
- grow aerobically or anaerobically?
- be catalase negative or positive?

2

Agents Targeting DNA

Antibacterial agents which target nucleic acids belong to the quinolones (DNA gyrase/DNA topoisomerase IV inhibitors), rifamycins (DNA-dependent RNA polymerase inhibitors), or nitroimidazoles.

2.1

Quinolone antibacterial agents

Key Points

- The quinolones are synthetic antibacterial agents which show greatest activity against Gram negative bacteria and are used in the treatment of urinary tract and respiratory infections.
- Resistance to quinolones arises due to alterations in the DNA gyrase (via mutations in the quinolone resistance-determining region (QRDR) of the *gyrA* gene), and similar mutations which decrease quinolone binding have been described in topoisomerase IV (in the *parC* gene).
- The quinolones are generally well tolerated, but some of the newer generations have been associated with serious adverse effects, such as hepatotoxicity and QT interval prolongation (cardiotoxicity).
- Some quinolones should not be routinely used in combination with theophylline because of the risk of developing theophylline toxicity.
- For further information, see Emmerson and Jones (2003) and Hooper (1998).

Antibacterial quinolones used clinically include nalidixic acid, ciprofloxacin, ofloxacin (racemic mixture of levofloxacin enantiomers), levofloxacin, norfloxacin, besifloxacin, and moxifloxacin; their structures are shown in Figure 2.1.1 and their therapeutic indications are listed in Table 2.1.1.

2.1.1 Discovery

Nalidixic acid was discovered in 1962, as a result of a drug discovery project at Sterling Winthrop, which used as the lead compound a quinolone impurity (Figure 2.1.2a) isolated during the synthesis and purification of the antimalarial agent, chloroquine (Figure 2.1.2b). Nalidixic acid, like other first-generation quinolones, was found to have weak antibacterial (bactericidal) activity and is now only used for the treatment of urinary tract infections (UTIs) caused by Gram negative bacteria, as is norfloxacin, a second-generation quinolone.

All second-generation and later quinolones contain, in addition to the 3-carboxyl-substituted quinolin-4-one essential for biological activity, a fluorine at the 6-position and a cyclic amine at the 7-position, both of

Antibacterial Agents: Chemistry, Mode of Action, Mechanisms of Resistance and Clinical Applications, First Edition.
Rosaleen J. Anderson, Paul W. Groundwater, Adam Todd and Alan J. Worsley.
© 2012 John Wiley & Sons, Ltd. Published 2012 by John Wiley & Sons, Ltd.

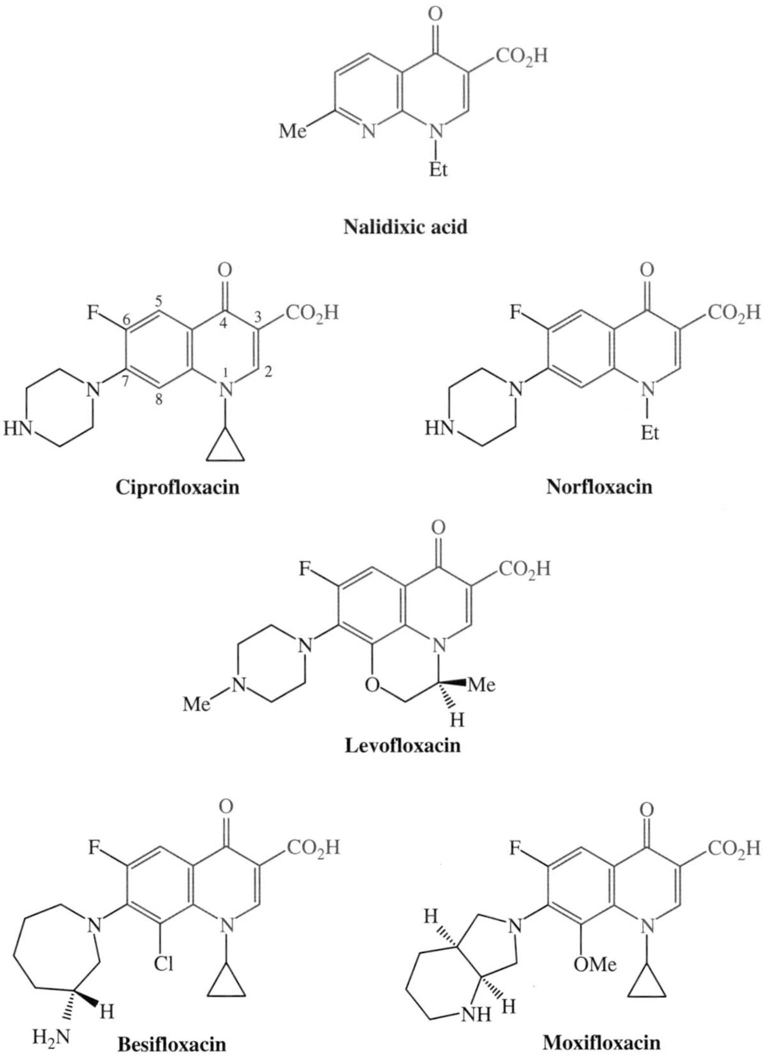

Nalidixic acid

Ciprofloxacin

Norfloxacin

Levofloxacin

Besifloxacin

Moxifloxacin

Figure 2.1.1 *Quinolone antibacterials: nalidixic acid (first generation), ciprofloxacin and norfloxacin (second generation), levofloxacin (third generation), and besifloxacin and moxifloxacin (fourth generation)*

Table 2.1.1 *Therapeutic indications for the quinolone antibacterials*

Quinolone antibacterial	Indications
Nalidixic acid	Urinary tract infections (UTIs)
Ciprofloxacin	Respiratory tract infections (RTIs), UTIs, anthrax treatment, gonorrhoea, chronic prostatitis, prophylaxis of meningitis
Ofloxacin	UTIs, RTIs, gonorrhoea, urethritis
Levofloxacin	Community-acquired pneumonia
Norfloxacin	UTIs
Besifloxacin	Bacterial conjunctivitis
Moxifloxacin	Sinusitis, community-acquired pneumonia, exacerbations of chronic bronchitis

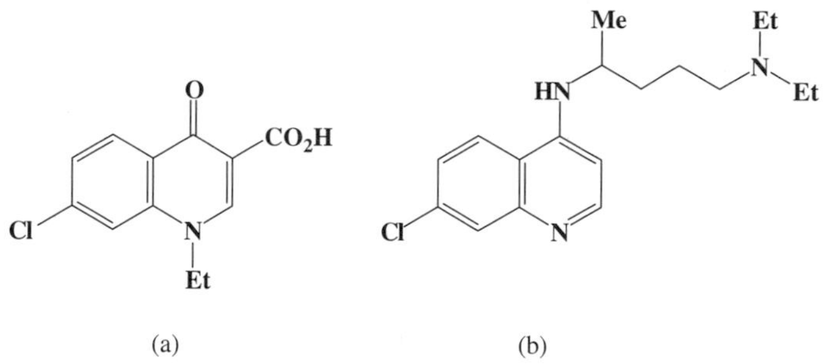

Figure 2.1.2   *(a) Quinolone impurity. (b) Chloroquine (Reprinted from J. M. Domagala, J. Antimicrob. Chemother., 33, 685–706, 1994, with permission of the Oxford University Press.)*

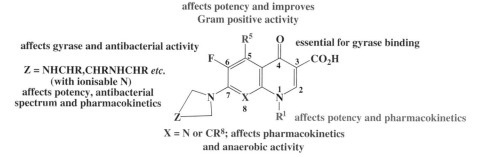

Figure 2.1.3   *Summary of structure-activity relationships (SARs) for the fluoroquinolones*

which were quickly discovered to increase the antibacterial activity of these agents during structure–activity relationship (SAR) studies (Domagala, 1994). The nature of the cyclic amine, along with the introduction of other substituents, influences the antibacterial spectrum of activity (including activity against Gram positive organisms), as well as the physicochemical and pharmacokinetic properties of these synthetic antibacterials (Figure 2.1.3); five- and six-membered rings confer greater activity (although besifloxacin, of course, has a seven-membered (azepine) ring at this position), with a six-membered ring improving the activity against Gram negative species, for example, the piperazine ring in ciprofloxacin, and ring alkylation improving both the Gram positive activity and the half-life. (For more on the discovery of the quinolones, see Rádl, 1996.)

2.1.2   Synthesis

As many thousands of quinolones have been synthesised and tested, there are a large number of literature syntheses (including patents) of this class of antibacterial agent and it is not within the scope of this section to discuss their various merits. One particularly elegant synthesis, however, which builds upon earlier synthetic strategies but on a solid phase support, is the use of DIVERSOMER technology by MacDonald *et al.* (1996), which is illustrated in Scheme 2.1.1 for the preparation of ciprofloxacin. This methodology has a number of advantages and a range of fluoroquinolones can be prepared from common intermediates **4** (from which various N^1-substituted derivatives can be obtained through variation of the amine used in the coupling step) and **6** (from which various cyclic amines can be introduced at position 7). Solid phase organic synthesis allows

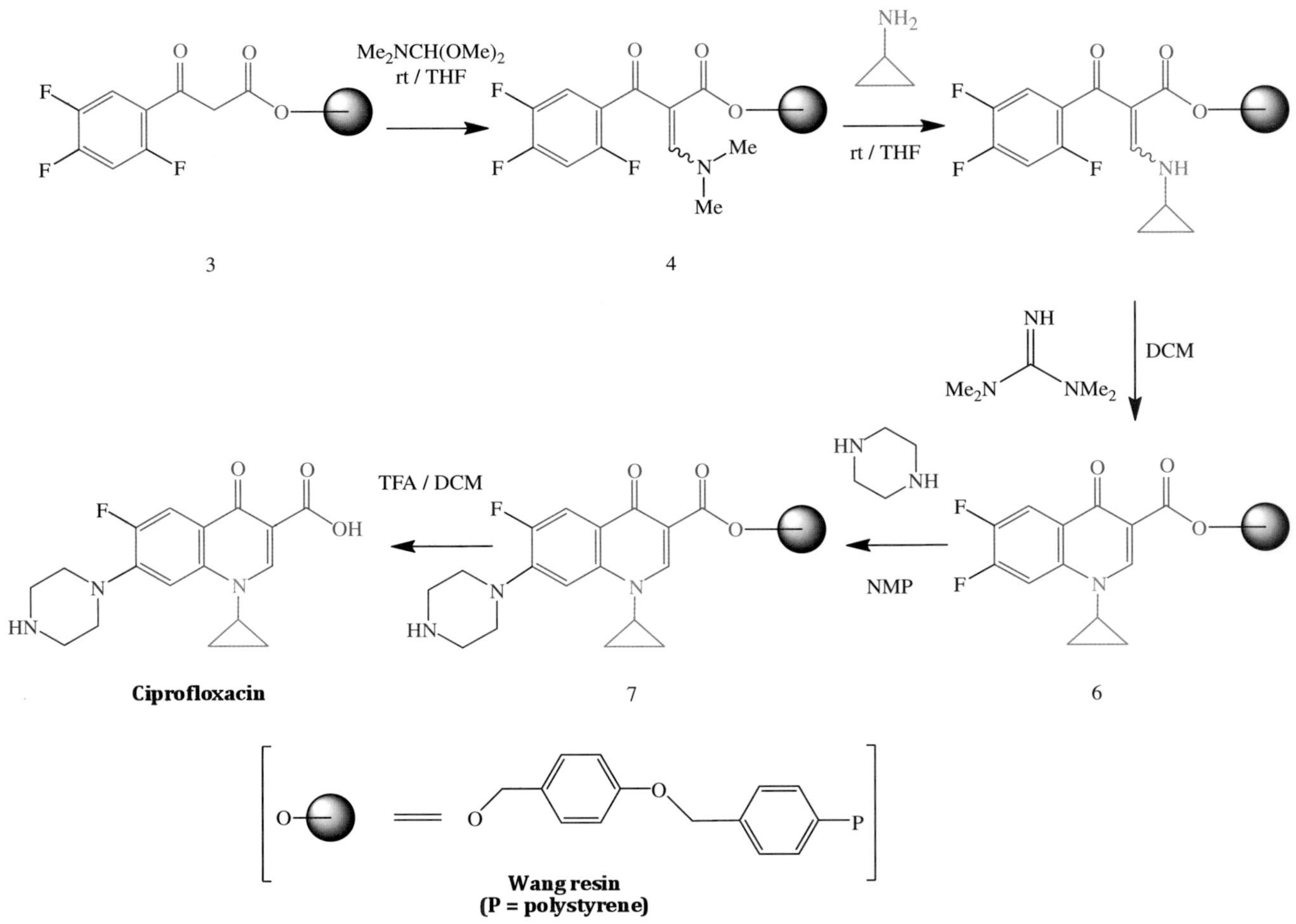

Scheme 2.1.1 Solid phase synthesis of ciprofloxacin (rt, room temperature; THF, tetrahydrofuran; DCM, dichloromethane; NMP, N-methylpyrrolidone; TFA, trifluoroacetic acid) (MacDonald et al., 1996)

optimal yields to be obtained at each stage through the use of excess reagents, which are simply washed away after each step, with the intermediates remaining linked to the Wang resin until the final cleavage of the resin-bound fluoroquinolone **7** with trifluoroacetic acid. In addition to the introduction of the N^1-substituent and the heterocycle (piperidine in this example) at position 7, other key steps are the reaction of the active methylene group in β-ketoester **3** with dimethylformamide dimethylacetal to introduce C-2 and an amine group, a transamination to the cyclopropylamine derivative **5**, the base-catalysed cyclisation of the amine **5** via a nucleophilic aromatic substitution (which is facilitated by the presence of the electron-withdrawing fluorine atoms and the *ortho*-carbonyl group on the aryl ring, as well as the fluoro leaving group), and the final cleavage step (Scheme 2.1.1).

As can be seen from Figure 2.1.1, the heterocyclic ring attached to position 7 of the quinolone nucleus increases in complexity in the structures of the third- and fourth-generation fluoroquinolones, with many of these later derivatives having a number of stereogenic centres. Ofloxacin is a mixture of enantiomers, but each of these optical isomers will have been synthesised and tested separately as part of the licensing application to the relevant agency – the European Medicines Agency (EMA), the Medicines and Healthcare products Regulatory Agency (MHRA), the Food and Drug Administration (FDA), the Australian Therapeutic Goods Administration (TGA), and so on. In general, the greater the number of stereogenic centres in a drug candidate, the more complex the synthesis, with each stereoisomer having to be prepared, isolated, characterised, and tested. As chiral synthesis is generally expensive (unless a route which utilises a relatively cheap and easily available chiral natural product can be employed), this all adds to the cost of the drug discovery process, and such an increase in the complexity of these heterocyclic side chains must be justified by an increase in the efficacy or antibacterial spectrum of the agent, or a decrease in their side effects.

2.1.3 Bioavailability

There have been many investigations into the bioavailability of the quinolones; much of this work has focussed upon ciprofloxacin, but others are currently under scrutiny as the newer-generation quinolones become more popular.

If you reinspect the general structure of an antibacterial quinolone (Figure 2.1.3) for ionisable groups that could affect the bioavailability, you will note the carboxylic acid at position 3, which is essential (as the negatively charged carboxylate) for binding to the target, and the ionisable amine group on position 7. Together, these groups confer a zwitterionic nature that leads to the quinolones being largely ionised in all physiological compartments. Despite the predominance of ionised species in the gastrointestinal (GI) tract, the quinolones are well absorbed after oral administration, with good overall bioavailability, typically 70% (e.g. ciprofloxacin) to 99% (e.g. levofloxacin) (Blondeau, 1999; Wright *et al.*, 2000; Rodvold and Neuhauser, 2001), with measured $logP_{app}$ values (octanol/buffer at pH 7.0) in the range of approximately -1.6 (norfloxacin, ciprofloxacin) to 2.5 (nalidixic acid) (Ross *et al.*, 1992; Bermejo, 1999; Piddock *et al.*, 2001; Perez *et al.*, 2002). While the majority of quinolones have negative apparent partition coefficients, the third- and fourth-generation quinolones are reported to be significantly more lipophilic and better absorbed, after oral, ophthalmic, or respiratory system delivery, presumably due to the presence in their structures of more lipophilic groups (Perletti *et al.*, 2009).

Absorption of the quinolones is known to be reduced in the presence of divalent or trivalent metal cations, such as Mg^{2+}, Ca^{2+}, Al^{3+}, or Fe^{3+}, due to chelation of the cation to the quinolone carbonyl and carboxylate groups; the bioavailability of the quinolones is, therefore, limited by co-administration with metal cation-rich products, such as antacid preparations and iron supplements (more on this later, in Subsection 2.1.8).

For these concentration-dependent antibiotics, the higher the serum concentration, the greater the efficacy of bactericidal action. The most important parameters for these agents are the serum peak concentration (C_{max}) and the area under the serum concentration–time curve (AUC) when compared to the minimum inhibitory concentration (MIC), alternatively expressed as the maximum concentration achieved and maintained over a specific time when compared to the concentration required for bactericidal effect. The half-life of the quinolone ($t_{1/2}$) impacts upon both of these parameters, as it represents the length of time the quinolone can be expected to exert its bactericidal effect.

The early quinolones, such as ciprofloxacin and norfloxacin, have relatively short half-lives (3–5 hours), while the third- and fourth-generation quinolones have extended serum half-lives (e.g. 7.4 hours for levofloxacin and 9.6 hours for moxifloxacin), which enables single daily dosing. The C_{max} values for quinolones vary from 1.1 to 8.7 mg/L and the AUC values lie in the range 10–75 mg/h/L. The quinolones show exquisite potency towards a wide range of Gram negative bacteria, with MICs in the ng/mL range, and reasonable activity against many Gram positive bacteria (MICs in the mg/mL range), although some of the newer quinolones, such as moxifloxacin, are significantly more potent against this group of bacteria. The excellent MICs and good C_{max} and AUC values lead to strong pharmacokinetic profiles for the quinolones overall and the newer quinolones tend to have enhanced parameters for improved delivery and efficacy (Blondeau, 1999; Wright *et al.*, 2000; Cheng *et al.*, 2007).

Quinolone distribution is influenced by protein binding, typically in the range of 20–70%. The volume of distribution (V_d) varies in the range 1.1–5.1 L/kg, newer quinolones usually having higher values and better tissue penetration, enabling their use across a wider range of infections. The quinolones are subject to comparatively little metabolism, most being excreted largely unchanged in the urine, although moxifloxacin exhibits significant faecal excretion ($\sim$10%) (Blondeau, 1999; Wright *et al.*, 2000).

The high urinary excretion of most quinolones, along with their short half-lives, highlights the treatment of UTIs as a possible indication. The third- and fourth-generation fluoroquinolones show greater versatility for the treatment of systemic infections, due to their superior activity across a wider range of Gram positive bacteria and their improved V_d and uptake into many cells, due to their increased lipophilicity, enabling, for example, the treatment of prostatic infections (Perletti *et al.*, 2009).

2.1.3.1 Prodrugs

As the bioavailability of the quinolones is generally very good, there has been little need for quinolone prodrugs, with the only one to have been licensed and used clinically being alatrofloxacin, a dialanyl-derivative of trovafloxacin, a broad-spectrum quinolone with excellent activity against a wide range of both Gram negative and Gram positive bacteria, including MRSA, but whose low aqueous solubility has limited its IV use. The two L-alanine amino acid residues confer considerable aqueous solubility to alatrofloxacin, thus allowing parenteral delivery of the prodrug, which undergoes rapid systemic hydrolysis to the parent, trovafloxacin (Scheme 2.1.2) (Melnik *et al.*, 1998; Vaudaux *et al.*, 2002).

Trovafloxacin and the prodrug alatrofloxacin were withdrawn from clinical use in Europe in 1999 (but are still available in the USA) due to unpredictable and severe hepatotoxicity, with a high risk of fatality (Pannu *et al.*, 2001).

Two successful phase III clinical trials of the prodrug prulifloxacin (Figure 2.1.4), designed for oral administration, were recently completed. Prulifloxacin delivers the parent drug, ulifloxacin, immediately and quantitatively after efficient absorption. This thiazeto-quinolone antibacterial agent has potent activity against a wide range of both Gram negative and Gram positive bacteria, with MICs in the ng/mL range.

Prulifloxacin is well-absorbed in the upper small intestine and metabolised to the parent, ulifloxacin, so rapidly that the prodrug is not detected in the systemic circulation. Interestingly, ulifloxacin accumulates in the

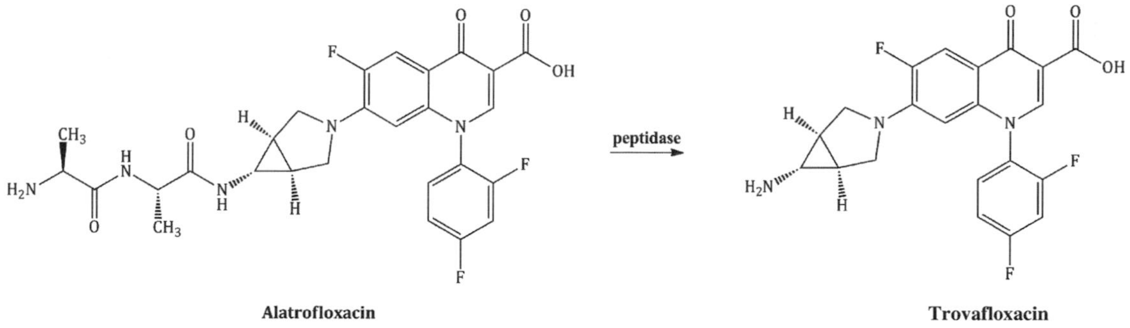

Scheme 2.1.2 *Structure of the prodrug alatrofloxacin and its hydrolysis to its parent, trovafloxacin*

liver and kidney, with a relatively high concentration in other organs, such as the spleen, pancreas, and lungs, yet a low distribution into the central nervous system (CNS) (Matera, 2006). The enzyme paraoxonase has a major role in the hydrolysis of this unusual prodrug (Tougou *et al.*, 1998), the postulated mechanism for which is shown in Scheme 2.1.3.

2.1.3.2   Bacterial uptake of quinolones

As we will see in the next subsection, the target for quinolones is found in the cytoplasm, so these agents must cross the bacterial cell wall and enter the cell to exert their antibacterial activity. They readily cross the cell wall of Gram positive bacteria, such as staphyloccocci, by passive diffusion and even though efflux mechanisms for hydrophilic quinolones in Gram positive bacteria have been identified, they are well accumulated rapidly into these bacteria, to a higher concentration than is achieved in Gram negative bacteria. The more complex cell wall of Gram negative bacteria is not amenable to passive diffusion and quinolones are taken into these bacteria through the outer membrance porins, facilitated for the more lipophilic examples by diffusion through the outer membrane. The physicochemical properties of each antibacterial agent, such as lipophilicity and molecular mass, have been correlated to the kinetics of uptake and, for some species, the antibacterial effect upon Gram negative bacteria (Piddock *et al.*, 2001).

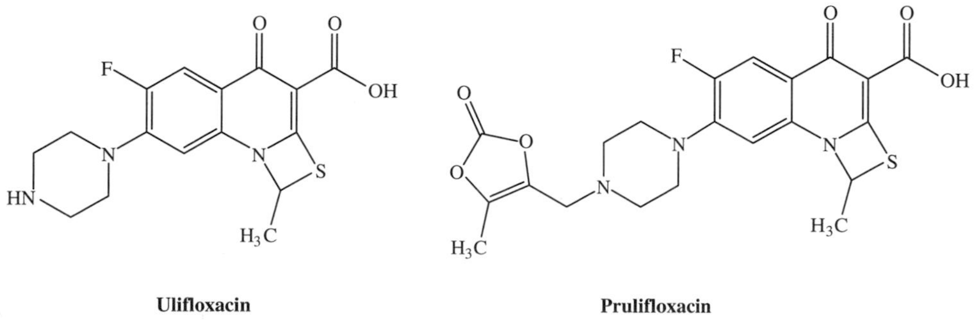

Figure 2.1.4 *Structure of ulifloxacin and its prodrug, prulifloxacin*

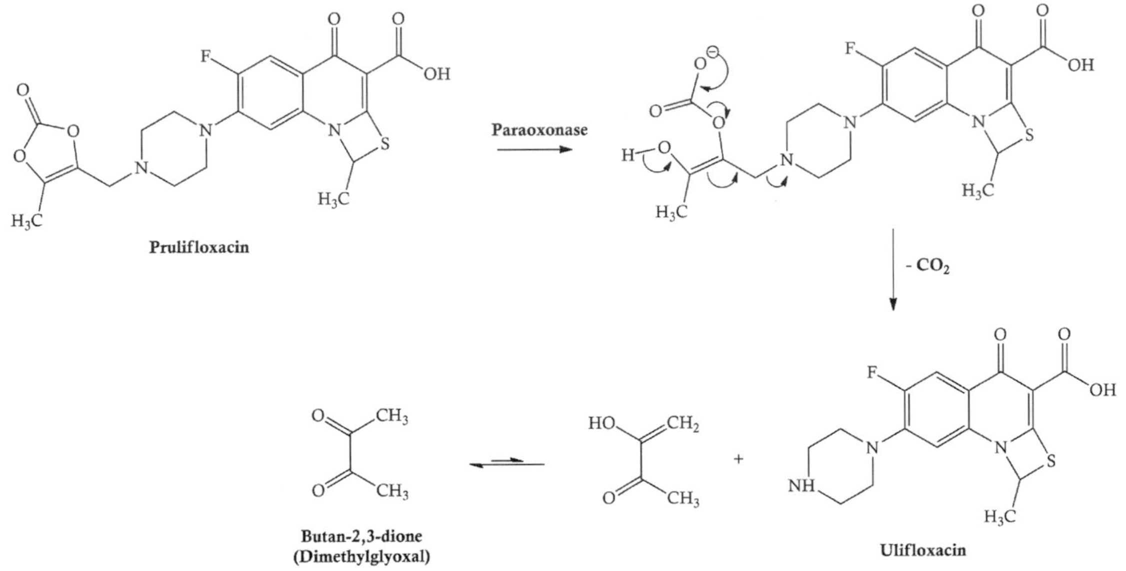

Scheme 2.1.3   *Postulated paraxonase hydrolysis of prulifloxacin (Tougou et al., 1998)*

2.1.4   Mode of action and selectivity

As discussed in the introductory section, if we know the mode of action of a class of antibacterial agents then we will be able to understand how their selectivity arises (i.e. what cellular process the agents target in prokaryotic cells and how this differs from eukaryotic cells) and how bacteria will develop resistance to them.

As you will have gathered from the title of this section, the quinolones exert their antibacterial activity by targeting DNA synthesis, so in order to understand how they do this, we need to look at the processes involved in DNA replication. We will also need to focus on how the quinolones can do this selectively in prokaryotic (bacterial) cells as, of course, this process will also be taking place in eukaryotic (mammalian) cells.

As we saw in Subsection 1.1.3, DNA replication is a complex process during which the two strands of the double helix separate and each strand acts as a template for the synthesis of complementary DNA strands. This process occurs at multiple, specific locations (origins) along the DNA strand, with each region of new DNA synthesis involving many proteins, which catalyse the individual steps involved in this process – the separation of the two strands at the origin to give a replication fork (*DNA helicase*); the synthesis and binding of a short primer DNA strand (*DNA primase*); DNA synthesis (*DNA polymerase*); and, finally, the proofreading and error-checking process to ensure the new DNA strand's fidelity – that is, that this strand (red) is exactly complementary to the template (black) strand (*DNA polymerase and endonucleases*) (Figure 1.1.4, Subsection 1.1.3). As you can see from Figure 1.1.4, the replication fork is a busy place, but for the purposes of the discussion of the mode of action of the quinolones, we can concentrate on the role of two of the enzymes shown – helicase and topoisomerase (Duguet, 1997).

DNA helicase is the enzyme which separates the DNA strands and in doing this, as a result of the right-handed helical nature of DNA, produces positive supercoils (knots) ahead of the replication site. In order for DNA replication to proceed, these supercoils must be removed by topoisomerases, relaxing the chain. By catalysing the formation of negative supercoils, through the cutting of the DNA chain(s) and the passing of one strand through the other, these enzymes remove the positive supercoils and give a tension-free DNA double helix, so that the replication process can continue. Type I topoisomerases relax DNA by cutting one

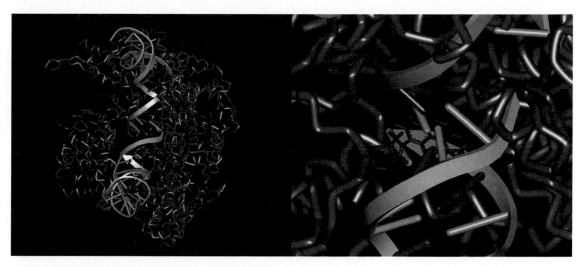

Figure 2.1.5 *A DNA-topoisomerase IV (blue) complex from* Streptococcus pneumoniae *stabilized by moxifloxacin (red); PDB code 3FOF (Laponogov et al., 2009)*

of the DNA strands, while, you've guessed it, type II cut both strands (Champoux, 2001). For the purpose of understanding the mode of action of the quinolones, it is the type II topoisomerases (DNA gyrase in Gram negative and topoisomerase IV in Gram positive bacteria) that are important (Ruiz, 2003). By binding to these enzymes, the quinolones interfere with bacterial DNA replication, and so with cell division (Figure 2.1.5). So far so good, but two questions you have probably already asked yourself are: given that mammalian cells must also relax supercoiled DNA, why are the quinolones selective antibacterial agents, and, if the quinolones affect cell division (i.e. the division of mature cells), why are they bactericidal rather than bacteriostatic? Good questions! First, mammalian cells do not have DNA gyrase or topoisomerase IV (they do have topoisomerases I and II, but quinolones do not bind to these enzymes), hence these agents have some selectivity as inhibitors of prokaryotic DNA replication. The fact that these agents are bactericidal rather than bacteriostatic is more difficult to address and is unlikely to be due to the inhibition of DNA synthesis, as the concentration of a quinolone required to block DNA synthesis is much lower than that required to kill cells.

One of the possible explanations for the bactericidal effect of quinolones is the generation of DNA ends (effectively DNA double-strand breaks) through the formation of reversible ternary 'cleavable' complexes (quinolone-topoisomerase-DNA) and the resultant release of the DNA ends from the constraints of the topoisomerase (Figure 2.1.6) (Froehlich-Ammon and Osheroff, 1995). Such double-strand DNA breaks, when generated by other means, are known to be lethal to cells (Drlica and Zhao, 1997).

2.1.5 Bacterial resistance

As we suggested previously, knowing the mode of action of an antibacterial agent will help us to understand the way in which bacteria develop resistance, as the specific resistance mechanisms will be related to interference with the processes targeted by the antibacterial agent. As described in the previous subsection, the cellular targets of the quinolones are the bacterial DNA gyrase and topoisomerase IV, both of which are hetero-tetrameric in structure; that is, formed from four subunits: A_2B_2 for DNA gyrase, C_2E_2 for topoisomerase IV.

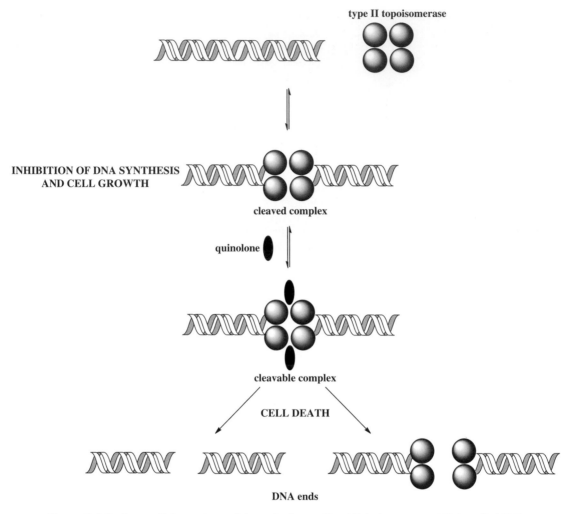

Figure 2.1.6   *Intracellular actions of the quinolones (Froehlich-Ammon and Osheroff, 1995)*

(The structure of these enzymes represents a key difference from the mammalian type I (monomer) and II (homodimer) topoisomerases and this may account for the binding of the quinolones to the prokaryotic, but not the eukaryotic, enzymes.) As might be expected, bacteria have developed resistance to the quinolones through alterations in the target enzymes – alterations to the DNA gyrase occur via mutations in the quinolone resistance-determining region (QRDR) of the *gyrA* gene, which encodes the two A subunits of the tetrameric enzyme (*gyrB* encodes the two B subunits), while similar mutations which decrease quinolone binding have been described in topoisomerase IV (in the *parC* gene, which codes for the two C subunits (*parE* codes for the two E subunits)) (Ruiz, 2003).

Both *Pseudomonas aeruginosa* (Gram negative) (Nikaido, 1996) and *Staphylococcus aureus* (Gram positive) exhibit well-characterised efflux pumps for the quinolones (Putman *et al.*, 2000), but bacteria have also developed resistance to the quinolones through decreased cellular uptake due to the impermeability of the cell membrane. Quinolones cross the outer membrane via specific porins (all quinolones) or by diffusion through the phospholipid bilayer (hydrophobic quinolones only). Porins are protein channels which allow passive diffusion of a specific agent across the cell membrane; for example, *E. coli* bacteria have three main porins, a decrease in the level of one of which (*OmpF*) is associated with an increase in resistance to the quinolones (Cohen *et al.*, 1989). In addition, the outer membrane of *P. aeruginosa* has very low permeability to small hydrophobic molecules, giving this bacterium some intrinsic resistance to the quinolones (Yoshimura and Nikaido, 1982).

2.1.6 Clinical applications

2.1.6.1 Spectrum of activity

The quinolones currently available generally have greatest activity against Gram negative bacteria, with the most susceptible organisms including members of *Enterobacteriaceae*, *Neisseria* species and *Haemophilus* species (Hooper, 1998); *Pseudomonas aeruginosa* and *Acinetobacter* species are also susceptible.

Activity against Gram positive bacteria has, over the years, been debatable, but the newer-generation quinolones, such as levofloxacin and moxifloxacin, appear to show improved activity; for example, moxi-floxacin shows good activity against *Staphylococcus aureus* (meticillin-susceptible) and *Streptococccus pneumoniae* (which are Gram positive bacteria).

2.1.6.2 Urinary tract infections (UTIs)

As mentioned previously, quinolones show good activity against Gram negative bacteria, with, for example, ciprofloxacin particularly active against *Salmonella*, *Shigella*, *Campylobacter*, *Neisseria*, and *Pseudomonas* species. This is important because UTIs, which are relatively common, especially in women, are frequently caused by Gram negative bacteria. It is estimated that 40–50% of women will experience a UTI in their lifetime; the infection is much less common in men, as bacteria can pass up the shorter urethra of a woman much more readily than that of a man. There is also evidence suggesting that prostatic fluid, present in men, has antibacterial properties and provides an additional defence mechanism against UTIs.

Generally, the first-line treatment for uncomplicated UTIs is either trimethoprim (a DHFR inhibitor; see Section 3.2) or nitrofurantoin (which is similar in mode of action to metronidazole; see Section 2.3) for three to seven days (three days treatment is usually sufficient), but if the first-line therapy fails, or bacterial culture suggests there is resistance to these agents, a quinolone can be used. Studies show that ciprofloxacin is highly active against the most common pathogens causing UTIs (e.g. *E. coli*) (Farrell *et al.*, 2003) and this activity appears to be evident for all of the quinolones, as a Cochrane review found no significant differences in clinical or microbiological activity between quinolones for the treatment of uncomplicated acute cystitis (Rafalsky *et al.*, 2006). UTIs in men, as discussed previously, are less common and generally occur as a result of an anatomical or functional problem with the genitourinary tract, so are considered more serious. UTIs in men are therefore often considered to be 'complicated' (as opposed to the uncomplicated UTIs that frequently occur in women) and may be treated with a two-week course of a quinolone (SIGN, 2006).

Quinolones from all generations, including nalidixic acid, ciprofloxacin, ofloxacin, levofloxacin and norfloxacin, are used in the management of UTIs. In elderly patients, when compliance is often a problem, a quinolone with a once-daily dosing regimen, such as ofloxacin, would be preferred over one with a twice-daily dosing regimen, such as ciprofloxacin. Moxifloxacin, a fourth-generation quinolone, is not used in the routine management of uncomplicated UTIs, because it has been associated with the risk of life-threatening hepatotoxicity (Verma *et al.*, 2009). As a result of this potentially serious adverse effect, moxifloxacin is reserved for the treatment of sinusitis, community-acquired pneumonia, or cases of chronic bronchitis that have failed to respond to other antibacterial therapies.

In addition to UTIs, quinolones can be used to treat bacterial prostatitis (inflammation of the prostate gland), which can present as either an acute or a chronic condition, with bacterial infection accounting for the majority of acute cases. Symptoms often include fever (a general sign of infection), dysuria (painful urination), low abdominal pain, pain on ejaculation, and urethral discharge. Ciprofloxacin (and other quinolones) are often preferred to other antibacterial agents for the treatment of bacterial prostatitis, because they penetrate well into the prostate tissue, a property few other antibacterial agents possess (Wolfson and Hooper, 1989). In the case of chronic bacterial prostatitis, extended courses of quinolones (in some cases up to six weeks) are often required.

2.1.6.3   Respiratory tract infections (RTIs)

Quinolones are also used clinically in the treatment of respiratory tract infections (RTIs). These infections, frequently acquired in the community, are often caused by pathogens such as *S. pneumoniae* (Gram positive) and *H. influenzae* (Gram negative). Ciprofloxacin and ofloxacin show good activity against the Gram negative *H. influenzae*, but reduced activity against the Gram positive *S. pneumoniae*, and these agents should therefore not be used in the treatment of pneumococcal pneumonia. The newer agents, such as levofloxacin (third generation) and moxifloxacin (fourth generation), show improved activity against *S. pneumoniae* and thus provide a useful therapeutic alternative to the earlier quinolones. A recent meta-analysis, concerning the treatment of exacerbations of chronic bronchitis, concluded that the clinical success rate was higher with moxifloxacin than for the standard antibiotic therapy (such as amoxicillin or clarithromycin) (Miravitlles *et al.*, 2007).

At present, quinolones are not the recommended first-line treatment for the management of uncomplicated community-acquired pneumonia, which is often managed successfully with amoxicillin (or erythromycin, if the patient is allergic to penicillin; see Section 4.2). In these cases, quinolones are not the recommended first-line treatment, in order to minimise the threat of bacterial resistance, allowing them to be kept in reserve for more serious infections.

In the case of hospital-acquired pneumonia, a quinolone such as ciprofloxacin, ofloxacin, levofloxacin, or moxifloxacin may be used. Hospital-acquired pneumonia can be defined as pneumonia which occurs 48 hours after admission to hospital and is often caused by *S. aureus* (including meticillin-resistant *Staphylococcus aureus* (MRSA)) or, in some cases, opportunistic Gram negative bacilli, such as *P. aeruginosa*. Unfortunately, as outlined previously, some strains of *P. aeruginosa* have developed resistance to the quinolones through the expression of efflux pumps, which limits their role in the treatment of some hospital-acquired pneumonias. Ultimately, the choice of agent for the treatment of hospital-acquired pneumonia is dependent on both bacterial sensitivity and the local antibiotic policy of the hospital (we will discuss this more in Subsection 5.1.6 when we outline the clinical uses of the β-lactams).

2.1.6.4   Sexually transmitted infections (STIs)

Some quinolones also have good activity against strains of *Neisseria gonorrhoeae* (the organism responsible for causing the sexually transmitted infection gonorrhoea). *N. gonorrhoeae* is a Gram negative organism which is found in the semen of infected men and the vaginal fluids of infected women. Around half of women infected with gonorrhoea can be asymptomatic, while men are much more likely to present with symptoms (British Association for Sexual Health and HIV, 2005).

One-off, single-dose treatments (as a tablet) of ciprofloxacin or ofloxacin are preferred in these instances to assure compliance with therapy, which is advantageous as it will help reduce the risk of further sexual transmission of the infection. There have also been reports suggesting that strains of *N. gonorrhoeae* are now beginning to show signs of resistance to ciprofloxacin and ofloxacin (Abeyewickereme *et al.*, 1996); if this is confirmed then cefixime, a third-generation cephalosporin, can be used as a single (oral) dose to treat the infection, although this is an unlicensed indication for cefixime.

2.1.6.5   Tuberculosis

Tuberculosis is caused by *Mycobacterium tuberculosis* and is a major public health concern, particularly in developing countries, where multidrug- (MDR-TB) and extensively drug-resistant (XDR-TB) strains have emerged. If left untreated or not treated appropriately, tuberculosis can be fatal, and it is claimed that tuberculosis causes more adult deaths worldwide than any other infection (Kochi, 1991). Quinolones have activity against *M. tuberculosis*, but until recently it has not been clear what role they have in the treatment of tuberculosis.

A recent, evidence-based Cochrane review concluded that the routine addition or substitution of quinolones to conventional antituberculosis chemotherapy is not recommended (Ziganshina and Squire, 2008). In addition, when ciprofloxacin was used to treat drug-sensitive tuberculosis, it was found to actually increase the relapse rate of infection when substituted into first-line antituberculosis chemotherapy (Ziganshina and Squire, 2008).

As with all infectious diseases, resistance to conventional antimicrobial therapy is becoming a significant clinical problem, as illustrated by the rise of multidrug-resistant tuberculosis (MDR-TB). When a strain of *M. tuberculosis* develops resistance to rifampicin and isoniazid, it is said to be multiple drug-resistant. MDR-TB is very costly and difficult to manage, even in countries with highly developed health care systems, and threatens to jeopardise overall tuberculosis control efforts (Mukherjee *et al.*, 2004). Quinolones appear to have a role in the management of MDR-TB, with the newer generations, such as moxifloxacin, showing increased *in vitro* activity compared to the older generations. Indeed, the quinolones have been included in MDR-TB drug regimens since the late 1980s and their unique mode of action, through DNA-gyrase/topoisomerase inhibition, would appear to offer a welcome additional therapeutic target to those of the drugs found in regular regimens. The exact role of quinolones in the management of TB remains controversial, however; for example, one small randomised controlled trial compared the standard antituberculosis drug regimens to the standard antituberculosis regimens plus levofloxacin in the management of TB (the trial included drug-resistant and drug-susceptible patients) (el-Sadr *et al.*, 1998) and found that there was no statistically significant difference in clinical outcomes between the two regimens. The recent Cochrane review, however, acknowledges that the quinolones have a role in the management of MDR-TB, but suggests that further trials, particularly involving studies on the newer-generation quinolones (such as moxifloxacin), are needed to establish exactly what that role is (Ziganshina and Squire, 2008).

2.1.6.6   Anthrax

Ciprofloxacin has also had an important role to play in the treatment of anthrax, which was used as a biological weapon in a relatively recent terrorist attack. This occurred just after the September 11th terrorist attacks in 2001; anthrax spores were sent through the US postal service to several media outlets and two US senators. As a result of this bioterrorist attack, at least five people died, and the number of victims (who required antibiotics) was as many as 68 people (Cymet and Kerkvliet, 2004). Anthrax, caused by *Bacillus anthracis*, is often fatal if left untreated. Ciprofloxacin shows activity against *B. anthracis* and, as a result, the FDA recommends the use of ciprofloxacin to treat anyone who has been exposed to anthrax spores (Doganay and Aydin, 1991), with a recommended dose of ciprofloxacin of 500 mg twice daily for 60 days (alternative agents are penicillins and doxycycline, but the drug of choice remains ciprofloxacin). The treatment period of 60 days remains a problem in terms of patient compliance and studies have demonstrated that eradication of the spore burden can be achieved after 35 days of treatment (95% of patients). Spore burden eradication was achieved in 99.5% of patients after 110 days' treatment, so 60 days remains a suitable compromise (Drusano *et al.*, 2008). Safety data are limited for the use of ciprofloxacin during pregnancy and lactation, and as result the US Centers for Disease Control and Prevention (CDC) recommended amoxicillin in pregnant and lactating mothers, when the strain is penicillin-sensitive. Retrospective safety data show that exposure of the developing foetus to norfloxacin, ciprofloxacin, and ofloxacin is not associated with an increased risk of malformations: 2.2% rate of malformations in the quinolone group compared to 2.6% in the control group (Loebstein *et al.*, 1998), although the rate of spontaneous abortions was higher in the quinolone-treated group of women, with a relative risk of 4.50. As data are still limited, the agent of choice in pregnant women remains penicillin.

2.1.6.7   Miscellaneous

Ciprofloxacin is often used to treat diarrhoea, which is one of the most common illnesses affecting travellers, particularly in developing countries. Travellers' diarrhoea is often primarily infective in origin, with *E. coli*, *Salmonella*, and *Campylobacter* species all implicated. A single 500 mg dose of ciprofloxacin (or levofloxacin) is sufficient to reduce the severity of symptoms and also the duration of diarrhoea (Salam *et al.*, 1994; Sanders *et al.*, 2007). While travellers' diarrhoea is usually self-limiting, such single-dose quinolone therapy helps to reduce the risk of developing complications, such as dehydration, which is particularly common in older people.

Quinolones are also used for the treatment of enteric fever (also known as typhoid or paratyphoid fever) caused by *Salmonella typhi* and *Salmonella paratyphi*. Quinolones are indicated for treating drug-sensitive and multiple drug-resistant (when there is resistance to chloramphenicol, amoxicillin, and co-trimoxazole) *S. typhi* and *S. paratyphi* in children and adults. Enteric fever is a major public health concern and has reached endemic status in many areas of the developing world. A recent meta-analysis published in the *British Medical Journal* and a Cochrane review concluded that quinolones may be better in helping to prevent clinical relapse of enteric fever than chloramphenicol, although data for recommending quinolones as first-line treatment in children are limited (Thaver *et al.*, 2008, 2009).

There have also been developments in expanding the therapeutic use of the quinolones already on the market. A good example of this is the recent development of ciprofloxacin eye drops and eye ointment and ofloxacin eye drops, marketed in the UK as Ciloxin and Oftaquix, respectively. Ciloxin eye drops are licensed for the treatment of corneal ulcers and superficial eye infections, while Ciloxin eye ointment is licensed to treat corneal ulcers, conjunctivitis, and blepharitis. These products are effective against *P. aeruginosa*, an organism that can cause corneal ulceration, which, if left untreated, can lead to blindness. Oftaquix is licensed for the topical treatment of bacterial external ocular infections caused by microorganisms susceptible to levofloxacin. All of these products are not recommended for use in children under one year of age, as the clinical experience with such products is limited.

In addition, quinolone **eye** drops are also used to treat chronic otitis media (an inner-ear infection). This use may seem surprising, but ciprofloxacin or ofloxacin ear drops are only available on a named-patient basis (and must be ordered from a specialised company and so are difficult to obtain), so the eye formulations are often used as an unlicensed alternative. Ciprofloxacin and ofloxacin ear drops (or eye drops if the ear drops cannot be obtained) also offer an alternative to aminoglycoside antibiotic ear drops in the management of chronic otitis media in patients who have a perforated tympanic membrane.

2.1.7   Adverse drug reactions

Quinolones are generally well tolerated but mild side effects such as nausea, vomiting, dyspepsia, abdominal pain, diarrhoea, headache, and rashes have been reported. Less frequent side effects include blood disorders, such as eosinophilia (abnormally high numbers of eosinophils), leucopenia (decreased total number of white blood cells), and thrombocytopenia (decreased number of platelets), as well as anorexia, sleep disturbances, confusion, anxiety, depression, and hallucinations.

Very rare side effects (those that occur in less than 1 in 10 000 patients) include tendon inflammation and damage (mainly affecting the Achilles tendon), psychoses, and convulsions. With regard to tendon damage and inflammation, the Committee on Safety of Medicines (CSM) has advised that quinolones are contra-indicated in patients with a history of tendon disorders associated with previous quinolone use. These adverse effects are rare, but are more common in elderly patients and in patients who have been treated with corticosteroids (Khaliq and Zhanel, 2005). Tendon damage is most common in the lower extremities, but has also been reported in the upper extremities. If a patient experiences any joint pain or inflammation, the quinolone should be discontinued immediately and care should be taken not to put excess weight on the affected extremity.

Quinolones should not be routinely used in children because of the hypothetical risk of arthropathy in weight-bearing joints. Although this risk has been established in immature animals but not, as yet, in humans, it is prudent to avoid the use of quinolones in children, unless the therapeutic benefit outweighs the potential risks associated with treatment. There are, however, some cases where the short-term use of a quinolone is warranted in children; for example, ciprofloxacin is often used to treat pseudomonal infections in cystic fibrosis patients over the age of five.

The side effects outlined above are all general to the quinolone family, but some side effects are specific to individual quinolones. For example, moxifloxacin (third generation) has been associated with life-threatening hepatotoxicity, resulting in fatalities (Verma *et al.*, 2009). This adverse effect is considered to be very rare, but because of the serious consequences, patients are advised to contact their doctor if signs of hepatotoxicity, such as developing asthenia (physical weakness) associated with jaundice, dark urine, a bleeding tendency, or hepatic encephalopathy develop (Avelox summary of product characteristics).

2.1.7.1 Effect on QTc wave and proarrhythmic potential

Some of the quinolones in clinical use demonstrate the adverse effect of prolongation of the QTc wave during administration, which can lead to torsade de pointes, a cardiac arrhythmia which can be life-threatening (Figure 2.1.7).

This effect has been clearly demonstrated for grepafloxacin, moxifloxacin, and gemifloxacin, the former of which was withdrawn from the market in 2003. Moxifloxacin, when administered at the recommended dose of 400 mg once daily, also prolonged the QTc wave interval (Tsikouris *et al.*, 2006). As a result of this, it is recommended that if signs of a cardiac arrhythmia occur during moxifloxacin therapy, treatment should be stopped and an ECG should be performed. Moxifloxacin is also contraindicated in patients with a history of QT interval prolongation, bradycardia, clinically relevant heart failure with reduced left-ventricular ejection fraction, electrolyte disturbances (particularly in uncorrected hypokalaemia), or symptomatic arrhythmias (Avelox summary of product characteristics).

In comparison, levofloxacin and ciprofloxacin, when administered at the recommended adult doses, did not exhibit the same effect upon the QTc wave, which led to further investigation of the structure activity-adverse event relationships of these drugs. Further retrospective analyses of quinolone use in the USA showed that QTc wave prolongation was demonstrated with levofloxacin (13 cases), gatifloxacin (8 cases), ciprofloxacin (2 cases), and ofloxacin (2 cases), with moxifloxacin not being associated with any cases (Frothingham, 2001). It would appear, therefore, that the proarrhythmic potential of the quinolones may depend upon the dose administered. However, structure activity-adverse event profiling has shown that some of the quinolones investigated affect the human cardiac potassium channel HERG

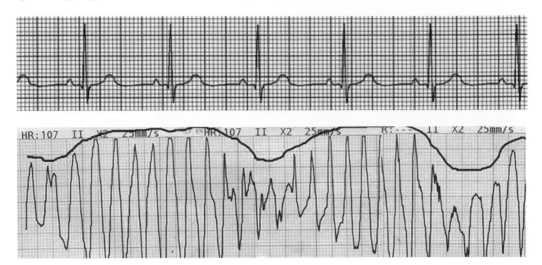

Figure 2.1.7   *Traces showing a normal electrocardiogram (above) and torsade de pointes (below). The blue line shows the characteristic 'twisting points' associated with torsade de pointes (Reproduced from 'Torsades de pointe', Wikipedia, Online, [http://fr.academic.ru/dic.nsf/frwiki/1645373].)*

(human ether a-go-go channel), which contributes to the electrical activity (and so rhythm) of the heart (Figure 2.1.8).

Several studies have illustrated that all fluoroquinolones inhibit HERG channel currents, but with differing potencies (Kang *et al.*, 2001). Inhibition of HERG channels occurs with grepafloxacin, moxifloxacin, and

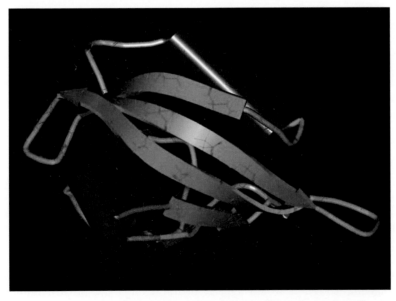

Figure 2.1.8   *The* N*-terminus of the HERG (human ether a-go-go channel); PDB code 1BYW (Morais Cabral et al., 1998)*

gatifloxacin at therapeutic plasma levels, whereas levofloxacin, ciprofloxacin, and ofloxacin require levels greater than those used clinically to demonstrate this effect. These data demonstrate that the effect upon HERG channel inhibition is not a class effect and that structural substitutions of the quinolone nucleus may affect affinity for this channel. Animal studies and analyses of clinical cases demonstrate that those agents most associated with HERG-channel blockade, namely sparfloxacin and grepafloxacin, have substituents in the C5 position of the quinolone nucleus (Figure 2.1.1) and those compounds that are unsubstituted at C5, such as levofloxacin, ciprofloxacin, and ofloxacin, exhibit reduced HERG blockade (Falagas *et al.*, 2007). Another possible structural interpretation is that the reduced HERG blockade occurs for those compounds with either unsubstituted C8 (ciprofloxacin) or substituents linked from C8 to N1, to create a conformationally restricted structure (levofloxacin, ofloxacin). Lastly, methoxy substitution at C8, as with gatifloxacin and moxifloxacin, exhibits HERG channel blockade at therapeutic doses. It is clear that further investigation is required in order to establish structure activity-adverse events relationships with the fluoroquinolones and QTc wave prolongation via the HERG channel, and it is also worth considering the pharmacokinetic profiles of the agents described, which may have an effect upon respective plasma concentrations. However, these structural differences may have an effect upon the development of future quinolones and may restrict their potential use.

2.1.7.2 Dermatological side effects

Another commonly observed adverse event with the fluoroquinolones is phototoxicity (Figure 2.1.9). This effect is particularly evident with those agents substituted with a halogen at the C8 position of the quinolone nucleus. These phototoxic effects have been demonstrated in both animal and human models, with particular reference to sparfloxacin (Domagala *et al.*, 1986; Domagala, 1994). Compounds that are methoxy-susbstituted at the C8 position (moxifloxacin and gatifloxacin) do not appear to demonstrate any significant phototoxic potential; this may be due to stabilisation of the quinolone nucleus with respect to ultraviolet light (Matsumoto *et al.*, 1992). Ultraviolet-halogenated quinolone toxicity has been associated with the generation of superoxide anions and free radicals, which in turn initiate prostaglandin release (such as prostaglandin E2) via tyrosine kinase and protein kinase C mechanisms (Shimoda, 1998; Viola *et al.*, 2004).

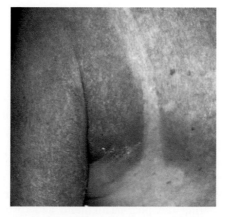

Figure 2.1.9 *Phototoxicity observed with quinolone use (Reprinted from T. Williams, 'Phototoxicity', Dermatology, Online, [http://drtedwilliams.net/kb/index.php?pagename=Dermatology] 2008.)*

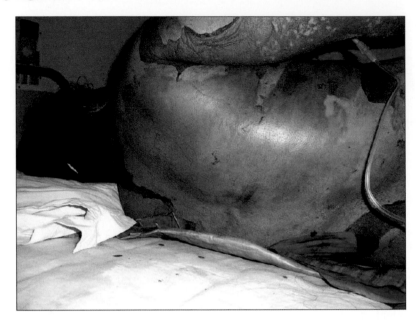

***Figure* 2.1.10**   *Clinical presentation of Lyell's syndrome (toxic epidermal necrolysis) caused by ciprofloxacin (Reprinted from G.M. Upadya and K. Ruxana, Indian J Med Sci. 63 (10) 461–3, 2009, with permission of the Board of the Indian Journal of Medical Sciences.)*

In addition, serious skin disorders, such as Stevens–Johnson syndrome (ulcerated skin lesions) or Lyell's syndrome (toxic epidermal necrolysis), have also been reported with quinolone use (such as ciprofloxacin or moxifloxacin) (Nori *et al.*, 2004). Stevens–Johnson and Lyell's syndromes are both potentially fatal, severe, bullous skin disorders that are most commonly drug-induced. Most cases are characterised by erythema (redness), blisters, necrosis, and eventual detachment of the epidermis and mucous membranes from the dermis (Figure 2.1.10), leaving the patient vulnerable to opportunistic infections, which can result in septicemia and possibly even death. Patients should thus be advised to contact their doctor if any skin reactions occur during treatment with a quinolone.

2.1.7.3   Central nervous system (CNS) effects

CNS effects represent the second most common group of events associated with the use of quinolone antibiotics (Paton and Reeves, 1992), with direct CNS effects of the quinolones including headache, dizziness, sleep disorders, agitation, and, in severe cases, convulsions. It has been postulated that these events are associated with the interaction of quinolones with central GABA receptors, causing direct CNS stimulation (Halliwell *et al.*, 1993). CNS effects are not commonly observed at therapeutic levels of administered quinolones; however, other, less-specific pharmacological mechanisms may exist with respect to CNS events. What is clear is that substitution of the quinolone nucleus at C7 with aliphatic side chains (e.g. enoxacin, norfloxacin, and ciprofloxacin) confers greater CNS adverse activity, probably due to greater affinity to the GABA receptors. Bulkier side chains (e.g. those in temafloxacin and sparfloxacin) appear to lead to reduced binding to GABA receptors. As with all adverse events, prediction of potential action is difficult without considering the pharmacokinetic and distribution profile of these agents, but it would appear that bulkier C7 substituents and lower lipophilicity result in fewer CNS adverse events.

As a result of these CNS effects, quinolones should be used with caution in patients with epilepsy, as there is the rare possibility of them causing seizures. In addition, the concomitant use of quinolones and NSAIDs (non-steroidal anti-inflammatory drugs) should be avoided because of the increased risk of seizures. This phenomenon was first observed when enoxacin was co-administered with fenbufen, which resulted in seizures in animals (Christ, 1990). The exact mechanism of this interaction has not been established, but it has been suggested that it may again be due to inhibition of central GABA receptors (Ito *et al.*, 1996).

2.1.8   Drug interactions

2.1.8.1   Antacids

As mentioned in Subsection 2.1.3, there is a well-documented drug interaction between the quinolones and antacids (or calcium-containing compounds) which impairs the absorption of the quinolones, causing the serum concentrations to fall below the MIC, leading to treatment failure and possibly also to the development of bacterial resistance. For example, when 750 mg of ciprofloxacin was given 2 hours after magnesium hydroxide and aluminum hydroxide (both of which are commonly present in antacids), the relative bioavailability was reduced to 23.2% of that when ciprofloxacin was administered without the antacids (i.e. around 80% of the dose of ciprofloxacin was not absorbed) (Nix *et al.*, 1989). It is thought that absorption of the quinolone is impaired through the formation of an insoluble chelate between the cationic antacid ions (such as magnesium ions or aluminum ions) and the 3-keto and 4-carboxyl functional groups of the quinolone (see Figure 2.1.1). As all of the quinolones currently available have these functional groups, they all interact with antacids to some extent, with the interaction being more significant with aluminum/magnesium-containing antacids than with calcium-containing antacids. It is important to note, however, that this drug interaction does not necessarily mean that combining the two drugs is contraindicated. Indeed, this interaction can be effectively managed by separating out the doses of the quinolone and antacid and it is generally recommended that the quinolone should be taken at least 2 hours before, and not less than 4–6 hours after, the antacid (Baxter, 2007).

2.1.8.2   Theophylline

Another clinically significant drug interaction is that between some quinolones and theophylline. Theophylline is a phosphodiesterase inhibitor which is used clinically in the management of respiratory diseases, such as chronic obstructive pulmonary disease (COPD) and asthma, although it is now rarely used due to its side effects and the discovery of the beta-2 agonists. Theophylline is metabolised through *N*-demethylation by cytochrome P450, CYP 1A2, and some quinolones appear to inhibit this enzyme, thus decreasing the clearance of theophylline and increasing the serum theophylline level. Molecular modeling studies and NMR binding studies have illustrated that the quinolone keto group, carboxylic acid, and *N*1 functionalities are probably the important groups for CYP 1A2 binding (Mizuki *et al.*, 1996). This interaction is significant for ciprofloxacin, norfloxacin, and ofloxacin, while no significant interaction has been documented for levofloxacin, moxifloxacin, or nalidixic acid. For example, one case describes a rise in theophylline serum levels by nearly sixfold when co-administered with ciprofloxacin (Baxter, 2007). Theophylline has a very narrow therapeutic index and such a dramatic increase in serum levels may cause toxicity, such as nausea, vomiting, agitation, tachycardia, and, more seriously, convulsions and supra-ventricular and ventricular arrhythmias. There have been a number of fatalities reported as a consequence of this drug interaction, particularly with ciprofloxacin (Baxter, 2007), which are well documented and of clinical significance. It appears to be most common in the elderly, or when a high dose of theophylline is used. The most sensible solution to this drug interaction would be to substitute the quinolone by a more suitable therapeutic alternative, such as a penicillin or tetracycline (if appropriate). If, however, this is not possible, a quinolone should be used that has no documented interaction with theophylline (such as levofloxacin). If, again, this is not possible, then a dose reduction of theophylline should be considered, combined with careful monitoring of serum theophylline levels – it has been suggested that the dose of theophylline should be reduced by up to 50% when ciprofloxacin treatment is started (Baxter, 2007).

2.1.8.3   Digoxin

Approximately 10% of the population have the Gram positive anaerobic bacilli *Eubacterium lentum* present within their normal gut flora. This organism has been shown *in vitro* to reduce digoxin to its inactive form, (20*R*)-dihydrodigoxin (Robertson *et al.*, 1986; Mathan *et al.*, 1989). Studies have demonstrated that the administration of erythromycin and tetracycline markedly reduced the levels of cardioinactive digoxin reduction products (DRPs), which implies that the bacteria normally present in the gut which inactivate digoxin were significantly eradicated, thus potentially increasing levels of absorbed 'active' digoxin (Lindenbaum *et al.*, 1981). It has been postulated that the administration of quinolones may reduce *E. lentum* levels in the gut and thus a reduction in the metabolism of digoxin to its inactive form may occur more readily, effectively increasing the potential for active digoxin to be absorbed and potentially leading to toxicity. However, several studies have demonstrated that certain quinolones, while decreasing gut flora growth (such as enterococci, *Corynebacterium*, lactobacilli, clostridia, and *Bacteroides*), actually allowed for the increased growth of *Eubacterium*, probably due to an increased potential environment (Nord *et al.*, 2003). Co-administration of quinolones and digoxin should still be treated with caution, as the antibacterial spectra of the quinolones do differ and some may reduce *E. lentum* growth. Careful monitoring of digoxin plasma should be performed when a patient is treated with a quinolone to ensure that the levels remain between 1.0 and 2.5 nmol/L.

2.1.9   Recent developments

The quinolones still remain attractive antimicrobial agents, retaining high potency, with a broad spectrum of activity. They have good bioavailability, attaining high serum concentrations, with a subsequently impressive V_d achieving high tissue concentrations.

The newer-generation quinolones appear to have a wider therapeutic activity than the older quinolones, but this advantage has been slightly compromised by an increased risk of side effects; for example, the addition of a fluorine atom at the 8 position of the quinolone nucleus has been linked to photosensitivity, while the substitution of an amine or methyl group at the 5 position has resulted in the potential to prolong the QTc wave and thus cause subsequent arrhythmias. Another problem encountered with the quinolones has been antimicrobial resistance. There are essentially three mechanisms of bacterial resistance towards the quinolone antibiotics, as described in Subsection 2.1.5. Despite the setbacks of resistance and toxicity, the success of the quinolones has encouraged further research into developing new quinolone derivatives, some of which are described in Table 2.1.2 and shown in Figure 2.1.11. The aim of the research is to develop novel quinolone antibacterials, which have a broad spectrum of activity, but are devoid of the adverse effects associated with the newer generations.

Table 2.1.2 A summary of the quinolones at various stages of development

Quinolone (generation)	Date registered and agency	Indication	Date withdrawn	Reasons for withdrawal	Reference
Gemifloxacin (fourth generation)	FDA approved in April 2003 MA submitted in March 2008 to European medicines agency	Community-acquired pneumonia, acute bacterial exacerbation of chronic bronchitis	Menarini International Operations Luxembourg SA withdrew MA application, June 2009	CHMP concluded that gemifloxacin did not have a positive benefit–risk balance; currently licensed in the USA	http://www.ema.europa.eu/humandocs/PDFs/EPAR/factive/38240809en.pdf
Nadifloxacin	Available in Japan	Used for the topical treatment of acne vulgaris	-	-	Jacobs and Appelbaum (2006)
Sitafloxacin (fourth generation)	Currently marketed in Japan as Gracevit				Anderson (2008)
Olamufloxacin	Currently under development		-	-	Jiraskova (2000)
DQ113	Currently under development	Shown activity against MRSA and other Gram positive bacteria	-	-	Tanaka (2002)
DW286	Currently under development	Shown activity against *Streptococcus pneumoniae* and MRSA	-	-	Yun *et al.* (2002)
Sparfloxacin (third generation)	December 1996, FDA		February 2001	QT interval prolongation	Rubinstein and Camm (2002)
Temafloxacin (third generation)	January 1992, FDA	UTIs, lower respiratory tract, skin and prostate infections	June 1992	Haemolytic anaemia (linked to several deaths)	Blum *et al.* (1994)

(*Continued*)

Table 2.1.2 (Continued)

Quinolone (generation)	Date registered and agency	Indication	Date withdrawn	Reasons for withdrawal	Reference
Grepafloxacin (third generation)	November 1997, FDA January 1998, MHRA	Broad-spectrum activity, often used for bronchitis	October 1999, worldwide	QT interval prolongation	Rubinstein and Camm (2002)
Gatifloxacin (fourth generation)	December 1999, FDA	RTIs	BMS stopped manufacture in May 2006; still available in USA and Canada as an ophthalmic solution	Hypoglycaemia, diabetes	Park-Wyllie et al. (2006)
Garenoxacin (fourth generation)	Not licensed	Community-acquired pneumonia, acute bacterial exacerbation of chronic bronchitis	Schering-Plough Europe withdrew MA application, July 2007	Schering-Plough Europe failed to provide information to the licensing authority	http://www.ema. europa.eu/ humandocs/PDFs/ EPAR/ garenoxacinmesy late/3411 7407en.pd
Tosufloxacin (third generation)	Only available in Japan		-	-	Niki (2002)

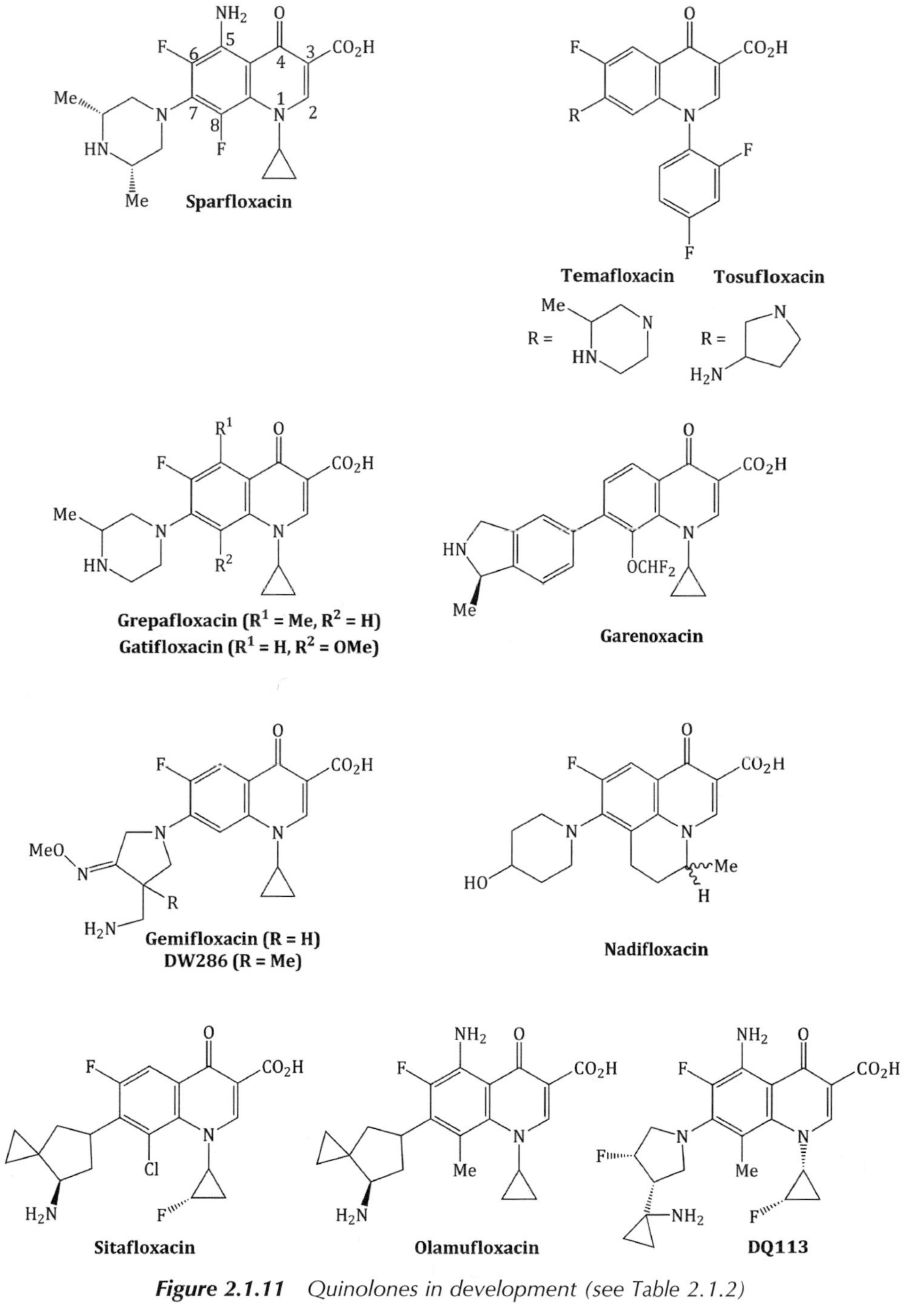

Figure 2.1.11 *Quinolones in development (see Table 2.1.2)*

References

I. Abeyewickereme, L. Senaratne, and V. B. Prithiviraj, *Genitourin. Med.*, 1996, **72**, 301–302V.

D. L. Anderson, *Drugs Today (Barc)*, 2008, **44**, 489–501.

Avelox summary of product characteristics (http://emc.medicines.org.uk, last accessed 7 January 2010).

K. Baxter (Ed.), *Stockley's Drug Interactions*, 8th Edn, Pharmaceutical Press, London, 2007.

J. M. Blondeau, *Clin Ther.*, 1999, **21**, 3–40.

M. D. Blum, D. J. Graham, and C. A. McCloskey, *Clin. Infect. Dis.*, 1994, **18**, 946–950.

British Association for Sexual Health and HIV, National guideline on the diagnosis and treatment of gonorrhoea in adults 2005, 2005 (http://www.bashh.org/documents/116/116.pdf, last accessed 14 March 2012).

J. J. Champoux, *Ann. Rev. Biochem.*, 2001, **70**, 369–413.

D. Cheng, W.-R. Xu, and C.-X. Liu, *World J. Gastroenterol.*, 2007, **13**, 2496–2503.

W. Christ, *J. Antimicrob. Chemother.*, 1990, **26**(Suppl B) 219–225.

S. P. Cohen, L. M. McMurry, D. C. Hooper, J. S. Wolfson, and S. B. Levy, *Antimicrob. Agents Chemother.*, 1989, **33**, 1318–1325.

T. C. Cymet and G. J. Kerkvliet, *J. Amer. Osteopathic Assoc.*, 2004, **104**, 452.

M. Doganay and N. Aydin, *Scand. J. Infect.*, 1991, **23**, 333–335.

J. M. Domagala, *J. Antimicrob. Chemother.*, 1994, **33**, 685–706.

J. M. Domagala, L. D. Hanna, C. L. Heifetz, M. P. Hutt, T. F. Mich, J. P. Sanchez, and M. Solomon, *J. Med. Chem.*, 1986, **29**, 394–404.

K. Drlica and X. Zhao, *Microbiol. Mol. Biol. Rev.*, 1997, **61**, 377–392.

G. L. Drusano, O. O. Okusanya, A. Okusanya, B. Van Scoy, D. L. Brown, R. Kulawy, F. Sorgel, H. S. Heine, and A. Louie, *Antimicrob. Agents Chemother.*, 2008, **52**, 3973–3979.

M. Duguet, *J. Cell Science*, 1997, **110**, 1345–1350.

W. M. el-Sadr, D. C. Perlman, J. P. Matts, E. T. Nelson, D. L. Cohn, N. Salomon, M. Olibrice, F. Medard, K.D. Chirgwin, D. Mildvan, B. E. Jones, E. E. Telzak, O. Klein, L. Heifets, and R. Hafner, *Clin. Infect. Dis.*, 1998, **26**, 1148–1158.

A. M. Emmerson and A. M. Jones, *J. Antimicrob. Chemother.*, 2003, **51**, 13–20.

M. E. Falagas, P. I. Rafaildis, and E. S. Rosmarakis, *Int. J. Antimicrob. Agents.*, 2007, **29**, 373–379.

D. J. Farrell, I. Morrissey, D. De Rubeis, M. Robbins, and D. Felmingham, *J. Infection.*, 2003, **46**, 94–100.

S. J. Froehlich-Ammon and N. Osheroff, *J. Biol. Chem.*, 1995, **270**, 21 429–21 432.

R. Frothingham, *Pharmacotherapy.*, 2001, **21**, 1468–1472.

R. F. Halliwell, P. G. Davey, and J. J. Lambert, *J. Antimicrob. Chemother.*, 1993, **31**, 457–462.

D. C. Hooper, *Biochim. Biophys. Acta.*, 1998, **1400**, 45–61.

Y. Ito, T. Miyasaka, H. Fukuda, K. Akahane, and Y. Kimura, *Neuropharmacology*, 1996, **35**, 1263–1269.

M. R. Jacobs and P. C. Appelbaum, *Expert Opin. Pharmacother.*, 2006, **7**, 1957–1966.

N. Jiraskova, *Curr. Opin. Investig. Drugs*, 2000, **1**, 31–34.

J. Kang, L. Wang, X.-L. Chen, D. J. Triggle, and D. Rampe, *Mol. Pharmacol.*, 2001, **59**, 122–126.

Y. Khaliq and G. G. Zhanel, *Clinics in Plastic Surgery*, 2005, **32**, 495–502.

A. Kochi, *Tubercle.*, 1991, **72**, 1–6.

I. Laponogov, M. K. Sohi, D. A. Veselkov, X. S. Pan, R. Sawhney, A. W. Thompson, K. E. McAuley, L. M. Fisher, and M. R. Sanderson, *Nature Struct. & Mol. Biol.*, 2009, **16**, 667–669.

J. Lindenbaum, D. G. Rund, V. P. Butler, D. Tse-Eng and J.R. Saha, *N. Eng. J. Med.*, 1981, **305**, 789–794.

R. Loebstein, A. Addis, E. Ho, R. Andreou, S. Sage, A. E. Donnenfeld, B. Schick, M. Bonati, M. Moretti, A. Lalkin, A. Pastuszak, and G. Koren, *Antimicrob. Agents Chemother.*, 1998, **42**, 1336–1339.

A. A. MacDonald, S. H. DeWitt, E. M. Hogan, and R. Ramage, *Tet. Lett.*, 1996, **37**, 4815–4818.

M.G. Matera, *Pulmonary Pharmacol. Ther.*, 2006, **19**, 20–29.

V. I. Mathan, J. Wiederman, and J. F. Dobkin, *Gut*, 1989, **30**, 971–977.

M. Matsumoto, K. Kojima, H. Nagano, S. Matsubara, and T. Yokata, *Antimicrob. Agents Chemother.*, 1992, **36**, 1715–1719.

G. Melnik, W. H. Schwesinger, L. C. Dogolo, R. Teng, and J. Vincent, *Am. J. Surgery*, 1998, **176**(Suppl 6A), 14S–17S.

M. Miravitlles, J. Molina, and M. Brosa, *Arch. Bronconeumol.*, 2007, **43**, 22–28.

Y. Mizuki, K. Yamamoto, T. Yamaguchi, T. Fujii, H. Miyazaki, and H. Ohmori, *Xenobiotica*, 1996, **26**, 1057–1066.

J. H. Morais Cabral, A. Lee, S. L. Cohen, B. T. Chait, M. Li, and R. Mackinnon, *Cell*, 1998, **95**, 649–655.

J. S. Mukherjee, M. L. Rich, A. R. Socci, J. K. Joseph, F. Alcántara Virú, S. S. Shin MD, J. J. Furin, M. C. Becerra, D. J. Barry, J. Y. Kim, J. Bayona, P. Farmer, M. C. Smith Fawzi, and K. J. Seung, *Lancet*, 2004, **363**, 474–481.

H. Nikaido, *J. Bacteriol.*, 1996, **178**, 5833–5839.

Y. Niki, *J. Infect. Chemother.*, 2002, **8**, 1–18.

D. E. Nix, W. A. Watson, M. E. Lener, R. W. Frost, G. Krol, H. Goldstein, J. Lettieri, and J. J. Schentag, *Clin. Pharmacol. Ther.*, 1989, **46**, 700–705.

C. E. Nord, L. Meurling, R. L. Russo, A. Bello, D. M. Grasela, and D. A. Gajjar, *J. Chemother.*, 2003, **15**, 244–247.

S. Nori, C. Nebesio, R. Brashear, and J. B. Travers, *Arch. Dermatol.*, 2004, **140**, 1537–1538.

L. Y. Park-Wyllie, D. N. Juurlink, A. Kopp, B. R. Shah, T. A. Stukel, C. Stumpo, L. Dresser, D. E. Low, and M. M. Mamdani, *N. Eng. J. Med.*, 2006, **354**, 1352–1361.

D. H. Paton and D. S. Reeves, *Adverse Drug Reaction Bulletin*, 1992, **153**, 575–578.

M. A. C. Perez, H. G. Dıaza, and C. F. Teruel, *Eur. J. Pharmaceut. Biopharmaceut.*, 2002, **53**, 317–325.

G. Perletti, F. M. E. Wagenlehner, K. G. Naber, and V. Magri, *Int. J. Antimicrob. Agents*, 2009, **33**, 206–210.

L. J. V. Piddock, Y. F. Jin, and D. J. Griggs, *J. Antimicrob. Chemother.*, 2001, **47**, 261–270.

M. Putman, H. W. Van Been, and W. L. Konings, *Microbiol. Mol. Biol. Rev.*, 2000, **64**, 672–693.

S. Rádl, *Arch. Pharm. Chem. Med. Chem.*, 1996, **329**, 115–119.

V. V. Rafalsky, I. V. Andreeva, and E. L. Rjabkova, *Cochrane Database of Systematic Reviews*, 2006, Issue 3.

L. W. Robertson, A. Chandrasekaran, R. H. Reuning, J. Hui, and B. D. Rawal, *Applied Environ. Microbiol.*, 1986, **51**, 1300–1303.

K. A. Rodvold and M. Neuhauser, *Pharmacotherapy*, 2001, **21**(10 Pt 2), 233S–252S.

D. L. Ross, S. K. Elkington, and C. M. Riley, *Int. J. Pharmaceutics*, 1992, **88**, 379–389.

E. Rubinstein and J. Camm, *J. Antimicrob. Chemother.*, 2002, **49**, 593–596.

J. Ruiz, *J. Antimicrob. Chemother.*, 2003, **51**, 1109.

I. Salam, P. Katelaris, S. Leigh-Smith, and M. J. Farthing, *Lancet*, 1994, **344**, 1537–1539.

J. W. Sanders, R. W. Frenck, S. D. Putnam, M. S. Riddle, J. R. Johnston, S. Ulukan, D. M. Rockabrand, M. R. Monteville, and D. R. Tribble, *Clin. Infect. Dis.*, 2007, **45**, 294–301.

K. Shimoda, *Toxicol Lett.*, 1998, **102**, 369–373.

SIGN, Management of suspected bacterial UTIs in adults, 2006 (http://www.sign.ac.uk/pdf/sign88.pdf, last accessed 14 March 2012).

M. Tanaka, E. Yamazaki, M. Chiba, K. Yoshihara, T. Akasaka, M. Takemura, and K. Sato, *Antimicrob. Agents Chemother.*, 2002, **46**, 904–908.

D. Thaver, A. K. M. Zaidi, J. Critchley, A. Azmatullah, S. A. Madni, and Z. A. Bhutta, *Cochrane Database of Systematic Reviews*, 2008, Issue 4.

D. Thaver, A. K. M. Zaidi, J. Critchley, A. Azmatullah, S. A. Madni, and Z. A. Bhutta, *Brit. Med. J.*, 2009, **338**, b1865.

K. Tougou, A. Nakamura, S. Watanabe, Y. Okuyama, and A. Morino, *Drug Metab. Disposition*, 1998, **26**, 355–359.

J. P. Tsikouris, M. J. Peeters, C. D. Cox, G. E. Meyerrose, and C. F. Seifert, *Ann. Noninvasive Electrocardiol.*, 2006, **11**, 52–56.

P. Vaudaux, P. Francois, C. Bisognano, J. Schrenzel, and D. P. Lew, *Antimicrob. Agents Chemother.*, 2002, **46**, 1503–1509.

R. Verma, R. Dhamija, D. Batts, S. C. Ross, and M. E. Loehrke, *Cases Journal*, 2009, **2**, 8063–8065.

G. Viola, L. Facciola, and M. Canton, *Chem and Biodiversity*, 2004, **1**, 782–801.

J. S. Wolfson and D. C. Hooper, *J. Antimicrob. Chemother*, 1989, **33**, 1655–1661.

D. H. Wright, G. H. Brown, M. L. Peterson, and J. C. Rotschafer, *J. Antimicrob. Chemother.*, 2000, **46**, 669–683.

F. Yoshimura and H. Nikaido, *J. Bacteriol.*, 1982, **152**, 636–642.

H.-J. Yun, Y.-H. Min, J.-A. Lim, J.-W. Kang, S.-Y. Kim, M.-J. Kim, J.-H. Jeong, Y.-J. Choi, H.-J. Kwon, Y.-H. Jung, M.-J. Shim, and E.-C. Choi, *Antimicrob. Agents Chemother.*, 2002, **46**, 371–374.

L. E. Ziganshina and S. B. Squire, *Cochrane Database of Systematic Reviews*, 2008, Issue 1.

2.2

Rifamycin antibacterial agents

Key Points

- The rifamycins are semisynthetic antibiotics which have broad-spectrum activity.
- Resistance to the rifamycins arises due to mutations in the *rpo*B gene, which encodes the β-subunit of RNA polymerase.
- Rifampicin is a potent enzyme inducer and is associated with many clinically significant drug interactions.
- For further information, see Mariani and Maffioli (2009) and Sensi (1983).

Antibacterial rifamycins used clinically are rifampicin (rifampin), rifabutin, rifapentine, and rifaximin; their structures are shown in Figure 2.2.1 and their therapeutic indications are listed in Table 2.2.1.

2.2.1 Discovery

The rifamycins are a class of the ansamycin antibiotics (Mariani and Maffioli, 2009; Floss and Yu, 2005), so called because an aliphatic chain links two nonadjacent carbons of a cyclic system, giving the structure a similarity to the handle of a basket (*ansa* being the Latin for 'handle'). In the case of the rifamycins, the aliphatic chain contains 17 atoms, with an ether (C-O) link at one end and an amide (CONH) at the other, while the cyclic system is a naphtho(2,1-*b*)furan (highlighted in red in the structure of rifamycin in Figure 2.2.1). The rifamycins were first isolated in 1957 from a soil sample, from the beach-side town of St Raphael in France, containing a microorganism which seems to have had some form of identity crisis, having first been classified as *Streptomyces mediterranei*, then *Nocardia mediterranea*, and finally *Amycolatopsis mediterranei* (despite the name changes, there are no prizes for guessing which coast of France St Raphael is on). The antibacterial extract contained a mixture of compounds, with only rifamycin B, the least biologically active

Antibacterial Agents: Chemistry, Mode of Action, Mechanisms of Resistance and Clinical Applications, First Edition.
Rosaleen J. Anderson, Paul W. Groundwater, Adam Todd and Alan J. Worsley.
© 2012 John Wiley & Sons, Ltd. Published 2012 by John Wiley & Sons, Ltd.

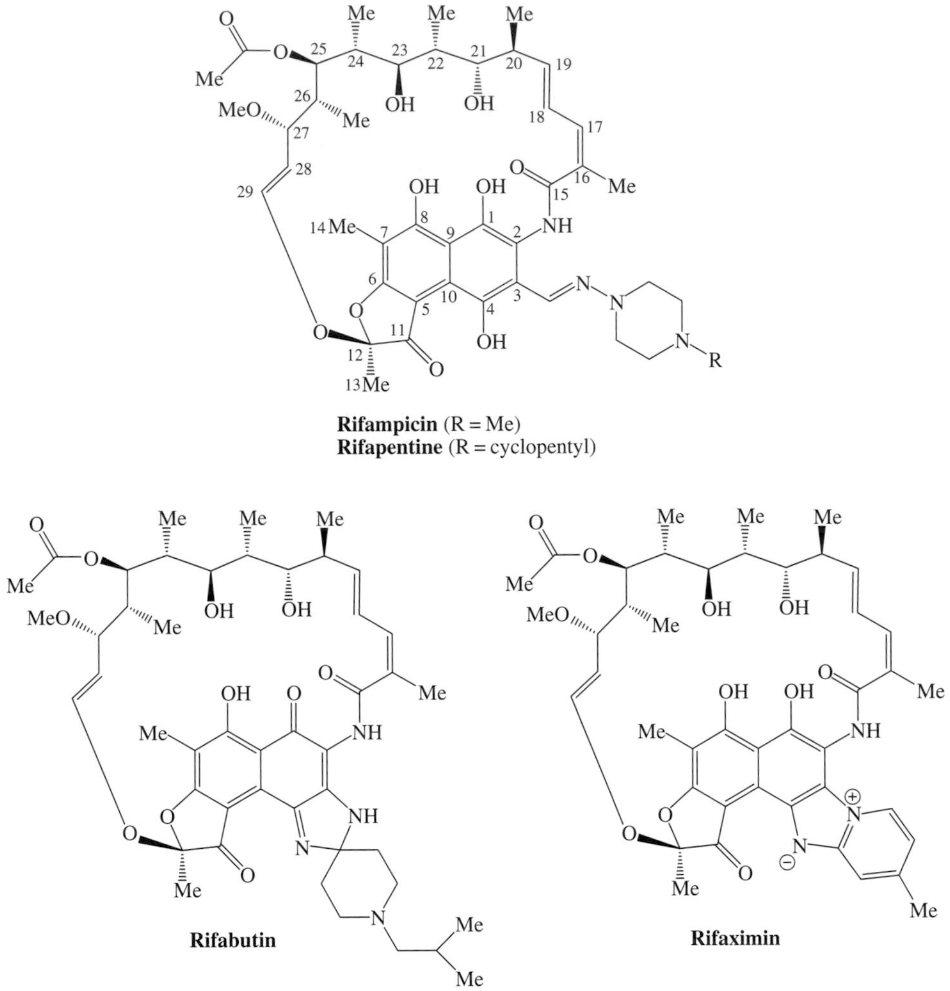

Rifampicin (R = Me)
Rifapentine (R = cyclopentyl)

Rifabutin

Rifaximin

Figure 2.2.1 *Rifamycin antibacterials*

component, being isolated in crystalline form. Although rifamycin B had very low activity, it is converted (through a sequence involving spontaneous cyclisation, hydrolysis, and reduction) on standing in solution to rifamycin SV, which has greatly enhanced antibacterial activity and was the first rifamycin introduced into the clinic (Scheme 2.2.1).

Table 2.2.1 *Therapeutic indications for the rifamycin antibacterials*

Rifamycin antibacterial	Indications
Rifampicin (rifampin)	Tuberculosis (TB), leprosy, other mycobacterial infections, meticillin-resistant *Staphylococcus aureus*, prophylaxis for meningococcal disease
Rifabutin	TB, prevention and treatment of *M. avium* complex (MAC)
Rifapentine	TB
Rifaximin	Travellers' diarrhoea

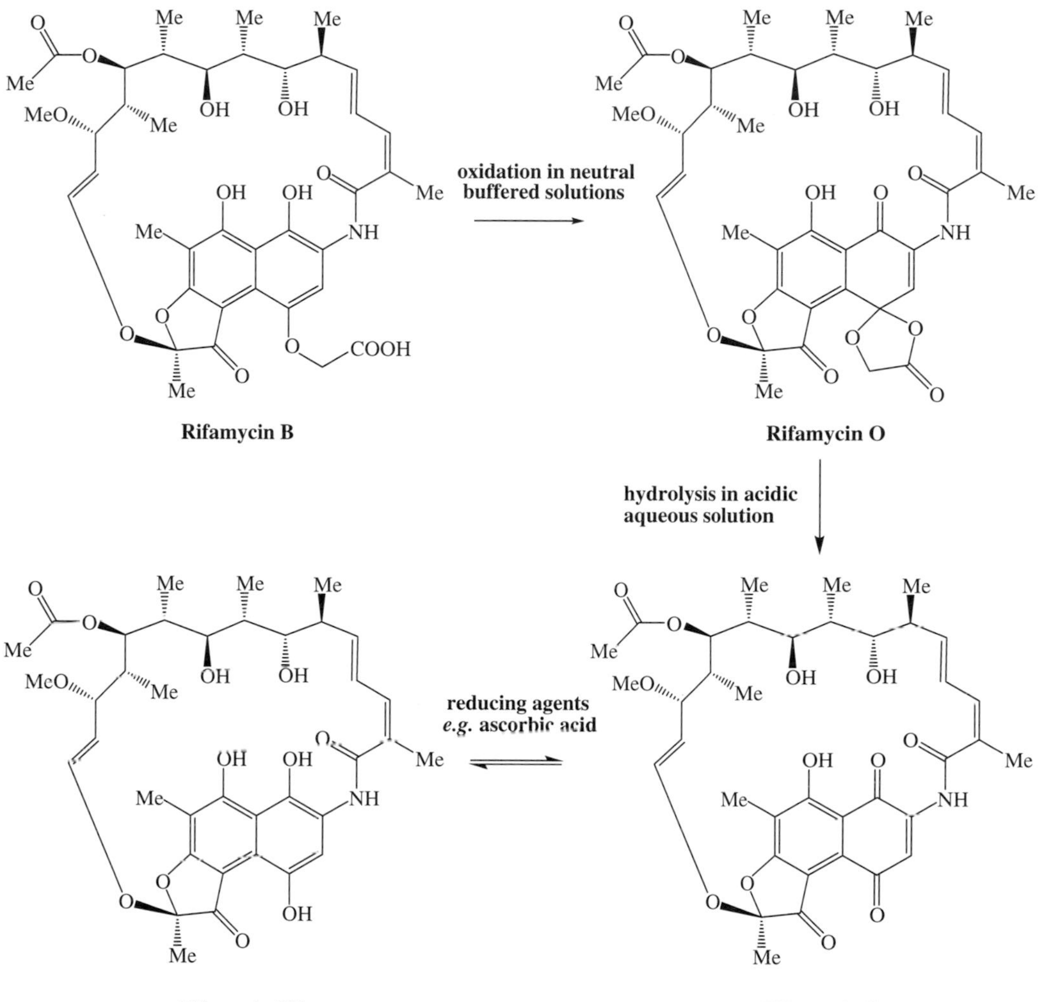

Scheme 2.2.1   *Conversion of rifamycin B to rifamycin SV (Prelog, 1964)*

2.2.2   Synthesis

It doesn't take a long look at the structure of the rifamycins to realise that a total synthesis, involving many functional group interconversions and the stereospecific synthesis of many stereogenic (chiral) centres, would be a supreme challenge. The total synthesis of rifamycin S has attracted the attention of some of the superstars of organic synthesis and the first total synthesis was reported by Kishi and coworkers as early as 1980, in 62 steps, with the ansa chain being derived from 3-benzyloxy-2-methylpropionaldehyde and the aromatic fragment from 2-methyresorcinol monomethyl ether in an elegant synthesis, but, unfortunately, with an overall yield of less than 0.01% (Scheme 2.2.2) (Kishi, 1981 and references cited therein; Paterson and Mansuri, 1985). The production of the rifamycins by such total syntheses is obviously impractical as the quantities obtained are too small, and such a sequence would be

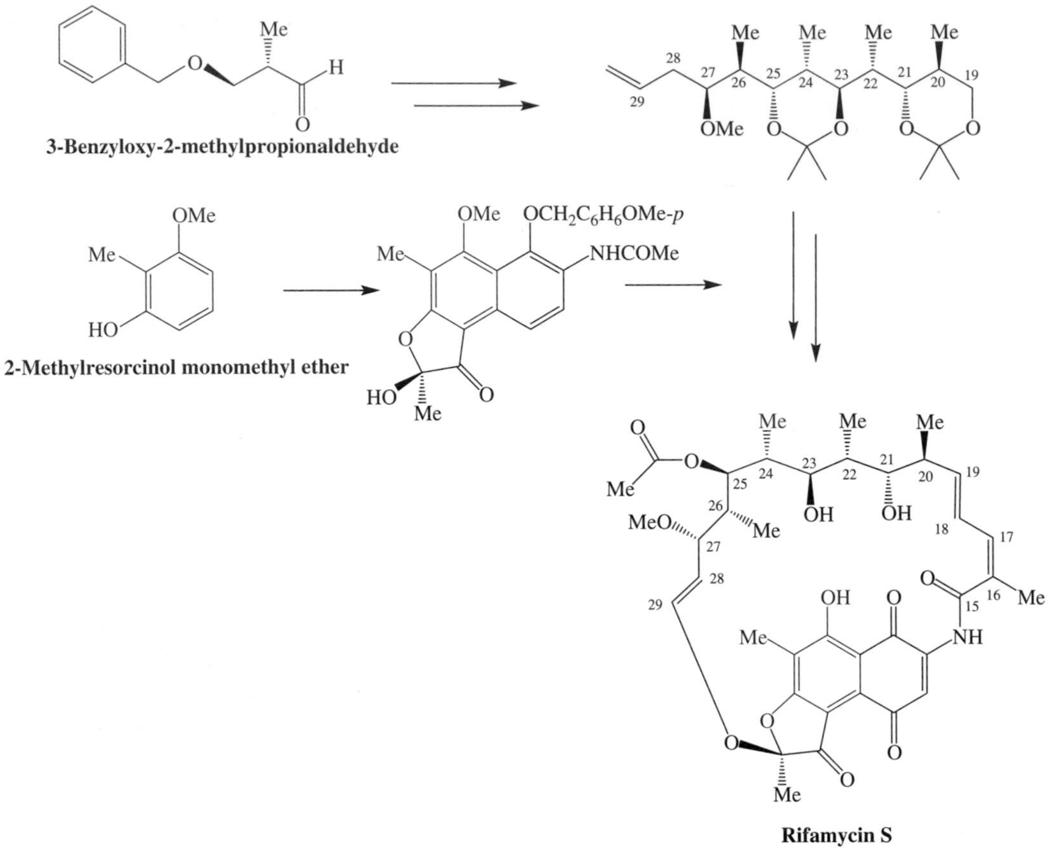

3-Benzyloxy-2-methylpropionaldehyde

2-Methylresorcinol monomethyl ether

Rifamycin S

Scheme 2.2.2   *Kishi's total synthesis of rifamycin S*

too costly, for these drugs ever to have made it to the clinic. How then are the rifamycins obtained? The rifamycins used clinically, like the β-lactams, which we will discuss later, are known as semi-synthetic antibacterial agents as they are prepared by a combination of fermentation and synthesis. Fermentation of a microbial broth produces the complex core structure, containing the ansa ring attached to the naphthofuran ring system, which is then elaborated to give the derivatives used clinically in a series of relatively simple chemical transformations. The key intermediate in the preparation of rifampicin and rifapentine is 3-formylrifamycin SV, which was originally obtained from the decomposition of some of the early, chemically unstable analogues which were synthesised for biological evaluation.

3-Formylrifamycin SV is now obtained by fermentation of *A. mediterranei* to give rifamycin B, which is converted to rifamycin SV (the chemical conversion gives a low overall yield, so an enzyme-based method has been developed using rifamycin B oxidase) (Lal *et al.*, 1995), and then 3-formylrifamycin SV via the mild oxidation of the Mannich base (Maggi and Sensi, 1967b) formed by reaction with paraformaldehyde and a secondary amine (Scheme 2.2.3). The synthesis of rifampicin and rifapentine then involves the simple condensation of a hydrazine with this aldehyde (Scheme 2.2.4) (Maggi and Sensi, 1967a), a chemical transformation which is very similar to one that many of you will have performed in school/university teaching laboratories: the preparation of yellow/orange crystalline derivatives of aldehydes and ketones through their condensation with Brady's reagent. As you can imagine, the fermentation of

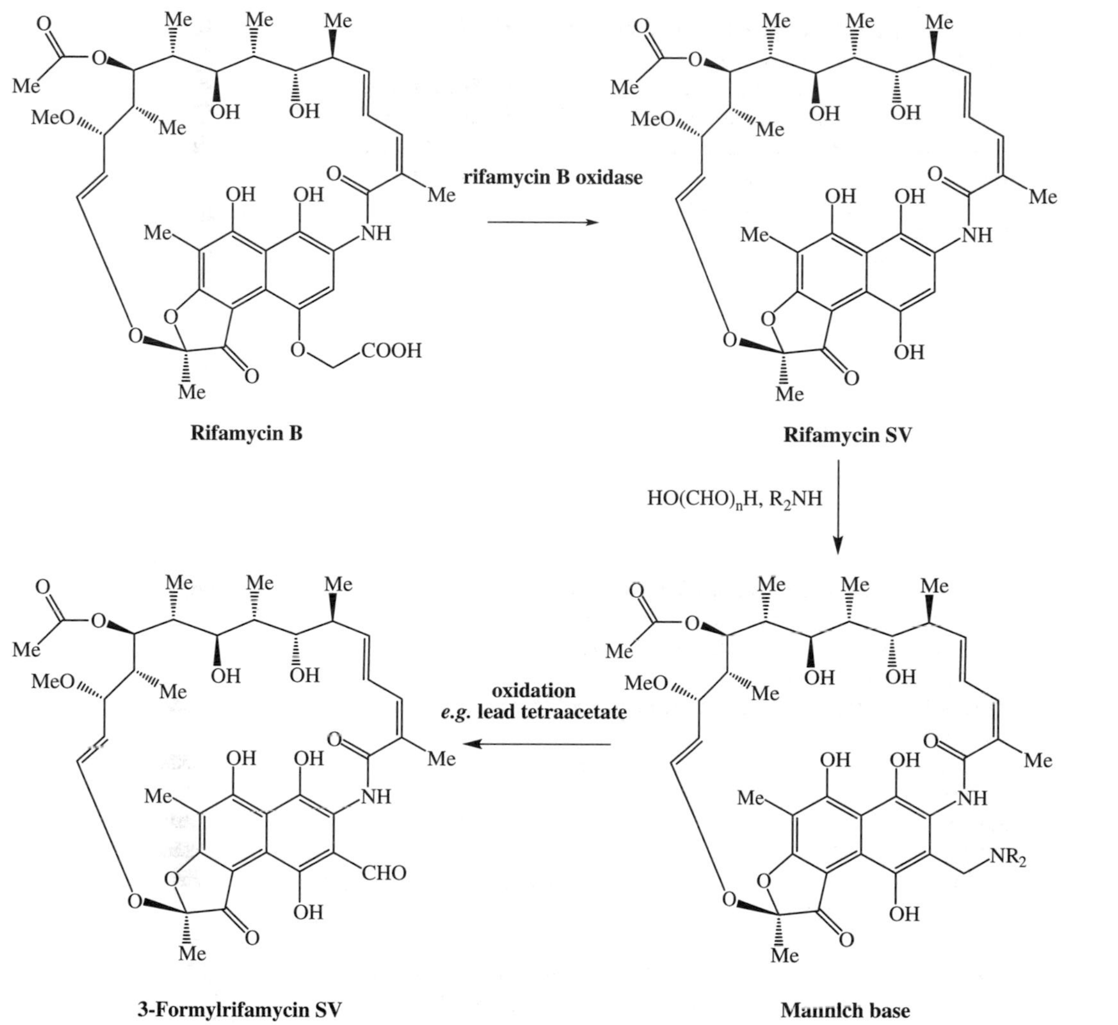

Scheme 2.2.3   *Synthesis of key semisynthetic intermediate 3-formylrifamcin SV*

rifamycin-producing microorganisms has been studied extensively, and while rifamycin SV can also be produced by fermentation, the yields are low so that it is more effective to utilise the two-stage route to rifamycin SV.

The rifamycins have been the subject of a number of structure–activity relationship (SAR) studies (Sensi and Lancini, 1990) and the key structural requirements for RNAP-inhibitory activity have been identified as:

- Oxygenated substituents at C1 and C8 (OH or C=O) of the aromatic system.
- Free hydroxyls at the C21 and C23 positions with the absolute stereochemical relationship shown in Figure 2.1.11.

Modifications at C3 and/or C4 of the aromatic system, as described above for rifampicin etc., are the most fruitful in terms of increasing the biological activity.

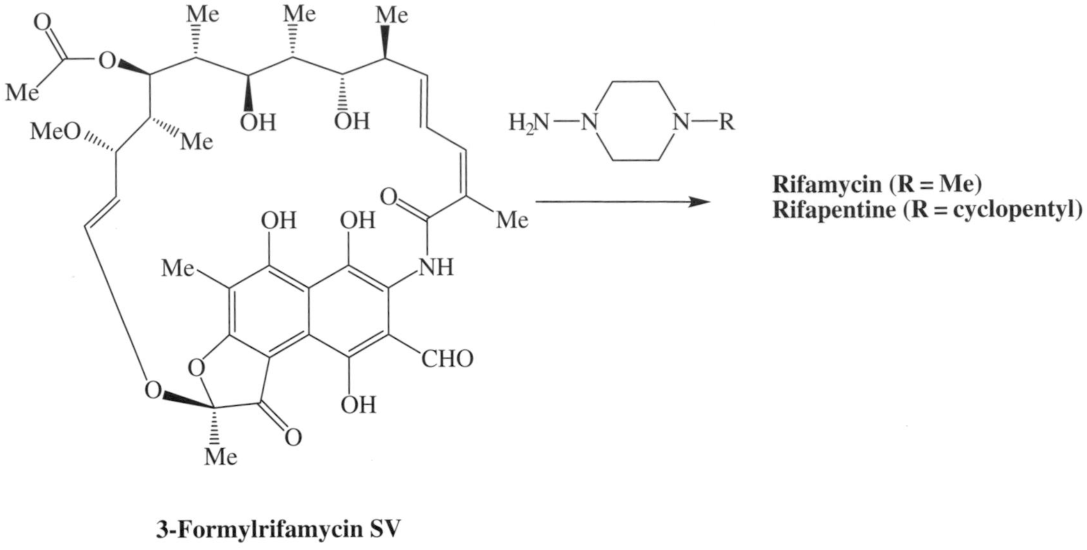

3-Formylrifamycin SV

Rifamycin (R = Me)
Rifapentine (R = cyclopentyl)

Scheme 2.2.4   *Synthesis of rifampicin and rifapentine (Maggi and Sensi, 1967a)*

2.2.3   Bioavailability

The rifamycins have different oral bioavailabilities and each is affected differently by co-administration with food. Rifampicin is the fastest and most completely absorbed, with an oral bioavailability of 68%; the presence of food in the GI tract has little effect upon its rate of absorption, but results in a decrease of the maximum serum concentration achieved (C_{max}). Rifabutin has lower bioavailability (20%) and the presence of food slows the absorption further. Although poorly studied, the absorption of rifapentine is known to be increased by food, resulting in significantly increased C_{max} and AUC values (Burman *et al.*, 2001).

In contrast, rifaximin was specifically designed to remain in the GI tract for the treatment of GI infections, due to its permanently ionised nature, and, as was the intention, is not significantly absorbed after oral administration (Scarpignato and Pelosini, 2005).

The half-lives of the rifamycins in plasma are also quite different and are not dose-dependent: 2–5 hours for rifampicin, 14–18 hours for rifapentine, and 32–67 hours for rifabutin. This property of the latter two allows for once- or twice-weekly administration, enabling directly observed therapy, which we will discuss in Subsection 2.2.6. The systemic rifamycins accumulate in neutrophils, granulocytes, and macrophages, giving these agents an advantage in the treatment of abscesses and bacterial infections involving direct invasion of the immune cells, such as tuberculosis (discussed in Subsection 2.2.6).

The metabolism of the rifamycins is also individual to each, although there are some underlying themes. All three of the absorbed rifamycins are deacetylated at the 25 position by hepatic enzymic action; however, while rifampicin and rifapentine are both hydrolysed to 3-formylrifamycin SV, rifabutin is not metabolised in this way – instead it is hydroxylated by cytochrome P_{450} isoform 3A (CYP3A). Interestingly, both rifampicin and rifabutin cause autoinduction of their own metabolism, which results in a decreased AUC until steady-state conditions are achieved after about a week. Lastly, all three of these rifamycins cause induction of CYP3A, although to different extents; rifampicin causes the most pronounced induction (and induces other hepatic enzymes too), while rifapentine causes moderate induction and rifabutin's inductive effect is weak. This CYP3A induction is of particular relevance to drug–drug interactions, of which more will be said later in this section. The rifamycins and their metabolites are largely excreted through the bile into the GI tract and faecally excreted. A relatively small proportion (10–25%) is excreted unchanged in the urine, turning it a red-orange colour[1] (Skinner and Blaschke, 1995; Burman *et al.*, 2001; Gumbo *et al.*, 2007).

There is considerable pharmacokinetic inter-individual variability observed in treatment with rifampicin, particularly in cases of comorbidity, such as HIV infection. With these agents, the AUC : MIC ratio is important for effective bactericidal action, but the rifamycins have complex pharmacodynamic parameters and are not described by a simple model. Once- or twice-weekly doses result in a relatively short exposure time to critical concentrations, yet effective antibacterial action can still be achieved. Rifampicin doses of 450 mg/day, rather than the usual 600 mg/day, have been shown to be suboptimal, while higher doses, such as 750 mg/day, are no more effective. Rifapentine, however, shows linear pharmacokinetics over a wide dose range (300–1200 mg) (Burman *et al.*, 2001; Davies and Nuermberger, 2008).

2.2.3.1　Prodrugs

The relatively high logP values of the rifamycins, except for rifaximin, and the acceptable absorption from the GI tract obviate the need for prodrug forms. However, the absorption, stability and antimycobacterial action of rifampicin and isoniazid has been shown to be improved by a polymeric micelle prodrug form of this combination therapy (Silva *et al.*, 2007).

2.2.3.2　Bacterial uptake of rifamycins

Rifampicin penetrates well into both Gram positive and Gram negative bacteria, which correlates with its broad spectrum of activity (see Section 2.2.6), and the effects of efflux pumps in reducing intracellular rifamycin concentrations can be overcome by a modest increase in antibacterial concentration (Williams and Piddock, 1998).

2.2.4　Mode of action and selectivity

As we shall see in later sections, the rifamycins (well, rifampicin in particular) are now the first-line treatments for *Mycobacterium tuberculosis* (TB) infections, but they have a very broad spectrum of activity and a high level of activity against Gram positive (e.g. *Staphylococcus aureus*), Gram negative (e.g. *Haemophilus influenzae*), and mycobacteria. The rifamycins are bactericidal and their broad spectrum of activity arises from their unique mode of action: the inhibition of bacterial DNA-dependent RNA polymerase (RNAP), a nucleotidyl transferase enzyme that is responsible for the process of transcription (Floss and Yu, 2005; Mariani and Maffioli, 2009). In bacteria, the transcription process can be divided into a number of distinct steps, in which the RNAP holoenzyme binds to duplex promoter DNA to form the RNAP-promoter complex, then a series of conformational changes leads to local unwinding of DNA to expose the transcription start site. RNAP can then initiate transcription, directing the synthesis of short RNA products.

As mentioned in Section 1.1.3.3, RNAP is a complex system, comprising five subunits ($\alpha_2\beta\beta'\omega$), each of which has a different function. The α subunits assemble the enzyme and bind regulatory factors, the β subunit contains the polymerase, the β' subunit binds nonspecifically to DNA, and the ω subunit promotes the assembly of the subunits and constrains the β' unit. The core structure of RNAP is thought to resemble a crab's

---

[1] Whenever dispensing a rifamycin, it is important to warn patients that, as a result of taking these antibiotics, their urine will turn red/orange.

claw – with the active centre on the floor of the cleft between the two 'pincers', the β and β′ subunits – and also contains a secondary channel, by which the nucleotide triphosphates access the active centre, and an RNA-exit channel (for a really good interactive tutorial showing the structure of RNAP, see http://www. pingrysmartteam.com/RPo/RPo.htm). Bacterial RNAP contains only these conserved subunits, while eukaryotic RNAP contains these and seven to nine other units (Ebright, 2000).

In bacteria, the transcription of a particular gene requires the binding of a further subunit, a σ factor (a transcription initiation factor), which increases the specificity of RNAP binding to a particular promoter region and so results in the transcription of a particular DNA sequence. Once the assembly process is complete, the holoenzyme (the active form containing all the subunits: $\alpha_2\beta\beta'\omega\sigma$) catalyses the synthesis of RNA, which is complementary to the DNA sequence characterised by the σ factor (Figure 2.2.2) (eukaryotic RNAP also requires the binding of transcription factors).

The precise RNAP-binding site of the rifamycins has been the matter of some debate and, despite having been in clinical use since the 1970s, has only just been unequivocally established (famous last words?). One of the mechanisms that has been proposed, the binding of rifamycin to RNAP at a site remote from the catalytic centre, resulting in allosteric modulation of the affinity for the catalytic Mg^{2+} bound at the active centre (Artsimovitch *et al.*, 2005), was recently discounted (Feklistov *et al.*, 2008). Crystal structures of rifamycin in complex with *Thermus aquaticus* RNAP, in addition to further mechanistic studies, indicate that rifamycin binds to the β-subunit and in doing so sterically blocks the extension of the RNA chain being formed, after the first or second condensation step, leading to the production of shortened RNA sequences (Zhang *et al.*, 1999).

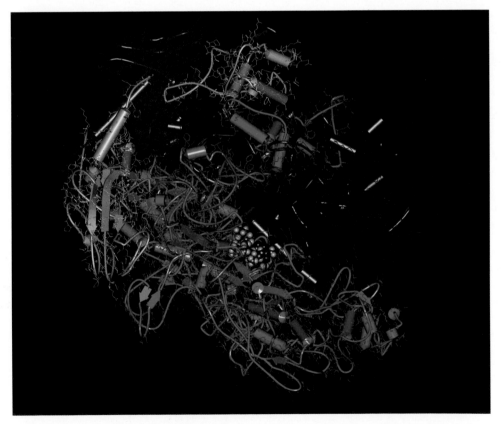

Figure 2.2.2   *Rifampicin (grey/red spheres) bound to RNAP; PDB code 1I6V (Campbell et al., 2001)*

Further support for this mechanism comes from the observation that bacteria are no longer sensitive to the rifamycins once RNA synthesis has produced an oligonucleotide chain.

As with all the antibacterial agents we will be studying, we should ask ourselves why antibacterial agents that target RNAP are selective, as transcription must also take place in eukaryotic cells. As mentioned briefly above, despite structural and functional similarities, the amino acid sequence of the bacterial enzyme differs sufficiently from that of the eukaryotic RNAP, which is also composed of different subunits to those in the prokaryotic RNAP and is present as several variants, each with a different function; for example, eukaryotic RNA polymerase I transcribes ribosomal RNA (rRNA) genes. These differences result in the eukaryotic enzymes being $100–10\,000\times$ less sensitive to the rifamycins than their bacterial counterparts (Ho *et al.*, 2009). We learned earlier that the rifamycins are broad-spectrum antibiotics, with activity against mycobacteria, as well as Gram negative and positive organisms, and this results from the fact that the amino acid sequences of their target, the β-subunit of bacterial RNAP, are highly conserved, meaning that rifamycin binding to all bacterial RNAP is equally likely. Differences in bacterial sensitivity, for example between Gram negative and positive species, may thus simply be due to differential permeation of cellular membranes by the rifamycin. For more on mode of action and selectivity, see Hartmann *et al.* (1985).

2.2.5 Bacterial resistance

In the previous subsection, we learned about the mode of action of the rifamycins and the reason why they are selective inhibitors of bacterial RNAP: bacterial and eukaryotic RNAP sequences are not highly conserved, resulting in selective toxicity. As we have seen, bacteria respond rapidly to the challenge of an antibiotic agent, and this is particularly true for the rifamycins, with pathogenic bacteria developing resistance at a high rate (typically $10^{-8}–10^{-9}$ per bacterium per cell division) (Gillespie, 2002). In order to prevent widespread resistance to these exceedingly useful antibiotics, the rifamycins are thus used in combination therapy, which effectively decreases the chances of resistance developing to any of the agents used in combination, since two, or more, cellular mechanisms are targeted. Use of the rifamycins is also restricted in many countries to the treatment of tuberculosis, or to clinical emergencies, in order to minimise the exposure of bacteria to these agents and so reduce the chances of bacterial resistance developing.

As the rifamycins exhibit their antibacterial effect through binding to the β-subunit of the RNAP, it will come as no surprise that bacterial resistance develops through modifications to the gene encoding this subunit, the *rpo*B gene, thus diminishing rifamycin-binding affinity for the polymerase. For example, acquired resistance to rifampicin in *M. tuberculosis* has been shown to arise due to mutations in only 27 codons near the centre of the *rpo*B gene (Chopra and Brennan, 1998), with more than 70% of mutations found in clinical isolates being due to either His526 to Tyr or Ser531 to Leu (Telenti *et al.*, 1993). These mutations appear to have little effect on the growth of *M. tuberculosis*, which naturally grows slowly, but result in decreased RNAP affinity for the drug.

In non-mycobacterial organisms, other resistance mechanisms are encountered, including chemical modification. For example, pathogenic *Nocardia* species are naturally rifamycin resistant and *Nocardia brasiliensis* converts rifampicin to its inactive 23-*O*-β-D-glucosyl derivative (the 21 and 23 hydroxy groups are essential for antibacterial activity) (Figure 2.2.3) (Morisaki *et al.*, 1993). For more on bacterial resistance, see Wehrli (1983), Heym and Cole (1997), and Chopra and Brennan (1998).

2.2.6 Clinical applications

As mentioned earlier, the target, bacterial RNAP, is well conserved, so the rifamycins have a wide spectrum of activity, with activity against Gram negative bacteria (e.g. *Neisseria meningitidis* and *Haemophilus*

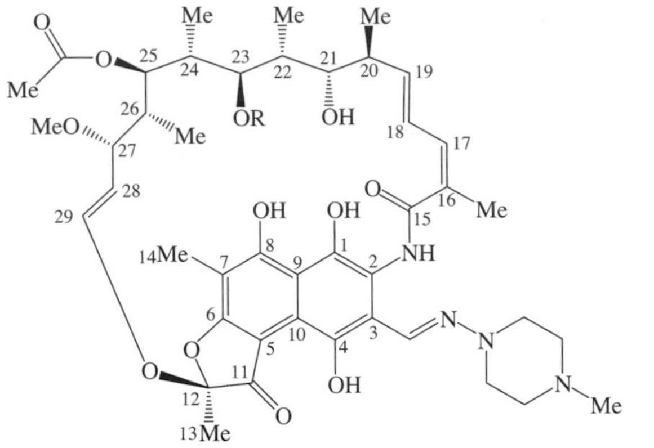

Rifampicin (R= H)
23-O-β-D-Glucosylrifampicin (R = β-D-glucopyranosyl)

Figure 2.2.3   *Inactivation of rifampicin by* N. brasiliensis

influenzae), Gram positive bacteria (e.g. *Staphylococcus aureus* – including, in some cases, MRSA), and mycobacteria (e.g. *M. tuberculosis*). Despite their spectrum of activity, the clinical applications of the rifamycins are generally limited to the treatment of serious (and sometimes life-threatening) infections in order to minimise the development of bacterial resistance.

2.2.6.1   Tuberculosis

The discovery of the rifamycins helped to transform the treatment of *M. tuberculosis* by reducing the duration of treatment, increasing cure rates, and decreasing relapse rates for this debilitating disease (Fox *et al.*, 2001). Prior to the rifamycins, treatment of active TB could take up to 18 months, leading to serious problems with patient compliance.

Rifampicin, the first clinically used rifamycin, has activity against both active and latent *M. tuberculosis* infection and is considered to be a key component of the initial and continuation phases of treatment. Rifampicin should therefore only be excluded from a treatment regimen if the patient has a specific contraindication or if resistance to rifampicin is suspected.

The treatment of tuberculosis is separated into two phases: an initial phase and a continuous phase. The initial phase usually lasts 2 months and is designed to quickly reduce the bacterial population, in order to minimise the possibility of the development of resistance. This is achieved by using four drugs in the initial treatment regimen, usually rifampicin, isoniazid, pyrazinamide, and ethambutol. Resistance to rifampicin normally occurs when treating active *M. tuberculosis* disease, rather than the latent infection, so it important that the initial phase of treatment is adhered to (Bishai and Chaisson, 1997). The continuous phase starts once the initial phase of treatment is completed, lasts 4 months, and consists of rifampicin and isoniazid.

If the treatment of TB is to be successful, it is critical that the prescribed drug regimen is strictly followed. This is sometimes unrealistic, even with the introduction of combined preparations—for example, Rifater is used in the UK (a preparation combining rifampicin, isoniazid, and pyrazinamide)—as the regimen is often complex, with a high pill burden. In an attempt to overcome this problem, a technique known as directly observed therapy (DOT) has been introduced. DOT involves an appointed person (usually a health worker, community volunteer, or family member) directly supervising patients taking their anti-TB drugs, in order to ensure the medication is taken appropriately. A typical DOT regimen is taken three times per week, while with the unsupervised regimen, the medications are taken daily. Despite the advantages of DOT, a recent Cochrane review concluded that there is insufficient evidence to recommend the routine use of DOT to improve cure rates or treatment completion in people with TB (Volmink and Garner, 2009). Further research is needed to determine in which circumstances DOT would be beneficial, as one small study reported that the use of DOT can actually decrease sucessful treatment outcomes when compared to self-administration (Zwarenstein *et al.*, 1998).

Rifabutin also has activity against *M. tuberculosis* and, interestingly, its *in vitro* activity is superior to that of rifampicin (MIC against *M. tuberculosis* reported as <0.06 mg/mL for rifabutin and 0.15 mg/mL for rifampicin), but rifampicin is still used in the first-line treatment of TB. Despite the positive data for rifabutin, there is still not enough evidence for it to replace rifampicin in the first-line treatment of TB (Davies *et al.*, 2007).

Rifabutin does, in some instances, still have an important role in the treatment of TB and was included in the list agents recommended by the WHO (World Health Organization, 2008). In addition, rifabutin has just been added to the WHO's list of essential medicines (World Health Organization, 2009b), because it is routinely prescribed (in place of rifampicin) for patients with HIV who are also taking protease inhibitors, since rifampicin is prone to causing significant drug interactions, typically with anti-HIV medication (such as protease inhibitors). Rifabutin has more favourable pharmacokinetic and pharmacodynamic properties and, as it is less likely to cause clinically significant drug interactions, is considered to be a good alternative to rifampicin in patients taking anti-HIV therapy.

Rifabutin may also have a role in the treatment of MDR-TB, which, as discussed in Subsection 2.1.6.5, is becoming a significant clinical problem. Studies have confirmed that some strains of rifampicin-resistant *M. tuberculosis* are still susceptible to rifabutin, although the evidence suggests that there is a high degree of cross-resistance to rifampicin and rifabutin. For example, one study examining the *in vitro* activity of rifabutin against *M. tuberculosis* with *rpo*B mutations (see Subsection 2.2.5) reported that, of the 41 isolates resistant to rifampicin, 35 were also resistant to rifabutin (Cavusoglu *et al.*, 2004). The WHO guidelines for the management of MDR-TB recommend that a drug should not be included in a treatment regimen if there is a high risk of cross-resistance (World Health Organization, 2008), so that, if a patient contracted MDR-TB resistant to rifampicin, it would not be good practice to include rifabutin in the treatment regimen. If, however, a patient could not tolerate rifampicin (due to side effects or drug interactions, for example) then rifabutin would clearly be an appropriate alternative.

Rifapentine, a long-acting rifamycin, is also used in the management of TB. Rifapentine is available in the USA under the propriety name of Priftin and is licensed for the treatment of pulmonary TB. When rifapentine was first licensed by the FDA in 1998, it was anticipated that its long half-life compared to rifampicin (14–18 hours versus 2–5 hours) would allow less-frequent dosing, or possibly even reduce the duration of treatment. Indeed, a trial published in the *Lancet* confirmed that once-weekly dosing of rifapentine is effective for the treatment of noncavitary pulmonary TB, although when compared to twice-weekly rifampicin, the rifapentine arm was associated with a higher failure rate (Benator *et al.*, 2002). Furthermore, it has been shown that rifapentine shortens the duration of TB treatment in a murine model when compared to rifampicin-containing regimens (Rosenthal *et al.*, 2007). While these data are encouraging, it is still not clear what role rifapentine has in the international management of TB, and interestingly, it is not yet recommended by the WHO as a first-line therapy (World Health Organization, 2009a). Currently, the American Thoracic Society recommends that rifapentine should be used once weekly (in combination with isoniazid) for the continuous phase of treatment in patients who are HIV sero-negative with non-cavitary, drug-susceptible pulmonary TB (American Thoracic Society, 2003).

2.2.6.2 Meningitis

Rifampicin is also used in the prevention of meningitis; a disease characterised by inflammation of the meninges, primarily caused by *Neisseria meningitidis* – a Gram negative, bean-shaped, diplococcal bacterium (Figure 2.2.4). Meningitis is a serious, often fatal, disease, and even if treated successfully, many patients will sustain severe neurological defects, including loss of feeling in limbs, mental retardation, and paralysis.

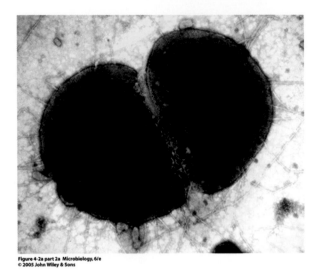

Figure 4-2a part 2a Microbiology, 6/e
© 2005 John Wiley & Sons

Figure 2.2.4   *A scanning electron micrograph of* Neisseria meningitidis *(Reprinted from J. G. Black, Microbiology: Principles and Explorations, 6th Edition, 2005, with permission of John Wiley & Sons.)*

Bacterial meningitis can also be caused by *Streptococcus pneumoniae* and *H. influenzae* type B, but the successful development and widespread use of vaccines against these bacteria have left *N. meningitidis* the most common cause of bacterial meningitis.

N. meningitidis is frequently found in the nasopharyngeal cavity of humans, with one study reporting that one in ten people carry the bacterium (Cartwright *et al.*, 1987). Thankfully, most of these carriers are asymptomatic, but they are, nonetheless, still thought to play a crucial role in the transmission of this debilitating disease. The eradication of nasopharyngeal *N. meningitidis* is used as a means of minimising the spread of meningitis. Currently, chemoprophylaxis is recommended for patients who have had prolonged close contact in a household setting during 7 days before the onset of the meningitis (Health Protection Agency, 2011).

A Cochrane review confirmed that rifampicin is effective at eradicating nasopharyngeal *N. meningitidis*, along with ciprofloxacin and ceftriaxone, and until recently it was considered the agent of choice in the prevention of the transmission of meningitis (Fraser *et al.*, 2009). Nowadays, however, ciprofloxacin is generally preferred over rifampicin as it is associated with less significant drug interactions, has a shorter regime duration, and is more widely available (Health Protection Agency, 2011).

Rifampicin is licensed for the prophylactic treatment of asymptomatic carriers of *N. meningitidis* and *H. influenzae* type B to minimise the transmission of meningitis. It should be taken for 2 days after exposure to meningococcal meningitis (i.e. meningitis caused by *N. meningitidis*), while for prophylaxis after *H. influenzae* type B exposure, it is usually taken for 4 days.

2.2.6.3   Leprosy

Rifampicin is also used to treat other infections caused by mycobacteria, such as leprosy. Leprosy is a chronic infection of the skin and peripheral nerves caused by *Mycobacterium leprae* or *Mycobacterium lepromatosis*. Patients with leprosy commonly present with skin lesions, such as macules and plaques (Figure 2.2.5), and if left untreated, it can cause permanent damage to peripheral nerves, resulting in impaired motor and sensory function, including blindness (Britton and Lockwood, 2004; Parikh *et al.*, 2009). Interestingly, *M. lepromatosis* has only recently been discovered and was first isolated from two patients who died of diffuse lepromatous leprosy (DLL) (Han *et al.*, 2008). Genetic sequencing analysis of the 16S rRNA gene suggests

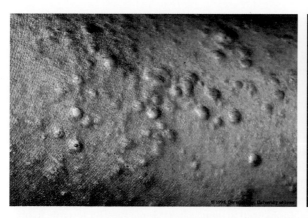
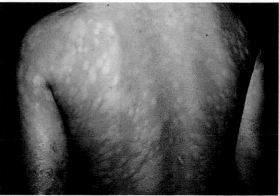

Figure 2.2.5 *Skin lesions associated with leprosy (Reprinted from 'Hansen's Disease (Leprosy) Lepromatous form - Left Arm' Dermatologic Image Database, Online, [http://www.healthcare.uiowa.edu/dermatology/DermImag. htm] 1995 with permission of T. Ray; Reprinted from 'Leprosy', Leprosy: Symptoms and Treatment, Online, [http://www.modernguidetohealth.com/conditions-diseases/leprosy.html], 2009.)*

that *M. lepromatosis* is a separate species to *M. leprae* and it is thought that both species diverged from a common ancestor around 10 million years ago (Han *et al.*, 2009).

Leprosy remains a significant public health problem in several developing countries, despite a leprosy elimination campaign led by the WHO in 1991, which focussed on eliminating leprosy as a public health problem by the year 2000 (elimination was defined as a prevalence rate of 1 case per 10 000 population) (Britton and Lockwood, 2004). The campaign was a great success and by the end of 2000, 108 out of the 122 countries previously listed as being leprosy endemic achieved their elimination target, with a further 8 countries reaching this goal by the end of 2005.

The WHO's staging process for classifying leprosy is based on skin smear results and often on the clinical picture of the patient as, in reality, skin smear testing is sometimes not available. The first stage, single-lesion paucibacillary leprosy (SLPB), is where the patient only has one skin lesion present and there is no damage to the nerve trunks. The second stage, paucibacillary leprosy (PB), is more serious and is characterised by the patient showing between two and five skin lesions with limited nerve damage; typically only one nerve trunk is involved. In the last stage, multibacillary leprosy (MB), there are typically more than five skin lesions present and damage occurs to many nerve trunks, resulting in loss of sensation.

The recent global success in trying to eliminate leprosy is largely due to the use of multidrug therapy (MDT), in which rifampicin forms a key component (Table 2.2.2). MDT was designed to minimise resistance problems

Table 2.2.2 *Recommended standard treatment regimens for MD and PP in adults (Britton and Lockwood, 2004) (Reprinted from T. Ray, 'Hansen's Disease (Leprosy) Lepromatous form - Left Arm' Dermatologic Image Database, [http://www.healthcare.uiowa.edu/dermatology/DermImag.htm] 1995 with permission of Thomas Ray.)*

Stage	Multibacillary leprosy (MB)	Paucibacillary leprosy (PB)	Single-lesion paucibacillary leprosy (SLPB)
Regimen	Rifampicin 600 mg once a month (supervised)	Rifampicin 600 mg once a month (supervised)	Rifampicin 600 mg
	Clofazimine 300 mg once a month (supervised) and 50 mg daily (self-administered)	Dapsone 100 mg daily (self administered)	Ofloxacin 400 mg
	Dapsone 100 mg daily (self-administered)		Minocycline 100 mg
Duration	12 months	6 months	Single dose

associated with single-agent dapsone therapy, and since the introduction of MDT, concerns over resistance have been alleviated, with very few patients who have undergone MDT having a relapse (Britton and Lockwood, 2004). The duration of treatment for MB was originally 24 months, but it was subsequently decided that it could be shortened to 12 months without increasing the risk of developing rifampicin resistance (World Health Organization, 1998).

Monthly rifampicin is highly effective at treating leprosy and offers the advantages that it is less toxic (i.e. has a decreased number of adverse effects, such as hepatitis), has a lower cost, and improves patient compliance when compared to daily therapy (Chandorkar *et al.*, 1984). Studies in nude mice confirm that rifampicin is highly bactericidal – it has been reported that three consecutive monthly doses of rifampicin kill more than 99.999% of viable *M. leprae* isolates (Ji *et al.*, 1996). In addition, rifabutin also appears to have significant activity against *M. leprae*, with some studies reporting that rifabutin has a lower MIC against *M. leprae* than rifampicin (Dhople and Williams, 1998; Dhople *et al.*, 1993). As a result of the success of rifampicin, and the lack of resistance to it, rifabutin is seldom used to treat leprosy, but it may have a role to play in the future if resistance to rifampicin becomes a significant clinical problem.

2.2.6.4   Miscellaneous

In addition to their activity against *M. tuberculosis*, the rifamycins are also used in the management of other mycobacterium infections, such as *Mycobacterium avium* complex (MAC) – an infection caused by several species of mycobacterium, namely *Mycobacterium avium* and *Mycobacterium intracellulare*. MAC, commonly found in soil, house dust, and water, rarely causes problems in patients with a normal immune response, but when a patient's immune response is impaired, a MAC infection can be serious or even life-threatening. For example, in patients infected with HIV, it is thought that the risk of developing MAC is significant if the CD4 count is less than 50 cells per μL (Nightingale, 1997).

In the UK, rifabutin is licensed for the prophylaxis of MAC in patients with HIV who have CD4 counts less than 75 cells per μL and also for the treatment of non-tuberculosis mycobacterial disease (such as that caused by MAC). Rifabutin, if given prophylactically to patients with AIDS, has been shown to reduce the incidence of MAC bacteraemia by one half when compared to placebo (Nightingale *et al.*, 1993), but appears to be less effective than clarithromycin (Benson *et al.*, 2000).

The American Thoracic Society recommends that rifabutin should be used to treat disseminated MAC in patients with HIV as part of triple combination therapy, alongside clarithromycin and ethambutol (Griffith *et al.*, 2007). The same guideline also recommends that rifabutin should not be used as first-line therapy for prophylaxis against MAC. If prophylaxis is indicated (for patients with a CD4 count of less than 50 cells per μL), azithromycin or clarithromycin is preferred and rifabutin should only be used if these agents cannot be tolerated (Griffith *et al.*, 2007).

Many of the clinical applications for the rifamycins described above are for serious infections. Rifaximin, a rifamycin that is very poorly absorbed (see Subsection 2.2.3), is used to manage less-severe infections, such as travellers' diarrhoea (Hill and Ryan, 2008). Rifaximin has comparable efficacy to ciprofloxacin in the treatment of *Escherichia coli*-associated diarrhoea, but is less effective than ciprofloxacin when invasive agents, such as *Campylobacter*, *Shigella*, or *Salmonella*, are implicated (Taylor *et al.*, 2006). It can also be used to prevent travellers' associated diarrhoea at a dose of 200 mg once or twice daily, although prophylactic use should only be initiated in selected patients when diarrhoea would be particularly problematic (after a colostomy, for example) (Hill and Ryan, 2008).

Rifaximin has also been approved for the treatment of hepatic encephalopathy by the FDA. A study published in the *New England Journal of Medicine* showed that, over a 6-month period, rifaximin maintained remission from hepatic encephalopathy more effectively than, and reduced the risk of patient hospitalisation when compared to, placebo (Bass *et al.*, 2010).

2.2.7   Adverse drug reactions

There are relatively few adverse effects of rifampicin and these are primarily hepatic and immunoallergic in character. Hepatic adverse effects are most commonly due to the dose of rifamycin used and pre-existing liver disease, whereas immunoallergic events are often associated with prolonged or intermittent therapy. The immunoallergic reactions tend to be cutaneous, 'flu-like', or related to the GI tract and, in severe cases, haemolytic anaemia and renal failure (Grosset and Leventis, 1983). Hepatotoxicity is often noted in combination with other antituberculosis drugs, with which rifampicin is most commonly used.

2.2.7.1   Hepatotoxicity

Several reports have demonstrated the hepatotoxic effects of rifampicin when used in conjunction with isoniazid used in the treatment of tuberculosis (Steele *et al.*, 1991; Askgaard *et al.*, 1995). The hepatotoxicity of the antituberculosis drugs is further increased by combination with pyrazinamide, but in general the hepatotoxicity of rifampicin is far less than that of isoniazid or pyrazinamide. One study demonstrated that the incidence of all major events was 1.48 per 100 person-months of exposure for pyrazinamide, 0.49 for isoniazid, and 0.43 for rifampicin (Yee *et al.*, 2003). With respect to the sole use of rifampicin, hepatotoxicity appears to be idiosyncratic and significantly less frequent than with combination therapy (Saukkonen *et al.*, 2006). Rifampicin hepatitis was first reported by Scheuer in a case series (Scheuer *et al.*, 1974) but there are few reports of sole rifampicin-induced hepatitis and the results of abnormal liver function tests are slightly confounded by the concomitant adminstration in several cases of other antituberculosis agents. Two cases were reported by Bachs *et al.* (1992), who observed hepatitis in two women being treated with rifampicin for primary biliary cirrhosis, while one later publication has suggested significant hepatic impairment with rifampicin monotherapy (Prince *et al.*, 2002). The mechanism of rifampicin hepatotoxicity is as yet unknown, but has been demonstrated to occur in rats upon administration of high doses (Gangadharan, 1986).

Agents that produce hepatic damage fall into two broad categories: those that have intrinsic properties or predictable hepatotoxicity and those that damage the liver unpredictably in uniquely susceptible hosts. The first group of intrinsically hepatotoxic agents is recognisable from the high rate of toxicity experienced in individuals, which is often dependent upon the dose, and can be further subdivided into those that lead to direct and to indirect effects. Those agents which produce hepatotoxicity in hosts that is not dose-dependent (and do not cause hepatotoxicity in animal studies) are generally recognised as idiosyncratic reactions. The mechanism of adverse effects associated with the rifamycins remains unclear; in some cases, pre-existing hepatic damage and high doses of rifampicin may cause some hepatotoxicity, suggesting intrinsic toxicity, whereas immunoallergenic reactions would indicate idiosyncratic toxicity. Several studies have demonstrated that a specified daily dose 'rhythm' – that is, the specific timing of doses – improves tolerance compared to the reduction of doses (Aquinas *et al.*, 1972; Girling, 1973).

It would appear that rifampicin, although less often associated with hepatotoxicity than isoniazid and pyrazinamide, may enhance toxicity when used in combination with these agents, due to enhanced CYP 2E1 activity and mRNA expression in the human hepatocyte (Shen *et al.*, 2008).

2.2.7.2   Gatrointestinal (GI) effects

The incidence of GI side effects is transitory, with associated risk factors being age, alcoholism, and existing hepatic disorders (Mandell and Sande, 1990). Adverse GI effects (such as nausea and vomiting) are exacerbated with the concomitant use of protease inhibitors in HIV patients (Haas *et al.*, 2009).

2.2.7.3   Blood and kidney

It has been suggested that two different antibodies to rifampicin may exist, one binding to the surface of red cells and the other to platelets (Blajchman *et al.*, 1970), and that complement fixation in those cells *in vivo* may cause intravascular lysis if all fractions of complement are activated or if intratissue lysis occurs in a number of organs, particularly in the Kupffer cells (Cooper and Brown, 1971). This theory accounts for adverse reactions to rifampicin in the form of haemolytic anaemia or thrombocytopenia purpura, but in most cases the adverse reaction presents as pyrexia, chills, myalgia, digestive disorders, and sometimes anuria. It is not known whether these symptoms are associated with the fixation on the red cell complexes that activate the complement and release degradation products or with the fixation of complexes on other cells (e.g. mastocytes or vascular endothelial cells) (Kleinknecht *et al.*, 1972; Decroix and Pujet, 1973). The classical explanation is that the pathogenesis of haemolysis due to rifampicin is via the immune complex system, as described above, and it is assumed that these complexes attach to erythrocytes in a non-specific manner, but further investigations have suggested that this complex formation is rather more specific. In a similar way to the specific blood group antigen interactions experienced with nomifensine, streptomycin, and latamoxef, it has been suggested that rifampicin forms immunocomplexes with anti-I specificity (the I antigen is found on the surface of the erythrocyte) (Pereia *et al.*, 1991). As yet, the mechanism of autoimmune haemolytic anaemia caused by rifampicin is unknown, or at best it is thought to be due to the formation of an unstable hapten.

2.2.7.4   Rifampicin and cholestasis treatment

Bile acids, steroids, and selective drugs activate the pregnane X receptor (PXR; also known as nuclear receptor subfamily 1, group I, member 2 (NR1I2)), which induces CYP 3A4 (responsible for drug metabolism) and inhibits CYP 7A1 (also known as cholesterol 7α-hydroxylase, which is responsible for bile acid synthesis in the liver) (Kliewer *et al.*, 2002). Rifampicin is a PXR agonist which inhibits bile acid synthesis and so has been used in the treatment of cholestasis and cholestatic jaundice (Li and Chiang, 2005).

Consequently, rifampicin is often used in the treatment of cholestasis, through the mediation of the PXR, which is utilised in cells (especially liver and GI tract cells) to detect the presence of foreign substances (i.e. drugs) and subsequently upregulates those proteins involved in the detoxification process. Substances interacting with this receptor may also include bile salts and steroids, as well as drugs, such as St John's Wort and the rifamycins (Hofmann, 2002).

2.2.8   Drug interactions

The ability of the rifamycins to induce hepatic enzymic activity leads to many important and clinically significant drug interactions. This is more evident in cases of comorbidity, and one classic example is when rifampicin is used in the treatment of TB when a patient also has HIV (and is using a protease inhibitor). As we have mentioned in Subsection 2.2.3, the rifamycins induce CYP isoenzymes, with rifampcin causing the most pronounced induction. Indeed, rifampicin can induce CYP 3A4, CYP 1A2, CYP 2C9, CYP 2C19, CYP 2D6 (Cupp and Tracy, 1998; Dilger *et al.*, 2000; Kanebratt *et al.*, 2008), leading to many clinically significant drug interactions, some of which we will discuss below. It is beyond the scope of this textbook to discuss every single documented drug interaction associated with the rifamycins (and in particular rifampicin), but, as a guide, if a patient is taking a drug that is metabolised by any of the CYP isoenzymes listed above, and they are also taking rifampicin, there is a fair chance you will observe a drug interaction.

2.2.8.1 Protease inhibitors and rifampicin/rifabutin

The drug interaction between protease inhibitors and rifampicin/rifabutin is well documented and significant in patients who are co-infected with TB and HIV (unfortunately, a large proportion of HIV patients do also have TB). Not suprisingly, the interaction is complex, and it is thought to be mediated by CYP isoenzymes. First, rifampcin causes induction of CYP 3A4 isoenzymes, which accelarates the metabolism of several protease inhibitors, meaning that, clinically, the effect of the protease inhibitor is substantially diminimsed. Given the seriousness of HIV infection, the use of rifampicin in combination with some protease inhibitors (e.g. amprenavir, indinavir, and nelfinavir) is not recommended, in order to avoid treatment failure. This is problematic as there are many millions of people around the world infected with both HIV and TB, and, as we have already discussed, rifampicin is a cornerstone in the management of TB.

The accelerated metabolism of protease inhibitors is not the only problem with taking a protease inhibitor in combination with rifampicin. Indeed, studies in healthy volunteers have also demonstrated increased GI adverse events (such as nausea and vomiting) and elevated liver enzymes with the combined use of protease inhibitors and rifampicin. Rifampicin undergoes enterohepatic circulation and is metabolised to 25-*O*-desacetylrifampicin. One postulated mechanism of interaction is that the inhibition of CYP 3A4 by ritonavir (and other protease inhibitors) affects the metabolism of 25-*O*-desacetylrifampicin, which may in itself exhibit some toxicity (Haas *et al.*, 2009).

This leads us nicely to the interaction between protease inhibitors and rifabutin. Rifabutin is often used in place of rifampicin when a patient is taking a protease inhibitor, as rifabutin induces CYP 3A4 only moderately (and less so than rifampcin), so has less of an effect on the metabolism of protease inhibitors. To give you an example, when rifampicin is given in combination with indinavir, the AUC of indinavir reduces by 89%, compared to only 32% when indinavir is used in combination with rifabutin (Finch *et al.*, 2002).

Unfortunately, however, some protease inhibitors also inhibit CYP 3A4, which is responsible for metabolising rifabutin. Therefore, when rifabutin is used in combination with a protease inhibitor, the dose of rifabutin should be reduced (in some cases, depending on the protease inhibitor, by up to 75%) to avoid rifabutin toxicity.

2.2.8.2 Hydroxymethylglutaryl coenzyme A reducatse inhibitors and rifampicin

As an inducer of both CYP 2C9 and CYP 3A4, which are responsible for the metabolism of hydroxymethylglutaryl coenzyme A (HMGCoA) inhibitors, rifampicin affects the metabolism of these inhibitors (commonly known as the statins: atorvastatin, simvastatin, lovastatin, and the withdrawn drug cerivastatin). In addition, rifampicin is also taken up into hepatocytes by the uptake transporter system OATP1B1 (organic anion transporting polypeptide 1B1), thus stimulating the PXR and so the the CYP isoenzyme system. Pravastatin, which is not metabolised via the CYP isoenzyme system, is a substrate of the canalicular multispecific organic anion transporter (cMOAT or MRP2), found in the intestines, liver and kidney (Sandusky *et al.*, 2002). This transporter affects the absorption, elimination, and tissue uptake of drugs. Rifampicin is known to induce the MRP2 protein in the intestine, reducing drug absorption, and P-glycoprotein, causing positive efflux and additional CYP 3A4 metabolism of HMGCoA inhibitors in the intestine. Both effects reduce the plasma levels of the HMGCoA inhibitors.

2.2.8.3 Rifampicin and warfarin

The anticoagulant drug warfarin consists of a racemic mixture of (*R*)- and (*S*)-enantiomers, with the latter being approximately five times more active than the former in inhibiting the vitamin K epoxide reductase enzyme, partly due to differential metabolism of the enantiomers (Kaminsky and Zhang, 1997). The

(*S*)-enantiomer is metabolised primarily by cytochrome P_{450} isoform (CYP 2C9), while the (*R*)-isomer is metabolised by several isoenzymes, including CYP 1A2, CYP 3A4, and CYP 2C19. Chronic use of rifampicin is known to induce CYP 2C9, and drug metabolism studies have demonstrated an increased metabolism of warfarin when used concomitantly with rifampicin (Heimark *et al.*, 1987). This clinically significant interaction has been shown to increase the overall clearance of warfarin, resulting in a reduction of the half-life from an average value of 36 hours to 0.47 hours in reported cases (Casner, 1996). The combined use of warfarin and rifampicin has consequently been shown to reduce the anticoagulant effect significantly (Michot *et al.*, 1970), with the subsequent recommendation of vigilant monitoring of INR[2] and significant increase in anticoagulant dose, approaching, in some cases, 20 mg daily (Romankiewiecz and Ehrman, 1975; Kim *et al.*, 2007).

2.2.8.4   Rifampicin and azole antifungals

The azole antifungal drugs interact with rifampicin via induction of CYP 3A4. Rifampicin, as described in the previous subsections, induces the PXR, thus mediating the *CYP 3A4* gene and consequently affecting the production of the isoenzyme. As a result, several azole antifungal derivatives have a reduced antifungal effect due to increased metabolism. This effect occurs with itraconazole, fluconazole, ketoconazole, voriconazole, and posaconazole (Gubbins *et al.*, 2001). Therefore, if these agents are given in combination, it is often necessary to increase the dose of the azole to counteract the increased metabolism.

A similar effect is also observed with rifabutin (i.e. the metabolism of the azole is increased through induction of CYP 3A4), and interestingly, some azole antifungals also increase serum rifabutin levels through inhibition of CYP 3A4 (remember that this is also observed with the protease inhibitors); for example, the manufacturers of rifabutin claim that when it is given in combination with itraconazole, the AUC of rifabutin increases by up to 75%. As a consequence of this interaction, a case of uveitis (inflammation of the middle layer of the eye) has been reported (a rare complication of rifabutin) in a patient taking both rifabutin and itraconazole; the symptoms resolved upon cessation of rifabutin (Lefort *et al.*, 1996).

In addition, when ketoconazole is given in combination with rifampicin, serum rifampicin can also be reduced (some reports suggest by up to 50%) – it is not really understood why this occurs, but it is suggested that ketoconazole may impair the absorption of rifampcin from the gut. As a result, if rifampicin is to be used with ketoconazole, it is recommended to stagger the dose of each agent by around 12 hours.

2.2.8.5   Rifampicin and calcium channel blockers

There are three classes of calcium channel antagonist used clinically: the dihydropyridine type (e.g. nifedipine), the diphenylalkylamine type (e.g. verapamil), and the benzothiazepine type (e.g. diltiazem). Rifampicin, through induction of CYP 3A4, would appear to interact with all three classes, resulting in reduced bioavailability of the calcium channel antagonist.

Interestingly, studies have shown that the bioavailability of intravenously administered nifedipine is not significantly affected by the concomitant use of rifampicin, in contrast to orally administered nifedipine, where bioavailability is severely reduced (Holtbecker *et al.*, 1996). The reduction in nifedipine bioavailabilty due to CYP 3A4 induction is most likely due to rifampicin-induced gut wall metabolism. Similarly, rifampicin-induced gut wall metabolism has also been reported with verapamil (Fromm *et al.*, 1998). Consequently, orally administered calcium channel blockers should be carefully monitored and increased doses utilised if given

---

[2] INR stands for International Normalised Ratio and is a technique used to check how quickly a patient's blood coagulates (or clots). INR monitoring is used to determine what dose of warfarin a patient should be prescribed; an INR higher than 1 indicates the blood takes longer to clot than normal. Generally, for a patient taking warfarin, their INR should be between 2.5 and 3.5, although this is dependent upon their medical condition.

concomitantly with rifampicin. Alternatively, if appropriate, another class of drug should be given that does not interact with rifampicin (such as a beta-blocker or ACE inhibitor, but this depends on what the calcium channel antagonist was originally indicated for).

2.2.9   Recent developments

A phase III clinical trial on the use of rifamycin SV MMX in the treatment of travellers' diarrhoea is currently underway (www.cosmopharmaceuticals.com).

References

American Thoracic Society/Centers for Disease Control and Prevention/Infectious Diseases Society of America: Treatment of Tuberculosis, *Am. J. Respir. Crit. Care Med.*, 2003, **167**, 603–662.

M. Aquinas, W. G. L. Allan, P. A. L. Horsfall, P. K. Jenkins, W. Hung-Yan, D. Girling, R. Tall, and W. Fox, *Brit. Med. J.*, 1972, **1**, 765–771.

I. Artsimovitch, M. N. Vassylyeva, D. Svetlov, V. Svetlov, A. Peredcrina, N. Igarishi, N. Matsuagki, S. Wakatsuki, T. H. Tahirov, and D. G. Vassylyev, *Cell*, 2005, **122**, 351–363.

D. S. Askgaard, T. Wilcke, and M. Dossing, *Thorax*, 1995, **50**, 213–214.

L. Bachs, A. Parés, M. Elena, C. Piera, and J. Rodés, *Gastroenterology*, 1992, **102**, 2077–2080.

N. M. Bass, K. D. Mullen, A. Sanyal, F. Poordad, G. Neff, C. B. Leevy, S. Sigal, M. Y. Sheikh, K. Beavers, T. Frederick, L. Teperman, D. Hillebrand, S. Huang, K. Merchant, A. Shaw, E. Bortey, and W. P. Forbes, *New Eng. J. Med.*, 2010, **362**, 1071–1081.

D. Benator, M. Bhattacharya, L. Bozeman, W. Burman, A. Cantazaro, R. Chaisson, F. Gordin, C. R. Horsburgh, J. Horton, A. Khan, C. Lahart, B. Metchock, C. Pachucki, L. Stanton, A. Vernon, M. E. Villarino, Y. C. Wang, M. Weiner, S. Weis: Tuberculosis Trials Consortium, *Lancet*, 2002, **360**, 528–534.

C. A. Benson, P. L. Williams, D. L. Cohn, S. Becker, P. Hojczyk, T. Nevin, J. A. Korvick, L. Heifets, C. C. Child, M. M. Lederman, R. C. Reichman, W. G. Powderly, G. F. Notario, B. A. Wynne, and R. Hafner, *J. Infect. Dis.*, 2000, **181**, 1289–1297.

R. W. Bishai and R. E. Chaisson, *Clin. Chest. Med.*, 1997, **18**, 115–122.

M. A. Blajchman, R. C. Lowry, J. E. Pettit, and P. Stradling, *Brit. Med. J.*, 1970, **3**, 24–26.

W. J. Britton and D. N. J. Lockwood, *Lancet*, 2004, **363**, 1209–1219.

W.J. Burman, K. Gallicano, and C. Peloquin, *Clin. Pharmacokinet.*, 2001, **40**, 327–341.

E. A. Campbell, N. Korzheva, A. Mustaev, K. Murakami, S. Nair, A. Goldfarb, and S. A. Darst, *Cell*, 2001, **104**, 901–912.

K. A. V. Cartwright, J. M. Stuart, D. M. Jones, and N. D. Noah, *Epidem. Inf.*, 1987, **99**, 591–601.

P. R. Casner, *Southern Med. J.*, 1996, **89**, 1200–1203.

C. Cavusoglu, Y. Karaca-Derici, and A. Bilgic, *Clin. Microbiol.Infec.*, 2004, **10**, 657–678.

G. Chandorkar, N. P. Burte, R. K. Gade, and P. M. Bulakh, *Indian J. Lepr.*, 1984, **56**, 63–70.

I. Chopra and P. Brennan, *Tubercule Lung Dis.*, 1998, **78**, 89–98.

A. G. Cooper and D. L. Brown, *Clin. Exp. Immunol.*, 1971, **9**, 99–110.

M. J. Cupp and T. S. Tracy, *Am Fam Physician*, 1998, **57**, 107–116.

G.R. Davies and E.L. Nuermberger, *Tuberculosis*, 2008, **88** (Suppl. 1), 565–574.

G. R. Davies, S. Cerri, and L. Richeldi, *Cochrane Database of Systematic Reviews*, 2007, Issue 4.

G. Decroix and J. C. Pujet, *Semaine des Hopitaux*, 1973, **49**, 2219–2224.

A. M. Dhople and S. L. Williams, *Int. J. Antimicrob. Agents*, 1998, **9**, 169–173.

A. M. Dhople, M. A. Ibanez, and G. D. Gardner, *Drug Res.*, 1993, **43**, 384–386.

K. Dilger, U. Hofmann, and U. Klotz, *Clin Pharmacol Ther.*, 2000, **67**, 512–520.

R. H. Ebright, *J. Mol. Biol.*, 2000, **304**, 687–698.

A. Feklistov, V. Mekle, Q. Jiang, L. F. Westblade, H. Irschik, Rolf Jansen, A. Mustaev, S. A. Darst, and R. H. Ebright, *Proc. Natl. Acad. Sci. USA*, 2008, **105**, 14 820–14 825.

C. K. Finch, C. R. Chrisman, A. M. Baciewicz, and T. H. Self. *Arch. Intern. Med.*, 2002, **162**, 985–992.

H. G. Floss and T.-W. Yu, *Chem. Rev.*, 2005, **105**, 621–632.

W. Fox, G. A. Ellard, and D. A. Mitchison, *Int. J. Tuberc. Lung Dis.*, 2001, **3**, 231–279.

A. Fraser, A. Gafter-Gvili, M. Paul, and L. Leibovici, *Cochrane Database of Systematic Reviews*, 2009, Issue 2.

M. F. Fromm, K. Dilger, D. Busse, H. K. Kroemer, M. Eichelbaum, and U. Klotz, *Br. J. Clin. Pharmacol.*, 1998, **45**, 247–255.

P. R. J. Gangadharan, *Ann. Rev. Respir. Dis.*, 1986, **133**, 963–965.

S. H. Gillespie, *Antimcrob. Agents Chemother.*, 2002, **46**, 267–274.

D. J. Girling, *Scand. J. Resp. Dis.*, 1973, **84**, 119–124.

D. E. Griffith, T. Aksamit, B. A. Brown-Elliott, A. Catanzaro, C. Daley, F. Gordin, S. M. Holland, R. Horsburgh, G. Huitt, M. F. Iademarco, M. Iseman, K. Olivier, S. Ruoss, C. von Reyn, R. J. Wallace, Jr, and Kevin Winthrop, *Am. J. Respir. Crit. Care Med.*, 2007, **175**, 367–416.

J. Grosset and S. Leventis, *Rev. Infect Dis.*, 1983, **5**, S440–S446.

P. O. Gubbins, S. A. McConnell, and S. R. Penzak, 'Antifungal agents', in S. C. Piscitelli and K. A. RodvoldKA (Eds), *Drug Interactions in Infectious Diseases*, Humana Press, Totowa, NJ, 2001.

T. Gumbo, A. Louie, M. R. Deziel, W. Liu, L. M. Parsons, M. Salfinger, and G. L. Drusano, *Antimicrob. Agents Chemother.*, 2007, **51**, 3781–3788.

D. W. Haas, S. L. Koletar, L. Laughlin, M. A. Kendall, C. Suckow, J. G. Gerber, A. R. Zolopa, R. Bertz, M. J. Child, L. Hosey, B. Alston-Smith, and E. P. Acosta, *J. Acquir. Immune. Defic. Syndr.*, 2009, **50**, 290–293.

X. Y. Han, Y. H. Seo, K. C. Sizer, T. Schoberle, G. S. May, J. S. Spensor, W. Li, and R. G. Nair, *Am. J. Clin. Pathol.*, 2008, **130**, 856–864.

X. Y. Han, K. C. Sizer, E. J. Thompson, J. Kabanja, J. Li, P. Hu, L. Gomez-Valero, and F. J. Silva, *J. Bacteriol.*, 2009, **191**, 6067–6074.

G. R. Hartmann, P. Heinrich, M. C. Kollenda, B. Skrobranek, M. Tropschug, and W. Weiß, *Angew. Chem. Intn. Edn. Engl.*, 1985, **24**, 1009–1074.

Health Protection Agency, Guidance for public health management of meningococcal disease in the UK, 2011 (http://www.hpa.org.uk/web/HPAwebFile/HPAweb_C/1194947389261, last accessed 9 March 2012).

L.D. Heimark, M. Gibaldi, W. F. Trager, R. A. O'Reilly, and D. A. Goulart, *Clin. Pharmacol. Ther.*, 1987, **42**, 388–394.

B. Heym and S.T. Cole, *Intn. J. Antimicrob. Agents*, 1997, **8**, 61–70.

D. R. Hill and E. T. Ryan, *Brit. Med. J.*, 2008, **337**, 863–867.

M. X. Ho, B. P. Hudson, K. Das, E. Arnold, and R. H. Ebright, *Curr. Opinion Struct. Biol.*, 2009, **19**, 715–723.

A.F. Hofmann, *Gut*, 2002, **51**, 756–757.

N. Holtbecker, M. F. Fromm, H. K. Kroemer, E. E. Ohnhaus, and H. Heidemann, *Drug Metab. Dispos.*, 1996, **24**, 1121–1123.

B. Ji, E. G. Perani, C. Petinom, and J. H. Grosset, *Antimicrob. Agents Chemother.*, 1996, **40**, 393–399.

L. S. Kaminsky and Z.-Y. Zhang, *Pharmacol. Therapeut.*, 1997, **73**, 67–74.

K. P. Kanebratt, U. Diczfalusy, T. Bäckström, E. Sparve, E. Bredberg, Y. Böttiger, T. B. Andersson, and L. Bertilsson, *Clin Pharmacol Ther.*, 2008, **84**, 589–594.

K. Y. Kim, K. Epplen, F. Foruhani, and H. Alexanropoulos, *Prog. Cardiovasc. Nursing*, 2007, **21**, 1–4.

Y. Kishi, *Pure Appl. Chem.*, 1981, **53**, 1163–1180.

D. Kleinknecht, J. C. Homberg, and G. Decroix, *Lancet*, 1972, **1**, 1238–1239.

S. Kliewer, B. Goodwin, and T. Wilson, *Endocr. Rev.*, 2002, **23**, 687–702.

R. Lal, M. Khanna, H. Kaur, N. Srivastava, K. K. Tripathi, and S. Lal, *Crit. Rev. Microbiol.*, 1995, **21**, 19–30.

A. Lefort, O. Launay, and C. Carbon, *Ann. Intern. Med.*, 1996, **125**, 939–940.

T. Li and J. Y. L. Chiang, *Am. J. Physiol. Gastrointest. Liver Physiol.*, 2005, **288**, G74–G84.

N. Maggi and P. Sensi,US Patent 3342810, 19 September 1967a.

N. Maggi and P. Sensi,US Patent 3349082, 24 October 1967b.

G. L. Mandell and M. A. Sande, 'Antimicrobial agents; drugs used in the chemotherapy of tuberculosis and leprosy' in Goodman and Gilman, *Goodman and Gilman's The Pharmacological Basis of Therapeutics*, 8th Edn, Pergamon Press, New York, NY, 1990, Chapter 49, p. 1146.

R. Mariani and S. I. Maffioli, *Curr. Med. Chem.*, 2009, **16**, 430–454.

F. Michot, M. Bürgi, and J. Büttner, *Schweiz. Med. Wochenschr.*, 1970, **100**, 583–584.

N. Morisaki, S. Iwasaki, K. Yazawa, Y. Mikami, and A. Maeda, *J. Antibiot.*, 1993, **46**, 1605–1610.

S. D. Nightingale, *Infection*, 1997, **25**, 67–70.

S. D. Nightingale, D. W. Cameron, F. M. Gordin, P. M. Sullam, D. L. Cohn, R. E. Chaisson, L. J. Eron, P. D. Sparti, B. Bihari, D. L. Kaufman, J. J. Stern, D. D. Pearce, W. G. Weinberg, A. LaMarca, and F. P. Siegal, *New Eng. J. Med.*, 1993, **329**, 828–833.

R. Parikh, S. Thomas, J. Muliyil, S. Parikh, and R. Thomas, *Ophthalmology*, 2009, **116**, 2051–2057.

I. Paterson and M. M. Mansuri, *Tetrahedron*, 1985, **41**, 3569–3624.

A. Pereia, C. Sanz, and R. Castilli, *Ann. Hematol.*, 1991, **63**, 56–58.

V. Prelog, *Angew. Chem., Int. Ed. Engl.*, 1964, **3**, 154–155.

M. I. Prince, A. D. Burt, and D. E. J. Jones, *Gut*, 2002, **50**, 436–439.

J. A. Romankiewiecz and M. Ehrman, *Ann. Int. Med.*, 1975, **82**, 224–225.

M. Rosenthal, M. Zhang, K. N. Williams, C. A. Peloquin, S. Tyagi, A. A. Vernon, W. R. Bishai, R. E. Chaisson, J. H. Grosset, and E. L. Nuermberger, *Public Library of Science Medicine*, 2007, **4**, 1931–1939.

G. E. Sandusky, K. S. Mintze, S. E. Pratt, and A. H. Dantzig, *Histopathology*, 2002, **41**, 65–74.

J. J. Saukkonen, D. L. Cohn, R. M. Jasmer, S. Schenker, J. A. Jereb, C. M. Nolan, C. A. Peloquin, F. M. Gordin, D. Nunes, D. B. Strader, J. Bernardo, R. Venkataramanan, and T. R. Sterling, *Am. J. Respir. Crit. Care Med.*, 2006, **174**, 935–952.

C. Scarpignato and I. Pelosini, *Chemotherapy*, 2005, **51** (Suppl. 1), 36–66.

P. J. Scheuer, S. Lal, J. A. Summerfield, and S. Sherlock, *Lancet*, 1974, **303**, 421–425.

P. Sensi, *Rev. Infect. Dis.*, 1983, **5**, S402–S406.

P. Sensi and G. Lancini, in C. Hansch (Ed.), *Comprehensive Medicinal Chemistry*, Vol. 2, Pergamon Press, New York, NY, 1990, p. 793.

C. Shen, Q. Meng, G. Zhang, and W. Hu, *Br. J. Pharmacol.*, 2008, **153**, 784–791.

M. H. Skinner and T. F. Blaschke, *Clin. Pharmacokin.*, 1995, **28**, 115–125.

M. A. Steele, R. F. Burk, and R. M. DesPrez, *Chest*, 1991, **99**, 465–471.

D. N. Taylor, A. L. Bourgeois, C. D. Ericsson, R. Steffen, Z. D. Jiang, J. Halpern, R. Haake, and H. Dupont, *Am. J. Trop. Med. Hyg.*, 2006, **74**, 1060–1066.

A. Telenti, P. Imboden, F. Marchesi, D. Lowrie, S. T. Cole, M. J. Colston, L. Matter, K. Schopfer, and T. Bodmer, *Lancet*, 1993, **341**, 647–650.

J. Volmink and P. Garner, *Cochrane Database of Systematic Reviews*, 2009, Issue 1.

W. Wehrli, *Rev. Infect. Dis.*, 1983, **5**, S407–S411.

K. J. Williams and L. J. V. Piddock, *J. Antimicrob. Chemother.*, 1998, **42**, 597–603.

World Health Organization, WHO Expert Committee on Leprosy, 7th Report, 1998, 1–43 (http://www.searo.who.int/LinkFiles/Reports_2-Leprosy_7th_Geneva1998.pdf, last accessed 9 March 2012).

World Health Organization, Guidelines for the Programmatic Management of Drug-resistant Tuberculosis, Emergency Update, 2008.

World Health Organization, Treatment of Tuberculosis Guidelines, 4th Edn, 2009a.

World Health Organisation, List of Essential Medicines, 16th List, 2009b.

D. Yee, C. Valiquette, M. Pelletier, I. Parisien, I. Rocher, and D. Menzies, *Am. J. Respir. Crit. Care Med.*, 2003. **167**, 1472.

G. Zhang, E. A. Campbell, L. Minakhin, C. Richter, K. Severinov, and S. A. Darst, *Cell*, 1999, **98**, 811–824.

M. Zwarenstein, J. H. Schoeman, C. Vundule, C. J. Lombard, and M. Tatley, *Lancet*, 1998, **352**, 1340–1343.

2.3

Nitroimidazole antibacterial agents

Key Points

- Nitroimidazoles are prodrugs which are reduced *in vivo* to give radical species that are toxic to cells. Resistance arises due to decreased reduction by the cellular enzymes.
- Owing to this mechanism of action, the nitroimidazoles show activity against anaerobic bacteria and are therefore used to treat anaerobic infections.
- The nitroimidazoles are generally well tolerated, but on rare occasions can cause severe side effects such as pancreatitis.
- Alcohol should be avoided when taking nitroimidazoles.

Antibacterial nitroimidazoles used clinically are metronidazole and tinidazole; their structures are shown in Figure 2.3.1 and their therapeutic indications are listed in Table 2.3.1.

2.3.1 Discovery

Nitroimidazole antibacterial activity was first discovered in extracts from *Nocardia mesenterica* (Maeda *et al.*, 1953) and *Streptomyces eurocidicus* (Osato *et al.*, 1955). The active agent, called azomycin, was identified as 2-nitroimidazole (Figure 2.3.2) (Nakamura *et al.*, 1955), a metabolite produced from arginine (at least in the latter of these two bacteria) (Nakane *et al.*, 1977). After many unsuccessful attempts to synthesise this simple molecule, researchers at Rhône-Poulenc turned their attention to the production and evaluation of the synthetically less challenging 5-nitroimidazole isomers, leading to the identification of metronidazole as a potent antiprotozoal agent with greater activity than azomycin and an acceptable toxicity profile (Cosar and Julou, 1959). The synthesis of 2-nitroimidazole was not achieved until 10 years later, and then only in low yield (Lancini and Lazzari, 1965). For more on the discovery of the nitroimidazoles, see Anon (1978).

Antibacterial Agents: Chemistry, Mode of Action, Mechanisms of Resistance and Clinical Applications, First Edition.
Rosaleen J. Anderson, Paul W. Groundwater, Adam Todd and Alan J. Worsley.
© 2012 John Wiley & Sons, Ltd. Published 2012 by John Wiley & Sons, Ltd.

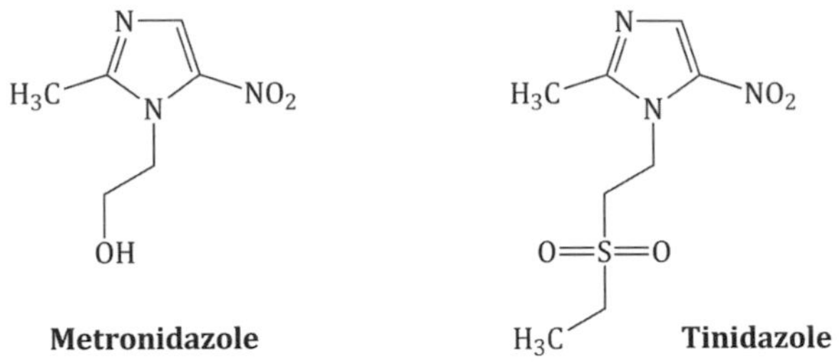

Metronidazole   **Tinidazole**

Figure 2.3.1   *Nitroimidazole antibacterials*

Table 2.3.1   *Therapeutic indications for the nitroimidazole antibacterials*

Nitroimidazole antibacterial	Indications
Metronidazole	Anaerobic infections, *Trichomonas vaginalis* infection, *Helicobacter pylori* eradication (part of triple therapy), rosacea, bacterial vaginosis, pelvic inflammatory disease, acute ulcerative gingivitis, acute oral infections, acute invasive *amoebic* dysentery, giardiasis
Tinidazole	Anaerobic infections, bacterial vaginosis, acute ulcerative gingivitis, abdominal surgery prophylaxis, acute invasive amoebic dysentery, giardiasis

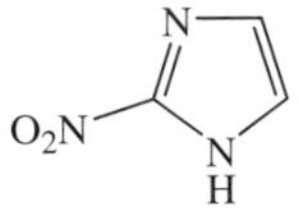

Figure 2.3.2   *Azomycin (2-nitroimidazole)*

2.3.2   Synthesis

The synthetic route to metronidazole patented by Rhône-Poulenc (Scheme 2.3.1) (Jacob *et al.*, 1960) starts with 2-methyl-5-nitroimidazole, which can be accessed from 2-methylimidazole, itself prepared by reacting ethylene diamine with acetic acid to obtain the salt, then heating at high temperature until dry, resulting in the diamide **1**. Deprotonation by calcium oxide initiates cyclisation to product **2**, which can then be oxidised to the aromatic imidazole using Raney-nickel. This classical chemistry is successful in producing the product, but poor yields are obtained for most steps. Despite this limitation, hundreds of derivatives have been synthesised and tested (Cosar and Julou, 1959; Cosar *et al.*, 1966) and there are many patents on the synthesis and activities of nitroimidazoles (mostly from the 1960s), the majority of which focus on the particularly low-yielding final step. Besides ethylene oxide, as depicted in Scheme 2.3.1, chloroethanol or ethylene carbonate were also commonly used. Later optimisation of the synthetic route improved the yields (Kraft *et al.*, 1989) such that metronidazole analogues are still being prepared by the same methodology (e.g. Atia, 2009) or else use metronidazole as the starting material (e.g. Mao *et al.*, 2009; Kumar *et al.*, 2010).

2.3.3   Bioavailability

The 5-nitroimidazoles are generally well-absorbed, with an oral bioavailability for metronidazole and tinidazole approaching 100% (Lau *et al.*, 1992; Freeman *et al.*, 1997; Fung and Doan, 2005). Following

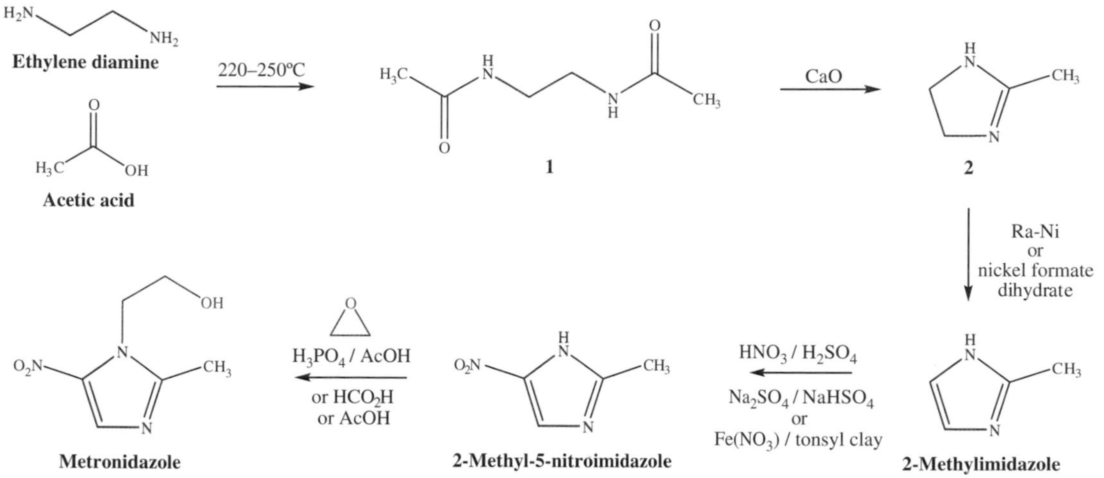

Scheme 2.3.1 *Synthesis of metronidazole*

oral administration, peak plasma concentrations occur after 1–2 hours, with 500 mg of metronidazole typically delivering C_{max} values of 8–13 mg/L after about 15 minutes to 4 hours. Tinidazole shows similar, if slightly more favourable, pharmacokinetic parameters and the absorption of neither drug is significantly affected by food. As the MIC of these concentration-dependent antibacterial agents is generally between 0.25 and 4 mg/L for susceptible species, appropriate therapeutic concentrations are readily achieved. Both metronidazole and tinidazole distribute well throughout a variety of tissues, including the CNS, with reported volumes of distribution of 0.53–0.96 L/kg for metronidazole and approximately 0.65 L/kg for tinidazole, thus providing excellent therapeutic potential for the treatment of anaerobic infections in most organs and tissues.

The plasma half-life of metronidazole is between 6 and 8 hours, while that of tinidazole is rather longer (12–14 hours), resulting in a requirement for 2–3 times daily dosing for metronidazole and 1–2 times daily for tinidazole. Both metronidazole and tinidazole are primarily metabolised in the liver by CYP 3A4; the metabolites are mostly excreted in the urine, while a small proportion (about 10–15%) of unchanged drug is excreted in the bile and in the urine, and a further 10% is excreted in the faeces.

2.3.3.1   Bacterial uptake of metronidazole

This small molecule can enter many bacteria by passive diffusion; however, the uptake of metronidazole into bacterial cells has also been linked to the presence of ferredoxin-linked electron transport or hydrogenase systems, which correlates with the antibacterial activity. As we will discuss in the next subsection, the mode of action of metronidazole involves its bioreduction by a ferredoxin-based enzyme or hydrogenase, which facilitates its transport into susceptible organisms.

2.3.4   Mode of action and selectivity

As far as possible, the following account of the mode of action and selectivity of the 5-nitroimidazoles relates to all products in the class, but metronidazole is the most commonly prescribed nitroimidazole and has been the most extensively studied for its mode of action, so it is often the focus of this summary.

During studies on the scope of activity and the mode of action, it was quickly evident that the nitro group is responsible for the bactericidal action of 5-nitroimidazoles, while the other substituents on the imidazole ring confer slightly different pharmacokinetic properties on each nitroimidazole, providing a choice of agents for different infections (Goldman, 1982).

Despite the clinical use of metronidazole over many years, the exact details of its mode of action remain unclear, but it is known to involve four steps: (1) the nitroimidazole must pass into the cell, before (2) the nitro group is bioreduced to a reactive radical species, which (3) can react with a cellular species, such as DNA, and lastly (4) inactivated products are released. A similar mode of action is believed to exist for the other 5-nitroimidazoles (Tocher, 1997).

Studies of susceptible bacteria and protozoa led to a link between the bactericidal activity and the presence of ferredoxin-linked electron transport or hydrogenase systems. Although not a perfect correlation, the presence and activity of ferredoxin (Edwards *et al.*, 1973) or a ferredoxin-linked hydrogenase (Church *et al.*, 1996) can generally be matched to susceptibility to the 5-nitroimidazoles. Furthermore, the cellular uptake of metronidazole is linked to the rate of reduction – as it is reduced intracellularly, a favourable concentration gradient results in more metronidazole being taken into the cell to be reduced, contributing to its efficiency. This was first noted in the protozoan species *Trichomonas vaginalis* (Ings *et al.*, 1974), but appears to be general across metronidazole-susceptible species, including bacteria (Edwards, 1980).

Once in the cell, the reduction of a nitro compound can progress through a number of stages to the ultimate production of the corresponding amino compound; the first one-electron reduction results in the reactive nitroimidazole radical anion, the reactive radical species referred to in Step (2) above, which is a key intermediate and has a number of possible fates (Scheme 2.3.2). It can:

- React with a cellular species, such as DNA or protein, and cause irreparable damage, leading to bactericidal activity: pathway (a).
- Accept a further electron to form the nitrosoimidazole product: pathway (b).

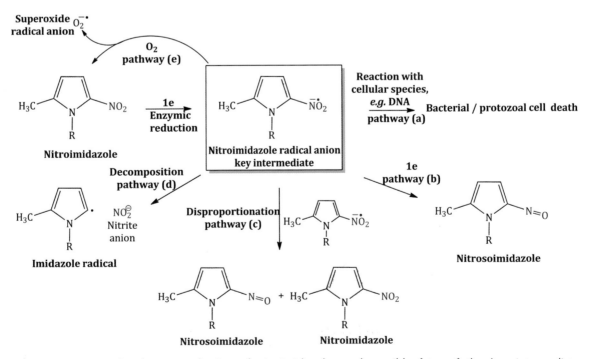

Scheme 2.3.2 *Single electron reduction of nitroimidazoles and possible fates of the key intermediate, nitroimidazole radical anion*

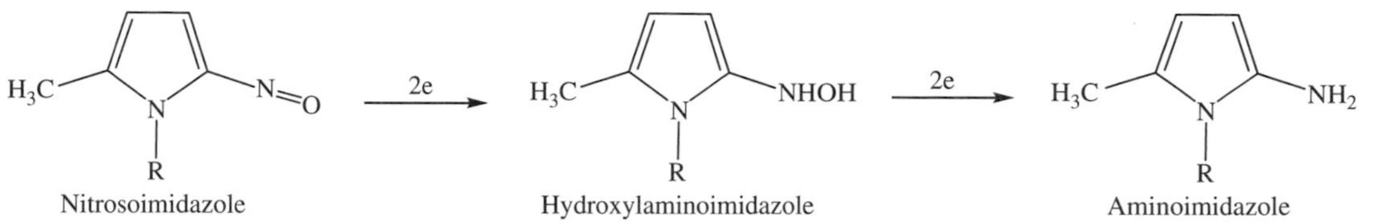

Scheme 2.3.3 *Further reduction of nitrosoimidazole to the aminoimidazole through the hydroxylaminoimidazole intermediate*

- React with another nitroimidazole radical anion in a disproportionation reaction to form a neutral nitrosoimidazole product and a molecule of nitroimidazole: pathway (c).
- Decompose to form an imidazole radical and nitrite anion: pathway (d).
- React with molecular oxygen, forming a superoxide radical anion and the starting nitroimidazole: pathway (e) (Goldman, 1982; Edwards, 1993; Tocher 1997).

There is some evidence that the nitroimidazole radical anion becomes protonated and in this form is more cytotoxic (Tocher, 1997).

The nitroimidazoles, then, are prodrugs that are bioactivated *in vivo* to the reactive nitroimidazole radical anion. In the presence of oxygen, such as that found in obligate aerobes, the cytotoxic nitroimidazole radical anion is rapidly inactivated, eliminating the bactericidal effect; only when reduced under anaerobic conditions, such as by obligate anaerobic bacteria, where the potential to trap and neutralise the reactive species is limited, does irreparable cellular damage result, mostly observed as DNA degradation (Knight *et al.*, 1978; Tocher and Edwards, 1994), leading to cell death.

Further two-electron reduction of the nitrosoimidazole yields the hydroxylamine, and a final two-electron reduction results in the amine (Scheme 2.3.3), but studies with these products suggest that they are non-cytotoxic and have no apparent role in the bactericidal effect (Tocher and Edwards, 1994).

2.3.4.1 Selective toxicity

Tests on the reduction of metronidazole in cultured mammalian cells showed significant cytotoxicity, yet there are few side effects to metronidazole treatment and it is commonly used during pregnancy. It seems likely that metronidazole is not reduced by mammalian enzymes and that its selectivity arises due to its reduction solely by intracellular bacterial and protozoal enzymes (probably due to favourable redox potentials in these species) (Edwards, 1993; Ehlhardt *et al.*, 1988), followed by rapid reaction of the reactive products with the bacterial or protozoal DNA, before the products are able to escape the cell into the mammalian host.

2.3.5 Mechanisms of resistance

As the mode of action involves reduction of the nitroimidazole to a bioactive form, it is probably not surprising to find that resistance arises due to a decrease in expression and/or activity of the reductive bacterial (or protozoal) enzyme, known to be ferredoxin in clostridia species (Edwards, 1980, 1993). Despite this obvious 'Achilles' heel' point of fragility, resistance to the 5-nitroimidazoles is rare, even though they have been used extensively worldwide over a 40-year period. Ferredoxin is part of the essential glucose metabolism pathway and the few bacteria which have developed resistance to the nitroimidazoles appear to have compensated for reduced ferredoxin/oxidoreductase activity through the upregulation of alternative enzymes for glucose metabolism (such as lactate dehydrogenase in a resistant strain of *Bacteroides fragilis*) (Edwards, 1980, 1993), and these resistant strains do not reduce and activate the nitroimidazole.

As we will see in the next subsection, the recent rise in resistance to metronidazole of *Helicobacter pylori*, again through the inactivation of the oxidoreductase enzyme(s) reponsible for the reduction and activation of the nitroimidazole, is a great concern (van der Wouden *et al.*, 2000; Jenks and Edwards, 2002).

The resistance of anaerobic protozoa has been studied more fully and there are several reviews of the topic (e.g. Land and Johnson, 1997, 1999; Upcroft and Upcroft, 2001; Dunne *et al.*, 2003), but this is beyond the scope of this book, which seeks to concentrate upon bacteria.

2.3.6   Clinical applications

2.3.6.1   Spectrum of activity

Metronidazole and tinidazole have activity against a wide range of pathogenic microorganisms, encompassing various protozoa (such as *Giardia lamblia* and *T. vaginalis*) and Gram negative (such as *B. fragilis* and *Fusobacterium necrophorum*) and Gram positive anaerobic bacteria (such as *Clostridium difficile* and *Peptostreptococcus anaerobius*). Another relatively recent development is that these agents also have activity against *H. pylori*, a Gram negative, microaerophillic bacterium found in the stomachs of humans, which is thought to cause peptic ulcers.

These drugs were originally introduced to treat protozoal infections, but were then subsequently found to have activity against certain pathogenic anaerobic bacteria. The combined protozoal and anaerobic activity of these agents allows them to be used in the treatment of a broad range of conditions.

As infections due to protozoa are outwith the scope of this book, we will mention these here only in brief.

2.3.6.2   Trichomoniasis

The introduction of metronidazole to the market in 1957 significantly improved the management of trichomoniasis infections. Single-dose therapy with metronidazole (2 g) is effective at treating trichomoniasis (Woodcock, 1972) (also sometimes known as 'trich'), a sexually transmitted infection (STI) affecting men and women caused by the flagellated, singled-celled, protozoan *T. vaginalis*. Trichomoniasis infection is a significant public health concern with around 170 million people estimated to be infected worldwide (World Health Organization, 2001). Indeed, it has been reported that the incidence of trichomoniasis infection is higher than that of other common non-viral STIs, such as chlamydia and gonorrhoea (Weinstock *et al.*, 2004). In women, the clinical manifestations of *T. vaginalis* infection can include green/yellow vaginal discharge, vulvar erthema, and dyspareunia (painful sexual intercourse), while in men they can include urethral discharge and occasionally urethritis. Symptoms of infection are more common in women than in men, but as many as 50% of women will not show any symptoms.

Single-dose therapy offers the advantage of improving patient compliance, but may be associated with higher side effects, such as nausea, when compared to lower-dose weekly regimens. Patients who cannot tolerate the adverse effects associated with the single-dose treatment may be offered a longer treatment regimen; for example, 200 mg every 8 hours for 7 days, or 400–500 mg every 12 hours for 5–7 days. Alternatively, if metronidazole is ineffective or not tolerated, tinidazole can be given as a one-off single 2 g dose to treat the infection. Metronidazole can also be given intravaginally, but this is not recommended, as parasitological cure rates are low compared to oral treatment (around 90% cure rate for oral treatment compared to around 50% cure rate for intravaginal treatment) (Forna and Gülmezoglu, 2003).

Unfortunately, clinical isolates resistant to metronidazole have been reported and their number appears to be increasing (Cudmore *et al.*, 2004). Refractory cases are often treated using an increased dose of metronidazole, or by switching to tinidazole, but because of the similar mode of action of these agents, cross-resistance is a real clinical concern.

A Cochrane review confirms that the nitroimidazoles, metronidazole and tinidazole, are effective at treating trichomoniasis, but the authors were unable to recommend the use of one drug over another due to lack of good-quality evidence comparing the two agents (Forna and Gülmezoglu, 2003).

2.3.6.3 Giardiasis

Symptoms of giardiasis infection include diarrhoea, abdominal cramps, nausea, and steatorrhea ('floaty poos'), and one risk of chronic infection, particularly in children, is malabsorption, which could ultimately lead to malnutrition and poor physical development.

Several years after the introduction of metronidazole for the treatment of *T. vaginalis* infections, in 1962, Darbon and coworkers reported that metronidazole could also be used to manage *Giardia lamblia* infections (Darbon *et al.*, 1962). Since then, other agents, such as the anthelmintics, albendazole and mebendazole, have also been shown to be effective, and these have a different mode of action (and spectrum of activity) to the nitroimidazoles.

A review of the drugs used to treat giardiasis showed that metronidazole was the most frequently tested drug (i.e. it had the largest number of published clinical trials in the literature). In addition, the data in this review showed that, in randomised controlled trials, metronidazole had a mean cure rate of 81.5%, while tinidazole, albendazole, and mebendazole had mean cure rates of 91.1, 73.4, and 65.6%, respectively (Busatti *et al.*, 2009). It appears that metronidazole is more effective at treating giardiasis when given at high doses over a prolonged period of time, while tinidazole remains effective when given as a one-off single dose, and these findings are reflected in the recommended dosage regimens for metronidazole and tinidazole: metronidazole should be given as 2 g daily for 3 days (or alternatively, 400 mg three times daily for 5 days, or 500 mg twice daily for 7–10 days), while tinidazole can be given as a 2 g single dose. Currently, metronidazole is the drug of choice in the treatment of giardiasis, and tinidazole should be used as an alternative.

2.3.6.4 Amoebiasis

Amoebiasis is a disease caused by the protozoan parasite *Entamoeba histolytica* and is responsible for a high number of mortalities annually worldwide (in excess of 50 000 deaths per year). Indeed, in terms of mortality caused by protozoan parasites, amoebiasis is second only to malaria, which gives an indication of the severity of this infection. Perhaps the most famous (and significant) outbreak of amoebiasis was during 1933, in Chicago, when a sewer pipe contaminated the water supply to a local hotel, causing more than 700 cases of amoebic dysentery to be reported (Bundesen *et al.*, 1934).

Metronidazole is considered the drug of choice for invasive intestinal amoebiasis and extraintestinal amoebiasis (including liver abcesses). Tinidazole is a useful alternative to metronidazole; it is more effective at reducing clinical failure when compared to metronidazole and is also associated with fewer side effects (Gonzales *et al.*, 2009).

In patients with invasive disease, metronidazole should first be prescribed (usually for 5 or 10 days) to eradicate trophozoites in the tissues (such as the liver). This should be followed by the administration of a luminal amoebicide (usually for 10 days) to eliminate any surviving organisms in the colon.

2.3.6.5 *Helicobacter pylori*

Metronidazole is also used as part of combination therapy to eradicate *H. pylori* – a Gram negative, micro-aerophillic bacterium, which is a leading cause of gastritis, peptic ulcer disease, and gastric cancer worldwide (around 95% of duodenal ulcers and 80% of gastric ulcers are thought to be caused by *H. pylori*). Indeed, because of the increased risk of developing gastric cancer associated with *H. pylori* infection, the WHO's International Agency for research on cancer has classified it as a type I carcinogen (i.e. carcinogenic to humans).

H. pylori infection is now universally accepted as being a leading cause of gastroduodenal disease. This, however, was not always the case – prior to 1982 it was thought that stomach and duodenal ulcers were caused by smoking, stress, and lifestyle factors. It was not until 1982 that two Australian researchers, Robin Warren and Barry Marshall, made the connection between *H. pylori* infection and peptic ulcer disease (Marshall and Warren, 1984). Despite this evidence, several attempts to culture the bacterium were unsuccessful and it wasn't until the culture plate was unintentionally left in the incubator for 5 days over the Easter holidays (we've all seen kitchens where something similar happens) that a culture of *H. pylori* was successfully obtained (Marshall *et al.*, 1984). Prior to this, the cultures were discarded after 48 hours if no growth had occurred. Once *H. pylori* had been isolated, the medical community still remained sceptical that a bacterium could cause gastroduodenal disease. Marshall then showed that ingestion of *H. pylori* could cause acute symptoms of gastritis, by drinking a suspension containing a 4-day-old culture of *H. pylori*, a dramatic (and brave) confirmation that *H. pylori* was, indeed, a pathogen.

By the mid-nineties, the medical communities had accepted Warren and Marshall's pioneering research, and in 2005 they were awarded the Nobel Prize in Physiology or Medicine for their discovery of 'the bacterium *Helicobacter pylori* and its role in gastritis and peptic ulcer disease'.

It soon became apparent that antibacterial drugs could be used to eradicate *H. pylori*, which proved to be an efficient and effective way of treating gastritis and peptic ulcer disease[3] (Marshall *et al.*, 1988). The use of triple therapy (two antibacterial agents plus an acid suppressant, such as ranitidine) proved to be particularly effective at eradicating *H. pylori* (Borody *et al.*, 1989). One study showed that ranitidine, metronidazole, and amoxicillin eradicated *H. pylori* in 89% of patients compared to 2% of patients taking ranitidine alone. Furthermore, the same study also showed that ulcers reoccurred in 2% of patients in whom *H. pylori* had been eradicated, compared to 85% in whom *H. pylori* had persisted (Hentschel *et al.*, 1993).

Nowadays, metronidazole forms a key component in many regimens used to eradicate *H. pylori* (Table 2.3.2). Tinidazole can also be used for eradication therapy, as an alternative to metronidazole, although

Table 2.3.2   *Recommended regimens used to eradicate H. pylori in adults. Regimens are generally taken for seven consecutive days (adapted from BNF)*

Regimen	Dose
Metronidazole	400 mg twice daily
Clarithromycin	250 mg twice daily
Lansoprazole	30 mg twice daily
Clarithromycin	500 mg twice daily
Amoxicillin	1 g twice daily
Lansoprazole	30 mg twice daily
Metronidazole	400 mg three times daily
Amoxicillin	500 mg three times daily
Omeprazole	20 mg twice daily
Amoxicillin	1 g twice daily
Clarithromycin	500 mg twice daily
Omeprazole	20 mg twice daily

---

[3] Up until the discovery of *H. pylori*, peptic ulcers were treated using H_2 receptor antagonists (e.g. cimetidine). Unfortunately, after H_2 receptor treatment was stopped, the ulcer almost always reoccurred, meaning that for many patients, lifelong therapy was required.

this is not commonplace. Unfortunately, it will come as no surprise to learn that, as with most bacterial infections, the development of resistance has had a significant impact on successful treatment. Resistance to metronidazole is common and can vary widely between geographical locations (Wang *et al.*, 2000; Kim *et al.*, 2001), with resistance rates as high as 80% reported in central Africa (Glupczynski *et al.*, 1990).

2.3.6.6 Anaerobic infections

Metronidazole has activity against a range of pathogenic anaerobic bacteria and is consequently considered the agent of choice to treat anaerobic infections (Löfmark *et al.*, 2010). Indeed, metronidazole has been successfully used to treat a variety of serious anaerobic infections, with examples including septicaemia caused by *F. necrophorum*, osteomyelitis caused by *B. fragilis*, and necrotising pneumonia caused by *Fusobacterium nucleatum* (Tally *et al.*, 1975; Boutoille *et al.*, 2003). It is also used to treat other infections from which pathogenic anaerobic bacteria have been isolated, such as brain abscess, bacteraemia, pelvic abscess, puerperal sepsis, and post-operative wound infections.

Metronidazole also has a role to play in the treatment of bacterial vaginosis, a condition commonly caused by overgrowth of *Gardnerella vaginalis*. Some women with bacterial vaginosis can remain asymptomatic, while others can show symptoms – commonly an abnormal vaginal discharge with an unpleasant 'fish-like' odour. In view of this, it was previously considered a minor infection with few complications, but there is now evidence to suggest that, in pregnancy, bacterial vaginosis can lead to a higher risk of preterm birth (Hillier *et al.*, 1995). Metronidazole is considered the agent of choice for the treatment of bacterial vaginosis; it may be given as a tablet, or if the patient prefers, an intravaginal gel – both formulations appear to be equally effective, with similar cure rates observed (Ferris *et al.*, 1995; Hanson *et al.*, 2000).

Another indication of metronidazole is for the treatment of *C. difficile* infection (CDI) (also known as 'C. diff') – a Gram positive anaerobe that has, over the last few years, received a great deal of media attention. Around 5% of healthy adults carry *C. difficile* in their colon without showing any signs of infection. However, if a patient is taking a broad-spectrum antibiotic[4] for, say, a UTI, this can disrupt the normal intestinal flora, which may facilitate the colonisation of *C. difficile* and ultimately lead to CDI. *C. difficile* produces two principal toxins – A and B – which cause diarrhoea, inflammation, and ultimately, *C. difficile*-associated diarrhoea (CDAD), which can be fatal (Starr, 2005). CDAD is diagnosed by the presence of diarrhoea and the positive detection of *C. difficile* toxins in the faeces. As diagnosis tends to focus on the presence of *C. difficile* toxins, laboratories tend not to culture it, although culturing techniques are available (Figure 2.3.3). This, however, has its disadvantages, as toxins are absent in a small number of confirmed cases, meaning that only testing for the presence of *C. difficile* toxin may not necessarily diagnose all cases of CDAD (Delmée *et al.*, 2005).

There are several risk factors established for CDAD, including long hospital in-patient stays, long antibiotic course durations, multiple antibiotic treatments,[5] increasing age, and treatment with anti-ulcer medications.

For a patient with CDAD, there are several treatment interventions that should be considered. First, if appropriate, the original antibiotic therapy should be stopped, and second, depending on the symptoms, antibiotic treatment should be started to eradicate *C. difficile* (Starr, 2005). The standard first-line therapy used to eradicate *C. difficile* is oral metronidazole, with oral vancomycin used as an alternative treatment (or when the patient is critically ill). Metronidazole appears to have equal efficacy to vancomycin when used for minor CDI, and is generally preferred because it is less expensive and allows vancomycin to be kept in reserve for more serious infections.

---

[4] Fluoroquinolones, clindamycin, and cephalosporins are most frequently associated with CDI.
[5] *H. pylori* eradication therapy, as discussed in the previous subsection, usually consists of two antibiotics and a proton pump inhibitor (an anti-ulcer agent) and is therefore associated with a risk of developing CDI.

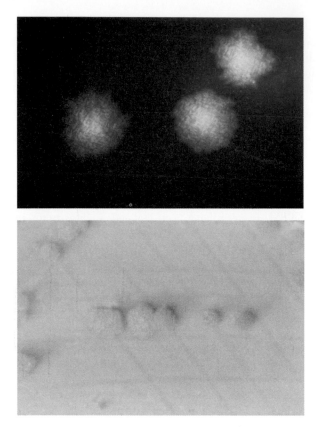

Figure 2.3.3   C. difficile *colonies grown on blood agar (top) and cycloserine mannitol agar (bottom); both after 48 hours (Image courtesy of Public Health Image Library, Image ID3647 and ID3649, Online, [http://www.phil.cdc.gov/phil/home.asp, last accessed 29th March 2012].)*

2.3.6.7   Rosacea

Metronidazole is also used for the treatment of rosacea, a skin condition commonly affecting the cheeks, nose, chin, and forehead, characterised by flushing, erythema, papules, and pustules. Rosacea, previously called acne rosacea, is distinct from the acne vulgaris commonly seen in teenagers. The cause of rosacea is unknown, but, unlike in acne vulgaris, sebum secretion and skin microbiology appear normal.

Topical metronidazole is effective for the treatment of minor rosacea (van Zuuren *et al.*, 2005), but, if the condition is severe, systemic antibiotics may be required. Interestingly, the efficacy of metronidazole appears to be due to its anti-inflammatory properties rather than its antimicrobial effects (Schmadel and McEvoy, 1990).

2.3.7   Adverse drug reactions

Metronidazole is not generally associated with any severe adverse events; however, several case studies have demonstrated limited toxicity in some individuals.

2.3.7.1   Pancreatitis

Small-scale studies have demonstrated a threefold increase in pancreatitis associated with metronidazole use (Lankisch *et al.*, 1995), and this increases eightfold if used in combination with other drugs, namely amoxicillin and omeprazole for the treatment of *H. pylori* (Norgaard *et al.*, 2005). A proposed mechanism of toxicity is that under aerobic conditions metronidazole could undergo redox cycling and might yield hydrogen peroxide, superoxide, and other free radicals that can damage pancreatic cells. No reports, to date, have demonstrated pancreatitis with tinidazole, but, as tinidazole and metronidazole are both derivatives of 5-nitroimidazole, caution should be demonstrated (Sanford *et al.*, 1988; Sura *et al.*, 2000).

The clinical implications of metronidazole-associated pancreatitis are that it is unpredictable and generally remits with standard pancreatitis treatment and removal of metronidazole. It is not recommended that metronidazole or any other 5-nitroimidazole be introduced to the same patient. This effect has also been reported with the related drug secnidazole (Slim *et al.*, 2010).

2.3.7.2   QT wave prolongation

There have been limited reports that metronidazole induces QT wave prolongation (Cohen *et al.*, 2008), which has been well documented with the structurally similar azole antifungal agents, including ketoconazole, itraconazole, and fluconazole. The former imidazoles may have an effect upon HERG receptors and potassium efflux into cardiac cells, thus affecting cardiac rhythm, and it is possible that metronidazole may have a similar action on HERG receptors, although further research is still required into this proposed theory.

Like the azole drugs, metronidazole has been shown to prolong QT wave duration when interacting with other QT wave-prolonging drugs (Cooke *et al.*, 1996; Fehri *et al.*, 2004; Kounas *et al.*, 2005).

2.3.7.3   Serotonin syndrome

Animal studies have demonstrated that metronidazole increases serotonin levels in Wistar rat brains and inhibits monoamine oxidase activity in bovine brains (Befani *et al.*, 2001; Karamanakos *et al.*, 2007; Karamanakos, 2008).

2.3.7.4   Encephalopathy

Single case reports have linked encephalopathy with the use of high-dose metronidazole, but the mechanism for this proposed neurotoxicity is undisclosed (Seok *et al.*, 2003; Groothoff *et al.*, 2010).

A few case studies have suggested that metronidazole may be associated with Wernicke's encephalopathy (WE) in non-alcoholic patients, but the evidence is suggestive rather than definitive (Kim *et al.*, 2007). It has been suggested that metronidazole is metabolised into a thiamine analogue, which may antagonise vitamin B1 and thus be responsible for the characteristcs of WE identified in nuclear imaging studies of patients (Zuccoli *et al.*, 2008).

2.3.8   Drug interactions

2.3.8.1   Alcohol

The commonest drug interaction experienced with metronidazole is that with ethanol, described by Edwards and Price (1967). There are six known classes of mammalian alcohol dehydrogenase (ADH), but only ADH I and IV are characterised as being involved in ethanol metabolism. Metronidazole is thought to block the ADH I enzyme, thus allowing for the accumulation of *in vivo* acetaldehyde, leading to very unpleasant adverse reactions, such as nausea, vomiting, and a throbbing headache (Visapaa *et al.*, 2002). It is likely that metronidazole increases intracolonic acetaldehyde levels, and this is reflected in animal studies (Tillonen *et al.*, 2000).

2.3.8.2   Warfarin

The first report of a warfarin–metronidazole interaction occurred in 1976, when the relative levels of the (*S*)-enantiomer of warfarin were elevated by concommitant administration of metronidazole (O'Reilly, 1976). Other reports demonstrated an increased INR due to the use of both drugs (Kazmier, 1976; Dean and Talbert, 1980).

Metronidazole inhibits cytochrome CYP 2C9 and consequently the metabolism of the (*S*)-enantiomer of warfarin is reduced, thus elevating (*S*)-warfarin levels and increasing the risk of haemorrhage. The clinical implication of the failure to monitor a patient's INR while receiving both drugs is an increased risk of intracerebral haemorrhage (Howard-Thompson *et al.*, 2008). The use of metronidazole with warfarin is thus discouraged, as is the use of tinidazole, but if these agents are deemed necessary, careful monitoring of INR during (and up to 8 days post) co-admininistration is recommended (http://dailymed.nlm.nih.gov/dailymed/drugInfo.cfm?id=4730 (Tinidazole), last accessed 30 March 2012).

2.3.8.3   Phenytoin/carbamazepine

Phenytoin and carbamazepine are drugs commonly used in the management of epilepsy; both induce CYP 450 enzymes and are therefore responsible for causing many clinically significant drug interactions. As phenytoin and carbamazepine are enzyme inducers (including CYP 3A4, which, as we mentioned in Subsection 2.3.3, is responsible for metabolising metronidazole/tinidazole), you would probably expect that when they are given concurrently with metronidazole/tinidazole, we would observe a drug interaction whereby the metronidazole/tinidazole is 'over-metabolised', resulting in reduced serum levels. This is actually not the case and both agents are often co-administered without the metronidazole/tinidazole being significantly affected. Surprisingly, the concomitant administration of metronidazole and intravenous phenytoin reduces the clearance and therefore prolongs the half-life of phenytoin (Jensen and Gugler, 1985). Similarly, the plasma levels of carbamazepine have been shown to be raised with metronidazole use (Wurden and Levy, 2002). The mechanism of this interaction is not fully understood and further work is necessary to establish the clinical significance.

2.3.9   Recent developments

There are many nitroimidazole antibacterial agents used worldwide – all with varying structures, activities, and potencies (Table 2.3.3 and Figure 2.3.4). The nitroimidazole that is often considered the gold standard is metronidazole. Other, newer, agents may offer advantages (such as increased potency or a better pharmacokinetic profile, enabling once-daily dosing) over metronidazole, but generally metronidazole is still the agent of choice, as it has the most clinical evidence. As a result of the widespread success of metronidazole and the fact that many thousands of nitroimidazole derivatives have already been evaluated for antibacterial activity, Big Pharma companies are not pursuing this avenue of research and are instead focussing their attention on more attractive (and novel) antibacterial drug targets (some of which we discuss in the coming sections).

Table 2.3.3 *Nitroimidazoles used in various parts of the world*

Nitroimidazole	Registered	Indication	R^1	R^2	Reference
Azanidazole	Licensed in Italy	Treatment of trichomoniasis	Me	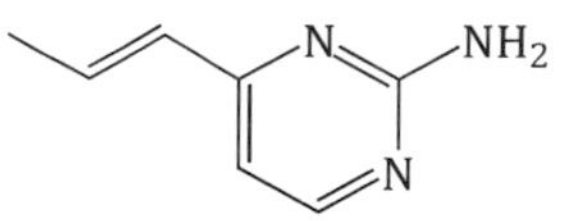 (2-aminopyrimidinyl-vinyl structure)	Marchionni *et al.* (1981)
Nimorazole	Licensed in Brazil, Hong Kong, and France	Treatment of protozoal infections	H	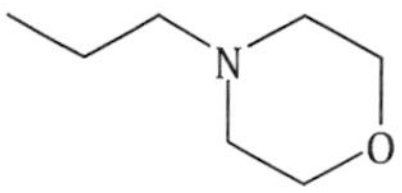 (ethyl-morpholine structure)	Mohanty and Deighton (1987)
Secnidazole	Licensed in Brazil, France, and Mexico	Treatment of protozoal infections	Me	$-CH_2CH(OH)Me$	Gillis and Wiseman (1996)
Ronidazole	Veterinary use	Controls trichomoniasis in cage birds and pigeons	Me	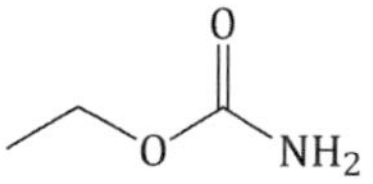 (carbamate ester structure)	Miller (1971)
Ornidazole	Licensed in France, Italy, and Mexico	Treatment of anaerobic and protozoal infections	Me	$-CH_2CHOHCH_2Cl$	Sköld *et al.* (1977)

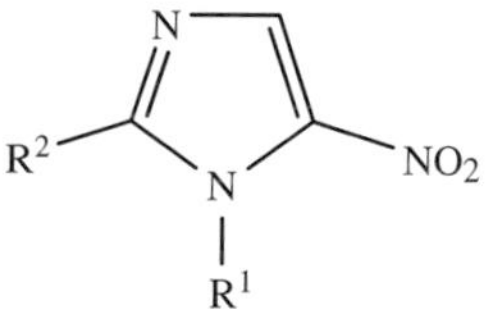

Figure 2.3.4 *Nitroimidazole general structure*

References

Anon (Editorial), *Brit. J. Vener. Dis.*, 1978, **54**, 69.

A. J. Kh. Atia, *Molecules*, 2009, **14**, 2431–2436.

O. Befani, E. Grippa, L. Saso, P. Turini, and B. Mondoi, *Inflamm. Res.*, 2001, **2** (Suppl.), S136–S137.

T. J. Borody, P. Cole, S. Noonan, A. Morgan, J. Lenne, L. Hyland, S. Brandl, E. G. Borody, and L. L. George, *Med. J. Aust.*, 1989, **151**, 431–435.

D. Boutoille, J. P. Talarmin, V. Prendki, and F. Raffi, *Eur. J. Intern. Med.*, 2003, **14**, 63–64.

H. N. Bundesen, F. O. Tonney, and I. D. Rawlings, *J. Am. Med. Assoc.*, 1934, **102**, 367–369.

B. G. N. O. Busatti, J. F. G. Santos, and M. A. Gomes, *Biologics: Targets & Therapy*, 2009, **3**, 273–287.

D. L. Church, R. D. Bryant, V. Sim, and E. J. Laishley, *Anaerobe*, 1996, **2**, 147–153.

O. Cohen, N. Saar, M. Swartzon, O. Kliuk-Ben-Bassat, and D. Justo, *Int. J. Antimicrob. Agents*, 2008, **31**, 180–181.

C. E. Cooke, G. E. Sklar, and J. M. Nappi, *Ann. Pharmacother.*, 1996, **30**, 364–366.

C. Cosar and L. Julou, *Ann. Inst. Pasteur*, 1959, **96**, 238.

C. Cosar, C. Crisan, R. Horclois, R. M. Jacob, J. Robert, S. Tchelitcheff, and R. Vaupre, *Arzneim. Forsch.*, 1966, **16**, 23.

S. L. Cudmore, K. L. Delgaty, S. F. Hayward-McClelland, D. P. Petrin, and G. E. Garber, *Clin. Microbiol. Rev.*, 2004, **17**, 783–793.

A. Darbon, A. Portal, L. Girier, J. Pantin, and C. Leclaire, *Presse. Med.*, 1962, **70**, 15–16.

R. P. Dean and R. L. Talbert, *Drug Intell. Clin. Pharm.*, 1980, **14**, 864–866.

M. Delmée, J. V. Broeck, A. Simon, M. Janssens, and V. Avesani, *J. Med. Microbiol.*, 2005, **54**, 187–191.

R. L. Dunne, L. A. Dunn, P. Upcroft, P. J. O'Donoghue, and J. A. Upcroft, *Cell Res.*, 2003, **13**, 239–249.

D. I. Edwards, *Br. J. Vener. Dis.*, 1980, **56**, 285–290.

D. I. Edwards, *J. Antimicrob. Chemother.*, 1993, **31**, 9–20.

J. A. Edwards and J. Price, *Biochem. Pharmacol.*, 1967, **16**, 2026–2027.

D. I. Edwards, M. Dye, and H. Carne, *J. Gen. Microbiol.*, 1973, **76**, 135–145.

W. J. Ehlhardt, B. B. Beaulieu, Jr, and P. Goldman, *Biochem. Pharmacol.*, 1988, **37**, 2603–2606.

W. Fehri, S. Abdessalem, Z. Smiri, H. Mhenni, N. Barakett, N. Rahal, and H. Haouala, *Tunis. Med.*, 2004, **82**, 867–874.

D. G. Ferris, M. S. Litaker, L. Woodward, D. Mathis, and J. Hendrich, *J. Fam. Pract.*, 1995, **41**, 443–449.

F. Forna and A. M. Gülmezoglu, *Cochrane Database of Systematic Reviews*, 2003, Issue 2.

C. D. Freeman, N. E. Klutman, and K. C. Lamp, *Drugs*, 1997, **54**, 679–708.

H. B. Fung and T. L. Doan, *Clin. Ther.*, 2005, **27**, 1859–1884.

J. C. Gillis and L. R. Wiseman, *Drugs*, 1996, **51**, 621–638.

Y. Glupczynski, A. Burette, E. Koster, J. F. Nyst, M. Deltenre, S. Cadranel, L. Bourdeaux, and D. D. Vos, *Lancet*, 1990, **335**, 976–977.

P. Goldman, *J. Antimicrob. Chemother.*, 1982, **10** (Suppl. A), 23–33.

M. L. M. Gonzales, L. F. Dans, and E. G. Martinez, *Cochrane Database of Systematic Reviews*, 2009, Issue 2.

M. V. R. Groothoff, J. Hofmeijer, M. A. Sikma, and J. Meulenbelt, *Clin. Ther.*, 2010, **32**, 60–64.

J. M. Hanson, J. A. McGregor, S. L. Hillier, D. A. Eschenbach, A. K. Kreutner, R. P. Galask, and M. Martens, *J. Reprod.*, 2000, **45**, 889–896.

E. Hentschel, G. Brandstatter, B. Dragosics, A. M. Hirschl, H. Nemec, K. Schutze, M. Taufer, and H. Wurzer, *N. Eng. J. Med.*, 1993, **328**, 308–312.

S. L. Hillier, R. P. Nugent, D. A. Eschenbach, M. A. Krohn, R. S. Gibbs, D. H. Martin, M. F. Cotch, R. Edelman, J. G. Pastorek, A. V. Rao, D. McNellis, J. A. Regan, J. C. Carey, and M. A. Klebanoff, *New Eng. J. Med.*, 1995, **333**, 1737–1742.

A. Howard-Thompson, A. C. Hurdle, L. B. Arnold, C. K. Finch, C. Sands, and T. H. Self, *Am. J. Geriat. Pharmacother.*, 2008, **6**, 33–36.

R.M. Jacob, G.L. Regnier, and C. Crisan, US Patent 2944061, 5 July 1960.

P. J. Jenks and D.I. Edwards, *Int. J. Antimicrob. Agents*, 2002, **19**, 1–7.

J. C. Jensen and R. Gugler, *Clin. Pharmacol. Ther.*, 1985, **37**, 407–410.

P. N. Karamanakos, *Minerva anestiologica*, 2008, **74**, 679.

P. N. Karamanakos, P. Pappas, V. A. Boumba, C. Thomas, M. Malamas, T. Vougiouklakis, and M. Marselos, *Int. J. Toxicol.*, 2007, **26**, 423–432.

F. J. Kazmier, *Mayo Clin. Proc.*, 1976, **51**, 782–784.

J. J. Kim, R. Reddy, M. Lee, J. G. Kim, F. A. K. El-Zaatari, M. S. Osato, D. Y. Graham, and D. H. Kwon, *J. Antimicrob. Chemother.*, 2001, **47**, 459–461.

E. Kim, D. G. Na, E. Y. Kim, J. H. Kim, K. R. Son, and K. H. Chang, *Am. J. Neuroradiol.*, 2007, **28**, 1652–1658.

R. C. Knight, I. M. Skolimowski, and D. I. Edwards, *Biochem. Pharmacol.*, 1978, **27**, 2089–2093.

S. P. Kounas, K. P. Letsas, A. Sideris, M. Efraimidis, and F. Kardaras, *Pacing Clin. Electrophysiol.*, 2005, **28**, 472–473.

M. Ya. Kraft, P. M. Kochergin, A. M. Tsyganova, and V. S. Shlikhunova, *Pharm. Chem. J.*, 1989, **23**, 861–863.

L. Kumar, A. Sarswat, N. Lal, V.L. Sharma, A. Jain, R. Kumar, V. Verma, J.P. Maikhuri, A. Kumar, P.K. Shukla, and G. Gupta, *Eur. J. Med. Chem.*, 2010, **45**, 817–824.

G.C. Lancini and E. Lazzari, *Cell. Mol. Life Sci.*, 1965, **21**, 83.

K. M. Land and P. J. Johnson, *Exp. Parasitol.*, 1997, **87**, 305–308.

K. M. Land and P. J. Johnson, *Drug Resistance Updates*, 1999, **2**, 289–294.

P. G. Lankisch, M. Droge, and F. Gottesleben, *Gut*, 1995, **37**, 565–567.

A. H. Lau, N. P. Lam, S. C. Piscitelli, L. Wilkes, and L. H. Danziger, *Clin. Pharmacokinet.*, 1992, **23**, 328–364.

K. Maeda, T. Osato, and H. Umezawa, *J. Antibiot. (Tokyo) Series A*, 1953, **6**, 182.

W.-J. Mao, P.-C. Lv, L. Shi, H.-Q. Li, and H.-L. Zhu, *Bioorg. Med. Chem.*, 2009, **17**, 7531.

M. Marchionni, A. D. Innocenti, andC. Penna, *Clin. Exp. Obstet. Gynecol.*, 1981, **8**, 18–20.

B. J. Marshall and J. R. Warren, *Lancet*, 1984, **1**, 1311–1315.

B. J. Marshall, H. Royce, D. I. Annear, C. S. Goodwin, J. W. Pearman, J. R. Warren, and J. A. Armstrong, *Microbios. Lett.*, 1984, **25**, 83–88.

B. J. Marshall, J. R. Warren, E. D. Blincow, M. Phillips, C. S. Goodwin, R. Murray, S. J. Blackbourn, T. E. Waters, and C. R. Sanderson, *Lancet*, 1988, **332**, 1437–1442.

A. K. Miller, *Appl. Microbiol.*, 1971, **22**, 480–481.

K. C. Mohanty and R. Deighton, *J. Antimicrob. Chemother.*, 1987, **19**, 393–399.

S. Nakamura, *Pharm. Bull. (Tokyo)*, 1955, **3**, 379.

A. Nakane, T. Nakamura, and Y. Eguchi, *J. Biol. Chem.*, 1977, **252**, 5267.

M. Norgaard, C. Ratanajamit, J. Jacobsen, M. V. Skriver, L. Pedersen, and H. T. Sorensen, *Aliment Pharmacol. Ther.*, 2005, **21**, 415–420.

R. A. O'Reilly, *New. Eng. J. Med.*, 1976, **295**, 354–357.

K. A. Sanford, J. E. Mayle, H. A. Dean, and D. S. Greenbaum, *Ann. Int. Med.*, 1988, **109**, 756–757.

L. K. Schmadel and G. K. McEvoy, *Clin. Pharm.*, 1990, **9**, 94–101.

J. I. Seok, H. Yi, Y. M. Song, and M. Y. Lee, *Arch. Neurol.*, 2003, **60**, 1796–1800.

M. Sköld, H Gnarpe, and L Hillström, *Br. J. Vener. Dis.*, 1977, **53**, 44–48.

R. Slim, C. Ben Salem, M. Zamy, N. Fathallah, J. J. Raynaud, K. Bouraoui, and M. Biour, *J Pancreas (Online)*, 2010, **11**, 85–86.

J. Starr, *Brit. Med. J.*, 2005, **331**, 498–501.

M. E. Sura, K. A. Heinrich, and M. Suseno, *Ann. Pharmacother.*, 2000, **34**, 1152–1155.

F. P. Tally, V. L. Sutter, and S. M. Finegold, *Antimicrob. Agents Chemother.*, 1975, **7**, 672–675.

J. Tillonen, S. Vaelevaeinen, V. Salaspuro, Y. Zhang, M. Rautio, H. Jousimies-Somer, K. Lindros, and M. Salaspuro, *Alcohol Clin. Exp Res.*, 2000, **24**, 570–575.

I. H. Tocher, *Gen. Pharmac.*, 1997, **28**, 485–487.

J. H. Tocher and D. I. Edwards, *Biochem. Pharmacol.*, 1994, **48**, 1089–1094.

P. Upcroft and J. A. Upcroft, *Clin. Microbiol. Rev.*, 2001, **14**, 150–164.

E. J. van der Wouden, J. C. Thijs, A. A. van Zwet, and J. H. Kleibeuker, *Aliment. Pharmacol. Ther.*, 2000, **14**, 7–14.

E. J. van Zuuren, M. A Graber, S. Hollis, M. M. M. C. Chaudhry, A. K. Gupta, and M. D. Gover, *Cochrane Database Syst. Rev.*, 2005, **3**.

J. P. Visapaa, J. S. Tillonen, P. S. Kaihovaara, and M. P. Salaspur, *Ann. Pharmacother.*, 2002, **36**, 971–974.

W. H. Wang, B. C. Y. Wong, A. K. Mukhopadhyay, D. E. Berg, C. H. Cho, K. C. Lai, W. H. C. Hu, F. M. Y. Fung, W. M. Hui, and S. K. Lam, *Ailment. Pharmacol. Ther.*, 2000, **14**, 901–910.

H. Weinstock, S. Berman, and W. Cates, Jr, *Perspect. Sex. Reprod. Health*, 2004, **36**, 6–10.

K. R. Woodcock, *Brit. J. Vener. Dis.*, 1972, **48**, 65–68.

World Health Organization, Global Prevalence and Incidence of Selected Curable Sexually Transmitted Infections, Overview and Estimates, 2001 (http://whqlibdoc.who.int/hq/2001/WHO_HIV_AIDS_2001.02.pdf, last accessed 9 March 2012).

C. J. Wurden and R. H. Levy, 'Carbamazepine. Interactions with other drugs' in R. H. Levy, R. H. Mattson, B. S. Meldrum, and E. Perucca (Eds.), *Antiepleptic Drugs*, 5th Edn, Lippincott Williams and Wilkins, Philidelphia, PA, 2002, pp. 247–261.

G. Zuccoli, N. Pipitone, and D Santa Cruz, *Am. J. Neuroradiol.*, 2008, **29**, E84.

Questions

(1) Explain why ciprofloxacin should not be used to manage MDR-TB.

(2) A doctor wishes to prescribe a quinolone antibiotic for the treatment of an acute exacerbation of chronic bronchitis. Unfortunately, the patient also has elevated liver function tests (greater than five times the upper limit of normal). Which quinolone antibiotic would you consider to be NOT suitable for prescribing to this patient?

(3) How many stereoisomers are possible for rifamycin?

(4) Starting from 3-bromorifamycin S, how might you prepare rifaximin?

3-Bromorifamycin S   **Rifaximin**

(E. Marchi and L. Montecchi, *US Patent* 4341785, 27 July 1982.)

(5) The emergence of MDR-TB is becoming a significant clinical problem. Describe the mechanism by which bacteria become resistant to rifampicin.

(6) What is DOT? Explain how this concept could help minimise bacterial resistance.

(7) What are the possible consequences of *H. pylori* infection? Outline a regimen used to eradicate *H. pylori* and discuss the possible risks and adverse effects associated with it.

3
Agents Targeting Metabolic Processes

Antibacterial agents that target metabolic processes include the sulfonamides and trimethoprim.

3.1

Sulfonamide antibacterial agents

Key Points

- The sulfonamide antibacterials are antifolates as they interfere with folic acid synthesis in bacterial cells.
- Resistance to sulfonamides arises due to mutations in the *dihydropteroate synthetase* (*folP*) gene, leading to alterations in the sulfonamide binding site of their enzyme target, dihydropteroate synthase.
- Co-trimoxazole and sulfadiazine should not be used routinely in the management of common infections (such as urinary tract infections and otitis media) due to the high risk of resistance and adverse effects.
- Silver sulfadiazine is used topically for the treatment and prevention of infection in burn wounds, although the clinical evidence for this is lacking.
- For further information, see Lesch (2007).

Sulfonamide antibiotics used clinically include sulfadiazine and sulfamethoxazole (in conjunction with trimethoprim (co-trimoxazole)); their structures are shown in Figure 3.1.1 and their therapeutic indications are listed in Table 3.1.1.

3.1.1 Discovery

Although there are only two sulfonamides currently in clinical use, and their indications are very specific, this class of drugs deserves detailed study as the sulfonamides were the first specific and synthetic antibacterials and, as we shall see, they have also helped shape the course of history. The sulfonamides were discovered in 1932 when a red azo dye (Prontosil), synthesised by the chemist Josef Klarer at the Bayer Corporation (IG Farbenindustrie), was found to prevent the multiplication of bacteria in animal experiments (but not in *in vitro* tests). Heinrich Hörlein had initiated a pharmaceutical research programme in the 1920s which was aimed at the treatment of bacterial and tropical diseases with synthetic chemicals, and by the end of the decade, more than 30 new synthetic chemicals were being tested per week against a streptococcal strain (Mietzsch and Klarer, 1935; Bentley, 2009).

Antibacterial Agents: Chemistry, Mode of Action, Mechanisms of Resistance and Clinical Applications, First Edition.
Rosaleen J. Anderson, Paul W. Groundwater, Adam Todd and Alan J. Worsley.
© 2012 John Wiley & Sons, Ltd. Published 2012 by John Wiley & Sons, Ltd.

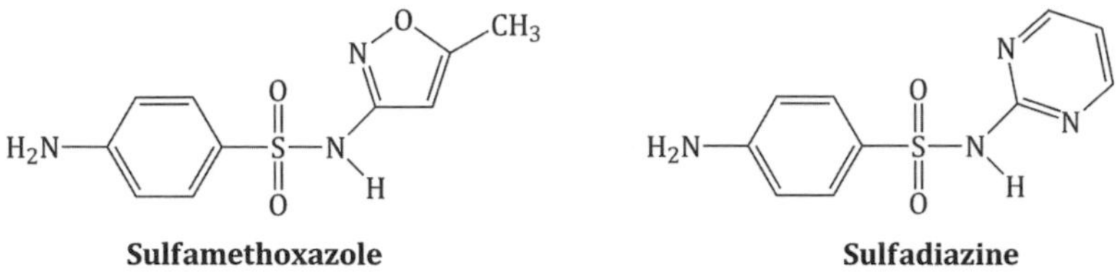

Figure 3.1.1 *Sulfonamide antibacterial agents*

Table 3.1.1 *Therapeutic indications for the sulfonamide antibiotics*

Sulfonamide antibiotic	Indications
Co-trimoxazole (sulfamethoxazole and trimethoprim)	Treatment prevention of *Pneumocystis jirovecii* pneumonia Treatment of toxoplasmosis Treatment of nocardiosis
Sulfadiazine	Prevention of rheumatic fever Treatment of toxoplasmosis Prevention and treatment of infection in burn wounds (used topically as the silver sulfadiazine salt)

Gerhard Domagk, who was responsible for the antibacterial testing and discovered the antibacterial effect of the sulfonamides (Domagk, 1935), was awarded the Nobel Prize in Medicine in 1939, but was forbidden from accepting the prize by Adolf Hitler (http://nobelprize.org/nobel_prizes/medicine/laureates/1939/domagk-lecture.pdf, last accessed 29 March 2012). After the war, Domagk was able to accept the prize, but was not given the monetary sum which went with it, as this was said to have expired (this is the kind of bad luck we can sympathise with) (Sköld, 2000).

As the authors can confirm, having taught on a medicinal chemistry module which involved their synthesis and analysis, the sulfonamides are not the most soluble of drugs, and an early attempt (1937) by Harold Cole Watkins to prepare a soluble form resulted in the tragic deaths of more than 100 patients who had taken Elixir Sulfanilamide[1] (Figure 3.1.2), as diethylene glycol ($HOCH_2CH_2OCH_2CH_2OH$), a highly toxic chemical which is used as an antifreeze, had been used as the solvent. This tragedy ultimately led to the increased regulation of drugs by the US Food and Drug Administration (FDA) (Wax, 1995).

On a more positive historical note, a sulfonamide, thought to be either sulfapyridine (then known as M&B 693)[2] or sulfathiazole (M&B 760) (Figure 3.1.2), was used to treat Winston Churchill, who had contracted pneumonia in 1943, en route to Tunisia after a tour of the Middle East to meet Allied leaders. Churchill's personal physician, Lord Moran, feared for the life of the Prime Minister (and the effect that his death would have on the morale of the British nation) and called in Brigadier D. E. Bedford, a heart specialist, as Churchill was also having heart fibrillations, thus giving Churchill his opportunity to joke that 'The M&B, which I may also call Moran and Bedford, did the job most effectively. There is no doubt that pneumonia is a very different illness from what it was before this marvellous drug was discovered' (Lesch, 2007).

Anyway, that's enough history for now; let's get back to the discovery of the sulfonamides. Prontosil was soon found to be inactive *in vitro*, as it is a prodrug which requires activation, via an azoreductase enzyme *in vivo* (Scheme 3.1.1). Using prontosil, Domagk was able to control streptococcal infections in mice and

---

[1] Watkins, a chemist at the Massengill Company (Bristol, TN, USA), committed suicide when he discovered the effect of his elixir.
[2] M&B stands for May and Baker, a British chemical company.

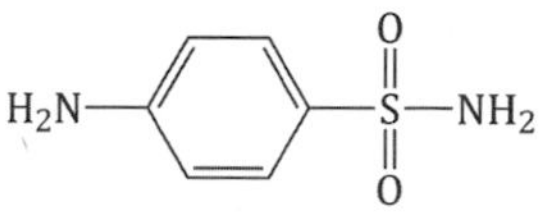

Sulfanilamide

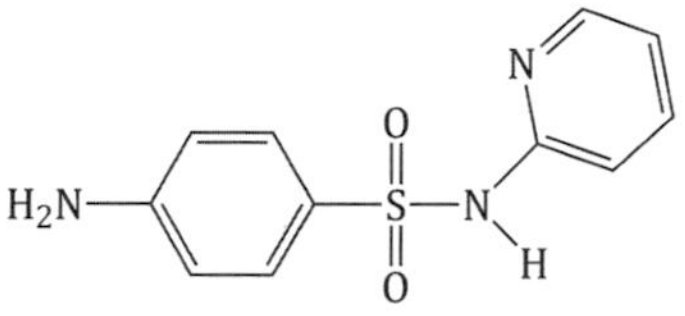

Sulfapyridine
(M&B 693)

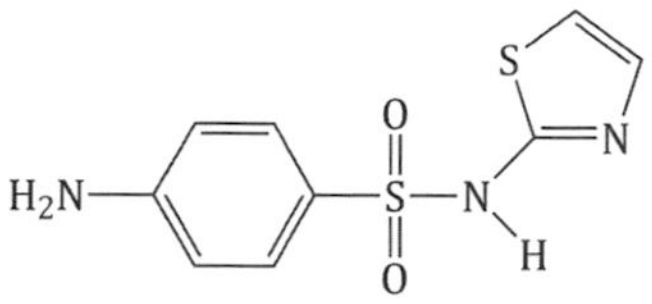

Sulfathiazole
(M&B 760)

Figure 3.1.2   *Sulfanilamide, sulfapyridine, and sulfathiazole*

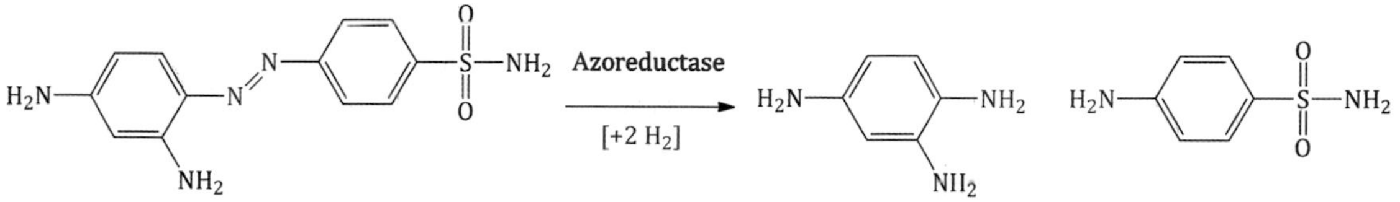

Prontosil   **1,2,4-Triaminobenzene (inactive)**   **Sulfanilamide**

Scheme 3.1.1   In vivo *metabolism of prontosil by azoreductase*

staphylococcal infections in rabbits. In 1935, Ernest Fourneau and his colleagues at the Institute Pasteur in Paris discovered the *in vivo* metabolism of prontosil to sulfanilamide, which soon became the first synthetic antibacterial agent and was on the market in the UK, France, and USA by 1937 (Tréfouël *et al.*, 1935).

3.1.2   Synthesis

Having discovered this exciting antibacterial activity, the research groups mentioned above were joined by pharmaceutical companies from all over the world, which began to synthesise sulfanilamide analogues, initially as probes of the structural requirements for activity, thereafter in attempts to prepare sulfonamides with more desirable properties: increased antibacterial activity, a broader spectrum of activity, reduced toxicity, longer half-life, and so on. Many thousands of analogues were tested and the sulfonamide group ($-SO_2NR_2$) and a free amino group at the *para*-position of an otherwise unsubstituted benzene ring were found to be essential (Figure 3.1.3) (Domagk, 1957; Miller *et al.*, 1972).

In summary, only variations in R^1 (on N^1) and R^4 (on N^4) lead to active sulfonamides; R^4 must be a group that can be hydrolysed *in vivo* to give the active free amino group, while variations in R^1 (especially the introduction of heterocyclic substituents) are responsible for the many sulfonamides which have been used clinically. Thankfully, there are a number of possible synthetic routes to suitably substituted sulfonamides, and we will mention two here.

A wonderfully simple looking route involves the nucleophilic substitution of commercially available *p*-acetamidobenzenesulfonyl chloride by an aromatic amine, followed by hydrolysis of the intermediate

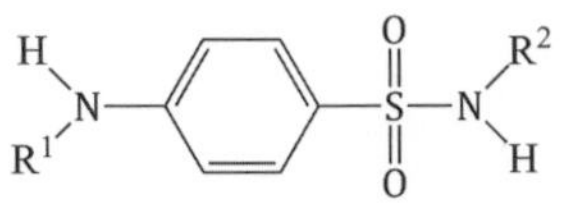

Figure 3.1.3   *Variations on the sulfonamide nucleus*

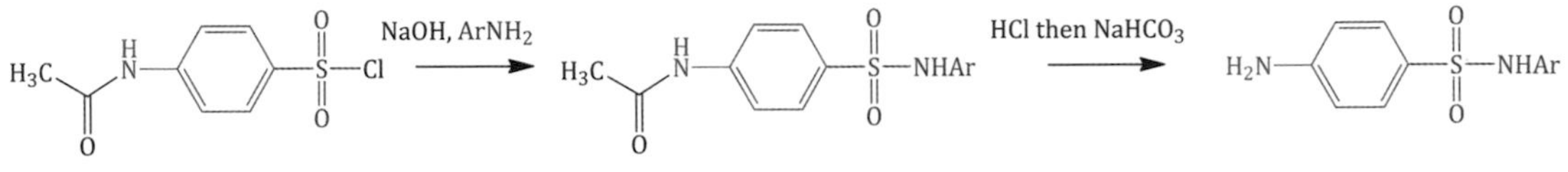

para- acetamidobenzenesulfonyl chloride   (Ar = aryl or heteroaryl [Het])

Scheme 3.1.2   *Synthesis of sulfonamides via nucleophilic substitution*

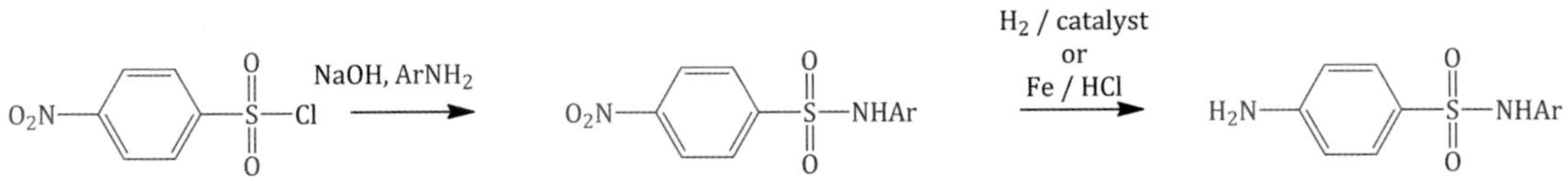

para- nitrobenzenesulfonyl chloride   (Ar = aryl or heteroaryl [Het])

Scheme 3.1.3   *Synthesis of sulfonamides via nucleophilic substitution followed by nitro-group reduction*

(Scheme 3.1.2) (Winnek and Roblin, 1947). The acetyl group in the starting material is being used as a protecting group, as it prevents any unwanted side reactions of the amino group and is hydrolysed in the second step to release the biologically active free amine. In theory, this route could be used to prepare any heteroaryl-substituted sulfonamide, as there are any number of aromatic amines which could be used in the nucleophilic substitution step. In practice, however, this route is limited by the solubilities of the intermediates and by the final products in the aqueous acidic, then basic, conditions employed in the hydrolysis step.

An equally attractive method involves the initial nucleophilic substitution of *p*-nitrobenzenesulfonyl chloride (Scheme 3.1.3), in which the nitro group cannot undergo any unwanted side reactions and will also facilitate the initial substitution through its strongly electron-withdrawing effect. In this case, the free amino group can be generated in the second step by reduction of the aromatic nitro group in a suitable solvent, a well-studied synthetic process (Senear *et al.*, 1946; Luk'yanov *et al.*, 1980).

3.1.3   Bioavailability

The oral bioavailability of the individual sulfonamides is not affected by co-administration with food, but, although generally good, it is dependent upon the acidity of the sulfonamide (pK_a),[3] as this determines the degree of ionisation at the various pHs of the GI tract and, consequently, the lipophilicity and absorption across the intestinal wall.

---

[3] $pK_a = -\log_{10}K_a$, where K_a is the acid dissociation constant (for $HA \rightleftharpoons H^+ + A^-$; $K_a = [H^+][A^-]/[HA]$. The stronger an acid (HA), the larger the K_a, and so the smaller the pK_a. When discussing bases (B), we still use the term pK_a, but this refers to the conjugate acid, BH^+, of the base ($BH^+ \rightleftharpoons H^+ + B$). The stronger the conjugate acid (smaller pK_a for BH^+), the weaker the base, and the weaker the conjugate acid (larger pK_a for BH^+), the stronger the base.

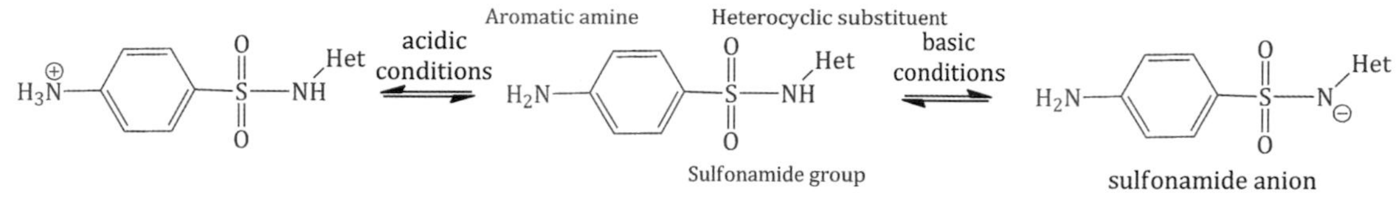

Scheme 3.1.4 *The functional groups involved in the acid–base equilibria of sulfonamides*

Table 3.1.2 *pK$_a$ values for some common sulfonamides (HA in parentheses)*

Sulfonamide	pK$_a$ values	Reference
Sulfamethoxazole	pK$_{a1}$ (NH$_3$$^+$) 1.85, pK$_{a2}$ (NH) 5.60	Qiang and Adams (2004)
Sulfadiazine	pK$_{a1}$ (NH$_3$$^+$) 2.10, pK$_{a2}$ (NH) 6.28	Lin *et al.* (1997)
Sulfathiazole	pK$_{a1}$ (NH$_3$$^+$) 2.01, pK$_{a2}$ (NH) 7.11	Qiang and Adams (2004)
Sulfadoxine	pK$_{a1}$ (NH$_3$$^+$) 1.59, pK$_{a2}$ (NH) 5.82	Ghafourian *et al.* (2006)

The general structure of a sulfonamide antibacterial agent is shown in Scheme 3.1.4. If you consider the structure, you can identify two sites that can take part in acid–base equilibria: the **aromatic amine** (pK$_{a1}$), which has low basicity due to resonance of the electron pair with the aromatic system and through to the sulfonyl group, and the **sulfonamide NH** (pK$_{a2}$), which is acidic, due to stabilisation of the conjugate anion by delocalisation with both the sulfonyl group and the heterocyclic substituent. The basicity of the aromatic amine is sufficiently low that it is poorly protonated, even in the stomach, and it varies little throughout the sulfonamide series, since, as we discovered in the previous subsection, the introduction of substituents on to the aromatic ring (which would affect the basicity) results in a loss of antibacterial activity. The acidity of the sulfonamide NH is more variable as it is dependent upon the nature of the heterocyclic substituent, as indicated in Table 3.1.2 (Hekster and Vree, 1982). It is sufficiently acidic in the majority of sulfonamides that many exist predominantly as the anion in most physiological compartments, which aids with their solubility and distribution.

The solubility of a molecule is affected by its inherent solubility, taking into account the number and nature of polar groups and the extent of ionisation at any given pH. The oral bioavailability is affected by the solubility, because some aqueous solubility is required in the GI tract in order to allow the sulfonamide to partition between the aqueous gastric contents and the lipophilic GI wall.

In the blood, those sulfonamides with a pK$_{a2}$ value below the blood pH will be ionised and exist as the sulfonamide anion and as a result will have good solubility and distribution, being well distributed throughout the body, with an apparent volume of distribution (V$_d$)[4] of approximately 0.1–0.25 L/kg in adults (Ghafourian *et al.*, 2006), reaching most tissues and compartments and conferring wide therapeutic application.

The aqueous solubility also affects the renal clearance of the sulfonamides; greater aqueous solubility, in general, increases the renal excretion. However, systemic sulfonamides with low aqueous solubility, such as sulfadiazine (see Table 3.1.3), can crystallise in the blood and kidneys, which causes irritation to the renal tubules. The co-administration of sodium bicarbonate results in a slight increase in the blood pH and an accompanying increase in sulfonamide solubility, reducing the risk of crystallisation (Hekster and Vree, 1982).

As a result of the variations in solubility, absorption, and excretion of the sulfonamides, they can be classified into separate groups for systemic use: ultrashort acting, short–medium acting, and long acting; as well as those for GI and topical use.

---

[4] The apparent volume of distribution (V$_d$) is the theoretical volume in which the total amount of a drug would need to be uniformly distributed in order to produce the desired blood concentration of a drug (V$_d$ = total amount of drug in body/drug concentration in plasma).

Table 3.1.3   *The solubility of some sulfonamides at different pH values and selected pharmacokinetic data*

Sulfonamide	pK_{a2}	Solubility at pH 5.5[a] (mg/L at 25°C)	Solubility at pH 7.0[a] (mg/L at 25°C)	$t_{1/2}$ (h)[a]	C_{max} (µg/mL)	$AUC_{(0-\infty)}$ (µg.h/mL)
Sulfamethoxazole	5.60	300	1900	10–12	20–40[b]	280–710[b]
Sulfadiazine	6.28	265	950	10–12	17–19[c]	340–382[c]
Sulfadoxine	5.82	186	2387	110	63–72[d]	12 394[d]

[a] Hekster and Vree (1982);
[b] 400 mg oral dose, co-administered with 80 mg trimethoprim, in healthy adults (Campero *et al.*, 2007);
[c] 500 mg oral dose in healthy adults (Meyer *et al.*, 1978);
[d] 500 mg oral dose, co-administered with 25 mg pyrimethamine, in healthy adults (Weidekamm *et al.*, 1982; Sarikabhuti *et al.*, 1988; Bhoir *et al.*, 2001)

The oral sulfonamides commonly used clinically, sulfamethoxazole (in co-trimoxazole) and sulfadiazine, fall into the category of short–medium acting agents that are well absorbed, with reported bioavailabilities of 100% after oral administration (Bertz and Granneman, 1997) and blood plasma half-life ($t_{1/2}$) values of 4–16 hours (Hekster and Vree, 1982). Peak plasma concentrations are seen within 1–3 hours after oral administration (Hekster and Vree, 1982). Long-acting agents, such as sulfadimethoxine, have been largely replaced by less-toxic sulfonamides, due to reports of severe allergic reactions (e.g. Rallison *et al.*, 1961; reviewed in Choquet-Kastylevsky *et al.*, 2002; Brackett 2007), not only in humans, but also in animals (e.g. Campbell, 1999; Trepanier, 1999). Sulfadoxine is an example of a long-acting sulfonamide ($t_{1/2}$ about 110 hours) that remains in clinical use – in combination with pyrimethamine, it is indicated for the treatment of malaria and toxoplasmosis. Its long half-life is believed to be appropriate for the slow life cycle of *Plasmodium falciparum*, the causative bacterium in malaria.

GI sulfonamides are those with low absorption from the GI tract; such agents have been used in sterilising the GI tract before surgery, but this is now usually achieved through the use of aminoglycosides. Sulfapyridine is the only sulfonamide from this group still used clinically (in the treatment of ulcerative colitis), as part of the prodrug sulfasalazine, which is reduced by bacterial azoreductase in the lower GI tract to release 5-aminosalicylic acid (which is considered to be the active agent) and sulfapyridine. Sulfonamides have been used for the topical treatment of skin and wound infections for many years, since the effectiveness of the combination of sulfamethoxazole/trimethoprim was observed by three colonels and a major in the US Army Medical Corps (Moncrieff *et al.*, 1966).

The relationship between pK_{a2} and the solubility of the sulfonamide is also important for uptake into the bacterial target – we will look at this in the last part of this subsection. The data in Table 3.1.3 show that the sulfonamides listed have greater aqueous solubility at pH 7 (when they are all ionised to some extent) than at pH 5.

Much work has gone into the study of the pharmacokinetic parameters of sulfonamides, especially in comparison to those of trimethoprim, to identify the optimal sulfonamide for combination therapy with trimethoprim (reviewed in Watson *et al.*, 1988 and in Hekster and Vree, 1982). In designing a combination therapy, it is essential that the two (or more) individual agents have compatible pharmacokinetics in order to ensure that therapeutic concentrations at the target site(s) are achieved simultaneously. If the combination agents do not act together at the site of infection, the synergistic action is lost and the development of resistance becomes a significant risk. Even though the combination of sulfamethoxazole and trimethoprim has been used for many years, and continues to be a key treatment for certain pneumonial infections, the differences in their respective bioavailability and distribution have caused some discussion and even criticism of their combined use (O'Grady, 1975; Reeves and Wilkinson, 1979; Watson *et al.*, 1981; Männistö *et al.*, 1982; Burman, 1986; Królicki *et al.*, 2004).

All of the current sulfonamide antibacterial agents are significantly metabolised to their corresponding N^4-acetyl derivatives, through acetylation by *N*-acetyl transferase (NAT) activity, predominantly in the liver.

The aromatic amine (N^4) must be unsubstituted for antibacterial activity, so N^4-acetylation leads to inactive metabolites, which are excreted in the urine, along with unchanged sulfonamide. Early pharmacogenetic analysis of NAT activity led to the classification of patients as 'slow' or 'fast' acetylators (Evans and White, 1964; Evans, 1969; Schroder and Evans, 1972); unfortunately, the rate of acetylation of each individual agent does not necessarily hold across the series of sulfonamides, preventing reliable clinical prediction (Hekster and Vree, 1982). For slow acetylators, there is the risk of toxicity due to increased levels of sulfonamide, while fast acetylators risk sub-therapeutic levels of the antibacterial agent and a less favourable response to treatment. Even in the early 1980s, the possibility of personalised medicine was raised in the context of sulfonamide treatment of bacterial infections (Vree *et al.*, 1980).

Like their sulfonamide parent drugs, the N^4-acetyl metabolites are more soluble at pH 7.0 than at pH 5.5, although they are usually less soluble than the sulfonamide parent. However, measurement and comparison of the urinary excretion of sulfonamides to that of their N^4-acetyl metabolites has shown that the direct relationship between pH and sulfonamide excretion was not mirrored by the N^4-acetyl compounds, which maintained a relatively constant urinary excretion that was almost independent of pH (this phenomenon is exemplified by the behaviour of sulfamethoxazole and N^4-acetylsulfamethoxazole) (Vree *et al.*, 1978). Urinary excretion accounts for about 90% or more of the route of elimination for sulfonamides and their N^4-acetyl metabolites (Hekster and Vree, 1982).

Other minor metabolic pathways result in N^1-glucuronidation and hydroxylation of the phenyl and/or heterocyclic rings. *O*- and *N*-dealkylation are also possible in appropriately substituted sulfonamides (Hekster and Vree, 1982).

3.1.3.1   Bacterial uptake of sulfonamides

The concentration of sulfonamide found inside a bacterial cell depends upon its ability to enter the cell (and be retained once inside), which depends upon the difference in pH between the extracellular medium and the cytoplasm of the bacterial cell, as this determines the extent of ionisation of the sulfonamide in each site (Tappe *et al.*, 2008; Zarfl *et al.*, 2008). In essence, sulfonamides with a pK_{a2} close to 5–7 are better accumulated in bacteria, since they predominantly exist in the unionised (neutral) form outside the cell, and can pass through the bacterial membrane by passive diffusion, then become partly ionised inside the bacterial cells, due to the slightly higher pH (usually around 7.6), leaving them trapped. Sulfonamides with a higher pK_{a2} value are ionised to a lesser extent in bacterial cells, as this would require more basic conditions than the intracellular pH provides; these sulfonamides can freely diffuse into and out of the cell, preventing their accumulation.

3.1.4   Mode of action and selectivity

During the early antibacterial studies, it was discovered that the sulfonamides (then referred to as sulpho-namides, sulpha, or sulfa drugs) were bacteriostatic, as they prevented the growth of bacteria, but did not kill them, and that, when *para*-aminobenzoic acid (PABA) was added to bacterial culture at the same time as sulfonamide, the drug had little or no effect.

As we learned in Subsection 1.1.3.2, bacteria require PABA (an essential metabolite) for the synthesis of folic acid and lack the protein required for folate uptake from their environment. Folic acid is an essential metabolite for mammals as it cannot be synthesised by mammalian cells and must be obtained from the mammalian diet (you can check how much folic acid you require daily by reading the ingredients on the side of your breakfast cereal packet). The sulfonamides are thus highly selective as they interfere with a key process in bacterial (prokaryotic) cells which does not take place in eukaryotic (mammalian) cells.

We now know that this interference in folic acid synthesis is the basis of the selectivity of the sulfonamides, but why is folic acid so important? Folic acid is indirectly involved in DNA synthesis as the enzyme co-factors which are required for the synthesis of the purine and pyrimidine bases of DNA are derivatives of folic acid (see Subsection 1.1.3.2). If the synthesis of folic acid is inhibited, the cellular supply of these co-factors will be diminished, and DNA synthesis will be prevented.

Bacterial synthesis of folic acid (actually dihydrofolic acid) involves a number of steps, with the key steps in terms of the mode of action of the sulfonamides shown in Scheme 3.1.5. Once again, a nucleophilic substitution is involved, in which the free amino group of PABA substitutes for the pyrophosphate group introduced on to 6-hydroxymethylpterin by the enzyme PPPK. In the final step, amide formation takes place between the free amino group of L-glutamic acid and the carboxylic acid group derived from PABA (Achari *et al.*, 1997).

Sulfanilamide and suitably substituted derivatives have similar molecular dimensions and properties to PABA (Figure 3.1.4), so are mistaken for it by the enzyme dihydropteroate synthase. These sulfonamides bind competitively to the PABA binding domain of the active site of dihydropteroate synthase (which is highly conserved between bacteria) (Figure 3.1.5), and some are incorporated into the macromolecule (Roland *et al.*, 1979). Once incorporated into the macromolecule, further reaction with L-glutamic acid will not produce folic acid (Scheme 3.1.6). The incorporation of the sulfonamide is, however, reversible, since the addition of excess PABA results in the formation of folic acid (Woods, 1940).

You will already have noticed why the free amino group is required in the sulfonamides: the introduction of any substituent other than hydrogen may well interfere (electronically or sterically) with the nucleophilic substitution which is key to this mode of action.

The sulfonamides are thus examples of antimetabolites (more properly, antifolates), which are defined as agents that are similar in structure to an essential metabolite but cannot take its place. The antimetabolite

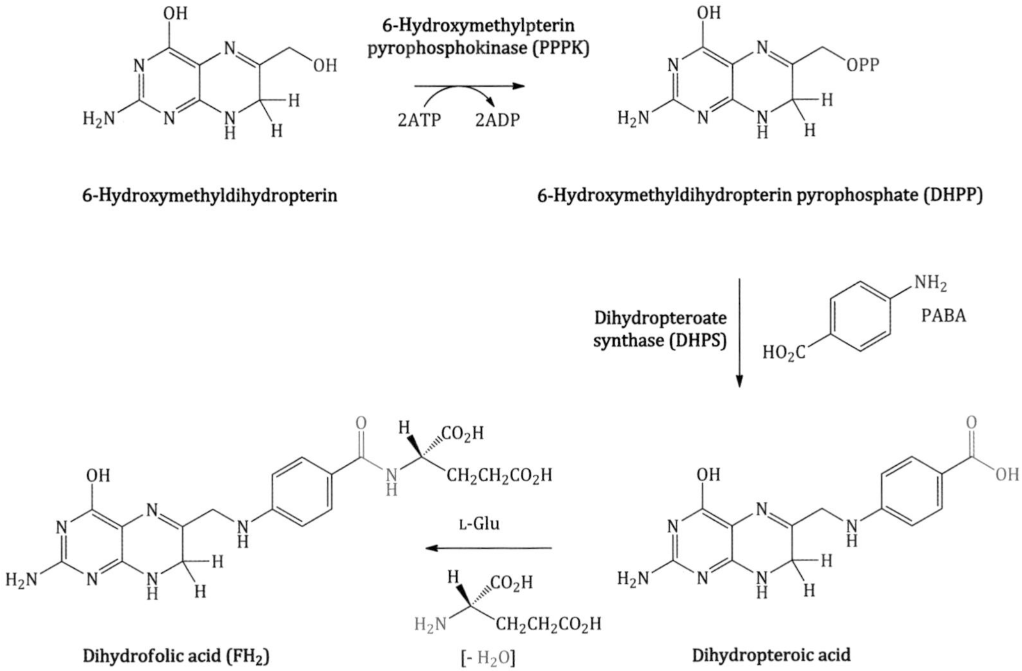

Scheme 3.1.5 *Bacterial synthesis of dihydrofolic acid*

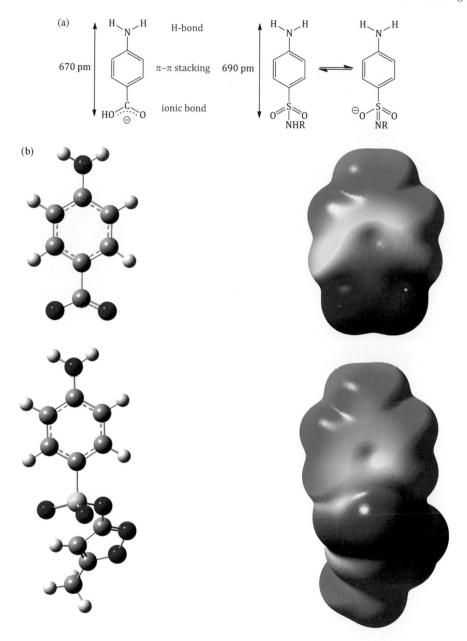

Figure 3.1.4 *(a) Comparison of molecular properties of PABA and sulfonamides. (b) The electrostatic similarities between PABA (top) and sulfamethoxazole (bottom)*

theory attributed to Woods (1940) and Fildes (1940) is most applicable to the sulfonamides, trimethoprim, and the anticancer agents 5-fluorouracil, 6-mercaptopurine, and methotrexate.

Look out for more on the mode of action of the sulfonamides in Section 3.2, where we will learn about the dihydrofolate reductase inhibitor trimethoprim and the combination of sulfamethoxazole and trimethoprim (co-trimoxazole).

Figure 3.1.5   *Dihydropteroate synthase (DHPS) from S. pneumoniae with bound DHPP; PDB code 2VEG (Levy et al., 2008)*

3.1.5   Bacterial resistance

As we have said a number of times so far, knowing the mode of action of an antibacterial agent gives us an idea of how bacteria will develop resistance, and it should, therefore, come as no surprise that altering the structure of dihydropteroate synthase, the enzyme that the sulfonamides bind to (in competition with PABA), resulting in some of them being incorporated into the folic acid precursor, is one of the ways in which bacterial cells overcome the effects of these agents.

Chromosomal resistance through mutations in the dihydropteroate synthase (*folP*) gene in *Escherichia coli*, *Staphylococcus aureus*, *Pneumocystis jirovecii* (fungal pneumonia infection), *Campylobacter jejuni* (gastroenteritis), and *Neisseria meningitidis* (bacterial meningitis) leads to alterations in the sulfonamide binding site of the DHPS. In one such *E. coli* mutant, the result of a single base-pair mutation leading to a Phe changing to Ile at position 28 of the amino acid sequence, led to a trade-off between resistance and enzyme efficiency, with the K_i for sulfathiazole being increased by 150 times and the K_M for PABA by 10 times[5] (Dallas *et al.*, 1992).

Sulfonamide resistance in Gram negative bacteria is plasmid-borne, with the two plasmid-borne genes *sul1* and *sul2*, found in roughly equal frequency among clinical isolates, again encoding drug-resistant DHPS

---

[5] K_i is the inhibition constant, the concentration of inhibitor required to decrease the maximal rate of reaction to half the uninhibited value; the lower the K_i, the lower the concentration required. K_M is the Michaelis constant, which is a measure of an enzyme's affinity for a substrate; the lower the K_M, the higher the affinity.

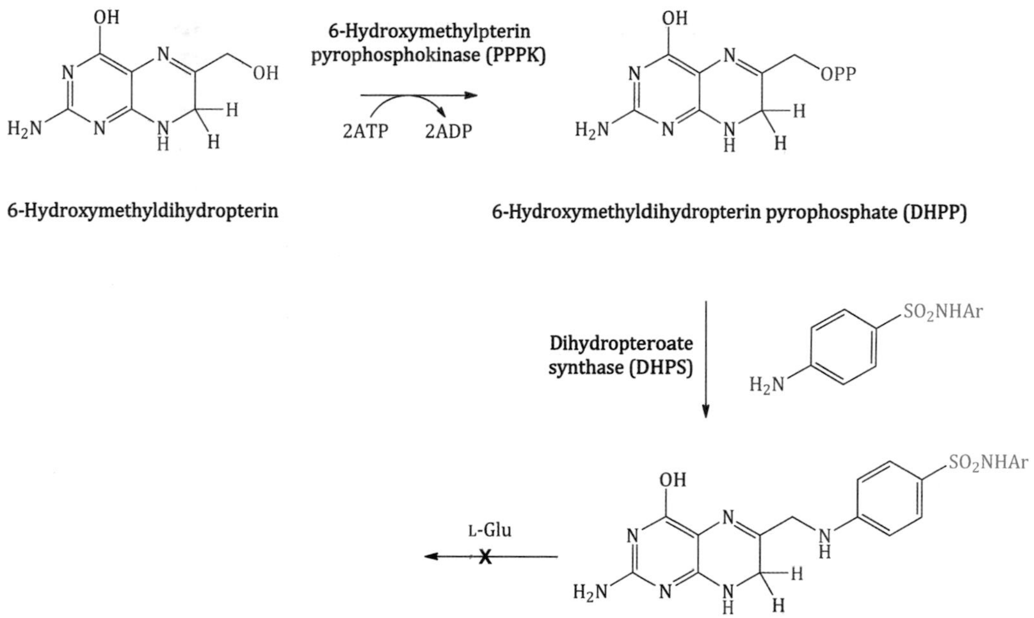

Scheme 3.1.6 *Sulfonamide incorporation into folic acid precursor by DHPS*

variants (Akiba *et al.*, 1960; Sköld, 1976). For more on the bacterial resistance of the sulfonamides, see Sköld (2000).

3.1.6 Clinical applications

3.1.6.1 Spectrum of activity

The sulfonamides are broad-spectrum antibiotics, with activity against both Gram negative and Gram positive bacteria, ranging from *Streptococcus pneumonia*, *Nocardia asteroides*, *E. coli*, *Brucella* species, and *Corynebacterium diphtheria*, to the protozoan *Toxoplasma gondii* (Allegra *et al.*, 1990). When first discovered, the sulfonamides were used to treat a wide range of clinical conditions, but nowadays, due to toxicity problems (e.g. blood dyscrasias and Stevens–Johnson syndrome) and increased resistance, they are only used as first-line treatment for a few clinical conditions (discussed below). The sulfonamides should only be used to treat common infections (such as urinary tract infections [UTIs], otitis media, and bronchitis) when there is good bacteriological evidence suggesting sensitivity and there is a clinical reason to use a sulfonamide in preference to another antibiotic. The sulfonamides lack activity against mycobacteria, while strains of *Pseudomonas aeruginosa* are often resistant (Poole, 2001; Livermore, 2002), and this limits their usefulness against diseases caused by these pathogens.

3.1.6.2 Protozoan infections

As infections due to protozoa and fungi are outwith the scope of this book, we will mention these here only in brief. Toxoplasmosis is a widespread parasitic disease caused by the protozoan parasite *Toxoplasma gondii*. A common source of *T. gondii* infection is from cats, as *T. gondii* replicates in the intestines of cats and is expelled in cats' faeces. As a result, any food or water ingested that is contaminated by cat faeces may result in the development of *T. gondii* infection (a good reason to keep cats off your vegetable patch) (Dabritz and Conrad, 2010).

When the patient has a healthy immune system, the infection does generally not cause any symptoms, but in cases where the immune system is compromised (such as in HIV), the infection can affect the brain, causing toxoplasmic encephalitis (TE), which can present as a multitude of neurological symptoms, including seizures, ataxia (loss of muscle coordination resulting in lack of balance), headaches, fever, and confusion. TE is an opportunistic infection, and if left untreated can be fatal, and it is a major cause of mortality in patients with HIV/AIDS, particularly in developing countries. Another scenario where *T. gondii* infection can be serious is if the infection is contracted during pregnancy, as *T. gondii* can cross the placenta to the foetus leading to foetal mental retardation and blindness, a condition known as congenital toxoplasmosis.

In those cases where infection with *T. gondii* can be serious (and result in complications), treatment can be prescribed to eradicate the parasite. Generally, the co-administration of sulfadiazine and pyrimethamine (an antimalarial drug which, like trimethoprim, is a DHFR inhibitor) is used as first-line treatment. Co-trimoxazole is used as an alternate therapy in resource-poor areas where sulfadiazine and pyrimethamine are not available. Treatment usually lasts for 3–6 weeks, but if the infection is associated with HIV/AIDS, the patient may have to take the medication for longer (sometimes for life) in order to keep it under control.

A Cochrane review failed to recommend any one regimen for the treatment of TE due to the lack of good-quality clinical trials, but from the data available, no significant differences in clinical response were found between co-trimoxazole alone and sulfadiazine and pyrimethamine together (Dedicoat and Livesley, 2006).

Pneumocystis jirovecii (formally known as *Pneumocystis carinii*) is a fungus that is commonly found in the respiratory tract of many mammals, including humans (Wakefield, 2002). The fungus rarely causes problems in healthy individuals, but can cause problems in immunocompromised individuals – the most common manifestation being pneumocystis pneumonia. Common symptoms of this condition include nonproductive cough, shortness of breath, fever, and tachypnoea (rapid breathing).

Historically, pneumocystis pneumonia has been associated with patients receiving chemotherapy or undergoing organ transplantation, but more recently it has become a common opportunistic infection in patients with HIV/AIDS, causing significant morbidity and mortality (with an estimated mortality rate of 10–20% in this patient group during the initial infection). Surprisingly, the mortality rate amongst non-HIV patients is higher, at a reported 30–60% (Thomas and Limper, 2004). Thankfully, because of the introduction of HAART (Highly Active Anti-Retro viral Therapy), the incidence of pneumocystis pneumonia appears to be decreasing in HIV patients, but it still remains the most common opportunistic infection amongst these patients (Thomas and Limper, 2004).

Despite this positive progression, the risk of contracting pneumocystis pneumonia must be taken seriously and therefore it is suggested that the following patient groups should be offered chemoprophylaxis to minimise the risk of developing infection: patients with a CD4 + count (T-helper cell count) of less than 200 cells/µL, those who have received organ transplantation (e.g. lung transplant), those who have immune-deficiency diseases or severe protein malnutrition, and those who are undergoing chemotherapy (Green *et al.*, 2007).

Co-trimoxazole is considered the agent of choice for the management of pneumocystis pneumonia, being used to treat active infection and to prevent infection in patients who are at risk (see above) (Grimwade and Swingler, 2003, 2006; Green *et al.*, 2007). For treatment of mild and moderate disease, high-dose co-trimoxazole is given orally, but in cases where the disease is severe and a rapid response to treatment is needed, high-dose co-trimoxazole is given by intravenous infusion. When used as prophylaxis, co-trimoxazole is given at a lower dose than when it is given to treat active infection. Typically, for an adult, a dose of 960 mg daily is used for prophylaxis,[6] but in cases where compliance is a problem

---

[6] As co-trimoxazole is a combined preparation, this dose equates to 800 mg of sulfamethoxazole and 160 mg of trimethoprim.

(which may often be the case in developing countries), the same dose may be taken on alternate days without significantly affecting the efficacy (Green *et al.*, 2007).

3.1.6.3   Nocardiosis

Nocardia are species of aerobic Gram positive bacteria commonly found in the soil, where they are involved in the decomposition of plant material. In some cases, however, certain species of *Nocardia* can be pathogenic to humans; for example, *N. asteroides* has been shown to cause pulmonary disease, while *Nocardia brasiliensis* can cause skin lesions (Saubolle and Sussland, 2003). These infections are most commonly associated with patients who have a compromised immune system, but can also occur in healthy patients who have a competent immune system – although this is less likely.

Nocardia can affect many organ systems, leading to different clinical forms of the disease. For example, pulmonary nocardiosis, the most common clinical form, is thought to be acquired via inhaling contaminated nocordial particles, while the introduction of *Nocardia* through the skin by a puncture wound or trauma (e.g. while gardening) can result in cutaneous nocardiosis (Figure 3.1.6). Other clinical forms of nocardiosis can include CNS infection (including brain abscesses), eye infection, and systemic (or disseminated) disease, which occurs when two or more body sites are affected.

Co-trimoxazole is effective at treating nocardiosis and is generally considered the agent of choice (Bayley *et al.*, 1981), but treatment regimens should always be tailored to suit the severity and location of infection. For brain abscesses, surgical excision may be used in combination with antibacterial therapy. The duration of treatment is variable, depending on the immune status of the patient and also the site of infection, but can last for anything up to 1 year. The outcome of treatment is generally dependent on the site of infection; high cure rates (approaching 100%) have been reported in patients with skin infection, whereas cure rates as low as 50% have been reported in patients with brain abscesses (Corti and Fioti, 2003).

In patients with AIDS, secondary prophylaxis with co-trimoxazole should also be considered to prevent re-infection.

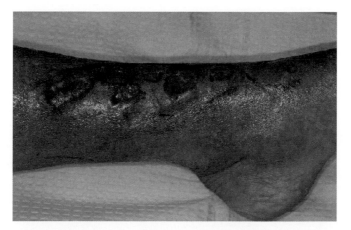

Figure 3.1.6   *A form of cutaneous nocardiosis caused by traumatic implantation of* Nocardia brasiliensis *in the foot (Reprinted from 'Lymphocutaneous Nocardiosis Caused by Nocardia brasiliensis', Dermatology Grand Round Cases, American Osteopathic College of Dermatology, Online, [http://www.aocd-grandrounds.org/case_46. shtml] 2006, with permission of J. Greg Brady.)*

3.1.6.4   Burn wounds

Sulfadiazine is also used (as the silver salt) for the prophylaxis and treatment of infection in burn wounds, which can be prone to infection. Depending on the severity, significant discomfort, disability, and sometimes even death can arise from the infection of a burn, so antimicrobial agents are used to prevent complications.

Silver sulfadiazine, formulated as a cream, has been used to reduce the risk of infection in burn wounds as it has activity against *P. aeruginosa* (Fox, 1968), as well as *in vitro* activity against a wide range of bacteria, including strains of *E. coli*, *Proteus mirabilis*, *Streptococcus pyogenes*, and *S. aureus* (Carr *et al.*, 1973). Interestingly, it is thought that the silver moiety (and *not* the sulfadiazine part) is directly responsible for the antibacterial activity of silver sulfadiazine, with only the silver moiety binding to cellular components; for example, the binding of silver to microbial DNA is related to the growth-inhibitory activity of silver sulfadiazine (Modak and Fox, 1973). The antibacterial activity of silver is well established and at this point you may be asking why other silver salts of acidic antibacterials have not been developed (silver penicillin, for example). This is a good question and can in part be explained by the rate of silver dissociation under physiological conditions. In the wound, silver sulfadiazine dissociates only moderately, over a period of time, and acts as a 'silver reservoir', while other silver salts either dissociate too rapidly (so the silver is rapidly removed) or too slowly (so there is not enough silver in the wound) (Fox and Modak, 1974).

Despite this clever mode of action, a Cochrane review failed to support the use of silver sulfadiazine in the management of superficial and partial-thickness burns, as a number of studies showed that it actually delays wound healing compared to other dressings (Wasiak *et al.*, 2008). For example, one randomised controlled trial, which investigated the healing times of residual burn wounds, showed that the mean healing times when silver-impregnated dressings were applied were significantly shorter than for the use of silver sulfadiazine (12 versus 16 days) (Li *et al.*, 2006).

3.1.6.5   Rheumatic fever

Acute rheumatic fever is an inflammatory disorder that can occur after a group A streptococcal pharyngeal infection (often known as 'strep throat') (Figure 3.1.7). The pathogenesis of this disorder is highly complex and not completely understood, but it is thought to result from a series of immunological events. Current

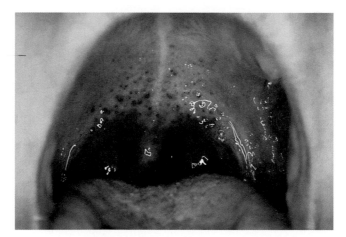

Figure 3.1.7   *'Strep throat', caused by group A* Streptococcus *bacteria, showing the characteristic petechiae (small red spots) on the soft palate (Image courtesy of the Public Health Image Library, Image ID 3186 [http://phil.cdc.gov/phil/details.asp?pid=3185, last accessed 29th March 2012].)*

research suggests that the immune response directed towards bacterial antigens also targets, through molecular mimicry, human tissues, thus causing inflammation and disease (Carapetis *et al.*, 2005a). Tissues affected can include large joints (causing arthritis), the heart (causing carditis), and the brain (causing chorea), so that acute rheumatic fever can present with a number of different symptoms. In view of this, a set of criteria, known as the Modified Jones Criteria, has been developed to assist in the diagnosis of acute rheumatic fever (Dajani *et al.*, 1992).

Acute rheumatic fever predominantly affects children aged between 5 and 14 years. The annual number of cases is estimated to be more than 471 000 worldwide, with 60% of these going on to develop rheumatic heart disease, a complication of acute rheumatic fever in which the heart valves (particularly the mitral and aortic valves) become damaged (Carapetis *et al.*, 2005b). Rheumatic heart disease can thus disrupt the normal blood flow through the heart and so is associated with significant morbidity and mortality, particularly in developing countries (Carapetis *et al.*, 2005b). The condition is more likely to occur after recurrent episodes of acute rheumatic fever (due to cumulative heart damage), but occasionally occurs after an initial episode.

There are a number of treatment options available to manage acute rheumatic fever but, because many approaches lack good-quality clinical evidence, the focus is often placed upon preventative therapy, with the prevention of an initial attack known as primary prophylaxis, and that of a recurrent attack known as secondary prophylaxis. Giving a patient an antibiotic to treat a sore throat caused by a streptococcal infection would thus be considered primary prophylaxis, while giving a patient an antibiotic after an episode of acute rheumatic fever to prevent further episodes (and thus decrease the risk of rheumatic heart disease) would be considered secondary prophylaxis. Generally, benzathine benzylpenicillin and penicillin V are considered the agents of choice for both primary and secondary prophylaxis (World Health Organization, 2001), but in cases where the patient is allergic to penicillins, oral erythromycin can be used for primary prophylaxis and oral sulfadiazine for secondary prophylaxis (World Health Organization, 2001). Sulfadiazine, although effective at preventing colonisation by group A streptococci in the upper respiratory tract (the goal of secondary prophylaxis), should *not* be used as primary prophylaxis due to problems associated with resistance.

The duration of secondary prophylaxis should always be tailored to suit the circumstances of the patient and is dependent on the degree of observed heart damage. For example, in cases where a patient has mild carditis, the duration of prophylaxis is 10 years after the last attack, or until the patient is 25 years of age (whichever is longer), while in cases when there is severe valvular disease, prophylaxis is recommended for life (World Health Organization, 2001).

3.1.7 Adverse drug reactions

Adverse drug reactions to the sulfonamide group of drugs occur at a rate of approximately 3% and this high incidence has resulted in the term 'sulfa allergy' (Choquet-Kastylevsky *et al.*, 2002; Slatore and Tilles, 2004). Structural differences between the sulfonamides and the nonantibiotic sulfur-containing drugs, namely the COX-II inhibitors, the triptan agents used for migraine (e.g. sumatriptan), the sulfonylureas (e.g. chlorpropamide, glipizide), and the diuretics (e.g. furosemide, bendroflumethiazide), mean that, generally speaking, only the antibiotic sulfonamides are commonly associated with sulfa allergy (Strom *et al.*, 2003). Some immunomodulatory drugs, such as sulfasalazine, are associated with sulfa allergy and should not be used in susceptible patients. Groups of patients who appear to be more susceptible to sulfa allergy include those who metabolise the sulfonamides slowly and patients with AIDs (in some cases up to 30% of this patient group shows some degree of sulfa allergy), who, as mentioned previously, are often treated with co-trimoxazole for *P. jirovecii* (Cribb *et al.*, 1996; Tiles, 2001).

Sulfonamide antibiotics have also been associated with hypoglycaemia due to their effect on insulin release, and with renal tubular acidosis and blood dyscrasias due to folate deficiency (Hollander, 1985; Murphy *et al.*, 1990; Strevel *et al.*, 2006).

3.1.7.1   Mechanism of toxicity

Despite it being currently accepted that cross-sensitivity between sulfonamide antibiotics and non-antibiotic sulfonamides is not significant (Strom *et al.*, 2003), increasing numbers of reports of allergenicity to sulfur-containing compounds have stimulated interest in determining a possible mechanism of sulfur allergenicity, including the serious cutaneous allergies observed with the sulfhydryl group (SH)-containing compounds, captopril and penicillamine, and the allergic reactions seen with intravenously administered furosemide (Hansbrough *et al.*, 1987; Kitamura *et al.*, 1990; Brackett *et al.*, 2004).

There does not appear to be an interaction between sulfa-containing drugs and the human immune system, so such cross-reactivity between non-antibiotic and antibiotic sulfonamide compounds appears to be unlikely. There are two structural characteristics that differ between the non-antibiotic- and antibiotic-containing sulfonamides: an amino group at the *para*-position of the benzene ring, and the presence of a five- or six-membered ring (usually heterocyclic) attached to the nitrogen of the sulfonamide group. Consequently, these N^4 and N^1 substitutions are useful predictors of an immunological response. In addition, non-type I hypersensitivity reactions to the sulfonamide antibiotics are generally suggested to be due to reactive metabolites of the sulfonamide, producing cytotoxic or immunoreactive responses.

Sulfonamide-containing antibiotics are associated with a range of immunologically mediated reactions, including:

- **Type I mediated responses** These involve immunoglobulin E (IgE) and include urticaria, hypotension, and anaphylaxis (often with cardiovascular collapse). These reactions occur within 1 hour of drug administration and constitute the most serious clinical reactions to the sulfonamides. Those patients reacting to sulfonamides with type I mediated response demonstrate increased levels of IgE antibodies towards the parent drug (Harle *et al.*, 1988). The interaction between IgE and sulfonamide antibiotics has been shown to be highly specific, with the N^1 heterocyclic ring being recognised immunologically, rather than the sulfonamide or *para*-aminophenyl (arylamine) groups (Knowles *et al.*, 2001).

- **Type II reactions** (cytotoxic) These involve IgG and IgM and are responsible for haemolytic reactions, neutropenias, and thrombocytopenias. Generally, drug-antigen and/or antigenic drug metabolites adhere to mature progenitor/stem cells, which are lysed by activated complement. Case studies have shown that sulfamethoxazole-trimethoprim administration can result in the production of antibodies (IgG and IgM) dependent upon sulfamethoxazole, while other patients have produced IgG and IgM in response to the metabolite N^4-acetyl-sulfamethoxazole (Kiefel *et al.*, 1987).

- **Type IV reactions** (cell-mediated immunity, also known as delayed hypersensitivity) These involve cytokines released by sensitised T cells. The low molecular weight of the sulfonamides would suggest that the parent compounds, when directly linked to an endogenous protein, should cause an immunological response. However, unmetabolised parent sulfonamides are typically unreactive, but may be immunogenic in a very small subset of patients (Brackett *et al.*, 2004), whereas their metabolites, such as hydroxylamines, are reactive, forming highly reactive covalent bonds with cellular proteins. For example, studies have indicated haptenation (the formation of a covalent bond with cellular proteins and drug/drug metabolite/metal ion, etc.) with the hydroxylamine derivative of sulfamethoxazole and the nitroso derivative of sulfamethoxazole, but not sulfamethoxazole itself (Manchandra *et al.*, 2002). Cytokines activate and subsequently attract macrophages, which release lytic enzymes, causing cutaneous reactions, such as Stevens–Johnson syndrome and toxic epidermal necrolysis. These reactions occur 48–72 hours post drug administration.

Other adverse drug reactions with sulfonamides include:

- **Skin reactions** The most common adverse drug reaction experienced with the sulfonamides is a skin reaction, ranging from a simple skin rash to the more serious Stevens–Johnson syndrome and toxic epidermal necrolysis. These reactions are caused by the whole range of immunological reactions (type I to IV).
- **Renal toxicity** Sulfonamides and their metabolites are excreted via the renal tubule and the incidence of renal toxicity associated with the first-generation sulfonamides (i.e. prontosil and M&B 693) was due to their pH-dependent solubility (as mentioned previously in Subsection 3.1.3) (Snyder, 1947). These drugs precipitated relatively easily within the renal tubules, causing obstruction and, in some cases, sulfadiazine haematuria, crystalluria, and renal colic (Arneil, 1958). Other mechanisms of toxicity, namely glomerulonephritis, are primarily due to **type III immunological reactions**, producing a soluble, transportable antigen immunoglobulin complex which infiltrates the renal tubule, with subsequent anaphylotoxin and histamine release. The immune complexes also deposit in the large joint spaces, in blood vessel membranes, and in glomeruli, resulting in arthritis, vasculitis, and glomerulonephritis, respectively. These reactions occur several weeks after drug administration.
- **Hepatotoxicity** This is, once again, primarily due to the poor solubility of the parent drug, metabolites, and any immune-drug complexes which form in the hepatocytes. The sulfonamides, being aromatic amines, are oxidised to hydroxylamines and then nitroso compounds, which undergo a redox cycle, leading to oxidative stress and free radical production, with subsequent hepatocellular damage. Hepatitis due to trimethoprim/sulfamethoxazole is relatively rare, with approximately 1 in 280 000 patients requiring hospitalisation (Gordin *et al.*, 1984; Beard *et al.*, 1986; Carson *et al.*, 1993). There is a 20% increase in the hepatitis rates in HIV patients, which is probably a result of their increased susceptibility to sulfonamides due to altered metabolism via CYP 450 pathways, leading to an increase in the oxidation of metabolites to toxic derivatives.
- **Blood reactions** The incidence of blood dyscrasias with sulfonamides has been reported to be approximately 3.3 per 100 000 person years, with a relative risk of 7.6 (Huerta and García Rodríguez, 2002). In severe cases, aplastic anaemia and agranulocytosis resulting from an adverse interaction between the drug and the haemopoietic pathway have been associated with sulfonamide use (Garvin, 1942). The incidence of agranulocytosis has been reported in one study to be approximately 0.19 cases per million per year, compared to β-lactams with an incidence of 0.56 per million per year, or an attributable risk of 5.4% and 12%, respectively (Ibáñez *et al.*, 2005). A combination of immunogenic reactions and oxidative stress is responsible for these blood dyscrasias.

The low incidence of cross-allergenicity of the sulfonamide antibiotics with other members of this class and also with non-antibiotic sulfa drugs is such that avoidance is often seen as unnecessary, but many clinicians avoid further use of sulfa drugs in those patients who have had a hypersensitivity reaction. This often complicates the clinical picture and may, in some instances, compromise patient care. The specificity of type I IgE-mediated allergic reactions suggests that cross-reactivity with other sulfonamides is highly improbable. It is recognised that it is the ring attached to N^1, and possibly the N^4 arylamine, that confers allergenicity in the mechanisms suggested above, and few other therapeutic agents contain such groups.

3.1.8   Drug interactions

Many of the drug interactions listed here refer to the combined use of trimethoprim and sulfamethoxazole, so it is difficult to apportion the toxicity reported in case studies to either of these agents alone.

3.1.8.1   Co-trimoxazole and ACE inhibitors

The combined use of co-trimoxazole and angiotensin-converting enzyme (ACE) inhibitors may result in hyperkalaemia[7] in elderly patients (Antoniou *et al.*, 2010). This interaction may be due to trimethoprim alone since its structure resembles that of the potassium-sparing diuretic amiloride and the concomitant use of potassium-sparing diuretics and ACE inhibitors is well known to cause hyperkalaemia.

3.1.8.2   Sulfonamides and anticoagulants

Several antibiotics, including sulfamethoxazole, inhibit CYP 2C9, and one very important drug interaction is with the *S*-warfarin enantiomer, which is also metabolised by CYP 2C9 (Jacobs, 2006). Indeed, one population-based study demonstrated that concomitant use of co-trimozaxole with warfarin resulted in a fourfold increase in hospitalisation of patients with bleeding in the genito-urinary tract (Fischer *et al.*, 2010), while data from another study showed that the incidence of GI bleeds increases with the combined use of warfarin and co-trimoxazole (Schellman *et al.*, 2008).

3.1.8.3   Sulfonamides and penicillin V

Penicillin V (phenoxymethylpenicillin) concentrations are increased with the concurrent use of several sulfonamides and it has been suggested that this elevation is due to the displacement of penicillin from protein-binding sites (Kunin, 1966).

3.1.8.4   Sulfonamides and methotrexate

One observational study and 17 case reports have shown an increased risk of cytopaenia with the concurrent use of methotrexate and co-trimoxazole (Bourré-Tessier and Haraoui, 2010).

3.1.8.5   Sulfonamides and phenytoin

Co-trimoxazole, and the concomitant use of sulfonamides and the hydantoins, have been reported to lead to hepatic damage (Ilario *et al.*, 2000). Both sets of drugs are metabolised in the liver and subsequently undergo conjugation (Cockerill and Edson, 1991). The proposed mechanism of interaction is combined toxicity, in that phenytoin initiates, in certain individuals, hepatic damage (as reflected by elevated alkaline phosphatase liver enzyme[8]), and co-trimoxazole contributes to fulminant hepatic failure, probably due to glucuronic acid depletion and thus the inability to glucuronidate toxic metabolites of all three agents.

3.1.8.6   Sulfonamides and oral hypoglycaemic agents

In vitro models have illustrated a possible interaction between tolbutamide and the sulfonamides. The enzyme involved in the metabolism of both agents is CYP 2C9. This study showed a 4.6- and 1.6-fold increase in the area under the curve (AUC) of tolbutamide (indicating enhanced bioavailability) when administered with

---

[7] Hyperkalaemia means there is a high level of potassium in the blood. Generally, levels of potassium of >6.5 mmol/L require urgent treatment, as there is a danger that the heart will stop. Indeed, large amounts of potassium can be very toxic and it is actually one of the drugs used in the lethal injection.

[8] Alkaline phosphatase (ALP) levels are often measured alongside other liver enzymes (such as alanine aminotransferase (ALT) and aspartate aminotransferase (AST)). These tests, known as liver function tests (or LFTs for short) are used to help detect liver disease or liver damage.

sulfaphenazole and with sulfamethizole, respectively (Komatsu *et al.*, 2000). Small-scale studies did not demonstrate similar effects upon glibenclamide concentrations, so there is no consistent pharmacokinetic interaction between these drugs (Sjöberg *et al.*, 1987).

Some researchers suggest that sulfonamides share a similarity to sulfonylureas in that they stimulate insulin release from the pancreas (Lee and Maddix, 1997). This possible effect has been demonstrated in patients receiving both insulin and co-trimoxazole (Hughes *et al.*, 2001; Strevel *et al.*, 2006), and with higher doses of co-trimoxazole (Johnson *et al.*, 1993).

3.1.9   Recent developments

The sulfonamides were once commonly used to treat a variety of infections caused by both Gram positive and Gram negative bacteria; however, due to the development of bacterial resistance and the serious adverse effects associated with sulfonamide use (such as the 'sulfa allergy'), they are now only considered as first-line therapy for a few conditions. Despite this problem, dihydropteroate synthase (DHPS), the target of the sulfonamides, still remains attractive for antibacterial drug design as it is absent in higher organisms. The development of a non-sulfonamide-based DHPS inhibitor would potentially overcome the problems associated with the current antibacterial sulfonamides.

For example, the pterin binding pocket of DHPS has been identified as a potential drug target (this is where the natural substrate 6-hydroxymethyldihydropterin pyrophosphate (DHPP) binds to DHPS) (see Scheme 3.1.5) (Hevener *et al.*, 2010). As we saw in Subsection 3.1.4, PABA and DHPP combine to form dihydropteroic acid, a process catalysed by DHPS. Rather than inhibit the PABA binding site of DHPS (as is the case for the sulfonamides), we could target the DHPP binding site, as the overall pharmacological effect would (theoretically) be the same (i.e. no dihydropteroic acid would be produced) but the inhibitor would lack the problems associated with the sulfonamides (isn't drug design easy?). Such compounds are in very early stages of development (see Figure 3.1.8 for examples) and are yet to undergo clinical trials, but in the future they may be a useful addition to the range of antibacterial agents available to practitioners. It will come as no surprise to you that the structures of the inhibitors are related to that of the natural substrate DHPP (Figure 3.1.8).

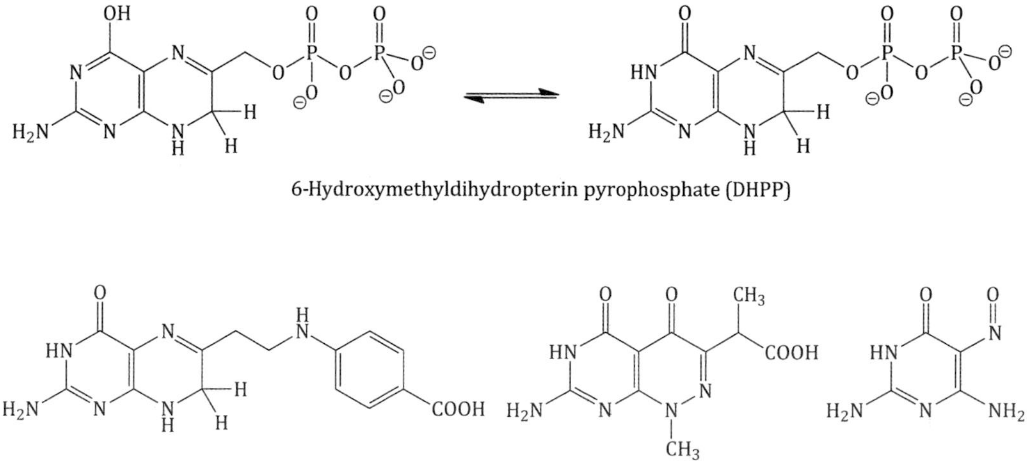

6-Hydroxymethyldihydropterin pyrophosphate (DHPP)

Figure 3.1.8   *DHPP and novel DHPS pterin inhibitors (Reprinted from K. E. Hevener, M. K. Yun, J. Qi, et al., J. Med. Chem., 53, 166–177, 2010, with permission of the American Chemical Society.)*

References

A. Achari, D. O. Somers, J. N. Champness, P. K. Bryant, J. Rosenmond, and D. K. Stammers, *Nature Struct. Biol.*, 1997, **4**, 490–497.

T. Akiba, K. Koyama, Y. Ishiki, S. Kimura, and T. Fukushima, *Jpn J. Microbiol.*, 1960, **4**, 219–227.

C. J. Allegra, D. Boarman, J. A. Kovacs, P. Morrison, J. Beaver, B. A. Chabner, and H. Masur, *J. Clin. Invest.*, 1990, **85**, 371–379.

T. Antoniou, T. Gomes, D. N. Juurlink, M. R. Loutfy, R. H. Glazier, and M. M. Mamdani, *Arch Intern Med.* 2010, **170**, 1045–1049.

G. C. Arneil, *Lancet.*, 1958, **1**, 826–827.

S. Bayley, P. S. Robinson, and S. J. Eykyn, *J. Infect.*, 1981, **3**, 230–233.

K. Beard, L. Belic, P. Aselton, D. R. Perera, and H. Jick, *J Clin Pharmacol.*, 1986, **26**, 633–637.

R. Bentley, *J. Ind. Microbiol. Biotechnol.*, 2009, **36**, 775–786.

R. J. Bertz and G. R. Granneman, *Clin. Pharmacokinet.*, 1997, **32**, 210–258.

S. I. Bhoir, I. C. Bhoir, A. M. Bhagwat, and M. Sundaresan, *J. Chromatog. B*, 2001, **757**, 39–47.

J. Bourré-Tessier and B. Haraoui, *J Rheumatol.*, 2010, **37**, 1416–1421.

C. C. Brackett, *Curr. Allergy Asthma Rep.*, 2007, **7**, 41–48.

C. C. Brackett, H. Singh, and J. H. Block, *Pharmacotherapy*, 2004, **24**, 856–870.

L. G. Burman, *Scand. J. Infect. Dis.*, 1986, **18**, 89–99.

K. L. Campbell, *Vet. Dermatol.*, 1999, **10**, 205–215.

R. A. Campero, R. B. Escudero, D. D. C. V. Alvarez, M. G. de la Parra, S. N. Montalvo, V. B. Fraga, R. S. Hernandez, and A. D. L. Acosta, *Clin. Therapeut.*, 2007, **29**, 326–333.

J. R. Carapetis, M. McDonald, and N. J. Wilson, *Lancet*, 2005a, **366**, 155–168.

J. R. Carapetis, A. C Steer, E. K. Mulholland, and M. Weber, *Lancet Infect. Dis.*, 2005b, **5**, 685–694.

H. S. Carr, T. J. Wlodkowski, and H. S. Rosenkranz, *Antimicrob. Agents Chemother.*, 1973, **4**, 585–587.

J. L. Carson, B. L. Strom, A. Duff, A. Gupta, M. Shaw, F. E. Lundin, and K. Das, *Ann. Intern. Med.*, 1993, **119**, 576–583.

G. Choquet-Kastylevsky, T. Vial, and J. Descotes, *Curr. Allergy Asthma Rep.*, 2002, **2**, 16–25.

F. R. Cockerill and R. S. Edson, *Mayo Clin Proc.*, 1991, **66**, 1260–1269.

M. E. Corti and M. F. V. Fioti, *Int. J. Infect. Dis.*, 2003, **7**, 243–250.

A. E. Cribb, B. L. Lee, L. A. Trepanier, and S. P. Spielberg, *Adverse Drug React. Toxicol. Rev.*, 1996, **15**, 9–50.

H. A. Dabritz and P. A Conrad, *Zoonoses Public Health*, 2010, **57**, 34–52.

A. S. Dajani, E. Ayoub, F. Z. Bierman, A. L. Bisno, F. W. Denny, D. T. Durack, P. Ferrieri, M. Freed, M. Gerber, E. L. Kaplan, A. W. Karchmer, M. Markowitz, S. H. Rahimtoola, S. T. Shulman, G. Stollerman, M. Takahashi, A. Taranta, K. A. Taubert, and W. Wilson, *JAMA*, 1992, **268**, 2069–2073.

W. S. Dallas, J. C. Rowan, P. H. Ray, M. J. Cox, and J. K. Dev, *J. Bacteriol.*, 1992, **174**, 5961–5970.

M. Dedicoat and N. Livesley, *Cochrane Database of Systematic Reviews*, 2006, Issue 3.

D. A. P. Evans, *J. Med. Genet.*, 1969, **6**, 405–407.

D. A. P. Evans and T. A. White, *J. Lab. Clin. Med.*, 1964, **63**, 394–403.

H. D. Fischer, D.N. Juurlink, M. M. Mamdani, A. Kopp, and A. Laupacis, *Arch. Intern. Med.*, 2010, **170**, 617–621.

C. A. Hughes, C. L. Chik, and G. D. Taylor, *Can. J. Infect. Dis.*, 2001, **12**, 314–316.

L. G. Jacobs, *Clin. Geriatr. Med.*, 2006, **22**, 17–32.

J. A. Johnson, J. E. Kappel, and M. N. Sharif, *Ann. Pharmcother.*, 1993, **27**, 304–306.

X. L. Li, Y. S. Huang, Y. Z. Peng, Z. J. Liao, G. A. Zhang, Q. Liu, J. Tang, X. S. Liu, and Q. Z. Luo, *Chinese Journal of Burns*, 2006, **22**, 15–18.

G. Domagk, *Deutsche Med. Wochenschr.*, 1935, **61**, 250–253.

G. Domagk, *Annal N. Y. Acad. Sci.*, 1957, 380–384.

P. Fildes, *Lancet*, 1940, **1**, 955–957.

C. L. Fox, Jr, *Arch. Surg.*, 1968, **96**, 184–188.

C. L. Fox, Jr., and S. M. Modak, *Antimicrob. Agents Chemother.*, 1974, **5**, 582–588.

C. F. Garvin, *Am. J. Orthodont. and Oral Surg.*, 1942, **28**, B102–108.

T. Ghafourian, M. Barzegar-Jalali, S. Dastmalchi, T. Khavari-Khorasani, N. Hakimiha, and A. Nokhodchi, *Int. J. Pharmaceut.*, 2006, **319**, 82–97.

F. M. Gordin, G. L. Simon, C. B. Wofsy, and J. Mills, *Ann. Intern. Med.*, 1984, **100**, 495–499.

H. Green, M. Paul, L. Vidal, and L. Leibovici, *Cochrane Database of Systematic Reviews*, 2007, Issue 3.

K. Grimwade and G. H. Swingler, *Cochrane Database of Systematic Reviews*, 2003, Issue 3.

K. Grimwade and G. H. Swingler, *Cochrane Database of Systematic Reviews*, 2006, Issue 1.

J. R. Hansbrough, H. J. Wedner, and D. D. Chaplin, *J. Allergy Clin. Immunol.*, 1987, **80**, 538–541.

D. G. Harle, B. A. Baldo, and J. V. Wells, *Molecular Immunol.*, 1988, **25**, 1347–1354.

C.A. Hekster and T.B. Vree, *Antibiot. Chemother.*, 1982, **31**, 22–118.

K. E. Hevener, M. K. Yun, J. Qi, I. D. Kerr, K. Babaoglu, J. G. Hurdle, K. Balakrishna, S. W. White, and R. E. Lee, *J. Med. Chem.*, 2010, **53**, 166–177.

H. Hollander, *Ann. Intern. Med.*, 1985, **102**, 138.

C. Huerta and L. A. García Rodríguez, *Pharmacotherapy*, 2002, **22**, 630–636.

L. Ibáñez, X. Vidal, E. Ballarín, and J. R. Laporte, *Arch. Intern. Med.*, 2005, **165**, 869–874.

M. J. Ilario, J. E. Ruiz, and C. A. Axiotis, *Arch. Pathol. Lab. Med.*, 2000, **124**, 1800–1803.

V. Kiefel, S. Santosos, S. Schmidt, A. Salama, C. Mueller-Eckhardt, *Transfusion*, 1987, **27**, 262–265.

K. Kitamura, M. Aihara, J. Osawa, S. Naito, and Z. Ikezawa, *J. Dermatol.*, 1990, **17**, 44–51.

S. Knowles, L. Shapiro, and N. Shear, *Drug Safety*, 2001, **24**, 239–247.

K. Komatsu, K. Ito, Y. Nakajima, S. Kanamitsu, S. Imaoka, Y. Funae, C. E. Green, C. A. Tyson, N. Shimada, and Y. Sugiyama, *Drug Metab. Dispos.*, 2000, **28**, 475–481.

A. Królicki, A. Klimowicz, S. Bielicka-Grzela, A. Nowak, and R. Maleszka, *Pol. J. Pharmacol.*, 2004, **56**, 257–263.

C. M. Kunin, *Clin. Pharmacol. Ther.*, 1966, **7**, 180–188.

A. J. Lee and D. S. Maddix, *Ann. Pharmacother.*, 1997, **31**, 727–732.

J. E. Lesch, The First Miracle Drugs: How the Sulfa Drugs Transformed Medicine, Oxford University Press, New York, NY, 2007.

C. Levy, D. Minnis, and J. P. Derrick, *Biochem. J.*, 2008, **412**, 379–388.

X. L. Li, Y. S. Huang, Y. Z. Peng, Z. J. Liao, G. A. Zhang, Q. Liu, J. Tang, X. S. Liu, and Q. Z. Luo, *Chinese Journal of Burns*, 2006, **22**, 15–18.

C.-F. Lin, C. C. Chang, and W.-C. Lin, *J. Chromatog. A*, 1997, **768**, 105–112.

D. M. Livermore, *Clin. Infect. Dis.*, 2002, **34**, 634–640.

A. V. Luk'yanov, V. S. Onoprienko, K. S. Borodina, V. A. Zasosov, E. V. Van'kovich, T. A. Gracheva, O. N. Volzhina, I. A. Kuznetsova, V. G. Potapava, Z. M. Klimonova, Yu. G. Zelinskii, I. N. Vinokurova, and V. S. Gershanovskii, *Khimiko-Farmatsevticheskii Zhurnal (Pharm. Chem. J.)* 1980, **14**, 640–644.

P. T. Männistö, R. Mäntylä, J. Mattila, S. Nykänen, and U. Lamminsivu, *J. Antimicrob. Chemother.*, 1982, **9**, 461–470.

T. Manchandra, D. Hess, L. Dale, S. G. Ferguson, and M. J. Rider, *Mol. Pharmacol.*, 2002, **62**, 1011–1026.

M. C. Meyer, A. B. Straughn, G. Ramachander, J. C. Cavagnol, and A. F. Biola Mabàdeje, *J. Pharm. Sci.*, 1978, **67**, 1659–1661.

F. Mietzsch and J. Klarer, *Ger. Patent*, **2** January 1935, 607–537.

G. H. Miller, P. H. Doukas, and J. K. Seydel, *J. Med. Chem.*, 1972, **15**, 700–706.

H. N. Mistri, A. G. Jangid, A. Pudage, A. Shah, and P. S. Srivastav, *Microchem. J.*, 2010, **94**, 130–138.

S. M. Modak and C. L. Fox, Jr, *Biochem. Pharmacol.*, 1973, **22**, 2391–2404.

Col. J. A. Moncrieff, Col. R. B. Lindberg, Col. W. E. Switzer, and Maj. B. A. Pruitt, Jr, *J. Trauma*, 1966, **6**, 407–419.

J. L. Murphy, W. R. Griswold, V. M. Reznik, and S. A. Mendoza, *Child Nephrol. Urol.*, 1990, **10**, 49–50.

F. O'Grady, *Can. Med. Assoc. J.*, 1975, **112**, 5S–7S.

K. Poole, *J. Mol. Microbiol. Biotechnol.*, 2001, **3**, 255–264.

Z. Qiang and C. Adams, *Water Res.*, 2004, **38**, 2874–2890.

M. L. Rallison, J. O'Brien, and R. A. Good, *Pediatrics*, 1961, **6**, 908–917.

D. S. Reeves and P. J. Wilkinson, *Infection*, 1979, **7**, S330–S341.

S. Roland, R. Ferone, R. J. Harvey, V. L. Styles, and R. W. Morrison, *J. Biol. Chem.*, 1979, **254**, 10 337–10 345.

M. A. Saubolle and D. Sussland, *J. Clin. Microbiol.*, 2003, **41**, 4497–4501.

B. Sarikabhuti, N. Keschamrus, S. Noeypatimanond, E. Weidekamm, R. Leimer, W. Wernsdorfer, and E. U. Kölle, *Acta Trop.*, 1988, **45**, 217–224.

H. Schellman, W. B. Bilker, C. M. Brensinger, X. Han, S. E. Kimmel, and S. Hennessy, *Clin. Pharm. Ther.*, 2008, **84**, 581–588.

A. E. Senear, M. M. Rapport, J. F. Mead, J. T. Maynard, and J. B. Koepfli, *J. Org. Chem.*, 1946, **11**, 378–383.

S. Sjöberg, B. E. Wiholm, R. Gunnarsson, H. Emilsson, E. Thunberg, I. Christenson, and J. Ostman, *Diabet. Med.*, 1987, **4**, 245–247.

O. Sköld, *Antimicrob. Agents Chemother.*, 1976, **9**, 49–54.

O. Sköld, *Drug Resistance Updates*, 2000, **3**, 155.

C. G. Slatore and S. A. Tilles, *Immunol. Allergy Clin. North Am.*, 2004, **24**, 477–490.

L. J. Snyder, *Proc. Annu. Meet. Cent. Soc. Clin. Res. US*, 1947, **20**, 70.

E. L. Strevel, A. Kuper, and W. L. Gold, *Lancet Inf. Dis.*, 2006, **6**, 178–182.

B. L. Strom, R. Schinnar, A. J. Apter, D. J. Margolis, E. Lautenbach, S. Hennessy, W. B. Bilker, and D. Pettitt, *N. Engl. J. Med.*, 2003, **349**, 1628–1635.

W. Tappe, C. Zarfl, S. Kummer, P. Burauel, H. Vereecken, and J. Groeneweg, *Chemosphere*, 2008, **72**, 836–843.

C. F. Thomas, Jr and A. H. Limper, *New. Eng. J. Med.*, 2004, **350**, 2487–2498.

S. A. Tiles, *South Med. J.*, 2001, **94**, 817–824.

J. Tréfouël, J. Tréfouël, F. Nitti, and D. Bovet, Comptes Rendus des Sceances Soc. *Biol Filail.*, 1935, **120**, 756–758.

L. A. Trepanier, *Vet. Dermatol.*, 1999, **10**, 241–248.

T. B. Vree, Y. A. Hekster, A. M. Baars, J. E. Damsma, and E. Van der Kleijn, *Clin. Pharmacokinet.*, 1978, **3**, 319–329.

T. B. Vree, W. J. O'Reilly, Y. A. Hekster, J. E. Damsma, and E. Van der Kleijn, *Clin. Pharmacokinet.*, 1980, **5**, 274–294.

A. E. Wakefield, *Br. Med. Bull.*, 2002, **61**, 175–188.

J. Wasiak, H. Cleland, and F. Campbell, *Cochrane Database of Systematic Reviews*, 2008, Issue 4.

I. D. Watson, H. N. Cohen, S. J. McIntosh, J. A. Thomson, and A. Shenkin, *Clin. Therapeut.*, 1981, **4**, 103–108.

I. D. Watson, M. J. Stewart, and D. J. Platt, *Clin. Pharmacokinet.*, 1988, **15**, 133–164.

P. M. Wax, *Ann. Internal Med.* 1995, **122**, 456–461.

E. Weidekamm, H. Plozza-Nottebrock, I. Forgo, and U. C. Dubach, *Bull. World Health Organ.*, 1982, **60**, 115–122.

World Health Organization, Rheumatic Fever and Rheumatic Heart Disease. Report of a WHO Expert Consultation, 2001 (http://www.who.int/cardiovascular_diseases/resources/en/cvd_trs923.pdf, last accessed 19 July 2010)

P. S. Winnek and R. O. Roblin, US Patent 2430439, 4 November 1947.

D. D. Woods, *Brit. J. Exp. Pathol.*, 1940, **21**, 74–90.

C. Zarfl, M. Mattheis, and J. Klasmeier, *Chemosphere*, 2008, **70**, 753–760.

3.2

Trimethoprim

Key Points

- Trimethoprim is an antifolate agent which is a **di**hydrofolate **r**eductase (DHFR) inhibitor.
- Trimethoprim is used in combination (synergistic) with sulfamethoxazole (as co-trimoxazole).
- Trimethoprim is most commonly used for the management of lower urinary tract infections.

3.2.1 Discovery

Trimethoprim (Figure 3.2.1) is an example of early rational drug design; it resulted from the work of George H. Hitchings and his group at Burroughs Wellcome in the USA during the 1940s, 1950s, and 1960s, when they were studying the cellular actions of purines and pyrimidines, with the rationale that interference in the processes associated with these biologically important heterocycles could result in therapeutic outcomes.

Using *Lactobacillus casei* as a representative organism, the group showed that metabolism of nucleic acids could be inhibited by selected purine and pyrimidine analogues and that this inhibition could be reversed by the addition of folic acid. In particular, 2,4-diaminopyrimidines were found to be competitive inhibitors of the folate cycle in *L. casei*. The synthesis of many substituted analogues of 2,4-diaminopyrimidine eventually led to the identification of trimethoprim as the optimal agent for the inhibition of this biochemical process in bacteria (Roth *et al.*, 1962; Burchall and Hitchings, 1965). In addition, the extensive studies on *L. casei* provided much important information on biochemical cycles, which led to the identification of similarities and differences between species, and aided the subsequent development of selective therapeutic agents.

The Hitchings group was highly successful and developed many active agents, including: the prophylactic antimalarial agent pyrimethamine; the antileukaemic/immunosuppressant 6-mercaptopurine; the xanthine oxidase inhibitor allopurinol, used to treat gout; and, of course, the antibacterial agent trimethoprim (Hitchings, 1969; Elion, 1989). In fact, Hitchings, along with his colleague Gertrude B. Elion, shared the Nobel Prize for Physiology or Medicine in 1998 with a British pharmacologist, Sir James Black; their work which led to the awarding of the Nobel Prize has been recently reviewed (Raviña, 2011).

Antibacterial Agents: Chemistry, Mode of Action, Mechanisms of Resistance and Clinical Applications, First Edition.
Rosaleen J. Anderson, Paul W. Groundwater, Adam Todd and Alan J. Worsley.
© 2012 John Wiley & Sons, Ltd. Published 2012 by John Wiley & Sons, Ltd.

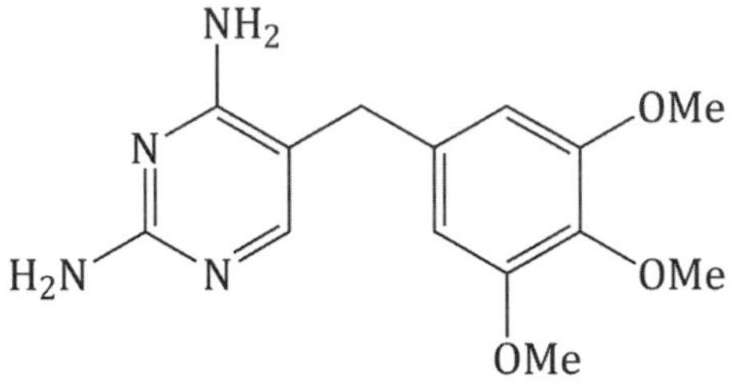

Figure 3.2.1 *Trimethoprim*

Strangely, although the *in vitro* and clinical results with trimethoprim were more than encouraging, it was several years after its discovery before it was adopted into regular clinical practice, despite further articles detailing its success across a range of infections (Cooper and Wald, 1964; Martin and Arnold, 1967; Darrell *et al.*, 1968; Grüneberg and Kolbe, 1969).

3.2.2 Synthesis

Trimethoprim was originally synthesised from the appropriate trimethoxy-substituted ethyl dihydrocinnamate **1** (Falco *et al.*, 1951). Formylation of the dihydrocinnamate **1**, with ethyl formate and sodium, gave the ethyl 3′,4′,5′-trimethoxybenzylmalonic semialdehyde **2** (Scheme 3.2.1), which tautomerises[9] to its enol form **3**. This mixture of tautomers was reacted with guanidine and sodium ethoxide to form 2-amino-4-hydroxy-5-(3′,4′,5′-trimethoxybenzyl)pyrimidine **4**. Substitution of chlorine for the hydroxyl group was effected by phosphorus oxychloride (a reaction that is still commonly used for introducing the chloro group, which is a much better leaving group than hydroxyl) and, finally, amination of the 4-chloro compound using ammonia resulted in trimethoprim (Hitchings and Roth, 1959; Roth *et al.*, 1962). The overall yield from the cinnamic acid precursor to the dihydrocinnamate **1** was poor (around 20%) and better syntheses were required for industrial-scale production.

Since then, there have been several synthetic routes developed, some of which show significant yield improvements over the others, and the most common routes have been summarised (Vardanyan and Hruby, 2006). One route worth mentioning allows for the synthesis of a wide range of analogues, although again in moderate yield. It employs an appropriately substituted aldehyde **5**, to which is added a 3-alkoxypropanenitrile **6**, under strongly basic conditions, to provide the corresponding cinnamonitrile **7** (or its enol ether **8**), followed by condensation with guanidine, as illustrated in Scheme 3.2.2 for trimethoprim (Stenbuck *et al.*, 1963).

Many newer substituted 2,4-diaminopyrimidines, such as iclaprim (which you will meet in Subsection 3.2.9) (Jaeger *et al.*, 2011), are accessed by a modification of this synthesis, which replaces the 3-alkoxypropionitrile **6** with an *N*-substituted 3-aminopropionitrile **9** and proceeds through the analogous β-cyano-*N*-substituted cinnamylamine intermediate **10**, followed by reaction with guanidine to give the appropriately substituted 5-benzyl-2,4-diaminopyrimidines (analogues of trimethoprim), in moderate to good yields (Scheme 3.2.3) (Poe and Ruyle, 1981).

Using this method, numerous analogues of trimethoprim have been prepared and used to explore the requirements of binding in the enzyme active site, with particular emphasis on understanding the selectivity (see, for example, Roth and Aig, 1987; Roth *et al.*, 1987; Davis *et al.*, 1989; Selassie *et al.*, 1998).

---

[9] Tautomerisation (here keto-enol tautomerism) involves the shift of a hydrogen (in this case that shown in red) from one atom to another to give an isomer that is in equilibrium with the original compound. We've already seen an example of this process in this section (amide-iminol). But where?

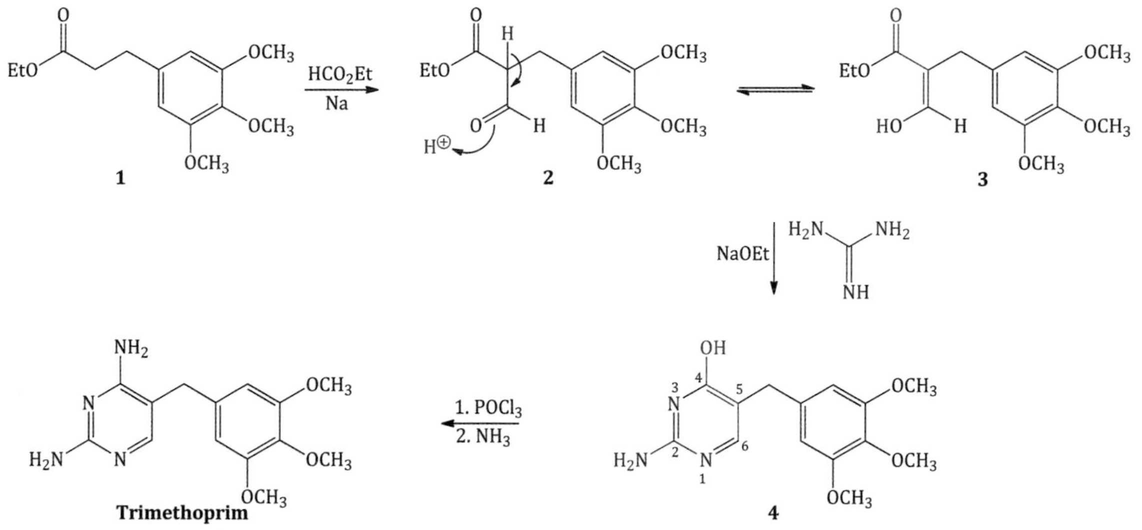

Scheme 3.2.1 *The original Hitchings synthesis of trimethoprim (Hitchings and Roth, 1959; Roth et al., 1962)*

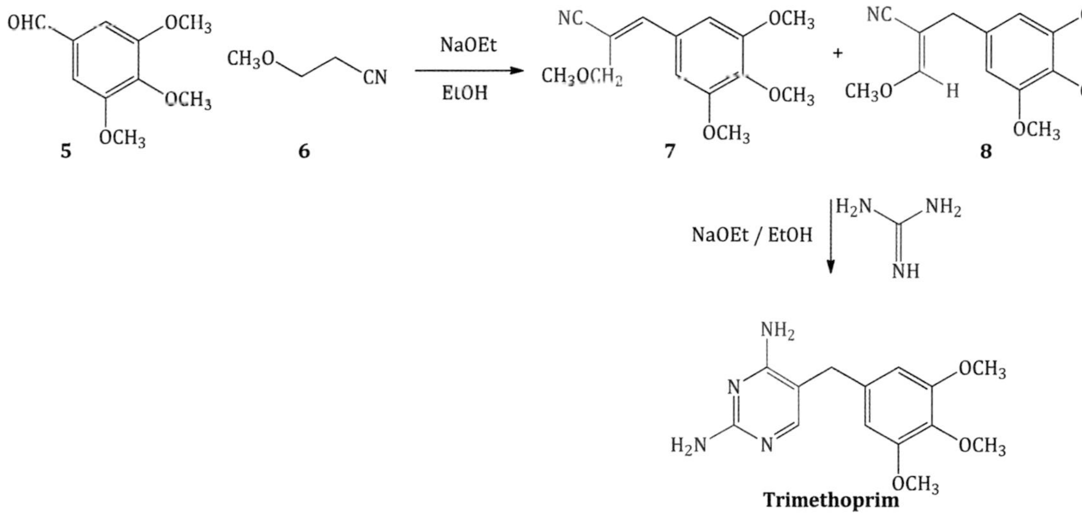

Scheme 3.2.2 *Synthesis of trimethoprim from the corresponding aldehyde **5** via cinnamonitrile **7***

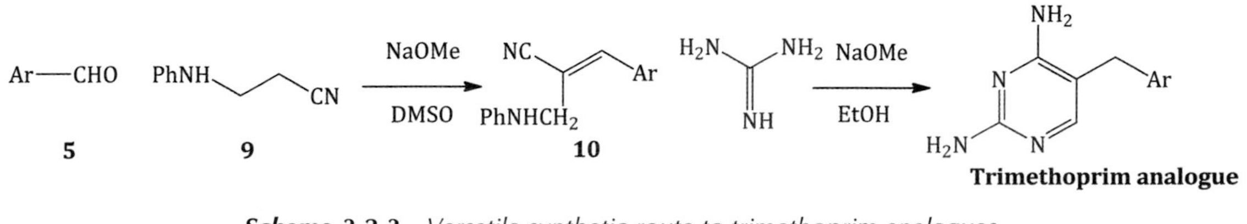

Scheme 3.2.3 *Versatile synthetic route to trimethoprim analogues*

We will consider some of these derivatives and their contribution to the knowledge of the mode of action in Subsection 3.2.4.

3.2.3   Bioavailability

The bioavailability of trimethoprim is almost 100%, with rapid absorption following oral administration and no impact of food upon the rate. It has a half-life of 10 hours, allowing twice-daily administration for acute infections and once-daily for its prophylactic action, although differences in the half-life were observed between adults and children when treated orally with trimethoprim (Hoppu and Arjomaa, 1984; Rylance *et al.*, 1985). The volume of distribution is considerable at 71–136 L/kg, so that it is widely distributed throughout tissues, organs, and fluids. For example, it is well distributed into lymph tissue, cerebrospinal fluid, prostatic fluid, saliva, sinus secretions, and breast milk, readily achieving therapeutic concentrations (Brogden *et al.*, 1982). Although trimethoprim is partly metabolised by demethylation and oxidation at the N^1 and N^3 positions (Sigel *et al.*, 1973), the majority is excreted unchanged in the urine – one reason why it is commonly used to treat urinary infections (Kasanen *et al.*, 1978; Libecco and Powell, 2004). As for the sulfonamides, the urinary excretion of trimethoprim is influenced by pH; if you look at the structure of trimethoprim, you will not be surprised to find that it is weakly basic, with a pK_a of about 7.3. In blood (pH 7.2–7.4) then, you would expect it to be approximately 50% ionised and 50% unionised.[10] A more acidic pH will increase ionisation (greater formation of N^1H^+) and so solubility, leading to greater excretion, while a higher pH will decrease ionisation and thus lead to lower excretion (Brogden *et al.*, 1982). The clinical pharmacokinetics of trimethoprim and trimethoprim in combination with a sulfonamide have been reviewed (Patel and Welling, 1980; Friesen *et al.*, 1981; Watson *et al.*, 1988).

3.2.3.1   Bacterial uptake of trimethoprim

Similarly to the sulfonamides, trimethoprim is sufficiently small and lipophilic that it can enter bacterial cells by passive diffusion. Poor concentrations achieved in Gram negative bacteria have been linked to an active efflux pump.

3.2.4   Mode of action and selectivity

Through their many experiments, as discussed at the beginning of this chapter, Hitchings and coworkers learned that trimethoprim was interfering with the folic acid cycle, but the actual bacterial enzyme target was not isolated and characterised until 1958. Like the sulfonamides, such as sulfamethoxazole, trimethoprim acts upon the folic acid cycle (Scheme 3.2.4).

As Scheme 3.2.4 shows, trimethoprim acts upon the enzyme dihydrofolate reductase (DHFR), EC 1.5.1.3. Details about the enzyme itself can be accessed from one of the excellent Internet resource sites, such as BRENDA (www.brenda-enzymes.info) or the Kyoto Encyclopedia of Genes and Genomes (KEGG; www.genome.jp/kegg), from which the folic acid pathway is also available.[11]

---

[10] From the Henderson–Hasselbalch equation, $pH = pK_a + \log([A^-]/[HA])$; if $pH = pK_a$ then the concentrations of the conjugate base $[A^-]$ and acid $[HA]$ must be equal (since $\log 1 = 0$).
[11] http://www.genome.jp/kegg-bin/show_pathway?map00790, (last accessed 30th March 2012), from which the hyperlink to enzyme EC 1.5.1.3 can be used to source further information about dihydrofolate reductase.

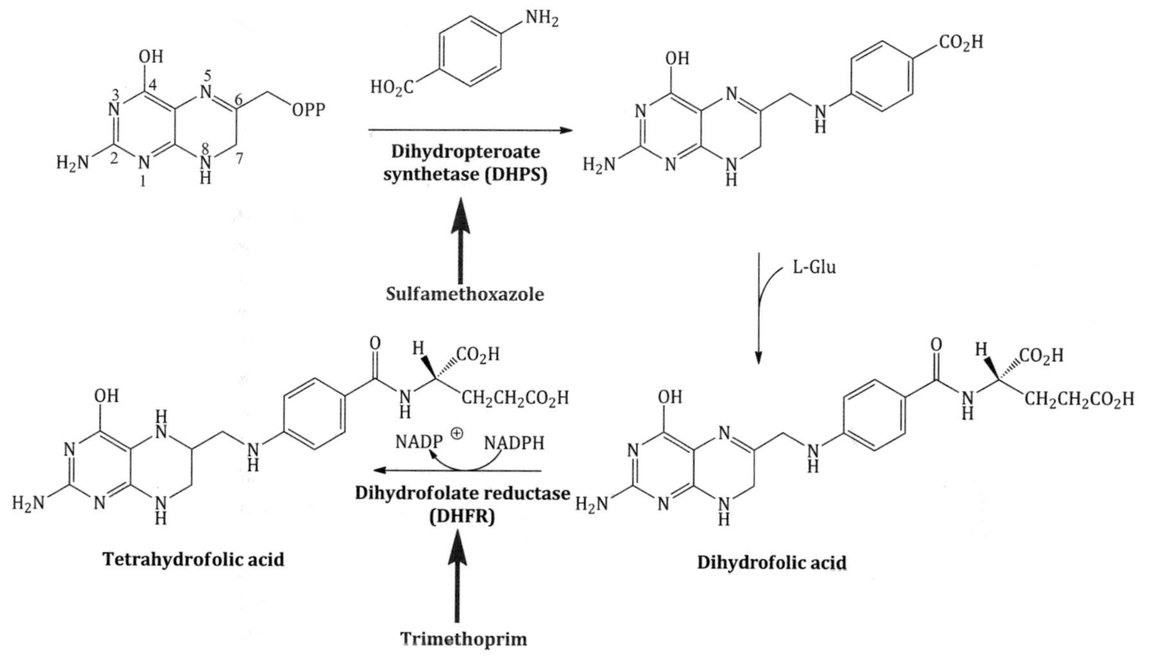

Scheme 3.2.4 *Targets for inhibition of the folic acid cycle by trimethoprim and the sulfonamides (such as sulfamethoxazole)*

Trimethoprim is a DHFR inhibitor and its bacteriostatic effect occurs through the disruption of the folate cycle and depletion of tetrahydrofolate (Quinlivan *et al.*, 2000), which, as we saw in Subsection 1.1.3.2, is required as a co-factor in the biosynthesis of thymidine (Kasanen *et al.*, 1978).

More than 200 crystal structures of dihydrofolate reductase are available on the RCSB protein crystal structure database (www.pdb.org), almost 100 of which are of DHFR from a range of bacterial species, either alone or in complex with an inhibitor.

Perhaps surprisingly for an enzyme that catalyses the same reaction across a wide range of species, there is only about 30% homology between the amino acid sequences of the DHFR from humans, chickens, mice, *Escherichia coli*, *L. casei*, and *Plasmodium falciparum*. When examined more closely using X-ray crystallography, however, it becomes apparent that the overall tertiary structure and 3D shape of the enzymes show much greater similarity, with relatively high conservation of the amino acids in and around the active sites of the mammalian and avian DHFR enzyme. The bacterial enzymes show similarity to one other but have key differences to the mammalian and avian forms (Selassie *et al.*, 1998; Schweitzer *et al.*, 1990).

If you look again at Scheme 3.2.4, you will see that dihydrofolate is reduced to tetrahydrofolate in the usual biochemical reaction catalysed by DHFR, using NADPH as a co-factor, which transfers a hydride (H^-) to C^6 (where the side chain is located) of dihydrofolate (Matthews *et al.*, 1986). The pteridine ring of the substrate binds close to the NADPH in the active site, presenting the imine ($N=C^6$) reduction site to the co-factor in order for hydride transfer to occur (Bolin *et al.*, 1982). Protonation of N^5, from a conserved water molecule located in the active site, follows, and this is assisted by Asp-27 in bacterial DHFR. Folic acid is also a substrate of DHFR (being reduced first to dihydrofolic acid, then tetrahydrofolic acid,[12] although it is reduced at around a 10–1000 times slower rate than the reduction of dihydrofolic acid and requires two transfers of hydride from

---

[12] Dihydrofolic acid is often referred to as dihydrofolate and tetrahydrofolic acid as tetrahydrofolate.

the co-factor) (Oefner *et al.*, 1988). In order for a reversible competitive inhibitor to be successful without the formation of a covalent bond to the enzyme or co-factor, it needs to bind more strongly to the enzyme than the natural substrate, which is effectively blocked from the active site. How does trimethoprim do this in bacterial DHFR?

Like the natural substrate, trimethoprim and other bacterial DHFR inhibitors bind to the enzyme along with NADPH in a ternary complex, in which the reducing end of the co-factor is sited near to the pyrimidine ring of trimethoprim. Much work has gone into studying the binding of trimethoprim to bacterial DHFR and identifying the key interactions that facilitate tight binding to the active site. NMR spectroscopy and X-ray crystallography indicate that N^1 is protonated in the bound state (Cocco *et al.*, 1983; Matthews *et al.*, 1985a), and X-ray crystallography also shows that the pyrimidine ring of trimethoprim locates in a deep cleft of the protein, with favourable non-covalent interactions helping to maintain its position. This binding mode enables ionic interactions between the protonated N^1 and the carboxylate group of the conserved Asp-27 residue, and hydrogen-bonding interactions (Figure 3.2.2) between:

- N^1 and Asp-27;
- 2-NH$_2$ and Asp-27 and a conserved water molecule;
- the same conserved water molecule and Thr-113;
- 4-NH$_2$ and Ile-5 and Ile-94.

Several favourable van der Waals interactions between the pyrimidine and hydrophobic groups in the active site, involving the side chains of Ile-5, Ala-7, Leu-28, Phe-31, Ile-94, and the backbone of Ala-6, Ala-7, and Ile-94, contribute to the strong binding of trimethoprim to the DHFR from *E. coli* and other bacteria. The trimethoxybenzyl group is orientated at an angle to the pyrimidine ring, extended towards the cleft opening, with van der Waals interactions to Leu-28, Phe-31, and Met-20 enhancing the binding (Matthews *et al.*, 1985a, 1985b; Champness *et al.*, 1986). Further interactions of the three methoxy groups are possible with the proximal amino acid residues and with the hydrophilic exterior, into which the 4-OMe group extends (Selassie *et al.*, 1998). You can imagine how this complex balance of intermolecular interactions, holding the

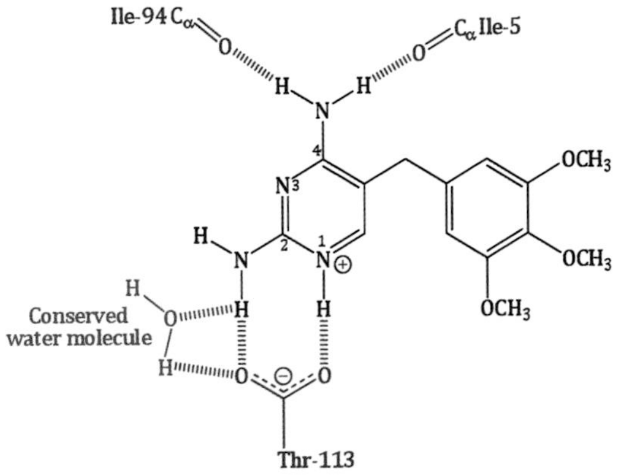

Figure 3.2.2 *Simplified illustration of hydrogen bonding between trimethoprim and the active site of E. coli DHFR. The DHFR residues are shown in blue, the conserved water molecule in pink, and the hydrogen bonds in red (adapted from Matthews et al., 1985a, 1985b; Champness et al., 1986.)*

structures tightly together, along with complementarity of size and shape, results in the potent inhibitory effect of trimethoprim upon bacterial DHFR.

The analysis of the structures of potential inhibitors and their activity against bacterial DHFR enzymes (in comparison to DHFR from other species) has been used to aid the design of improved inhibitors (Roth *et al.*, 1988; Johnson *et al.*, 1989; Rauckman *et al.*, 1989; Chan and Roth, 1991; Ohemeng and Roth, 1991; Selassie *et al.*, 1998), while the structures of various DHFR enzymes have been compared, and their binding to trimethoprim analysed (Schweitzer *et al.*, 1990).

The selectivity of trimethoprim for the bacterial enzyme over the mammalian enzyme has been appreciated for over 50 years (Hitchings, 1969, 1988; Elion, 1989). The mammalian and avian DHFR enzymes were isolated in 1958 (Osborn and Huennekens, 1958; Peters and Greenberg, 1958), followed by a bacterial form (Blakley and McDougall, 1961), but it was some years before it was confirmed that trimethoprim, pyrimethamine, and methotrexate all act as inhibitors of the same enzyme, with the selective inhibition of a particular DHFR enzyme dictating their pharmaceutical use (Burchall and Hitchings, 1965). The structures of these three classic DHFR inhibitors, all of which are still used clinically, are shown in Figure 3.2.3, alongside the natural substrate, dihydrofolic acid, with which they compete for the DHFR active site. You can see that the 2,4-aminopyrimidine group of these inhibitors resembles part of the pteridine nucleus of dihydrofolate, while the differently substituted benzene rings make a major contribution to the selective binding to the various DHFR enzymes.

If you consider the data in Table 3.2.1 (Burchall and Hitchings, 1965; Ferone *et al.*, 1969), you can see that trimethoprim inhibits the bacterial enzyme with great selectivity over mammalian enzymes, while pyrimethamine is selective for the protozoal enzyme. Although methotrexate has good inhibitory activity against the bacterial, protozoal, and mammalian DHFR enzymes, and you would be forgiven for thinking that it could have wide antimicrobial application, it has poor uptake into bacterial cells, which lack the membrane-bound folate transporter used by methotrexate to enter mammalian cells, so it has very little antibacterial activity *in vivo* (Allegra *et al.*, 1987).

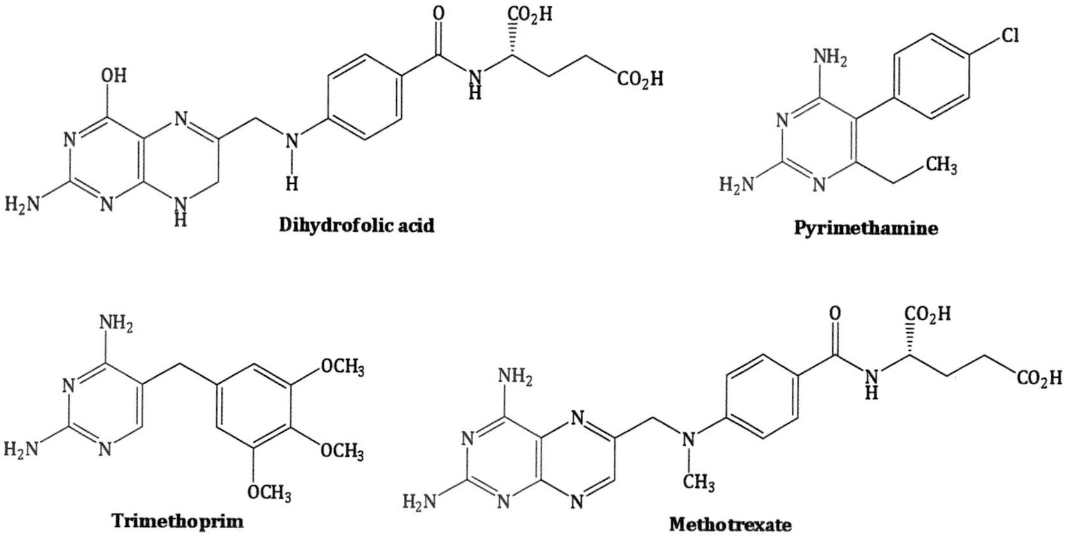

Figure 3.2.3 *The structures of dihydrofolic acid, trimethoprim, pyrimethamine, and methotrexate*

Table 3.2.1   *Selective inhibition of dihydrofolate reductase from varying sources by trimethoprim, pyrimethamine, and methotrexate (Burchall and Hitchings, 1965)*

Inhibitor	Species source of DHFR					
	Human liver	Rat liver	Rabbit liver	*Escherichia coli*	*Staphylococcus aureus*	*Plasmodium berghei*[a]
	Concentration required for 50% inhibition of activity ($IC_{50} \times 10^8$ M)					
Trimethoprim	30 000	26 000	37 000	0.5	0.5	7.0
Pyrimethamine	180	70	50	2500	300	0.05
Methotrexate	9	9	6	0.06	0.1	0.07

[a] Ferone *et al.* (1969).

A detailed analysis of the structure–activity relationship for DHFR inhibitors was made by Hansch[13] and coworkers in the 1980s (Blaney *et al.*, 1984), when more than 1500 DHFR inhibitors were analysed for their structural and electronic properties, to arrive at a number of conclusions:

- Substitution of the 4-hydroxy group in dihydrofolate (corresponding to the 4-oxo group in folate) by the 4-NH_2 group in 2,4-diaminopteridines and 2,4-diaminopyrimidines is the most important feature for the switch from agonist to antagonist activity.
- Binding of NADPH enhances inhibitor binding to the DHFR active site, and increased hydrophobicity in the inhibitor increases binding to the bacterial enzyme.
- Bacterial DHFR enzymes are the most inhibited by 5-benzyl-2,4-diaminopyrimidines and the greatest selectivity is achieved by the 3',4',5'-trisubstituted derivatives.
- In contrast to the mammalian enzymes, the size of the substituents on the benzyl ring is crucial for bacterial DHFR inhibition (Blaney *et al.*, 1984).

Since then, there has been a lot of effort directed at delineating the requirements for potent bacterial inhibition, while retaining the excellent selectivity displayed by trimethoprim. You would expect that the similarity of the active sites of DHFR across the species would make it difficult to design selectivity for one over another, but X-ray crystallography has shown that there are two possible binding sites for the trimethoxybenzyl in the active-site cleft: a lower, deeper cleft, and another sited higher. In non-bacterial DHFR, both of these active-site clefts are slightly wider, by a mere 1.5–2.0 Å, than that of the bacterial enzyme, possibly due to a three-amino-acid insertion found in all vertebrate DHFR enzymes near the active site (Matthews *et al.*, 1985b). These differences mean that trimethoprim cannot bind into both lower and upper binding sites, occupying the lower site alone, and lead to a conformational change in trimethoprim when it binds to mammalian DHFR, causing rotation of the trimethoxybenzyl group into a less favourable conformation for trimethoprim, as well as a close contact with part of the active site, disturbing one of the pyrimidine hydrogen bonds (4-NH_2 to Ile-94) (Matthews *et al.*, 1985b; Schweitzer *et al.*, 1990). In combination, these differences in the binding of trimethoprim to bacterial and mammalian DHFR lead to the selectivity observed.

Considering all of the information above, isn't it remarkable that Hitchings and his coworkers achieved such impressive selectivity for the DHFR of different species with little detailed knowledge of the

---

[13] Corwin Hansch is renowned for developing mathematical methodology for deriving quantitative structure–activity relationships that take into account factors such as hydrophobicity, steric size, and electronic properties to explain and predict the biological activity of a series of molecules. The approach, Hansch analysis, bears his name and is an important method available to medicinal chemists for rational drug design.

structural requirements? Indeed, much effort since then (reviewed in Then, 2004; da Cunha *et al.*, 2005; Kompis *et al.*, 2005; Chan and Anderson, 2006; Hawser *et al.*, 2006) has focussed on the design of new selective DHFR inhibitors, with only a few notable successes, despite the addition of X-ray crystal structure information (Baccanari and Kuyper, 1993; Beierlein *et al.*, 2008), molecular modelling (Roth, 1986; Ohemeng and Roth, 1991), and the use of NMR to probe the mechanism and structural requirements of binding (Cocco *et al.*, 1983; Roberts 1991; Kovalevskaya *et al.*, 2005, 2007). A considerable amount of research continues into improving DHFR activity and selectivity across species, such as *Bacillus anthrax* (Beierlein *et al.*, 2008; Bourne *et al.*, 2009), *Pneumocytis carinii*, *Toxoplasma gondii*, and *Mycobacterium avium* (these last three are of particular clinical relevance as opportunistic infections in AIDS patients) (Forsch *et al.*, 2004), and in overcoming resistance (Dale *et al.*, 1997), which is discussed in more detail in Subsection 3.2.5.

Another discovery of great clinical relevance was made in 1947, when the synergy between a pterin (based on the structure of folic acid) and sulfonamides was observed (Daniel and Norris, 1947); others followed with investigations into the use of early diaminopyrimidines with sulfonamides (Elion *et al.*, 1954; Hitchings and Burchall, 1965), and eventually the antibacterial effects of trimethoprim were combined with a range of different sulfonamides in order to try to identify the optimal combination and doses (Bushby and Hitchings, 1968). Although the synergy is generally believed to be due to sulfonamide and trimethoprim inhibition of the same pathway at two different points, it has also been suggested that both sulfonamide and trimethoprim can bind together to the active site of DHFR when present in sub-therapeutic concentrations, raising the possibility that at least some sulfonamides can act at two stages of the folate pathway (Lacey, 1982). The combined use of trimethoprim and sulfonamides has been under discussion almost since it was first used, as there is some doubt that the synergy is achieved routinely in clinical cases (Watson *et al.*, 1988) and concern that the side effects may outweigh the benefits (Brumfitt and Hamilton-Miller, 1994; Howe and Spencer, 1996).

How did clinicians know which of the many sulfonamides to combine with trimethoprim? A great deal of effort went into evaluating the right combination, as the components must have similar clinical pharmacokinetics in order to achieve the required therapeutic concentrations simultaneously. Several factors were investigated, including the penetration of the drugs into the active site, their volumes of distribution, and their elimination half-lives (Ortengren *et al.*, 1979; Bernstein, 1982). The two sulfonamides with the closest clinical pharmacokinetics to those of trimethoprim were found to be sulfadiazine and sulfamethoxazole (Vree *et al.*, 1978; Patel and Welling, 1980; Watson *et al.*, 1988). Both have been used in combination therapy with trimethoprim, and the combinations studied for their clinical pharmacokinetics and outcomes across a range of infections and complications (Bergan *et al.*, 1986; Chin *et al.*, 1995; Klepser *et al.*, 1996; Pokrajac *et al.*, 1998; Nyleen, *et al.*, 1999; Mistri *et al.*, 2010). Although both possible combinations are still available in some parts of the world, as co-trimazine (trimethoprim with sulfadiazine) and co-trimoxazole (trimethoprim with sulfamethoxazole), the former is not licensed in all countries and the latter is the more commonly used. Despite the fact that considerable effort was directed at matching the pharmacokinetics of the two components, it is necessary to administer sulfamethoxazole and trimethoprim in the ratio 5 : 1 in order to achieve appropriate concentrations *in vivo*. The differences in the rates of absorption and their abilities to distribute throughout various body compartments have also led to controversy about the combined use of sulfamethoxazole and trimethoprim (O'Grady, 1975; Watson *et al.*, 1988).

The observation of synergy was of great clinical importance, because it allowed both agents to be used at lower doses than is required in monotherapy, and the combined effect is that of a bactericidal agent against many different pathogenic bacteria, rather than the bacteriostatic effect of each agent separately. Another benefit is the lower risk of resistance developing due to a mutation in the enzyme targets, as the probability of a resistance-causing mutation arising in both enzymes (DHPS and DHFR) concurrently is very small (Bushby, 1975; Lacey, 1982).

3.2.5   Bacterial resistance

Within a few years of the clinical use of trimethoprim, resistance was reported across a range of bacteria (Datta and Hedges, 1972; Fleming *et al.*, 1972; May and Davies, 1972), although for many years it was restricted to Gram negative species. The incidence of resistance has continued to increase but has not yet resulted in the redundancy of trimethoprim as an antibacterial agent (Amyes and Towner, 1990; Huovinen, 2001; Roberts, 2002).

Low-level resistance to trimethoprim has been observed, and is linked to spontaneous chromosomal mutations, leading to decreased uptake or alterations to the target enzyme, DHFR (Goldstein, 1977). It is the high-level resistance, however, that causes clinical concerns, leading to increased resistance in the majority of regions and a disturbingly high incidence of resistant bacteria in some countries (Murray *et al.*, 1985; Mathai *et al.*, 2004; Guneysel *et al.*, 2009). In Gram negative bacteria, high-level resistance to trimethoprim is primarily due to the transmission of plasmids encoding a resistant form of DHFR (Amyes and Towner, 1990). In 1987, nine different plasmid-encoded trimethoprim-resistant DHFR enzymes and/or associated genes were known (Huovinen, 1987); by 2010, 30 different trimethoprim-resistant forms of the gene encoding DHFR (named *dfr*) had been described, most of which were associated with integrons (Brolund *et al.*, 2010). In Gram positive bacteria, resistance is largely due to chromosomal mutations, which lead to point mutations of amino acids in the DHFR enzyme; plasmid-borne *dfr* genes can also be observed in these species and generally give rise to similar resistant enzymes with point mutations (Huovinen, 1987; Bourne *et al.*, 2009).

The effect of mutation in a *dfr* gene can be seen, for example, in *Staphylococcus aureus*, in which Phe-98 is replaced by Tyr in the enzyme active site of S1 DHFR, a trimethoprim-resistant enzyme encoded by plasmid *dfrA*, located on transposon *Tn4003*, leading to high-level resistance to trimethoprim. The same mutation is seen in the chromosomally encoded DHFR and leads to intermediate-level resistance (Dale *et al.*, 1997). This same mutation was observed across a range of trimethoprim-resistant *S. aureus* clinical isolates, sometimes alongside other mutations in which His-30 was replaced by Asn (or His-149 was replaced by Arg), and may cause the loss of a hydrogen bond between trimethoprim and a conserved water molecule in the active site (Frey *et al.*, 2010). Mutations due to a variety of active-site amino acids have been observed in trimethoprim-resistant bacteria (DeGroot *et al.*, 1991; Adrian and Klugman, 1997; Coque *et al.*, 1999; Woodford, 2005).

3.2.6   Clinical applications

3.2.6.1   Spectrum of activity

Trimethoprim has a range of activity against UTI-causing pathogens, such as Gram negative bacteria (including *E. coli*, *Proteus mirabilis*, and *Klebsiella pneumoniae*) and some Gram positive bacteria, including *Staphylococcus saprophyticus*. Unfortunately, trimethoprim shows poor activity against *Pseudomonas aeruginosa* and anaerobic bacteria (Rosenblatt and Stewart, 1974; Then and Angehrn, 1979; Köhler *et al.*, 1996).

3.2.6.2   Urinary tract infections (UTIs)

The most common therapeutic use of trimethoprim is in the treatment of lower urinary tract infections (LUTIs) in women. LUTIs are very common, with symptoms including dysuria (pain when urinating) and polyuria (passage of large volumes of urine). In the case of upper urinary tract infections (UUTIs), symptoms are often suggestive of pyelonephritis (bacterial infection of the kidney) and can include fever and back pain. UUTIs are considered more serious than LUTIs, as they can be accompanied by bacteraemia, which is a potentially life-threatening condition.

E. coli is estimated to cause around 80% of UTIs, with other organisms such as *S. saprophyticus* and *P. mirabilis* also playing a role in some cases. Generally, empirical[14] antibacterial treatment is used to treat uncomplicated UTIs; for example, to treat a LUTI in a non-pregnant woman, a 3-day course of trimethoprim is recommended. In cases where the UTI is resistant to trimethoprim (or the symptoms suggest an UUTI), a quinolone (such as ciprofloxacin) is sometimes used.

Three-day courses of antibiotic therapy are recommended because the evidence presented in a Cochrane systematic review suggests that this duration of antibiotic therapy is similar to 5–10 days in achieving symptomatic cures for uncomplicated UTIs in women (Milo *et al.*, 2005). In addition, the 3-day course of antibiotic therapy offers the advantage of causing fewer adverse effects and decreases antibiotic exposure, thereby minimising the risk of developing resistance.

If a woman presents with symptoms of a LUTI and pregnancy is suspected, a pregnancy test should be done, as trimethoprim is not recommended due to its possible teratogenic risk (because of its antifolate effects). Although several studies have failed to demonstrate the increased risk of foetal abnormalities associated with trimethoprim use (Brumfitt and Pursell, 1972, 1973), others have shown that trimethoprim can cause birth defects in animals (Briggs *et al.*, 2008). Trimethoprim should thus be avoided in pregnancy, unless the potential therapeutic benefit outweighs the possible risks.[15] If pregnancy is confirmed, and the patient has a LUTI, an alternative agent, such as amoxicillin, should be prescribed.

The use of antibacterial agents in the treatment of UTIs highlights an interesting point concerning the prescribing of antibacterial drugs: if a patient presents with signs of an acute UTI, some doctors will prescribe trimethoprim without taking a urine sample. Some doctors, however, will take a urine sample for bacterial culture and sensitivity testing prior to starting trimethoprim therapy. If the culture from the urine sample later confirms the presence of a pathogen that is resistant to trimethoprim, the patient will be quickly contacted and their treatment will be altered accordingly. This latter method is preferred, as the threat of antibacterial resistance will be decreased by minimising the inappropriate prescribing of antibacterial drugs.

In cases where a patient is suffering from recurrent UTIs (typically more than three episodes in 12 months), the use of prophylactic trimethoprim should be considered to reduce the number of recurrences. When used for UTI prophylaxis, trimethoprim is typically given as a single dose at night and is taken at a lower dose than that used to treat acute infections.

3.2.6.3   Miscellaneous

Trimethoprim can also be used to treat acne vulgaris. Generally, systemic antibiotics are used in the management of acne when topical treatment has either been ineffective, not tolerated, or the acne is predominately inflammatory. Trimethoprim, at a dose of 300 mg twice daily (which is higher than that used for treatment of acute UTIs), is effective in the management of acne vulgaris (Bottomley and Cunliffe, 1993). However, as the long-term use of trimethoprim can be associated with severe adverse effects (see Subsection 3.2.7), it is only regarded as a third-line treatment. Consequently, for this indication, trimethoprim should only be initiated by specialists for the management of acne that is resistant to other antibacterial agents.

Trimethoprim also has a role in the treatment of acute and chronic prostatitis (inflammation of the prostate gland). Acute prostatitis is an uncommon complication of UTIs; it is considered a medical emergency and is managed using empirical antibiotic therapy. In contrast, however, chronic prostatitis often presents with a

---

[14] In this case, 'empirical treatment' means that the treatment is based on clinical signs and symptoms which are unconfirmed by urine culture.

[15] If trimethoprim therapy is considered essential during pregnancy, the patient should take a folic acid supplement – usually 400 micrograms per day, but in patients who are at high risk of neural tube defects, 5 mg per day is recommended.

history of recurrent UTIs and, interestingly, it is thought that the bacteria that cause the UTIs (such as *E. coli*) also cause the chronic prostatitis. The quinolones are generally used first-line to treat both of these disorders, but if this is not suitable (for example, if the patient is taking an interacting medication, such as theophylline), trimethoprim may be used. The quinolones are preferred over trimethoprim as they have a broader spectrum of activity against the common urinary pathogens, but if trimethoprim is indicated for either acute or chronic prostatitis it should be taken twice daily for 28 days.

3.2.7   Adverse drug reactions

Until recently trimethoprim was considered to be a relatively safe drug – with a similar structure to the potassium-sparing diuretic, amiloride, it is rapidly excreted and has been reported to cause hyperkalaemia in some patients (this can be more significant if a patient is already taking a drug which elevates potassium levels, such as, say, an ACE inhibitor). Relatively recent publications have, however, demonstrated that trimethoprim is associated with skin reactions, toxic epidermal necrolysis, and neutropenic reactions (Das *et al.*, 1988; Hawkins *et al.*, 1993).

3.2.8   Drug interactions

Trimethoprim is primarily excreted unchanged in the renal tubule; however, approximately 20% of the drug is subject to drug metabolism by the CYP 450 system (Gleckman *et al.*, 1981), so the half-life of trimethoprim can be extended by up to 50% in patients with a degree of hepatic failure (Rieder and Schwartz, 1975).

Trimethoprim has also been shown to inhibit CYP 2C8 (Wen *et al.*, 2002), a recently identified isoenzyme, which it is expected has both arachidonic acid and retinoic acid as natural substrates. Other drug substrates known to be metabolised, wholly or in part, by CYP 2C8 are zopiclone, the 'glitazones', repaglinide, amodiaquine, amiodarone, chloroquine, dapsone, loperamide, retinoic acid derivatives, and paclitaxel (Lebovitz, 2000; Ohyama *et al.*, 2000; Cresteil *et al.*, 2002; Niemi *et al.*, 2003a; Gil and Gil Berglund, 2007; Scheen, 2007). As trimethoprim inhibits CYP 2C8, one would expect to see a significant number of drug interactions, but so far, clinically at least, this is not the case.

3.2.8.1   Pioglitazone

An *in vitro* study predicted the possible interaction of trimethoprim with pioglitazone via inhibition of CYP 2C8*3 (Tornio *et al.*, 2008), while the plasma levels of the antidiabetic drug repaglinide were raised by the concomitant use of trimethoprim (Niemi *et al.*, 2003b).

3.2.8.2   Methotrexate

The concomitant use of either co-trimoxazole or trimethoprim with methotrexate has been associated with pancytopaenia (Groenendal and Rampen, 1990; Steur and Gumpel, 1998).

3.2.8.3   Lamivudine

The co-administration of trimethoprim increases the bioavailability of lamivudine by 40%, raising the possibility that trimethoprim could be exploited in HIV treatment regimens (Hudson and Nash, 1996).

3.2.8.4 Memantine

A clinical case report indicated that the co-administration of memantine (used in the treatment of Alzheimer's disease) and trimethoprim caused myoclonus and exacerbated delirium in a 74-year-old patient (Moellentin *et al.*, 2008). The structure of memantine resembles that of amantidine, which has also been shown to interact with trimethoprim (Speeg *et al.*, 1989).

3.2.8.5 Digoxin

One report has demonstrated an up to 22% increase in serum digoxin levels when co-administered with trimethoprim (Petersen *et al.*, 1985), and it has been suggested that serum digoxin levels are raised due to the decreased renal excretion of the drug and not due to extrarenal drug clearance.

3.2.9 Recent developments

The exploitation of selectivity towards bacterial DHFR enzymes is still a target for rational drug design (Hawser *et al.*, 2006) and it is hoped that new DHFR inhibitors, such as those shown in Figure 3.2.4 and

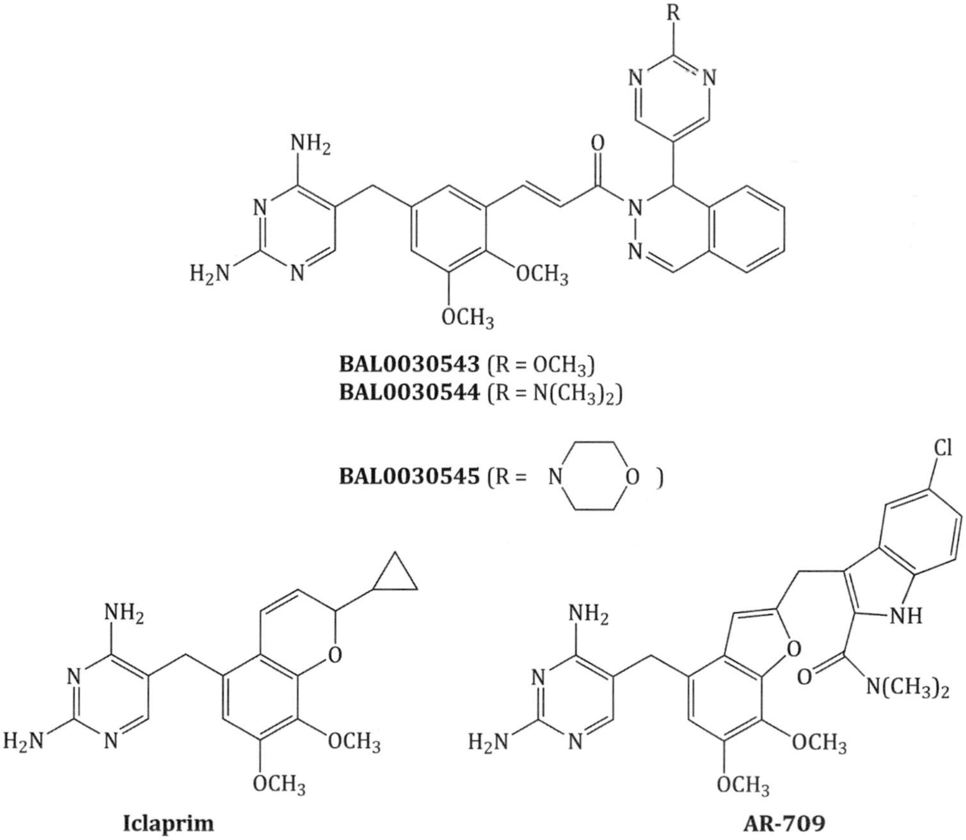

BAL0030543 (R = OCH$_3$)
BAL0030544 (R = N(CH$_3$)$_2$)

BAL0030545 (R = NO)

Iclaprim

AR-709

Figure 3.2.4 *Bacterial DHFR inhibitors in development*

Table 3.2.2 *Status of development of the bacterial DHFR inhibitors*

DHFR Inhibitor	Status	Indication	Reference
BAL0030543 BAL0030544 BAL0030545	In early preclinical development	Improved potency against Gram positive bacteria (compared to trimethoprim)	Bowker *et al.* (2009)
Iclaprim	Recently completed phase III clinical trials	Treatment of complicated skin and skin-structure infections	Krievins *et al.* (2009)
AR-709	Initial studies demonstrate *in vitro* activity against *Streptococcus pneumoniae*	Treatment of community-acquired respiratory tract infections	Jansen *et al.* (2008)

Table 3.2.2, will exhibit improved bactericidal activity, extend the spectrum of antibacterial activity, and overcome the problems associated with trimethoprim resistance.

References

P. V. Adrian and K. P. Klugman, *Antimicrob. Agents Chemother.*, 1997, **41**, 2406–2413.
C. J. Allegra, J. A. Kovacs, J. C. Drake, J. C. Swan, B. A. Chabner, and H. Masur, *J. Exp. Med.*, 1987, **165**, 926–931.
S. G. B. Amyes and K. J. Towner (Eds.), J. Med. Microbiol., 1990, **31**, 1–19.
D. P. Baccanari and L. F. Kuyper, *J. Chemother.*, 1993, **5**, 393–399.
J. Beierlein, K. Frey, D. Bolstad, P. Pelphrey, T. Joska, A. Smith, N. Priestley, D. Wright, and A. Anderson, *J. Med. Chem.*, 2008, **51**, 7532–7540.
T. Bergan, B. Ortengren, and D. Westerlund, *Clin. Pharmacokin.*, 1986, **11**, 372–386.
L. S. Bernstein, *Rev. Infect. Dis.*, 1982, **4**, 411–418.
R. L. Blakley and B. M. McDougall, *J. Biol. Chem.*, 1961, **236**, 1163–1167.
J. M. Blaney, C. Hansch, C. Silipo, and A. Vitorria, *Chem. Rev.*, 1984, **84**, 333–407.
J. T. Bolin, D. J. Filman, D. A. Matthews, R. C. Hamlin, and J. Kraut, *J. Biol. Chem.*, 1982, **257**, 13 650–13 662.
W. W. Bottomley and W. J. Cunliffe, *Dermatology*, 1993, **187**, 193–196.
C. R. Bourne, R. A. Bunce, P. C. Bourne, K. D. Berlin, E. W. Barrow, and W. W. Barrow, *Antimicrob. Agents Chemother.*, 2009. **53**, 3065–3073.
K. E. Bowker, P. Caspers, B. Gaucher and A. P. MacGowan, *Antimicrob Agents Chemother.*, 2009, **53**, 4949–4952.
G. G. Briggs, R. K. Freeman, and S. J. Yaffe, Drugs in Pregnancy and Lactation, 8th Edn, Lippincott Williams & Wilkins, Philadelphia, PA, 2008, 1866–1870.
R. N. Brogden, A. A. Carmine, R. C. Heel, T. M. Speight, and G.S. Avery, *Drugs*, 1982, **23**, 405–430.
A. Brolund, M. Sundqvist, G. Kahlmeter, and M. Grape, *PLoS One*, 2010, **5**, e9233.
W. Brumfitt and J. M. Hamilton-Miller, *J. Chemother.*, 1994, **6**, 3–11.
W. Brumfitt and R. Pursell, *Brit. Med. J.*, 1972, **2**, 673–676.
W. Brumfitt and R. Pursell, *J. Infect. Dis.*, 1973, **128**, S657–S665.
J. J. Burchall and G. H. Hitchings, *Mol. Pharmacol.*, 1965, **1**, 126–136.
S. R. M. Bushby, *CMA Journal*, 1975, **112**, 63S–66S.
S. R. M. Bushby and G. H. Hitchings, *Br. J. Pharmac. Chemother.*, 1968, **33**, 72–90.
J. N. Champness, D. K. Stammers, and C. R. Beddell, *FEBS Letters*, 1986, **199**, 61–67.
D. C. Chan and A. C. Anderson, *Curr. Med. Chem.*, 2006, **13**, 377–398.
J. H. Chan and B. Roth, *J. Med. Chem.*, 1991, **34**, 550–555.
T. W. F. Chin, A. Vandenbrouke, and I. W. Fong, *Antimicrob. Agents Chemother.*, 1995, **39**, 28–33.
L. Cocco, B. Roth, C. Temple, Jr, J. A. Montgomery, R. E. London, and R. L. Blakley, *Arch. Biochem. Biophys.*, 1983. **226**, 567–577.

R. G. Cooper and M. Wald, *Med. J. Aust.*, 1964, **18**, 93–96.

T. M. Coque, K. V. Singh, G. M. Weinstock, and B. E. Murray, *Antimicrob. Agents Chemother.*, 1999, **43**, 141–147.

T. Cresteil, B. Monsarrat, J. Dubois, M. Sonnier, P. Alvinerie, and F. Gueritte, *Drug Metab. Dispos.*, 2002, **30**, 438–445.

E. F. F. da Cunha, T. C. Ramalho, E. R. Maia, and R. B. de Alencastro, *Expert Opin. Ther. Patents*, 2005, **15**, 967–986.

G. E. Dale, C. Broger, A. D'Arcy, P. G. Hartman, R. DeHoogt, S. A. Jolidon, I. Kompis, A. M. Labhardt, H. Langen, H. Locher, M. G. P. Page, D. Stüber, R. L. Then, B. Wipf, and C. Oefner, *J. Mol. Biol.*, 1997, **226**, 23–30.

L. J. Daniel and L. C. Norris, *J. Biol. Chem.*, 1947, **170**, 747–756.

J. H. Darrell, L. P. Garrod, and P. M. Waterworth, *J. Clin Path.*, 1968, **21**, 202–209.

G. Das, M. J. Bailey, and J. E. Wickham, *Br. Med. J.*, 1988, **296**, 1604–1605.

N. Datta and R.W. Hedges, *J. Gen. Microbiol.*, 1972, **72**, 349–355.

R. DeGroot, D. O. Chaffin, M. Kuehn, and A.L. Smith, *Biochem. J.*, 1991, **274**, 657–662.

G. B. Elion, *Science*, 1989, **244**, 41–47.

G. B. Elion, S. Singer, and G. H. Hitchings, *J. Biol. Chem.*, 1954, **208**, 477–488.

E. A. Falco, S. DuBreuil, and G. H Hitchings, *J. Amer. Chem. Soc*, 1951, **73**, 3758–3762.

R. Ferone, J. J. Burchall, and G. H. Hitchings, *Mol. Pharmacol.*, 1969, **5**, 49–59.

M. P. Fleming, N. Datta, and R. N. Grüneberg, *Br. Med. J.*, 1972, **1**, 726–728.

R. A. Forsch, S. F. Queener, and A. Rosowsky, *Bioorg. Med. Chem. Lett.*, 2004, **14**, 1811–1815.

K. M. Frey, M. N. Lombardo, D. L. Wright, and A. C. Anderson, *J. Struct. Biol.*, 2010, **170**, 93–97.

W. T. Friesen, Y. A. Hekster, and T. B. Vree, *Drug Intell. Clin. Pharm.*, 1981, **15**, 325–330.

J. P. Gil and E. Gil Berglund, *Pharmacogenomics*, 2007, **8**, 187–198.

R. Gleckman, N. Blagg, and D. W. Joubert, *Pharmacotherapy*, 1981, **1**, 14–20.

F. W. Goldstein, *Bull. Inst. Pasteur*, 1977, **75**, 109–139.

H. Groenendal and F. H. J. Rampen, *Clin. Exp. Dermatol.*, 1990, **15**, 358–360.

O. Guneysel, O. Onur, M. Erdede, and A. Denizbasi, *J. Emerg. Med.*, 2009, **36**, 338–341.

T. Hawkins, J. M. Carter, K. R Romeril, S. R. Jackson, and G. J. Green, *N. Z. Med. J.*, 1993, **106**, 251–252.

S. Hawser, S. Lociuro, and K. Islam, *Biochem. Pharmacol.*, 2006, **71**, 941–948.

G. H. Hitchings, *Cancer Res.*, 1969, **29**, 1895–1903.

G. H. Hitchings, *In Vitro Cell. Dev. Biol.*, 1988, **25**, 303–310.

G. H. Hitchings and J. J. Burchall, 'Inhibition of folate biosynthesis and function as a basis for chemotherapy', in F.F. Nord (Ed.), *Advances in Enzymology*, vol. **27**, John Wiley and Sons, Inc., New York, NY, 1965, pp. 417–468.

G. H. Hitchings and B. Roth,US Patent 2909522, 20 October 1959.

K. Hoppu and P. Arjomaa, *Chemother.*, 1984, **30**, 283–287.

R. A. Howe and R. C. Spencer, *Drug Saf.*, 1996, **14**, 213–218.

M. Hudson and C. Nash, 1996, *AMA*, 1996 **276**, 1140.

P. Huovinen, *Antimicrob. Agents Chemother.*, 1987, **31**, 1451–1456.

P. Huovinen, *Clin. Infect. Dis.*, 2001, **32**, 1608–1614.

J. Jaeger, K. Burri, S. Greiveldinger-Poenaru, and J. Hoffner,US Patent 7893262, 22 February 2011.

W. T. M. Jansen, A. Verel, J. Verhoef, and D. Milatovic, *Antimicrob. Agents Chemother.*, 2008, **52**, 1182–1183.

J. V. Johnson, B. S. Rauckman, D. P. Baccanari, and B. Roth, *J. Med. Chem.*, 1989, **32**, 1942–1949.

A. Kasanen, M. Anttila, R. Elfving, P. Kahela, H. Saarimaa, H. Sundquist, and R. Tikkanen, *Ann. Clin. Resis.*, 1978, **10**, 5–39.

M. E. Klepser, Z. Zhu, D. P. Nicolau, M. A. Belliveau, J. W. Ross, L. Broisman, R. Quintiliani, and C. H. Nightingale, *Pharmacother.*, 1996, **16**, 656–662.

T. Köhler, M. Kok, M. Michea-Hamzehpour, P. Plesiat, N. Gotoh, T. Nishino, L. K. Curty, and J. C. Pechere, *Antimicrob. Agents Chemother.*, 1996, **40**, 2288–2290.

I. M. Kompis, K. Islam, and R. L Then, *Chem. Rev.*, 2005, **105**, 593–620.

N. V. Kovalevskaya, Y. D. Smurnyy, V. I. Polshakov, B. Birdsall, A. F. Bradbury, T. Frenkiel, and J. Feeney, *J. Biomol. NMR*, 2005, **33**, 69–72.

N. V. Kovalevskaya, E. D. Smurnyi, B. Birdsall, J. Feeney, and V.I. Polshakov, *Pharm. Chem. J.*, 2007, **41**, 350–353.

D. Krievins, R. Brandt, S. Hawser, P. Hadvary, and K. Islam, *Antimicrob. Agents Chemother.*, 2009, **53**, 2834–2840.

R. W. Lacey, *J. Med. Microbiol.*, 1982, **15**, 402–427.

H. E. Lebovitz, *Diabetes Metab. Res. Rev.*, 2000, **18**, 23–29.

J. A. Libecco and K. R. Powell, *Pediatr. Rev.*, 2004, **25**, 375–380.

D. C. Martin and J. D. Arnold, *J. Clin. Pharmacol. J. New Drugs*, 1967, **7**, 336–341.

J. R. May and J. Davies, *Br. Med. J.*, 1972, **3**, 376–377.

E. Mathai, M. Grape, and G. Kronvall, *APIMS*, 2004, **112**, 159–164.

D. A. Matthews, J. T. Bolin, J. M. Burridge, D. J. Filman, K. W. Volz, B. T. Kaufman, C. R. Beddell, J. N. Champness, D. K. Stammers, and J. Kraut, *J. Biol. Chem.*, 1985a, **260**, 381–391.

D. A. Matthews, J. T. Bolin, J. M. Burridge, D. J. Filman, K. W. Volz, and J. Kraut, *J. Biol. Chem.*, 1985b, **260**, 392–399.

D. A. Matthews, S. L. Smith, D. P. Baccanari, J. J. Burchall, S. J. Oatley, and J. Kraut, *Biochemistry*, 1986, **25**, 4194–4204.

G. Milo, E. Katchman, M. Paul, T. Christiaens, A. Baerheim, and L. Leibovici, *Cochrane Database of Systematic Reviews*, 2005, Issue 2.

H. N. Mistri, A. G. Jangid, A. Pudage, A. Shah, and P. S. Shrivastav, *Microchem. J.*, 2010, **94**, 130–136.

D. Moellentin, C. Picone, and E. Leadbetter, *Ann. Pharmacother.*, 2008, **42**, 433–437.

B. E. Murray, T. Alvarado, K.-H. Kim, M. Vorachit, P. Jayanetrap, M. M. Levine, I. Prenzel, M. Fling, L. Elwell, G. H. McCracken, G. Madrigal, C. Odio, and L.R. Trabulsi, *J. Infect. Dis.*, 1985, **152**, 1107–1113.

M. Niemi, J. T. Backman, M. Neuvonen, and P. J. Neuvonen, *Diabetologia.*, 2003a, **46**, 347–351.

M. Niemi, L. I. Kajosaari, M. Neuvonen, J. T. Backman, and P. J. Neuvonen, *B. J. Clin. Pharmacol.*, 2003b, **57**, 441–447.

C. Oefner, A. D'Arcy, and F.K. Winkler, *Eur. J. Biochem.*, 1988, **174**, 377–385.

F. O'Grady, *CMA J.*, 1975, **112**, 5S–7S.

K. A. Ohemeng and B. Roth, *J. Med. Chem.*, 1991, **34**, 1383–1394.

K. Ohyama, M. Nakajima, S. Nakamura, N. Shimada, H. Yamazaki, and T. Yokoi, *Drug Metab. Dispos.*, 2000, **28**, 1303–1310.

B. Ortengren, H. Fellner, and T. Bergan, *Infection*, 1979, **7**, S367–S370.

R. B. Patel and P. G. Welling, *Clin. Pharmacokin.*, 1980, **5**, 405–423.

J. M. Peters and D. M. Greenberg, *Nature*, 1958, **181**, 1669–1670.

P. Petersen, J. Kastrup, R. Bartram, and J. Mølholm Hansen, *Acta Med. Scand.*, 1985, **217**, 423–427.

M. Poe and W. V. Ruyle, US Patent 4258045, 24 March 1981.

M. Pokrajac, B. Miljkovic, B. Brzakovic, and A. Galetin, *Pharmazie*, 1998, **53**, 470–472.

E. P. Quinlivan, J. McPartlin, D. G. Weir, and J. Scott, *FASEB J.*, 2000, **14**, 2519–2524.

B. S. Rauckman, M. Y. Tidwell, J. V. Johnson, and B. Roth, *J. Med. Chem.*, 1989, **32**, 1927–1935.

E. Raviña, The Evolution of Drug Discovery: From Traditional Medicines to Modern Drugs, Wiley-VCH, 2011, pp. 362–393.

V. J. Rieder and D. E. Schwartz, *Arzneimittelforschung*, 1975, **25**, 656–666.

G. C. Roberts, *Ciba Found. Symp.*, 1991, **158**, 169–182.

M. C. Roberts, *Mol. Biotechnol.*, 2002, **20**, 261–283.

J. E. Rosenblatt and P. R. Stewart, *Antimicrob. Agents Chemother.*, 1974, **6**, 93–97.

B. Roth, *Fed. Proc.*, 1986, **45**, 2765–2772.

B. Roth and E. Aig, *J. Med. Chem.*, 1987, **30**, 1998–2004.

B. Roth, E. A. Falco, G. H. Hitchings, and S. R. Bushby, *J. Med. Pharm. Chem.*, 1962, **5**, 1103–1123.

B. Roth, B. S. Rauckman, R. Ferone, D. P. Baccanari, J. N. Champness, and R. M. Hyde, *J. Med. Chem.*, 1987, **30**, 348–356.

B. Roth, D. P. Baccanari, C. W. Sigel, J. P. Hubbell, J. Eaddy, J. C. Kao, M. E. Grace, and B. S. Rauckman, *J. Med. Chem.*, 1988, **31**, 122–129.

G. W. Rylance, R. H. George, D. E. Healing, and D. G. V. Roberts, *Arch. Dis. Childhood*, 1985, **60**, 29–33.

A. J. Scheen, *Clin. Pharmacokinet.*, 2007, **46**, 1–12.

B. I. Schweitzer, A. P. Dicker, and J. R. Bertino, *FASEB J.*, 1990, **4**, 2441–2452.

C. D. Selassie, W.-X. Gan, L. S. Kallander, and T. E. Klein, *J. Med. Chem.*, 1998, **41**, 4261–4272.

C. W. Sigel, M. E. Grace, and C. A. Nichol, *J. Infect. Dis.*, 1973, **128**, S580–S583.

K. V. Speeg, J. A. Leighton, and A. L. Maldonado, *Am. J. Med. Sci.*, 1989, **298**, 410–412.

P. Stenbuck, R. Baltzly, and H. M. Hood, *J. Org. Chem.*, 1963, **28**, 1983–1988.

A. Steur and J. M. Gumpel, *Br. J. Rheuamtol.*, 1998, **37**, 105–106.

R. L. Then, *J. Chemother.*, 2004, **16**, 3–12.

R. L. Then and P. Angehrn, *Antimicrob. Agents Chemother.*, 1979, **15**, 1–6.

A. Tornio, M. Niemi, P. J. Neuvonen, and J. T. Backman, *Drug Metabol. Dispos.*, 2008, **36**, 73–80.

R. Vardanyan and V. J. Hruby, Synthesis of Essential Drugs, Elsevier BV, 2006, pp. 510–511.

T. B. Vree, Y. A. Hekster, A. M. Baars, J. E. Damsma, and E. van der Kleijn, *Clin. Pharmacokin.*, 1978, **3**, 319–329.

I. D. Watson, M. J. Stewart, and D. J. Platt, *Clin. Pharmacokin.*, 1988, **15**, 133–164.

X. Wen, J-S. Wang, J. T. Backman, J. Laitlia, and P. J. Neuvonen, *Drug Metab. Dispos.*, 2002, **30**, 631–635.

N. Woodford, *Clin. Microbiol. Infect.*, 2005, **11**, 2–21.

Questions

(1) Explain why the combination of sulfamethoxazole and trimethoprim (commonly known as co-trimoxazole) shows synergistic activity.

(2) Which group(s) of patients should be prescribed co-trimoxazole for prophylaxis of pneumocystis pneumonia?

(3) A female patient presents with symptoms of dysuria and polyuria, which suggest a UTI. However, the patient also complains of fever and back pain. Briefly explain which antibiotic you would prescribe for this patient and why.

(4) Refer back to Scheme 3.2.2 in Subsection 3.2.2. By working through the mechanism of the reaction of guanidine with both the conjugated cinnamonitrile **7** and the tautomeric enol ether **8**, confirm that the same product, trimethoprim, is formed.

(5) What evidence would you look for in the UV spectrum of compound **7** (Scheme 3.2.2) to support nitrile conjugation with the aromatic ring?

(6) Draw the structures of the following metabolites of trimethoprim:
- $4'$-OH
- N^1-oxide
- $C\alpha$-OH-trimethoprim

(7) Draw the structures of all the resonance forms of the **anion** of sulfadiazine.

Sulfadiazine

(8) Which of the following statements about sulfonamide antibacterials is **incorrect**?
- (A) Sulfanilamide is only active *in vivo*.
- (B) They are antimetabolites and mimic the essential metabolite PABA.
- (C) They can be used in combination with trimethoprim to treat *Pneumocystis carinii* (*jiroveci*) pneumonia in HIV patients.
- (D) They are basic due to the presence of the amine and sulfonamide groups.

4

Agents Targeting Protein Synthesis

Antibacterial agents that target protein synthesis include the aminoglycosides, macrolides, tetracyclines, chloramphenicol, and the oxazolidinones.

4.1

Aminoglycoside antibiotics

Key Points

- The aminoglycoside antibiotics are potent bactericidal agents that are rapidly lethal to bacteria in a concentration-dependent manner; their therapeutic potential for widespread use is restricted by ototoxicity and nephrotoxicity.
- Aminoglycosides have a wide spectrum of activity and are used for the treatment of many Gram negative and Gram positive bacterial infections; they are sometimes used in combination with β-lactam antibiotics.
- For further information, see Vakulenko and Mobashery (2003).

Aminoglycoside antibiotics used clinically include gentamicin, neomycin, streptomycin, tobramycin, and amikacin; their structures are shown in Figure 4.1.1 and their therapeutic indications are listed in Table 4.1.1.

4.1.1   Discovery

The first aminoglycoside antibiotic was discovered as the result of a systematic study of soil microbes by Selman Waksman and his co-workers, who were searching for an agent with activity against tuberculosis (TB). It had previously been noticed that certain microorganisms predominated in soil and it was suggested that these species must be producing an antibiotic agent in order to afford them survival superiority. Waksman and co-workers carried out over 10 000 careful culture experiments before they discovered the aminoglycoside agent, streptomycin, which is produced by an actinomycete mould, *Streptomyces griseus*, and is active against the TB-causing bacterium *Mycobacterium tuberculosis* (Schatz and Waksman, 1944; Schatz *et al.*, 1944). Streptomycin was first used clinically in 1944, when its use led to the curing of a young woman with pulmonary TB; at a time when TB was essentially a slow death sentence, this cure was astonishing.

   Waksman was awarded the Nobel Prize in Physiology or Medicine in 1952 'for his discovery of streptomycin, the first antibiotic effective against tuberculosis.' Sadly, this award was the subject of some

---

Antibacterial Agents: Chemistry, Mode of Action, Mechanisms of Resistance and Clinical Applications, First Edition.
Rosaleen J. Anderson, Paul W. Groundwater, Adam Todd and Alan J. Worsley.
© 2012 John Wiley & Sons, Ltd. Published 2012 by John Wiley & Sons, Ltd.

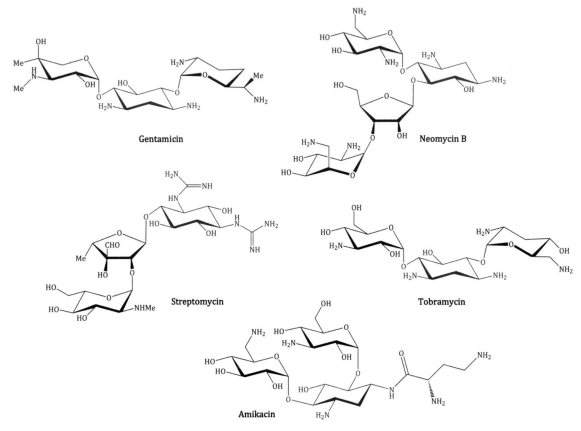

Figure 4.1.1   *Aminoglycoside antibiotics (aminocyclitol ring highlighted in red)*

controversy, as Albert Schatz, who had carried out the experiments that resulted in the discovery of streptomycin while working as a PhD student under Waksman's guidance, argued that the Nobel Prize should also have been awarded to him for his part in its discovery (Kingston, 2000, 2004; Ainsworth, 2006).

The search for new antibiotics for the treatment of bacterial infections, particularly TB, was continued by Waksman and his co-workers and other international groups. The *Streptomyces* family proved to be a rich source of antibiotics, resulting in the discovery of antibacterial agents belonging to several structural classes. Besides streptomycin, other key aminoglycoside discoveries were: neomycin from *S. fradiae* in 1949 (Waksman and Lechevalier, 1949; Waksman *et al.*, 1949); paromomycin from *Streptomyces rimosus* in

Table 4.1.1   *Therapeutic indications for the aminoglycoside antibiotics*

Aminoglycoside antibiotic	Indications
Gentamicin	Septicaemia, burns, endocarditis, used topically to treat eye and ear infections, pneumonia in hospital patients, Listeria meningitis
Neomycin	Suppression of intestinal bacteria prior to colonic surgery
Streptomycin	*M. tuberculosis*
Tobramycin	Chronic pulmonary *P. aeruginosa* infection in cystic fibrosis
Amikacin	Serious infections caused by Gram negative bacteria resistant to gentamicin

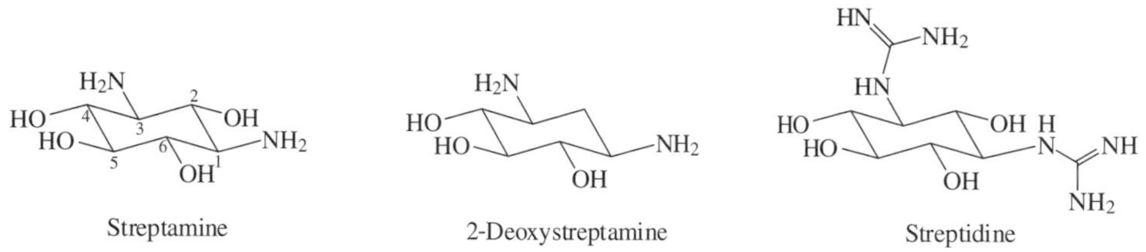

Figure 4.1.2 *The aminocyclitol ring systems found in the aminoglycoside antibiotics, 2-deoxystreptamine being the most common in clinical aminoglycosides*

1959 (Coffey *et al.*, 1959); kanamycin from *Streptomyces kanamyceticus* in 1957 (Maeda *et al.*, 1957; Takeuchi *et al.*, 1957; Umezawa *et al.*, 1957); and tobramycin – initially from *Streptomyces tenebrarius* in 1968 (Stark *et al.*, 1968), but which can also be obtained through the synthetic modification of kanamycin B (Takagi *et al.*, 1973, 1976).

Another aminoglycoside, gentamicin, was isolated in 1963 (Weinstein *et al.*, 1963) from various species of *Micromonospora* bacteria, as a complex of several structurally related agents. There are three forms, of which the most often encountered clinically is gentamicin C, consisting of gentamicins C_1, C_{1a}, C_2, and C_{2b} (Wagman *et al.*, 1967). The discovery of sisomicin, isolated from *Micromonospora inyoensis* in 1970, added to the aminoglycoside arsenal (Weinstein *et al.*, 1970), while further aminoglycosides have been made by synthetic modifications of the original natural aminoglycosides (more about this later). Of these, amikacin was found to have improved pharmacokinetic and pharmacodynamic parameters (Gooding *et al.*, 1976), in addition to excellent clinical efficacy and a lower incidence of bacterial resistance, and so was adopted for clinical use.

You may be wondering why some aminoglycosides end in –mycin and others with –micin. Well, it relates to the microorganism which is the source of each agent; those from a *Streptomyces* species are given the ending –mycin, while those agents from a *Micromonospora* species have –micin at the end of their name. While this is a useful guide to the origin of an aminoglycoside, it does not necessarily extend to the naming of other microorganism-derived agents and there are many examples of non-aminoglycoside agents from *Streptomyces* species that are not named by the same system, such as chloramphenicol, novobiocin, cycloserine, and clavulanic acid.

Structurally, the antibiotic aminoglycosides are glycosidic polycyclic structures, incorporating an aminocyclitol ring (streptamine, streptidine, or 2-deoxystreptamine) (Figure 4.1.2), of which the latter is most commonly found in clinical aminoglycosides, usually attached to two or three aminosugar rings. If you look back at Figure 4.1.1, you can see the aglycone[1] aminocyclitol ring in each structure – it is easily identifiable as it does not have an O atom as part of the ring (see Figure 4.1.2); the other cyclic parts of each molecule are sugars, recognisable as the five- or six-membered cyclic systems that include an O atom in the ring.

The aminoglycosides can be categorised into related groups based upon the glycosidic linkages of the cyclic aminosugar components to the central aminocyclitol ring: the 4,5-disubstituted group includes neomycin and paromomycin; the 4,6-disubstituted group includes the kanamycins, gentamicins and tobramycin; while the original aminoglycoside, streptomycin, has a six-membered monosubstituted aminocyclitol ring.

The possibility that the genetic sequencing of many bacteria may lead to the discovery of further aminoglycosides (and other classes of antibiotics) has been recognised, but no new agents have resulted from such genetic studies thus far (Paradkar *et al.*, 2003; Kudo and Eguchi, 2009).

Have another look at the structures in Figure 4.1.1 and identify how many ionisable groups each aminoglycoside possesses. How many do you think will be ionised and what will be the overall charge on

---

[1] An aglycone is a non-sugar component of a glycosidic molecule. In this case, the aglycone, aminocyclitol, is attached to two or more sugar molecules to form an aminoglycoside.

each molecule at pH 7.0? We will come back to these questions in Subsections 4.1.3 and 4.1.4, when we consider the bioavailability and the mode of action of these agents, which are both linked to ionisation.

4.1.2   Synthesis

In addition to ionisable groups, how many stereogenic, or chiral, centres can you find in each structure in Figure 4.1.1?[2] Having many chiral centres, each aminoglycoside represents a significant synthetic challenge. As these compounds were originally isolated from microorganisms, a biotechnological approach to their production, similar to that employed in the production of other particularly complex antibiotics (for examples, see Subsections 2.2.2 and 5.1.2), is the preferred method. In fact, Waksman and Schatz patented their original method for the production of streptomycin and licensed it to Merck (Waksman and Schatz, 1948). Many other patents followed, for example those from Pfizer (Ratajak and Nubel, 1959) and Ayerst (now Wyeth) (Trussel, 1951). Fermentation conditions have been well studied and remain the subject of research aimed at optimising the yields and purity of the aminoglycosides produced (Sánchez *et al.*, 2010). Full knowledge of the biosynthetic pathways would aid optimal fermentation, but they remain unclear in parts, despite considerable study (Llewellyn and Spencer, 2006; Flatt and Mahmud, 2007). The semi-synthetic aminoglycosides, such as amikacin, are produced commercially by chemical modification of the natural precursor, in this case kanamycin A, which is obtained from fermentation (Mangia, 1998).

As we have previously seen, some chemists view a complex chemical structure, such as those of the aminoglycosides, as a synthetic challenge to be overcome. You will probably not be surprised then to learn that the total synthesis of several members of the aminoglycoside family has been accomplished, although never in sufficient quantities to represent a viable production route. You've probably already realised, by considering the structures of the aminoglycosides in Figure 4.1.1, that with so many chiral centres and amino and hydroxyl groups, careful synthetic and protection strategies were required in these total syntheses. If those challenges were not enough, to make it even more difficult, each aminoglycoside consists of a different arrangement of substituents around the aminocyclitol ring, different glycosidic components attached to the aminocyclitol ring, and different substituents, with various relative substitution sites and stereochemistries, on the aminosugar rings. It is rather impressive that structures of such complexity were synthesised in the late 1960s and 1970s, when many of the elegant synthetic methods now available had not even been thought of. Much of the total synthesis of the aminoglycosides was carried out by Umezawa and co-workers and resulted in a number of synthetic routes that allowed the synthesis of the main clinical aminoglycosides and the evaluation of their essential features (Umezawa, 1974, 1978, 1979).

To understand fully the difficulty of these total syntheses, we have to consider some of the problems faced by carbohydrate chemists, which are varied and considerable in number and complexity. Any simple carbohydrate, such as a monosaccharide like glucose, has several hydroxyl groups that must be differentiated in order to react specifically at any particular position; this is normally achieved by a selective protection strategy, in which the relatively close spatial arrangement of particular hydroxyl groups is exploited to protect those groups, leaving others as the free OH groups. The example below will help to illuminate the problems that must be overcome, but for simplicity and brevity, a significant amount of detail that can be found in the original reviews is omitted (Umezawa, 1974, 1978, 1979).

Figure 4.1.3 highlights the component parts of streptomycin. The glycosidic part of streptomycin, streptose, linked to L-glucosamine, is attached at the C4-OH group of the aminocyclitol ring of streptidine. When starting

---

[2] You should find 13 chiral carbon atoms in gentamicin, 19 in neomycin B, 15 in streptomycin, 14 in tobramycin, and 16 in amikacin; do you agree?

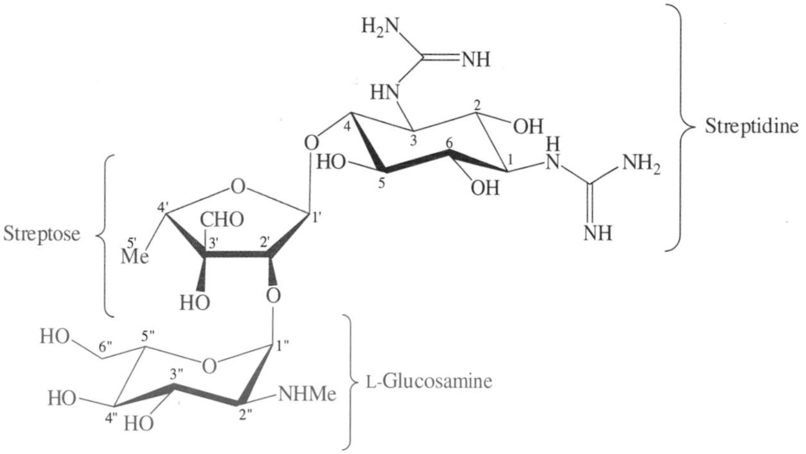

Figure 4.1.3 *The components of streptomycin: streptidine (aglycone), streptose, and L-glucosamine*

from streptamine, how can this linkage be made specifically at that position without attaching the streptose portion at the C2-OH, C5-OH, or C6-OH positions?

Note that the carbons numbered with a 'prime' in Figure 4.1.3 (with ring atom numbers $1'$, $2'$, $3'$, etc.) refer to the sugar (streptose) ring attached to the C4 position of the aminocyclitol systems. It is a general feature of aminoglycosides that the atoms of the ring attached to the C4 position are numbered $1'$, $2'$, $3'$, etc.

The same question may be asked about the linkage between L-glucosamine and streptose: how can it be made specifically between $C2'$-OH of streptose (as numbered in Figure 4.1.3) and the anomeric position of L-glucosamine ($C1''$)? For both questions, the answer involves a careful protection strategy to allow only the desired groups to react:

- Addition of acetyl and carbobenzyloxy carbamate (Cbz) protecting groups to each of the guanidine groups, and a cyclic acetal to the C5-OH and C6-OH groups of streptidine, through several steps, produces selectively protected streptidine **1**, ready for reaction with the streptose–glucosamine glycosidic unit (Scheme 4.1.1). You will note that protected streptidine **1** has both C2-OH and C4-OH available for reaction when forming the bond to streptose, which could cause a problem and a mixture of products. Although this

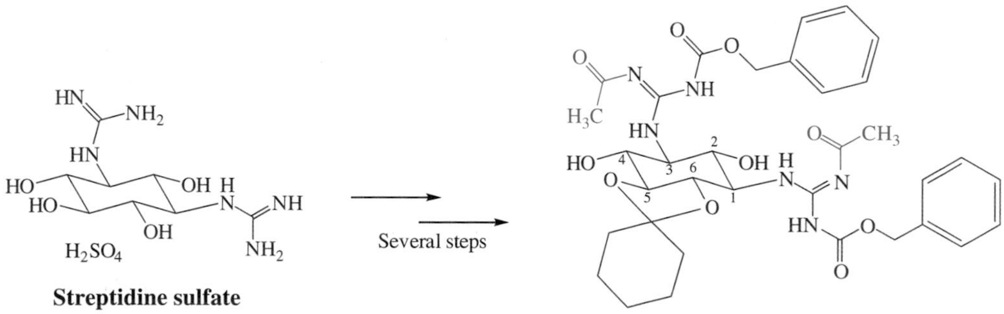

Scheme 4.1.1 *Selective protection of streptidine (Umezawa, 1979)*

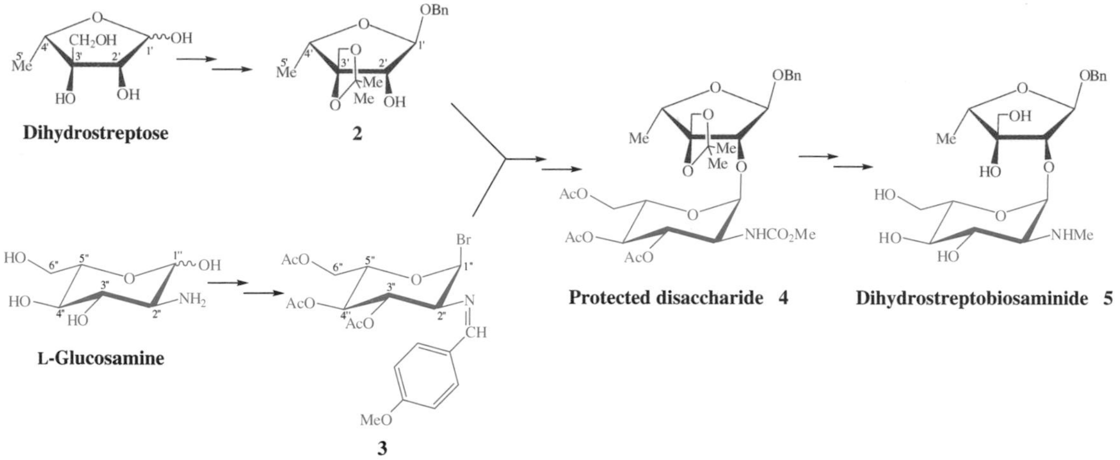

Scheme 4.1.2   *Selective protection of dihydrostreptose and ʟ-glucosamine, followed by glycoside bond formation to obtain the key intermediate dihydrostreptobiosaminide **5** (Umezawa, 1979)*

can and does happen, reaction at C4-OH is preferred, presumably due to steric hindrance at C2-OH caused by the two Cbz groups on the guanidines, so that the major product has the correct substitution at C4-OH and can be readily separated and purified.

- The glycosidic part requires substantially more preparatory work before it can be linked to the protected streptidine **1**. To simplify the atom numbering and make it easier to track each atom and its substituents through the synthesis, we will retain the numbering system used in Figure 4.1.3. First, the aldehyde group at C3$'$ of streptose complicates the protection and synthesis, so the reduced derivative, dihydrostreptose, is used, in which the C3$'$-CHO group has been reduced to C3$'$-CH$_2$OH. Dihydrostreptose is selectively protected by benzylation (CH$_2$Ph, Bn) of the anomeric hydroxyl (C1$'$), which is more reactive than the other hydroxyl groups, then acetalisation of the C3$'$-OH and C3$'$-CH$_2$OH groups forms the protected dihydrostreptose derivative **2**, with a sole hydroxyl group at C2$'$ ready for reaction (Scheme 4.1.2).

- The protected ʟ-glucosamine **3** can be prepared from ʟ-glucosamine in several steps; again, the anomeric centre is easily and selectively protected first. The amino group at C2$''$ is protected by Schiff's base formation, using *para*-methoxybenzaldehyde, before esterification of the C3$''$-OH, C4$''$-OH, and C5$''$-OH groups. Having protected the other functional groups, the anomeric centre can be deprotected and activated, and the α-bromo derivative **3** can be formed, which is ready for reaction with a nucleophilic hydroxyl group, to form the specific α-glycosidic bond between C2$'$ of dihydrostreptose and C1$''$ of glucosamine. Hydrolysis of the Schiff's base and *N*-methylation, via a carbamate in protected disaccharide **4**, gives the required N^1-methylated disaccharide, which is then deprotected, except for the benzyl group on C1$'$-OH of the dihydrostreptose molecule, to give a protected dihydrostreptobiosaminide **5**: a key intermediate (Scheme 4.1.2).

- Several more steps are still required to prepare the disaccharide for coupling to the protected streptamine **1**. These reactions follow a sequence of full protection, selective deprotection, and elaboration, before glycosyl chloride **6** is formed, which is reacted with protected streptamine **1**, followed by deprotection to dihydrostreptomycin **7**. Oxidation of the CH$_2$OH group on C3$'$ of dihydrostreptose requires yet more protection and deprotection steps before the aldehyde can be reliably formed and streptomycin is finally achieved (Scheme 4.1.3).

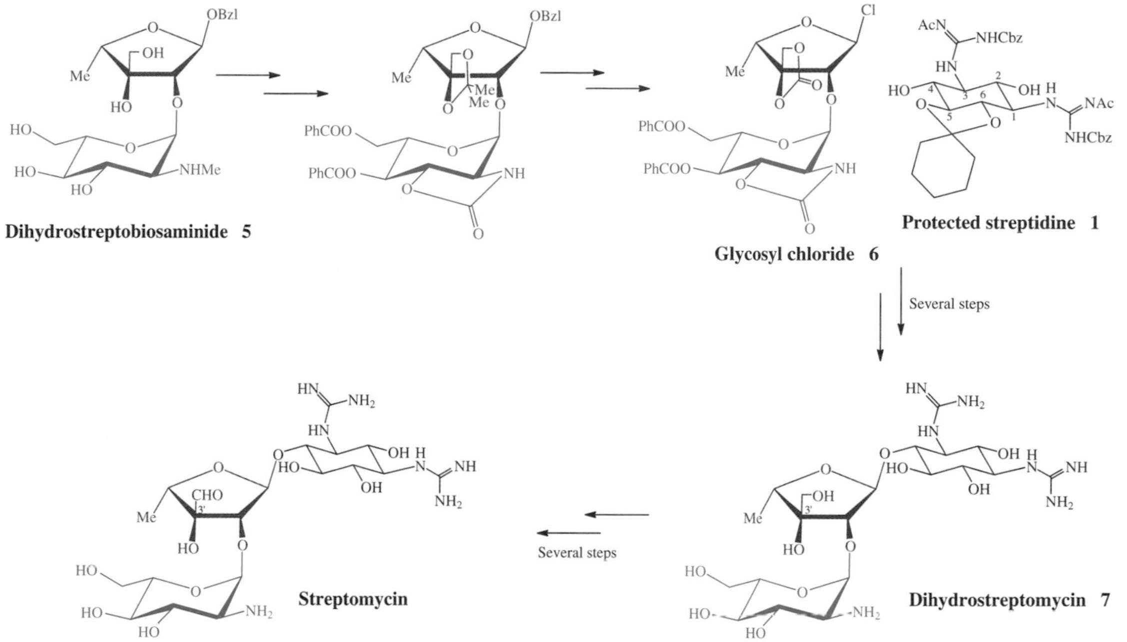

Scheme 4.1.3   *Final steps in the total synthesis of streptomycin via dihydrostreptomycin 7 (Umezawa, 1979)*

Considering this example was very much abbreviated, you can see that the total synthesis of aminoglycosides is far from a simple task and requires many sequential reactions. As we said earlier, it is thus easier to use the microorganisms that form these species to produce the commercial quantities needed for clinical use.

To investigate the essential structural features for antibacterial activity, selectivity, and efficacy, the synthetic routes outlined above have been further developed to allow the effect of modifications to be examined and pharmacophores to be identified (see, for example, Davies *et al.*, 1978; Igarashi *et al.*, 1982; Hanessian *et al.*, 2010; Yan *et al.*, 2011).

Following the identification of bacterial resistance pathways (more on these later, in Subsection 4.1.5) via conjugation to specific N or O atoms of an aminoglycoside, there has been much research on the modification of the target N and O atoms in order to reduce their susceptibility to metabolism and to avoid resistance; for example, the clinical aminoglycoside agent amikacin is a successful example of synthetic modification to N^1 of kanamycin A (Kawaguchi 1976), involving the addition of the 2S-hydroxy-4-aminobutyryl group (achieved by another selective protection strategy (Mangia, 1998; Hanessian *et al.*, 2003)), while netilmicin was synthesised from sisomicin (from the gentamicin family of aminoglycosides) by ethylation of N^1 (Scheme 4.1.4) (Wright, 1975; Miller *et al.*, 1976). The latter agents are not currently in clinical use in most countries, due to observed variations in efficacy, toxicity, and pharmacokinetics in different patient groups, and generally greater toxicity than the other aminoglycosides (Price, 1986).

More recently, there has been a considerable amount of research directed at the systematic exploration of structural requirements for activity and for avoiding resistance, particularly when it is due to metabolism (e.g. *N*-acetylation). The development of synthetic combinatorial methods suitable for the synthesis of many analogous structures has allowed libraries of aminoglycoside derivatives to be created and these methods, the resultant aminoglycosides, and their structure–activity relationships have been investigated (Silva and Carvalho, 2007; Zhou *et al.*, 2007).

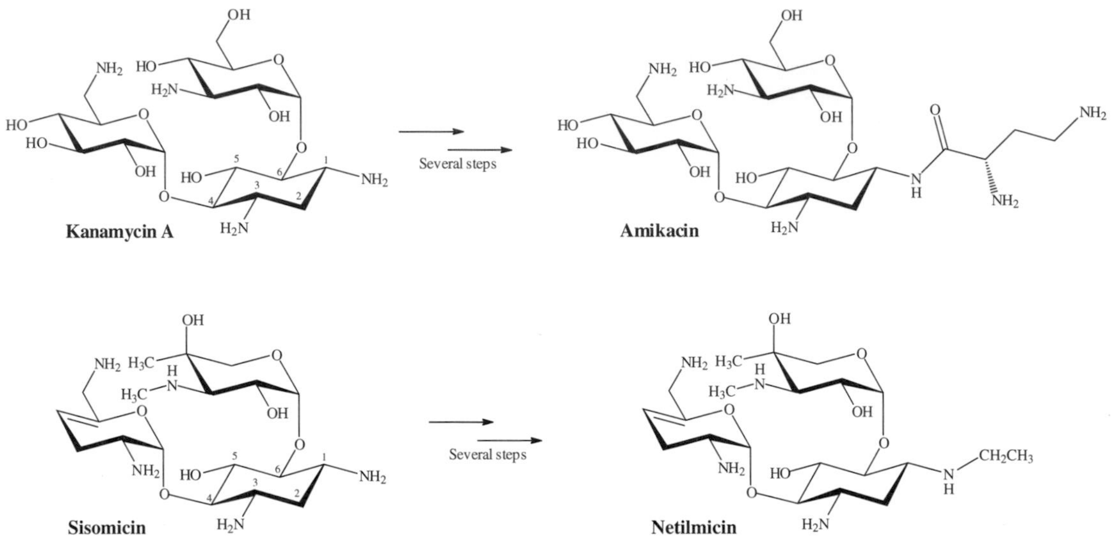

Scheme 4.1.4   *Semi-synthetic aminoglycosides: amikacin from kanamycin A and netilmicin from sisomicin*

4.1.3   Bioavailability

At the end of Subsection 4.1.1, we asked you to consider the structures of the five representative aminoglycosides in Figure 4.1.1 and to decide which were ionisable, which would be ionised at pH 7.0, and what the overall charge on the molecule would be at this pH. You will have found that they all have a number of hydroxyl (OH) and amino (NH_2) groups, while streptomycin also has two guanidine groups (-NH-C(=NH)-NH_2). Of these, at physiological pH only the amino and guanidino groups are ionisable, each group becoming protonated when the pH is lower than the pK_a of that individual amino or guanidino group. Remember that, aliphatic hydroxyl groups ($pK_a \sim 18$–19) are essentially neutral at physiological pH. We would therefore expect the overall charge on each molecule at physiological pH (7.0) to be: gentamicin $+5$, neomycin B $+6$, streptomycin $+3$, tobramycin $+5$, and amikacin $+5$, which reflects the number of protonated amino/guanidino groups on each at this pH. In reality, one of the amino groups in each molecule, N^3 of the aminocyclitol ring, has a pK_a of around 6, such that it is only partly ionised at pH 7, making the overall average charge slightly less than the integer values above (Walter *et al.*, 1999; Jin *et al.*, 2000; Kaul and Pilch, 2002).

When considering their oral bioavailabilities with respect to their state of ionisation, you could predict, from their polycationic character, that the expected absorption of the aminoglycosides after oral administration is likely to be fairly low. In fact, less than 1% of these antibiotics is absorbed across the gastrointestinal (GI) tract following oral administration (Forge and Schacht, 2000), so parenteral routes are commonly used. The poor absorption across the GI tract is very useful, however, in pre-operative bowel cleansing, when these agents are used to limit the risk of post-operative infection due to commensal or opportunistic bacteria.

Systemic aminoglycoside antibacterial treatment is usually achieved through carefully controlled IV infusion, which results in complete absorption and predictable peak serum levels. The latter parameter is particularly important, as infusions that are given too rapidly can result in high peak levels, which increases the risk of ototoxicity, when the cochlear and hair cell binding sites become saturated (Allegaert *et al.*, 2008). Intramuscular injection also results in high bioavailability, but as these very polar molecules do not easily cross membranes, they need to be administered directly to the site of infection. The clinical use of aminoglycosides with respect to their pharmacokinetics has been reviewed recently and recommendations have been made for their optimal use (Avent *et al.*, 2011).

As a result of the structures of these bactericidal agents (Figure 4.1.1), we might expect them to concentrate in the serum and extracellular fluid, with little distribution across membranes into cells and other compartments. In fact, the apparent volume of distribution of aminoglycosides approximates to the extracellular fluid volume, which is about 25% of the total body weight (Duff, 1992).

The use of these agents for the treatment of infections in readily accessible organs, such as their use in the treatment of infective endocarditis (see Subsection 4.1.6.2 below), seems obvious, but how can these agents be used for the treatment of meningitis, which requires therapeutic concentrations to be achieved in the cerebrospinal fluid (CSF)? While the highly polar nature of aminoglycosides suggests that they would not be able to cross the blood–brain barrier, and this is usually the case, this membrane is compromised and more permeable when inflamed in bacterial meningitis. The concentrations of aminoglycosides achieved in the CSF as a result are variable (Lutsar *et al.*, 1998), but appear to be sufficiently high for excellent therapeutic results (Gaillard *et al.*, 1995). Neonates have a more permeable blood–brain barrier and aminoglycoside therapy has been successfully combined with ampicillin for many years (McCracken, 1984). Interestingly, the introduction of a vaccine for *Haemophilus influenzae* B (Hib), a common cause of meningitis, has changed the epidemiology of this threatening infection and may result in the need for different antibiotic combinations as the infective agents change (Kim, 2009). Intralumbar, intrathecal, and intraventricular administration are alternatives for aminoglycoside treatment of meningitis, resulting in more rapid and higher aminoglycoside levels being achieved (Kaiser and McGee, 1975). New methods to improve the delivery of aminoglycosides are under investigation; one that has promising results is the use of liposomes to enhance uptake and delivery (Schiffelers *et al.*, 2001; Rukholm *et al.*, 2006; Alipour *et al.*, 2009).

The polycationic character of aminoglycosides imparts low serum protein binding, making the majority of the dose available for its therapeutic effect, and leads to efficient urinary excretion via the kidneys, with greater than 90% of an aminoglycoside dose being excreted by this route over 24 hours. For susceptible bacteria, the minimum inhibitory concentration (MIC) for most aminoglycosides lies in the range 0.5–5 µg/mL, which is evidence of their considerable potency, while the serum half-lives ($t_{1/2}$) of the aminoglycosides are around 2–3 hours (Turnidge, 2003). Due to their concentration-dependent effect, an important pharmacodynamic parameter for aminoglycoside efficacy is the ratio of C_{max} to the MIC, which should be around 10–12 for optimal therapeutic benefit (Filho *et al.*, 2007; Turnidge and Paterson, 2007). When used in this ratio, a large, once-daily dose is effective, despite the short half-life of these agents; this efficiency may be linked to the considerable post-antibiotic effect, which suppresses bacterial growth for over 8 hours after aminoglycoside concentrations fall below the MIC, and so enhances the bactericidal effect of the once-daily dosing (Begg and Barclay, 1995).

We should also consider the uptake of aminoglycosides into bacteria in order to exert their bactericidal action. How can sufficient amounts of a molecule so polar cross the complex bacterial cell wall of a Gram negative bacterium to access the ribosomal RNA? Studies have shown that there are several key aspects:

- First, a small amount of aminoglycoside enters the bacterial cells against a concentration gradient by an energy-requiring process (called energy-dependent phase I, EDP-I) during an essential induction phase. The energy is generated by a respiratory chain located on the membrane, which explains why anaerobic bacteria (with poor electron transport systems) are resistant to aminoglycosides and facultative anaerobes can resist low concentrations.
- The internalised molecules then bind to RNA in a second energy-dependent process (energy-dependent phase II, EDP-II) and cause misreading during protein synthesis, which results in catastrophic changes to the

cell wall construction, and a second, rapid-uptake phase of aminoglycoside accumulation, which is lethal to the bacterial cells (Vakulenko and Mobashery, 2003).

4.1.4   Mode of action and selectivity

Aminoglycosides result in rapid and complete lethality to cells, which is almost independent of the bacterial load, providing that they achieve the minimum bactericidal levels required. They are most active against aerobic bacteria, particularly Gram negative strains, with particular aminoglycosides displaying distinct bactericidal profiles, including activity against some Gram positive strains (Vakulenko and Mobashery, 2003; Magnet and Blanchard, 2005; Jana and Deb, 2006; Shakil *et al.*, 2008).

The aminoglycoside antibiotics act by binding to a part of the 30S subunit of prokaryotic ribosomes, to a region of the highly conserved 16S ribosomal RNA. Most in clinical use are believed to bind at the A site, which accepts the charged tRNAs during the process of translation and is a vital element of the translation mechanism in bacteria. At this point, if you need a reminder, you can refer back to Subsection 1.1.3.3, where the process of translation, which results in protein synthesis, is outlined in schematic form.

In the normal process of translation of mRNA, the correct (termed 'cognate') tRNA is selected through two key interactions. First, the codon on the mRNA has to match with the anticodon on the tRNA, and, second, when the cognate tRNA binds to the A site, subtle favourable conformational changes result and enable important interactions between the codon and anticodon RNA double helices, which discriminate between the correct and incorrect (non-cognate) tRNA. An incorrect tRNA, whether non-cognate or near-cognate (with a similar, but incorrect, codon sequence) cannot initiate the same favourable interactions; in this way, only the cognate tRNA is selected and proof-reading is achieved. In particular, two conserved adenine residues, A1492 and A1493, are known to be involved in the normal decoding process; before translation commences, these two adenine residues can be found stacked in the interior of the RNA double helix (in helix 44). Binding of the cognate tRNA causes a conformational change that 'flips' the two adenine residues away from their initial position, resulting in the formation of hydrogen bonds from N^1 of the two adenines to the tRNA residues in the first two positions of the codon–anticodon triplet. It is these intermolecular interactions that are thought to be responsible for differentiation between cognate and non-cognate tRNAs, as the latter cannot form the same favourable interactions, so are energetically less beneficial (Vakulenko and Mobashery, 2003).

Aminoglycoside binding to the ribosome is initially encouraged by the favourable interactions possible between the positively charged aminoglycoside and the negatively charged phosphate groups of the RNA double helix; tobramycin has been shown to bind as a trication (Jin *et al.*, 2000), while the neomycins have between three and five protonated groups on binding (Kaul and Pilch, 2002; Kaul *et al.*, 2003). These electrostatic interactions are important thermodynamic contributors to the binding. Each class of aminoglycoside has a structure complementary to slightly different sites on the ribosomal RNA, depending upon the exact nature and position of the substituents, resulting in a subtly altered outcome. For example, the 2-deoxystreptamine-based aminoglycosides, such as gentamicin, neomycin, tobramycin, and amikacin, are believed to bind to the A site and adversely affect the accuracy of translation. Streptomycin also induces poor accuracy, but appears to achieve this result in a slightly different way.

How does the binding of an aminoglycoside to the bacterial ribosome lead to misreading of the mRNA (Chittapragada *et al.*, 2009)?

- It has been shown that the 2-deoxystreptamine aminoglycosides bind in the major groove of double-stranded (ds) RNA (Jin *et al.*, 2000), near to A1492 and A1493 in helix 44 (which borders the A site), resulting in displacement of these adenine residues out of the interior of helix 44 and towards the minor groove (Kaul and Pilch, 2002). Several energetically favourable interactions encourage this conformational change.

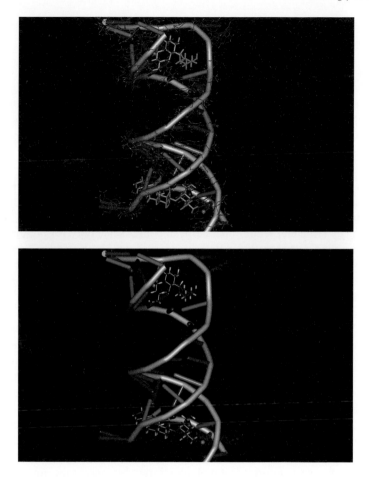

Figure 4.1.4   *Complex of two molecules of tobramycin with a synthetic putative ribosomal A binding site: (a) dsRNA with* tobramycin; *(b) with tobramycin colour-coded by atom*

- The 2-deoxystreptamine ring and the sugar moiety attached at C4 form highly conserved hydrogen bond interactions between N^1 and the carbonyl of uracil, U1495, N^3 and guanine, G1494, and between the ring O and C6'-NH$_2$ of the 'primed ring'[3] and adenine, A1408. Two conserved electrostatic interactions are also observed between the ionised N^3 and the phosphate groups of A1493 and G1494 (Chittapragada *et al.*, 2009).
- The new positions of A1492 and A1493 are stabilised by these interactions with the aminoglycoside, which compensates for the less favourable binding of non-cognate tRNA molecules, allowing the incorporation of incorrect amino acids and introducing misreading of the mRNA and reduced proof-reading of the sequence (Vakulenko and Mobashery, 2003; Yang *et al.*, 2006). The binding of tobramycin to a synthetic dsRNA portion of the 30S subunit is illustrated above, where the bases guanine, uracil, cytosine, and adenine can be seen close to the aminoglycoside (Figure 4.1.4). In this

---

[3] In Figure 4.1.3, we saw that the 'primed ring' is the one attached to C4 of the aminocyclitol.

model, the binding of tobramycin causes distortion of the dsRNA and mismatching of bases in the sequence used.

Specific aminoglycosides can form additional favourable interactions that further enhance the affinity; for example, tobramycin forms two extra hydrogen bonds from one of its carbohydrate rings to G1405 (Magnet and Blanchard, 2005), and amikacin's N^1-side chain extends alongside U1495 (Hobbie *et al.*, 2005).

Having discussed the high-affinity binding of aminoglycosides to the 30S RNA subunit, the question of selectivity arises. How can these agents, with such efficient inhibition of bacterial protein synthesis, cause no alteration to protein synthesis in eukaryotic cells?

- Firstly, their highly ionised structures make them too polar to diffuse across the plasma membrane of a mammalian cell to any great extent; if this were to occur, more severe adverse effects would be likely, due to non-selective binding to eukaryotic ribosomes.
- Secondly, there is a key structural feature that helps to confer selectivity for prokaryotic ribosomes over eukaryotic ribosomes: the C6'-NH$_2$ group. All of the aminoglycosides based on 2-deoxystreptamine – that is, gentamicin, neomycin B, tobramycin, and amikacin – possess an amino group at the C6' position. The analogues with a C6'-OH are less selective, also inhibiting protein synthesis in eukaryotic cells (Mingeot-Leclercq *et al.*, 1999).

Several research groups have studied the key differences in the structural interactions that cause high-affinity binding of aminoglycosides to prokaryotic ribosomes and weak, non-selective binding to the eukaryotic structures (Blount and Tor, 2006; Kondo *et al.*, 2007), not only to the ribosomal A site, but also to other target sites, such as helix 69 (Scheunemann *et al.*, 2010). Aminoglycoside binding to helix 69 inhibits dissociation, metabolism, and reprocessing of the bacterial ribosomes. In each case, the energetics of aminoglycoside binding to the prokaryotic ribosome were more favourable than binding to the eukaryotic ribosome. The interaction between the conserved A1408 and the aminoglycoside in the prokaryotic ribosome was found to be crucial for high-affinity selective binding. This base is replaced by guanine in eukaryotic ribosomes, in which the guanine lies closer to the major groove, occupying a slightly different position to its adenine counterpart, and blocking tight binding of the aminoglycoside (Lynch and Puglisi, 2001a, 2001b). This interaction of aminoglycoside and A1408 is considered to be key in differentiating binding to prokaryotic and eukaryotic ribosomes. Interestingly, a single RNA base mutation from adenine to guanine (at position 1408 in the *Escherichia coli* ribosome) confers resistance to the C6'-NH$_2$ aminoglycosides (Recht *et al.*, 1999).

Our knowledge of aminoglycoside binding to RNA is being used to predict structural modifications that would enhance activity and selectivity (Hermann, 2002; Setny and Trylska, 2009; Tran and Disney, 2010), providing the possibility of improved aminoglycoside agents in the future.

4.1.5   Bacterial resistance

Resistance to streptomycin was recorded as early as 1953 (Wyss and Schaiberger, 1953), when it was linked to differences in expression or activity of certain enzymes. There are several different mechanisms of resistance to aminoglycosides, including:

- Decreased uptake into bacterial cells, due to decreased cell membrane permeability or increased efflux (Poole, 2005b).

- Mutations to the ribosome A site (for example, A1408 to G1408) or to a ribosomal protein.
- Methylation of the ribosome, particularly at A1408, which dramatically reduces the binding affinity for most clinical 2-deoxystreptamine aminoglycosides (Morić *et al.*, 2010). The methyltransferases (often abbreviated to MTases) have initiated a flurry of research activity due to their detrimental role in the effectiveness of clinical aminoglycosides, as this is the most common form of aminoglycoside resistance in pathogenic bacteria (Macmaster *et al.*, 2010; Morić *et al.*, 2010). These enzymes methylate the key adenine base residue in the major groove that is responsible for the tight binding of the $C6'$-NH_2 aminoglycosides and discourages the interaction; in fact, this mechanism is essential for aminoglycoside-producing microorganisms (to ensure that they do not fall victim to their own toxic products).
- The production of aminoglycoside-modifying enzymes – aminoglycoside *N*-acetyltransferases, *O*-nucleotidyltransferases, and *O*-phosphotransferases (Wright, 1999; Ramirez and Tolmasky, 2010). These enzymes are found in both Gram positive and Gram negative bacteria and are transferred by plasmids, transposons, and integrons, which carry multiple genes to confer resistance to a wide range of antibacterial agents, and have spread worldwide during the last decade (Tolmasky, 2000; Yamane *et al.*, 2005; Davis *et al.*, 2010). They follow a common nomenclature of a three-letter code – AAC for aminoglycoside acetyltransferase, ANT for the nucleotidyltransferase, and APH for the phosphotransferase – followed by a number in parentheses that indicates the site on the aminoglycoside at which modification takes place. Examples include AAC(6′), which acetylates the $C6'$-NH_2; ANT(6), which catalyses the transfer of an adenine monophosphate to C6-OH; and APH(6), which phosphorylates C6-OH, particularly that of streptomycin (Wright and Thompson, 1999; Ramirez and Tolmasky, 2010). An unusual dual enzyme has been identified that has both AAC and APH activity; it is responsible for extensive resistance in staphylococci and enterococci (Daigle *et al.*, 1999).

The modification of aminoglycosides at the N and O atoms commonly used for conjugation and inactivation is one approach to overcoming resistance (Li *et al.*, 2007; Hanessian *et al.*, 2010; Yan *et al.*, 2011). Alternative strategies have been suggested, such as the inhibition of aminoglycoside-modifying enzymes by non-glycosidic, specifically targeted inhibitors (Welch *et al.*, 2005), the use of protein kinase inhibitors, which target the ATP-binding site of the relevant enzymes (Burk and Berghuis, 2002), and an antisense approach (Jana and Deb, 2005). For further information on bacterial resistance to the aminoglycosides, see Poole (2005a).

4.1.6　Clinical applications

4.1.6.1　Spectrum of activity

The aminoglycosides are primarily effective against Gram negative bacteria, but also show some activity against Gram positive bacteria (e.g. *Staphylococcus aureus*); resistance is frequently reported with species of streptococci and enterococci, while anaerobes are also resistant (Collatz *et al.*, 1984; van Asselt *et al.*, 1992; Chow, 2000). Amikacin, gentamicin, and tobramycin are active against *Pseudomonas aeruginosa*, but resistance to these agents has been reported in most parts of the world and could lead to problems in the not-so-distant future (Poole, 2005a). Streptomycin, amikacin, and kanamycin also show activity against *M. tuberculosis* and are used clinically to treat MDR-TB – of these, amikacin appears to have superior *in vitro* activity (Sanders *et al.*, 1982), while high rates of *M. tuberculosis* resistance have been reported for streptomycin.

4.1.6.2   Infective endocarditis

Infective endocarditis (IE) is a serious, often life-threatening condition that affects the endocardium (the tissue that lines the inside of the heart chamber). In most cases the infection is bacterial in origin – caused by a variety of different Gram positive (e.g. *S. aureus*, viridans streptococci,[4] or *Enterococcus faecalis*) or Gram negative bacteria (e.g. HACEK[5] organisms) – but it can also be caused by fungi (e.g. *Candida*). Patients with IE often present with fever, poor appetite, weight loss, and, in up to 80% of cases, heart murmurs (Beynon *et al.*, 2006).

The pathophysiology of IE is complex and is thought to occur by several processes. First, bacteria are introduced into the bloodstream (causing bacteraemia) via a number of possible mechanisms, including the spread of an initial infection (such as a gum or skin infection) or the use of an unsterilised needle to inject a drug into the bloodstream – this is the reason why IE is more common in the IV drug-using community (typically around 1–5% of patients (Beynon *et al.*, 2006)). The next stage concerns the attachment of bacteria to the endothelium surface. This surface is normally resistant to colonisation, but if the endothelium becomes damaged or inflamed, certain proteins are expressed or exposed (such as extracellular matrix proteins or integrins), promoting the attachment of bacteria. The bacteria can then attach to the surface and, along with other substances, such as fibrin and platelets, combine to form a 'vegetation'. Vegetations can prevent heart valves from working properly, or can break off and form emboli, which, if they travel to the brain, can lead to a stroke (a complication of IE).

Aminoglycosides (typically gentamicin) are the mainstay treatment for IE, but surgery can also be used in severe cases to repair/replace damaged heart valves. Both streptococcal and enterococcal species can cause IE, and if you look at Subsection 4.1.6.1, you will notice that these bacteria often show resistance to them. At this point, you may wonder why these agents are used to treat an infection where resistance is frequently observed (it would seem to be clinically inappropriate to prescribe them if resistance is common, as surely this will just increase the incidence of resistance). This is a good question and can be explained by the fact that the aminoglycosides are often used in combination with penicillin antibiotics to exert a synergistic antibacterial effect.[6] The penicillins target the bacterial cell wall (see Section 5.1) and, by doing so, allow greater penetration of the aminoglycoside into the bacterium (thereby allowing greater access to the ribosome, the target of the aminoglycosides) and thus increase antibacterial activity (Davis, 1982; Giamarellou, 1986). The clinical benefit of using this combination, however, is open to debate, as one meta-analysis failed to support the addition of an aminoglycoside to a penicillin for treatment of IE, and even showed that nephrotoxicity was more common with combination therapy when compared to single-agent penicillin therapy (Falagas *et al.*, 2006). Recent clinical reviews of the practice of combining an aminoglycoside with another agent have suggested that the increased risks of greater adverse events outweigh the potential benefits (Leibovici *et al.*, 2009; Marcus *et al.*, 2011).

Despite this debate, the combination of a penicillin and aminoglycoside is still recommended when treating bacterial IE, as summarised in Table 4.1.2.

4.1.6.3   Listeria meningitis

Gentamicin is also used in the management of another life-threatening infection, Listeria meningitis, which is caused by *Listeria monocytogenes* and is the third most common cause of bacterial meningitis (behind *Streptococcus pneumoniae* and *Neisseria meningitides*). Listeria meningitis is principally spread by food

---

[4] This term does not refer to one individual species, but rather a large group of streptococcal bacteria and includes *Streptococcus anginosus, Streptococcus constellatus, Streptococcus mitis, Streptococcus mutans, Streptococcus sanguis,* and *Streptococcus sobrinus*.

[5] HACEK organisms include **H**aemophilus, **A**ctinobacillus, **C**ardiobacterium, **E**ikenella, and **K**ingella species (hence the name).

[6] Remember that 'synergism' is a term used to describe a situation where the combined effect of two agents is greater than the sum of their individual effects.

Table 4.1.2 *Examples of antibiotic regimens used to treat infective endocarditis (see Habib et al., 2009 for full detailed guidance)*

Causative agent	Antibiotic regimen	Duration
Enterococci species	Amoxicillin with gentamicin[a]	4–6 weeks
'HACEK' organisms	Ampicillin with gentamicin (or in the case of penicillin resistance, ceftriaxone)	4 weeks
Meticillin-susceptible Staphylococci	Flucloxacillin with gentamicin	4–6 weeks for the flucloxacillin and 3–5 days for the gentamicin
Streptococci fully susceptible to penicillin (MIC < 0.125 mg/L)	Benzyl penicillin with gentamicin	2 weeks
Streptococci relatively resistant to penicillin (MIC 0.125–2 mg/L)	Benzyl penicillin with gentamicin	4–6 weeks for the benzyl penicillin and 2 weeks for the gentamicin

[a] If there is high level of resistance to gentamicin (MIC > 500 mg/L), streptomycin can be used as an alternative (providing the organism shows sensitivity to streptomycin).

contaminated with *L. monocytogenes*, so, thanks mainly to improved sanitisation and increased awareness of food hygiene, this condition is not very common in developed countries, with one study calculating an annual incidence as 0.07 cases per 100 000 adults (Brouwer *et al.*, 2006), and tends to occur among immunocompromised and elderly individuals. It presents with signs and symptoms (such as headache, fever, neck stiffness, and altered mental status) consistent with other forms of bacterial meningitis.

If a positive diagnosis of Listeria meningitis is made (after culturing CSF, for example), gentamicin can be used in combination with ampicillin to treat the infection (Tunkel *et al.*, 2004). The recommended duration of treatment is 21 days, although consideration can be given to stopping the gentamicin after 7 days (this should be contrasted with the other common forms of bacterial meningitis, where treatment usually lasts for only 7–14 days). It should be noted, however, that the recommended duration of treatment is only a guide and that treatment duration should always be individualised according to the patient's clinical response.

The combination of gentamicin and ampicillin for the treatment of Listeria meningitis has recently been questioned as a 3-year prospective observational study concluded that this combination is associated with a higher mortality rate and less favourable patient outcome when compared to other regimens (Amaya-Villar *et al.*, 2010), although further work in this area is required.

4.1.6.4 Chronic pulmonary infection in cystic fibrosis

As mentioned in Subsection 4.1.6.1, the aminoglycosides show activity against a wide range of Gram negative bacteria, including *P. aeruginosa*, with tobramycin and amikacin having the greatest activity (Durante-Mangoni *et al.*, 2009).

Tobramycin is used in patients with cystic fibrosis to manage chronic *P. aeruginosa* infection. These patients are prone to pulmonary infections as they have a mutation in the *CFTR* (cystic fibrosis transmembrane conductance regulator) gene, which codes for an ion channel that helps regulate the movement of salt (and water) across cells. Patients with this mutation tend to produce thick 'sticky' mucus (for example, in the airways of the lungs) and have high sodium and chloride sweat levels – a sign that is used as a diagnostic test for cystic fibrosis (levels of >60 mmol/L indicate cystic fibrosis). Consequently, these patients cannot clear mucus effectively from their lungs, and this results in the accumulation of inhaled bacteria, which can lead to chronic infection and inflammation, with common pathogens including *P. aeruginosa*, *H. influenzae*, and *S. aureus* (Coutinho *et al.*, 2008). This chronic infection and inflammation can, over time, cause irreversible airway damage, eventually leading to respiratory failure – a leading cause of death in cystic fibrosis patients (Davies *et al.*, 2007).

Managing lung infections in cystic fibrosis patients is thought to help maintain pulmonary function, and parenteral tobramycin has been used to treat *P. aeruginosa* infection in these patients. This route requires large doses of the aminoglycoside to achieve adequate drug concentrations at the site of infection[7] and, because the drug is systemically administered, can also cause nephro- and ototoxicity, a major disadvantage of parenteral therapy. Nowadays, however, tobramycin can also be administered by inhalation, using a nebuliser, to treat chronic *P. aeruginosa* infection. This route offers the advantage that it delivers high concentrations of antibiotic directly to the site of infection, while minimising systemic absorption, thus overcoming the adverse effects associated with parenteral administration. Tobramycin is given on a cyclical basis, with 28 days of antibiotic therapy followed by a 28-day 'off period' where no antibiotic is administered (Geller *et al.*, 2002). When administered in this way, tobramycin appears to be well tolerated, to improve pulmonary function, to decrease the density of *P. aeruginosa* in sputum, and also to decrease the risk of hospitalisation in patients with cystic fibrosis (Ramsey *et al.*, 1999).

Disadvantages of inhaled therapy include the cost (with 1 month's supply of tobramycin costing in excess of £1000), while the delivery of the drug can vary according to the type of nebuliser used.[8] Tobramycin is not the only antibiotic that is delivered by the inhalation route to manage chronic *P. aeruginosa* infection in cystic fibrosis patients, as aztreonam, a β-lactam antibiotic, which we will meet in Section 5.1, can also be used.

4.1.6.5   Miscellaneous

We have discussed some different drugs that can be used to treat TB (quinolones and rifamycins in Section 2; isoniazid and cycloserine will be covered in Section 5), and the aminoglycosides also have a role in the management of this disease. They are not used to treat drug-sensitive TB, but have a role in the management of MDR-TB and are listed by the WHO as Group 2 agents[9] (World Health Organization, 2010). The Group 2 aminoglycosides include kanamycin, amikacin, and streptomycin, and these agents should be used to treat MDR-TB when drug susceptibility has been documented or is suspected. Amikacin or kanamycin should be used in preference to streptomycin, as high rates of streptomycin resistance have been reported in MDR-TB. Other indications for the aminoglycosides include the treatment of burn and wound infections, septicaemia, abscesses, and respiratory tract infections (RTIs), particularly when these are caused by Gram negative organisms.

Until now, we have concentrated on the use of the aminoglycosides for the treatment of serious (in some cases life-threatening) infections, but they also have a role in the management of less-serious conditions, and are used topically to treat superficial ear and eye infections. In the management of superficial eye infections, chloramphenicol is generally considered to be the agent of choice (see Subsection 4.4.6), but when the infection is caused by *P. aeruginosa*, gentamicin eye drops may be more effective. Formulations of gentamicin ear drops are also available and are occasionally used to treat otitis externa (inflammation of the external ear canal) when *P. aeruginosa* is implicated. You will see in the Subsection 4.1.7 that one of the adverse effects of the aminoglycosides is ototoxicity, and you may therefore wonder if it is safe to use this class of drug as an ear

---

[7] Levels of aminoglycoside required typically exceed the MIC by 10–25-fold, as components in cystic fibrosis sputum are thought to bind to the aminoglycosides and reduce their bioactivity. See Mendelman *et al.* (1985) for more details.

[8] Early trials used an ultrasonic nebuliser at a dose of 600 mg three times daily to deliver the tobramycin. Nowadays, the more efficient jet nebuliser PARI LC PLUS is used, at a dose of 300 mg twice daily.

[9] Group 1 agents include pyrazinamide, ethambutol, and rifabutin and should be used for MDR-TB whenever possible.

drop. Generally speaking, the risk of ototoxicity when using gentamicin ear drops for the treatment of otitis externa is very low, providing the patient's eardrum is intact. In cases when the eardrum has been perforated, however, they should not be used due to the increased risk of ototoxicity (Wong and Rutka, 1997; Bath *et al.*, 1999). In patients with a perforated eardrum who present with either otitis media (inflammation of the middle ear) or otitis externa that has not responded to treatment, specialists may, however, use topical aminoglycoside ear drops as a last resort. In these cases it is considered that the presence of pus in the middle ear poses a higher risk of ototoxicity than the aminoglycoside itself and this is a good example of the benefit of treatment outweighing the risk of a potential adverse effect (a phrase you will commonly hear if you go on to work as a healthcare professional).

4.1.7 Adverse drug reactions

The problems of toxicity relating to the use of aminoglycosides have been known for more than 70 years, with the ototoxicity associated with streptomycin use having been detected in the 1940s.

The two main toxicities observed with all the aminoglycosides are nephrotoxicity and ototoxicity (which is related in some way to mitochondrial ribosomal mutations in certain individuals) (Conrad *et al.*, 2008). All the aminoglycosides share these adverse effects, so we will discuss their toxicity without referring to them individually.

4.1.7.1 Ototoxicity

Ototoxicity occurs in approximately 4–20% of patients (Lane *et al.*, 1977; Brummett and Morrison, 1990) and is generally irreversible – the use of aminoglycosides can result in permanent high-frequency hearing loss and temporary vestibular hypofunction (Guthrie, 2008). This damage can occur within as little as 4 hours of drug administration, and eventually progresses to affect lower-range frequencies (Boettcher *et al.*, 1987). Damage to the hair cells (sensing movement), in the organ of corti, and neurons of the cochlea is the basis of the permanence of damage with the aminoglycosides, which are taken up into the organ of corti by the protein megalin (Mizuta *et al.*, 1999).

Interpatient variability in the extent of ototoxicity has provoked suggestions that the aminoglycosides are converted in the sera of vulnerable subjects to an unidentified cytotoxin, the mechanism of the toxicity of which has yet to be determined (Crann *et al.*, 1992; Crann and Schacht, 1996; Wang *et al.*, 1999). In addition, it has been shown that the aminoglycoside molecule itself is not toxic, but actually requires the redox potential of a transition metal ion in order to induce ototoxicity (Schacht, 1993). Being poly-hydroxylated, the aminoglycosides can chelate transition metal ions and the complexes formed are redox active, generating reactive oxygen species that cause oxidative damage to surrounding biomolecules (Jezowska-Bojczuk *et al.*, 1998a, 1998b). It is believed that an iron-aminoglycoside complex potentiates cell damage in the inner ear by reactive oxygen species (Lopez-Gonzalez *et al.*, 1999), and animal studies have shown that iron chelators (such as desferrioxamine) and free radical scavengers (such as allopurinol, mannitol, dimethyl sulfoxide) reduce the degree of damage to the inner ear (Song and Schacht, 1996; Forge and Li, 2000).

4.1.7.2 Nephrotoxicity

The exact mechanism of the nephrotoxicity of the aminoglycosides is still unclear, despite many studies. Animal models are often used to demonstrate similarities to the human model, but the doses used are often far higher than those used in clinical practice.

As stated earlier, the aminoglycosides are primarily eliminated by glomerular filtration and excretion into the urine – in the case of gentamicin, approximately 15% of the filtered drug is re-absorbed into the proximal tubule (Contrepois *et al.*, 1985). An accumulation of the aminoglycoside may take place in the renal cortex tissue and this has a direct influence on nephrotoxicty, as excretion from this area may take many days (Schentag *et al.*, 1978). Being polycationic, the aminoglycosides are absorbed by a process known as pinocytosis, into the phospholipid membranes of the renal cortical tissue and are then transferred to the lysosomes, where their high affinity for phospholipids results in a decrease in lysosomal phospholipase activity. Large numbers of lysosomes are generally found in kidneys which have had high aminoglycoside exposure (Janknegt, 1990). There are two suggested mechanisms for tubular necrosis: the first assumes that the local concentration of the aminoglycoside has a direct relationship to its toxicity, with the lysosomes the key areas of damage and toxicity, while the second suggests that aminoglycosides become toxic after they have been released from the lysosome (once a critical concentration has been reached). In either case, the aminoglycosides chelate with mitochondrial iron, forming an Fe(II)-aminoglycoside complex which acts as a reactive oxygen species, as described for the mechanisms involved in ototoxicity.

The initial point of contact in the kidney tubular cells with the aminoglycosides is the acidic phospholipids of the brush borders. The aminoglycosides then transfer to a transmembrane protein called megalin (which, you will remember, is also found in the ear) and become internalised to an endosome (Moestrup *et al.*, 1995). The endosomes containing the aminoglycoside are then transferred to lysosomes, via a process of fusion – during this process the aminoglycoside molecule becomes fully protonated and is attracted to, and binds tightly to, the phospholipids of the tubular cells. This process causes phospholipid aggregation and the inhibition of phospholipases. These processes result in the accumulation of myeloid bodies – these steps in themselves do not explain kidney cell death, but they do contribute to overall toxicity.

Nephrotoxicity is sometimes reversible, especially if full hydration of the kidneys is maintained during antibiotic treatment.

4.1.7.3   Dosing schedule

Generally, once-daily dosing is preferred over multiple-daily dosing, as it is more convenient and provides adequate serum concentrations, although there are certain indications when once-daily dosing is not recommended (e.g. endocarditis or severe burns). However, due to the narrow therapeutic index of the aminoglycosides, care must always be taken when administering them systemically. This is particularly important if a patient has any degree of renal failure, as this could lead to accumulation of aminoglycoside, which may cause further renal failure and possibly even ototoxicity. It has been suggested that the uptake of aminoglycoside into the renal tubule is saturable (Giuliano *et al.*, 1986), so that, irrespective of the dose, if the renal tubular cells are saturated with aminoglycoside they cannot reabsorb any further drug once a particular concentration has been reached. This was found to be the case with once-daily drug dosing compared to the regimen of a thrice-daily dosing (Gilbert, 1997).

When an aminoglycoside is administered, it is therefore essential to assess the patient's renal function and adjust the dose, if appropriate. It is also important to measure the serum concentration of the aminoglycoside when administered parenterally in order to minimise the risk of developing toxicity, while also avoiding subtherapeutic concentrations. This is done after a dose is given (the 'peak' concentration) and just before the next dose is due (the 'trough' concentration); for example, when giving gentamicin in multiple doses, the peak concentration should not exceed 10 µg/mL (as ototoxicity has been observed above this concentration), while the trough concentration should not exceed 2 µg/mL; for once-daily doses, trough concentrations should not normally exceed 1 µg/mL.

4.1.8   Drug interactions

Because the aminoglycosides are not significantly metabolised by cytochrome P450 enzymes, drug interactions involving these antibiotics are restricted to those agents which reduce the excretion of the aminoglycosides, thereby increasing plasma concentrations of these drugs (and so the likelihood of ototoxicity and nephrotoxicity).

4.1.8.1   Gentamicin and furosemide

Case study reports have shown that the combined use of furosemide and gentamicin may increase the risk of ototoxicity, and this is most probably due to the synergistic effects of these drugs. One mechanism which has been suggested involves the interaction of the aminoglycosides with cell membranes in the inner ear, increasing their permeability and so allowing furosemide to penetrate into these cells in higher concentrations, thereby contributing to the ototoxicity (Bates *et al.*, 2002).

4.1.8.2   Aminoglycoside antibiotics and β-lactams

In vitro studies have demonstrated an interaction between carbenicillin and gentamicin, the clinical significance of which remains unclear (Holt *et al.*, 1976). It has been postulated that the β-lactam ring is opened by a nucleophilic attack of an amino group of the aminoglycoside, forming an inactive amide (Waitz *et al.*, 1972).

4.1.8.3   Aminoglycosides and NSAIDs

The combination of non-steroidal anti-inflammatory drugs (NSAIDs) and aminoglycosides has been shown to increase renal toxicity in animal studies (Assael *et al.*, 1985). Inhibition of prostaglandin E2 formation (which has a renal vasodilatory effect) by NSAIDs reduces the overall glomerular filtration rate and so will increase aminoglycoside nephrotoxicity. Indomethacin, which is commonly employed in symptomatic *patent ductus arteriosus* in pre-term infants, has been shown to reduce the glomerular filtration rate, with potential increases in aminoglycoside concentration (Kuehl *et al.*, 1985), while case reports have shown aminoglycoside toxicity in infants treated simultaneously with ibuprofen and gentamicin (Kovesi *et al.*, 1998).

   The effect of aspirin, when used in combination with the aminoglycosides, has been to attenuate ototoxicity (Chen *et al.*, 2007).

4.1.8.4   Aminoglycosides and general anaesthetics (volatile)

The combined use of aminoglycosides and enflurane has been associated with an increased risk of nephrotoxicity (Christensen *et al.*, 1993).

4.1.8.5   Aminoglycosides and neuromuscular blocking agents

One study has shown that tobramycin and gentamicin do not present, at therapeutic doses, any neuromuscular blocking effect or sub-clinical relaxant-potentiating effects (Lippmann *et al.*, 1982), but when used in combination with tubocurarine, enhancement of the neuromuscular blockade was experienced with tobramycin, ribostamycin, and dibekacin (Hashimoto *et al.*, 1975). The effect of the aminoglycosides upon skeletal muscle involves the inhibition of calcium influx into motor nerve terminals via the P/Q and N-type channels (Pichler *et al.*, 1996), and this effect has been shown to range in potency as follows (Knaus *et al.*, 1987):

$$\text{neomycin} > \text{gentamicin} = \text{tobramycin} > \text{streptomycin} > \text{amikacin} > \text{kanamycin}$$

Despite the suggested effects of the aminoglycosides themselves on neuromuscular blockade, the synergistic interaction with licensed neuromuscular blocking agents has been reported for atracurium and vecuronium in combination with gentamicin and tobramycin (Dupuis *et al.*, 1989), with a prolongation of neuromuscular blockade. The neuromuscular blockade associated with the aminoglycosides is enhanced when used in combination with calcium channel-blocking drugs, such as verapamil, but the clinical manifestation of this observation has not been realised (Paradelis *et al.*, 1988).

4.1.9   Recent developments

As we have seen a number of times, the aminoglycosides have potent bactericidal activity against a wide range of Gram negative and Gram positive pathogenic bacteria and, as a result, there is great deal of research underway to expand this existing class of antibiotics. One candidate, ACHN-490, known as a neoglycoside or a 'next-generation' aminoglycoside, is currently undergoing development (Figure 4.1.5). Encouragingly, this compound is active against many bacteria that possess aminoglycoside-modifying enzymes and are thus resistant to conventional aminoglycosides (Aggen *et al.*, 2010). ACHN-490 is currently being investigated for the treatment of complicated UTIs and acute pyelonephritis.

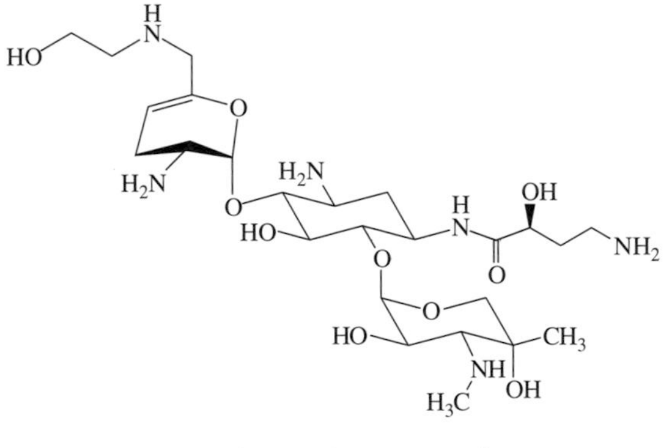

Figure 4.1.5   *Chemical structure of ACHN-490*

References

J. B. Aggen, E. S. Armstrong, A. A. Goldblum, P. Dozzo, M. S. Linsell, M. J. Gliedt, D. J. Hildebrandt, L. A. Feeney, A. Kubo, R. D. Matias, S. Lopez, M. Gomez, K. B. Wlasichuk, R. Diokno, G. H. Miller, and H. E. Moser, *Antimicrob. Agents Chemother.*, 2010, **54**, 4636–4642.

S. Ainsworth, *Pharm. J.*, 2006, **276**, 237–238.

M. Alipour, Z. E. Suntres, and A. Omri, *J. Antimicrob. Chemother.*, 2009, **64**, 317–325.

K. Allegaert, I. Scheers, E. Adams, G. Brajanoski, V. Cossey, and B. J. Anderson, *Antimicrob. Agents Chemother.*, 2008, **52**, 1934–1939.

R. Amaya-Villar, E. García-Cabrera, E. Sulleiro-Igual, P. Fernández-Viladrich, D. Fontanals-Aymerich, P. Catalán-Alonso, C. Rodrigo-Gonzalo de Liria, A. Coloma-Conde, F. Grill-Díaz, A. Guerrero-Espejo, J. Pachón, and G. Prats-Pastor, *BMC Infect. Dis.*, 2010, **10**, 324.

M. L. Avent, B. A. Rogers, A. C. Cheng, and D. L. Paterson, *Intern. Med. J.*, 2011, **41**, 441–449.

D. E. Bates, S. J. Beaumont, and B. W. Baylis, *Ann. Pharmacother.*, 2002, **36**, 446–451.

A. P. Bath, R. M. Walsh, M. L. Bance, and J. A. Rutka, *Laryngoscope*, 1999, **109**, 1088–1093.

E. J. Begg and M. L. Barclay, *Br. J. Clin. Pharmacol.*, 1995, **39**, 597–603.

R. P. Beynon, V. K. Bahl, and B. D. Prendergast, *Brit. Med. J.*, 2006, **333**, 334–339.

K. F. Blount and Y. Tor, *Chembiochem*, 2006, **7**, 1612–1621.

F. A. Boettcher, D. Henderson, M. A. Gratton, R. W. Daneilson, and C. D. Byrne, *Ear Hear.*, 1987, **8**, 192–212.

M. C. Brouwer, D. van de Beek, S. G. Heckenberg, L. Spanjaard, and J. de Gans, *Clin. Infect. Dis.*, 2006, **43**, 1233–1238.

D. L. Burk and A. M. Berghuis, *Pharmacol. Therapeut.*, 2002, **93**, 283–292.

Y. Chen, W. G. Huang, D. J. Zha, J. H. Qiu, J. L. Wang, S. H. Sha, and J. Schacht, *Hear. Res.*, 2007, **226**, 178–182.

M. Chittapragada, S. Roberts, and Y. W. Ham, *Perspect. Medicin. Chem.*, 2009, **3**, 21–37.

J. W. Chow, *Clin. Infect. Dis.*, 2000, **31**, 586–589.

L. Q. Christensen, J. Bonde, and J. P. Kampmann, *Acta. Anaesth. Scand.*, 1993, **37**, 231–244.

G. L. Coffey, L. E. Anderson, M. W. Fisher, M. M. Galbraith, A. B. Hillegas, D. L. Kohberger, P. E. Thompson, K. S. Weston, and J. Ehrlich, *Antibiot. Chemother.*, 1959, **9**, 730–738.

E. Collatz, C. Carlier, and P. Courvalin, *J. Gen. Microbiol.*, 1984, **130**, 1665–1671.

D. J. Conrad, A. E. Stenbit, and E. M. Zettner, *Pharmacogenet. Genomics.*, 2008, **18**, 1095–1102.

A. Contrepois, N. Brion, J. J. Garaud, F. Faurisson, F. Delatour, J. C. Levy, J. C. Deybach, and C. Carbon, *Antimicrob. Agents Chemother.*, 1985, **27**, 520–524.

H. D. Coutinho, V. S. Falcão-Silva, and G. F. Gonçalves, *Int. Arch. Med.*, 2008, **1**, 24.

D. M. Daigle, D. W. Hughes, and G. D. Wright, *Chem. Biol.*, 1999, **6**, 99–110.

D. H. Davies, A. K. Mallams, M. Counelis, D. Loebenberg, E. L. Moss, Jr, and J. A. Waitz, *J. Med. Chem.*, 1978, **21**, 189–193.

J. C. Davies, E. W. F. W. Alton, and A. Bush, *Brit. Med. J.*, 2007, **335**, 1255–1259.

B. D. Davis, *Rev. Infect. Dis.*, 1982, **4**, 237–245.

M. A. Davis, K. N. K. Baker, L. H. Orfe, D. H. Shah, T. E. Besser, and D. R. Call, *Antimicrob. Agents Chemother.*, 2010, **54**, 2666–2669.

P. Duff, *Obstet. Gynecol. Clinics N. America*, 1992, **19**, 511–517.

J. Y. Dupuis, R. Martin, and J. P. Tetrault, *Can. J. Anaeth.*, 1989, **36**, 407–411.

E. Durante-Mangoni, A. Grammatikos, R. Utili, and M. E. Falagas, *Int. J. Antimicrob. Agents*, 2009, **33**, 201–205.

P. M. Flatt and T. Mahmud, *Nat. Prod. Rep.*, 2007, **24**, 358–392.

M. E. Falagas, D. K. Matthaiou and I. A. Bliziotis. *J Antimicrob Chemother.*, 2006, **57**, 639–647.

A. Forge and L. Li, *Hear. Res.*, 2000, **139**, 97–115.

A. Forge and J. Schacht, *Audiol. Neurootol.*, 2000, **5**, 3–22.

J.-L. Gaillard, C. Silly, A. Le Masne, B. Mahut, F. Lacaille, G. Cheron, V. Abadie, P. Hubert, V. Matha, C. Coustere, and J. Fermanian, *Antimicrob. Agents Chemother.*, 1995, **39**, 253–255.

D. E. Geller, W. H. Pitlick, P. A. Nardella, W. G. Tracewell, and B. W. Ramsey, *Chest*, 2002, **122**, 219–226.

H. Giamarellou, *Am. J. Med.*, 1986, **80**, 126–137.

D. N. Gilbert, *Clin. Infect. Dis.*, 1997, **24**, 816–819.

R. A. Giuliano, G. A. Verpooten, L. Verbist, R. Weeden, M. E. De Broe, *J. Pharmacol. Exp. Ther.*, 1986, **236**, 470–475.

P. G. Gooding, E. Berman, A. Z. Lane, and K. Agre, *J. Infect. Dis.*, 1976, *134(Suppl.)*, S441–S452.

O. W. Guthrie, *Toxicology*, 2008, **249**, 91.

S. Hanessian, A. Kornienko, and E. E. Swayze, *Tetrahedron*, 2003, **59**, 995–1007.

S. Hanessian, K. Pachamuthu, J. Szychowski, A. Giguère, E. E. Swayze, M. T. Migawa, B. François, J. Kondo, and E. Westhof, *Bioorg. Med. Chem. Lett.*, 2010, **20**, 7097–7101.

Y. Hashimoto, T. Shima, S. Matsukawa, and K. Iwatsuki, *Nouv. Presse. Med.*, 1975, **55**, 1379–1390.

T. Hermann, *Biochimie*, 2002, **84**, 869–875.

S. N. Hobbie, P. Pfister, C. Brüll, E. Westhof, and E. C. Böttger, *Antimicrob. Agents Chemother.*, 2005, **49**, 5112–5118.

H. A. Holt, J. M. Broughall, M. McCarthy, and D. S. Reeves, *Infection*, 1976, **4**, 107–109.

K. Igarashi, T. Sugawara, T. Honma, Y. Tada, H. Miyazaki, H. Nagata, M. Mayama, and T. Kubota, *Carbohyd. Res.*, 1982, **109**, 73–88.

S. Jana and J. K. Deb, *Curr. Drug Targets*, 2005, **6**, 353–361.

S. Jana and J. K. Deb, *Appl. Microbiol. Biotechnol.*, 2006, **70**, 140–150.

R. Janknegt, *Pharm. World. Sci.*, 1990, **12**, 81.

M. Jezowska-Bojczuk, W. Bal, and H. Koztowski, *Inorg. Chim. Acta*, 1998a, **275**, 541–545.

M. Jezowska-Bojczuk, W. Bal, and H. Koztowski, *Carbohydr. Res.*, 1998b, **313**, 265–269.

E. Jin, V. Katritch, W. K. Olson, M. Kharatisvill, R. Abagyan, and D. S. Pilch, *J. Mol. Biol.*, 2000, **298**, 95–110.

A. B. Kaiser and Z. A. McGee, *New Engl. J. Med.*, 1975, **293**, 1215–1220.

M. Kaul and D. S. Pilch, *Biochemistry*, 2002, **41**, 7695–7706.

M. Kaul, C. M. Barbieri, J. E. Kerrigan, and D. S. Pilch, *J. Mol. Biol.*, 2003, **326**, 1373–1387.

H. Kawaguchi, *J. Infect. Dis.*, 1976, **134** (Suppl.), S242–S248.

K. S. Kim, *Minerva Pedriatr.*, 2009, **61**, 531–548.

W. Kingston, *Res. Policy*, 2000, **29**, 679–710.

W. Kingston, *J. Hist. Med. Allied Sci.*, 2004, **59**, 441–462.

H-G. Knaus, J. Striessnig, A. Koza, and H. Glossman, *Naunyn-Schmideberg's Arc. Pharmacol.*, 1987, **336**, 583–586.

J. Kondo, M. Hainrichson, I. Nidelman, D. Shallom-Shezifi, C. M. Barbieri, D. S. Pilch, E. Westhof, and T. Baasov, *Chembiochem*, 2007, **8**, 1700–1709.

T. A. Kovesi, R. Swartz, and N. MacDonald, *N. Engl. J. Med.*, 1998, **338**, 65–66.

F. Kudo and T. Eguchi, *J. Antibiot.*, 2009, **62**, 471–481.

A. Z. Lane, G. E. Wright, and D. Blair, *Am. J. Med.*, 1977, **78**, 911–918.

L. Leibovici, L. Vidal, and M. Paul, *J. Antimicrob. Chemother.*, 2009, **63**, 246–251.

J. Li, F.-I. Chiang, H.-N. Chen, and C.-W. T. Chang, *Bioorg. Med. Chem.*, 2007, **15**, 7711–7719.

M. Lippmann, E. Yang, E. Au, and C. Lee, *Anaeth. Analg.*, 1982, **61**, 767–770.

N. M. Llewellyn and J. B. Spencer, *Nat. Prod. Rep.*, 2006, **23**, 864–874.

M. A. Lopez-Gonzalez, F. Delgado, and M. Luca, *Hear. Res.*, 1999, **136**, 165–168.

I. Lutsar, G. H. McCracken, Jr, and I. R. Friedland, *Clin. Infect. Dis.*, 1998, **27**, 1117–1129.

S. R. Lynch and J. D. Puglisi, *J. Mol. Biol.*, 2001a, **306**, 1023–1035.

S. R. Lynch and J. D. Puglisi, *J. Mol. Biol.*, 2001b, **306**, 1037–1058.

R. Macmaster, N. Zelinskaya, M. Savic, C. R. Rankin, and G. L. Conn, *Nucleic Acids Res.*, 2010, **38**, 7791–7799.

K. Maeda, M. Ueda, K. Yagishita, S. Kawaji, S. Kondo, M. Murase, T. Takeuchi, Y. Okami, and H. Umezawa, *J. Antibiot. (Tokyo)*, 1957, **10**, 228–231.

S. Magnet and J. S. Blanchard, *Chem. Rev.*, 2005, **105**, 477–497.

A. Mangia,US Patent 5763587, 1998.

R. Marcus, M. Paul, H. Elphick, and L. Leibovici, *Int. J. Antimicrob. Agents*, 2011, **37**, 491–503.

G. H. McCracken, Jr, *Am. J. Med.*, 1984, **76**, 215–223.

P. M. Mendelman, A. L. Smith, J. Levy, A. Weber, B. Ramsey, and R. L. Davis, *Am. Rev. Respir. Dis.*, 1985, **132**, 761–765.

G. H. Miller, G. Arcieri, M. J. Weinstein, and J. A. Waitz, *Antimicrob. Agents Chemother.*, 1976, **10**, 827–836.

M.-P. Mingeot-Leclercq, Y. Glupczynski, and P. M. Tulkens, *Antimicrob. Agents Chemother.*, 1999, **43**, 727–737.

K. Mizuta, A. Saito, and T. Watanabe, *Hear. Res.*, 1999, **129**, 83–91.

S. K. Moestrup, S. Cui, H. Vorum, C. Bregengard, S. E. Bjorn, K. Norris, and E. I. Christiensen, *J. Clin. Invest.*, 1995, **96**, 1404–1413.

I. Morić, M. Savić, T. Ilić-Tomić, S. Vojnović, S. Bajkić, and B. Vasiljević, *J. Med. Biochem.*, 2010, **29**, 165–174.

A. G. Paradelis, C. J. Triantaphyllidis, M. Mirondou, L. G. Crassaris, D. N. Karachalios, and M. M. Giala, *Method. Find. Exp. Clin. Pharmacol.*, 1988, **10**, 687–690.

A. Paradkar, A. Trefzer, R. Chakraburtty, and D. Stassi, *Crit. Rev. Biotechnol.*, 2003, **23**, 1–27.

M. Pichler, Z. Wang, C. Grabner-Weiss, D. Reimer, S. Hering, M. Grabner, H. Glossmann, and J. Striessnig, *Biochemistry*, 1996, **35**, 14 659–14 664.

K. Poole, *Antimicrob. Agents Chemother.*, 2005a, **49**, 479–487.

K. Poole, *Antimicrob. Chemother.*, 2005b, **56**, 20–51.

K. E. Price, *Antimicrob. Agents Chemother.*, 1986, **29**, 543–548.

M. S. Ramirez and M. E. Tolmasky, *Drug Resis. Updates*, 2010, **13**, 151–171.

B. W. Ramsey, M. S. Pepe, J. M. Quan, K. L. Otto, A. B. Montgomery, J. Williams-Warren, M. Vasiljev-K, D. Borowitz, C. M. Bowman, B. C. Marshall, S. Marshall, and A. L. Smith, *N. Engl. J. Med.*, 1999, **340**, 23–30.

E. J. Ratajak and R. C. Nubel,US Patent 2875136, 1959.

M. I. Recht, S. Douthwaite and J. D. Puglisi, *EMBO J.*, 1999, **18**, 3133–3138.

G. Rukholm, C. Mugabe, A. O. Azghani, and A. Omri, *Int. J. Antimicrob. Agents*, 2006, **27**, 247–252.

S. Sánchez, A. Chávez, A. Forero, Y. García-Huante, A. Romero, M. Sánchez, D. Rocha, B. Sánchez, M. Ávalos, S. Guzmán-Trampe, R. Rodríguez-Sanoja, E. Langley, and B. Ruiz, *J. Antibiot.*, 2010, **63**, 442–459.

W. E. Sanders, Jr, C. Hartwig, N. Schneider, R. Cacciatore, and H. Valdez, *Tubercle*, 1982, **63**, 201–208.

J. Schacht, *Otolaryngol. Clin. North. Am.*, 1993, **26**, 845–856.

A. Schatz and S. A. Waksman, *Proc. Soc. Exper. Biol. Med.*, 1944, **57**, 244–248.

A. Schatz, E. Bugie, and S. A. Waksman, *Proc. Soc. Exper. Biol. Med.*, 1944, **55**, 66–69.

J. J. Schentag, G. Lasezkay, and T. J. Cumbo, *Antimicrob. Agent. Chemother.*, 1978, **13**, 649–656.

A. Scheunemann, W. D. Graham, F. A. P. Vendeix, and P. F. Agris, *Nucleic Acids Res.*, 2010, **38**, 3094–3105.

R. Schiffelers, G. Storm, and I. Bakker-Woudenberg, *J. Antimicrob. Chemother.*, 2001, **48**, 333–344.

P. Setny and J. Trylska, *J. Chem. Inf. Model*, 2009, **49**, 390–400.

S.-H. Sha and J. Schacht, *Lab. Invest.*, 1999, **79**, 807–813.

S. Shakil, R. Khan, R. Zarrilli, and A. U. Khan, *J. Biomed. Sci.*, 2008, **15**, 5–14.

J. G. Silva and I. Carvalho, *Curr. Med. Chem.*, 2007, **14**, 1101–1119.

Y. Takagi, T. Miyake, T. Tsuchiya, S. Umezawa, and H. Umezawa, *J. Antibiot.*, 1973, **26**, 403–406.

Y. Takagi, T. Miyake, T. Tsuchiya, S. Umezawa, and H. Umezawa, *Bull. Chem. Soc. Jpn.*, 1976, **49**, 3649–3651.

T. Takeuchi, T. Hikiji, K. Nitta, S. Yamazaki, S. Abe, H. Takayama, and H. Umezawa, *J. Antibiot. (Tokyo)*, 1957, **10**, 107–114.

T. Tran and M. D. Disney, *Biochemistry*, 2010, **49**, 1833–1842.

P. C. Trussel,US Patent 2541726, 13 February 1951.

A. R. Tunkel, B. J. Hartman, S. L. Kaplan, B. A. Kaufman, K. L. Roos, W. M. Scheld, and R. J. Whitley, *Clin. Infect. Dis.*, 2004, **39**, 1267–1284.

J. Turnidge, *Infect. Dis. Clin. N. Am.*, 2003, **17**, 503–528.

J. Turnidge and D. L. Paterson, *Clin. Microbiol. Rev.*, 2007, **20**, 391–408.

H. Umezawa, M. Ueda, K. Maeda, K. Yagishita, S. Kondo, Y. Okami, R. Utahara, Y. Osato, K. Nitta, and T. Takeuchi, *J. Antibiot. (Tokyo)*, 1957, **10**, 181–188.

S. Umezawa, *Adv. Carbohydr. Chem. Biochem.*, 1974, **30**, 111–182.

S. Umezawa, *Pure Appl. Chem.*, 1978, **50**, 1453–1476.

S. Umezawa, *Jpn. J. Antibiot.*, 1979, **32(Suppl.)**, S60–S72.

S. B. Vakulenko and S. Mobashery, *Clin. Microbiol. Rev.*, 2003, **16**, 430–450.

G. J. van Asselt, J. S. Vliegenthart, P. L. Petit, J. A. van de Klundert, and R. P. Mouton, *J. Antimicrob. Chemother.*, 1992, **30**, 651–659.

Q. Vicens and E. Westhof, *Chem. Biol.*, 2002, **9**, 747–755.

G. H. Wagman, J. A. Marquez, and M. J. Weinstein, *J. Chromatog. A*, 1967, **34**, 210–215.

J. A. Waitz, C. G. Drube. E. L. Moss, E. M. Oden, and J. V. Bailey, *J. Antibiot.*, 1972, **25**, 219–225.

S. A. Waksman and H. A. Lechevalier, *Science*, 1949, **25**, 305–307.

S. A. Waksman and A. Schatz,US Patent 2449866, 21 September 1948.

S. A. Waksman, J. Frankel, and O. Graessle, *J. Bacteriol.*, 1949, **58**, 229–237.

F. Walter, Q. Vicens, and E. Westhof, *Curr. Opin. Chem. Biol.*, 1999, **3**, 694–704.

S. Wang, Q. Bian, Z. Liu, and Y. Feng, *Hear. Res.*, 1999, **137**, 1–7.

M. J. Weinstein, G. M. Luedemann, E. M. Oden, G. H. Wagman, J. P. Rosselet, J. A. Marquez, C. T. Coniglio, W. Charney, H. L. Herzog, and J. Black, *J. Med. Chem.*, 1963, **6**, 463–464.

M. J. Weinstein, J. A. Marquez, R. T. Testa, G. H. Wagman, E. M. Oden, and J. A. Waitz, *J. Antibiot. (Jpn.)*, 1970, **23**, 551–554.

K. T. Welch, K. G. Virga, N. A. Whittlemore, C. Özen, E. Wright, C. L. Brown, R. E. Lee, and E. H. Serpersu, *Bioorg. Med. Chem.*, 2005, **13**, 6252–6263.

World Health Organization, Treatment of Tuberculosis Guidelines, 4th Edn, 2010 (http://whqlibdoc.who.int/publications/2010/9789241547833_eng.pdf, last accessed 16 March 2012).

D. L. Wong and J. A. Rutka, *Otolaryngol. Head Neck Surg.*, 1997, **116**, 404–410.

J. J. Wright, *J. Chem. Soc., Chem. Commun.*, 1975, **6**, 206–208.

G. D. Wright, *Curr. Opinion Microbiol.*, 1999, **2**, 499–503.

G. D. Wright and P. R. Thompson, *Frontiers Biosci.*, 1999, **4**, d9–21.

O. Wyss and G. E. Schaiberger, *J. Bacteriol.*, 1953, **66**, 49–51.

K. Yamane, J.-I. Wachino, Y. Doi, H. Kurokawa, and Y. Arakawa, *Emerg. Infect. Dis.*, 2005, **11**, 951–953.

R.-B. Yan, M. Yuan, Y. Wu, X. You, and X.-S. Ye, *Bioorg. Med. Chem.*, 2011, **19**, 30–40.

G. Yang, J. Trylska, Y. Tor, and J. A. McCammon, *J. Med. Chem.*, 2006, **49**, 5478–5490.

J. Zhou, G. Wang, L. H. Zhang, and X. S. Ye, *Med. Res. Rev.*, 2007, **27**, 279–316.

4.2

Macrolide antibiotics

Key Points

- Erythromycin, a commonly used macrolide antibiotic, has a broad spectrum of antibacterial activity which is similar to that of the penicillins, and so is frequently the agent of choice for patients who have penicillin allergies.
- Macrolides inhibit protein synthesis by blocking the exit of the nascent protein tunnel (in the 23S ribosomal RNA component of the large ribosome subunit) by binding to a high-affinity site inside the tunnel. The main nucleotide component of this binding pocket is adenine in prokaryotes, and the selectivity of the macrolides is due to the fact that they do not bind to guanine, which is the corresponding nucleotide in eukaryotes.
- For further information, see Zhanel *et al.* (2001).

Macrolide antibiotics used clinically include azithromycin, clarithromycin, erythromycin, and roxithromycin; their structures are shown in Figure 4.2.1 and their therapeutic indications are listed in Table 4.2.1.

4.2.1 Discovery

As we shall see on a number of occasions throughout this book, the discovery of the penicillins sparked worldwide interest in the antibacterial activity of microbial extracts. Samples from various locations and origins were tested in microbiology laboratories and those which exhibited antibacterial activity were selected for further study, involving the separation of the often-complex mixtures and the identification of the active component(s). The macrolide antibiotics are a product of just such a drug-discovery programme, with the agent which has been most used clinically, erythromycin, having been discovered in 1949 and introduced into the clinic in 1952. Erythromycin was discovered when researchers at Eli Lilly investigated the antibacterial activity of the metabolites of *Streptomyces erythreus* (now reclassified as *Saccharopolyspora erythreus*) isolated from a soil sample from the Iloilo region of the Philippines (for this reason, erythromycin was originally referred to as Ilotycin) (Bunch and McGuire, 1953).

Antibacterial Agents: Chemistry, Mode of Action, Mechanisms of Resistance and Clinical Applications, First Edition.
Rosaleen J. Anderson, Paul W. Groundwater, Adam Todd and Alan J. Worsley.
© 2012 John Wiley & Sons, Ltd. Published 2012 by John Wiley & Sons, Ltd.

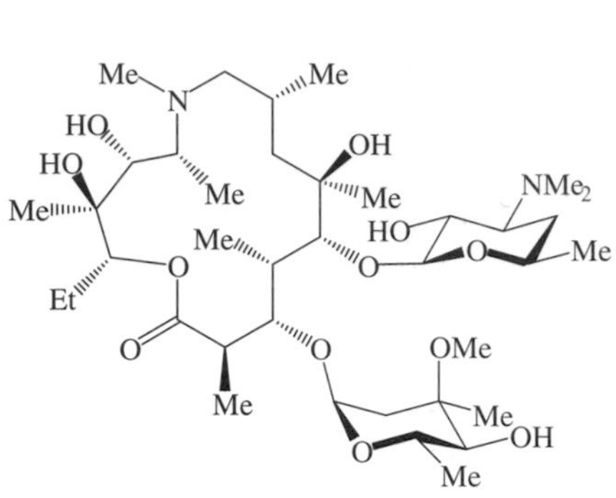

Azithromycin

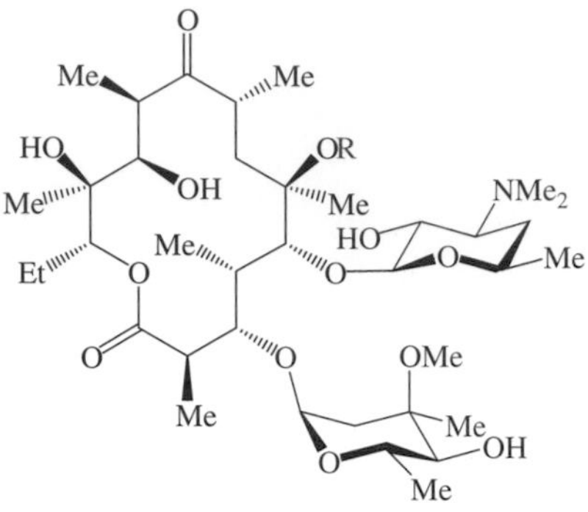

Clarithromycin (R = Me)
Erythromycin (R = H)

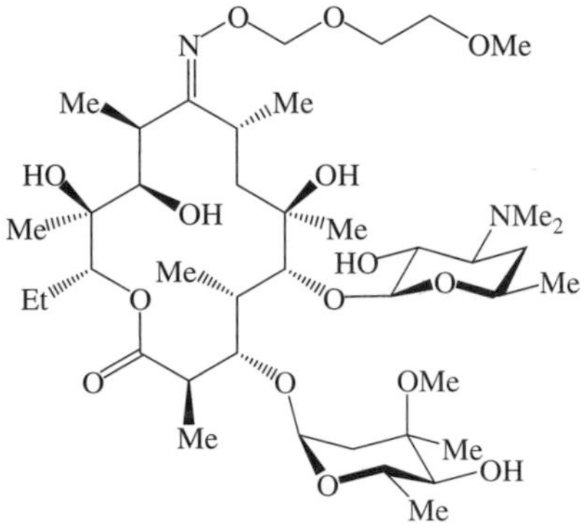

Roxithromycin

Figure 4.2.1 *Macrolide antibiotics*

Even today, with a vast array of spectroscopic techniques at our disposal, the characterisation of an unknown metabolite isolated from a natural product extract can be very challenging. How then did the research groups working, in the 1950s, on the structure of erythromycin manage to determine that it contained a 14-membered lactone (ester) ring, as well as two sugar units? (The term macrolide was coined by R. B. Woodward to denote a compound which has a macrocyclic (containing 12 or more atoms) lactone ring.)

Table 4.2.1 *Therapeutic indications for the macrolide antibiotics*

Macrolide antibiotic	Indications
Azithromycin	Chlamydial infections, e.g. urethritis, cervicitis, community-acquired pneumonia, streptococcal pharyngitis, prevention and treatment of *Mycobacterium avium* complex (MAC), donovanosis, typhoid, prevention and treatment of pertussis
Clarithromycin	Prevention and treatment of MAC, eradication of *Helicobacter pylori*, prevention and treatment of pertussis
Erythromycin	Upper and lower RTIs, rheumatic fever prophylaxis (in penicillin allergy), legionnaires' disease
Roxithromycin	Upper and lower RTIs, community-acquired pneumonia

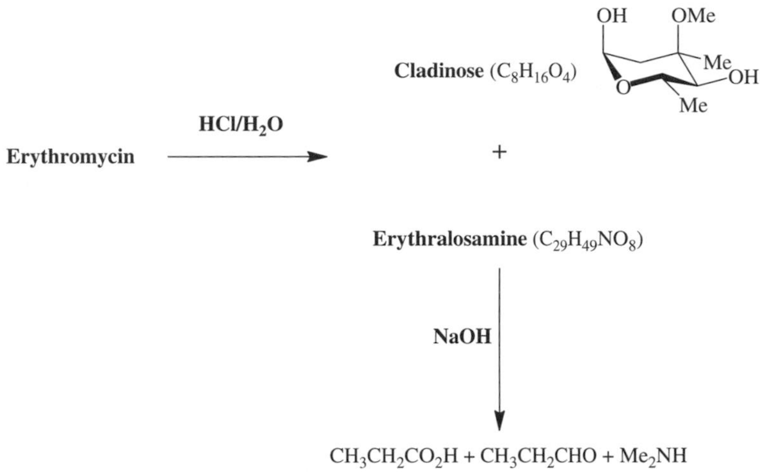

Cladinose ($C_8H_{16}O_4$)

OH   OMe

Me
OH
Me

Erythromycin $\xrightarrow{\text{HCl/H}_2\text{O}}$ +

Erythralosamine ($C_{29}H_{49}NO_8$)

NaOH

$CH_3CH_2CO_2H + CH_3CH_2CHO + Me_2NH$

Scheme 4.2.1   *Examples of chemical reactions used in the structure elucidation of erythromycin*

The absolute configuration of erythromycin had to wait until 1965 (Harris *et al.*, 1965), when an X-ray crystallographic analysis established the absolute stereochemistry at each of the stereogenic centres, but the structural units had been determined a decade earlier (well before the use of NMR spectroscopy became commonplace) through extensive degradation studies in which the macrocylic ring and sugar units were converted into smaller fragments that could be identified by the comparison of their physical properties, such as melting point, with those of previously prepared and characterised authentic standards (Scheme 4.2.1) (Flynn *et al.*, 1954). Note that, at the time this work was performed, the research group did not know the exact molecular formula for erythromycin. The products from the initial acid hydrolysis, cladinose and erythralosamine, had to be subjected to further reactions in order to determine their chemical constitution, with, for example, the dimethylamine produced by the basic hydrolysis of erythralosamine originating from the dimethylamino group in the desosamine sugar unit. You will probably also have noticed that erythromycin was introduced into the clinic before its chemical structure had been fully worked out, something which would not be possible today, given the very strict characterisation requirements of the various drug administrations around the world.

4.2.2   Synthesis

The total synthesis of the erythromycins (Figure 4.2.2) poses a supreme challenge and has attracted the attention of some of the world's most eminent synthetic chemists, leading to many elegant examples of the total synthesis of complex natural products. The total synthesis of the erythronolide A aglycone (lacking the sugar units) was first reported by E. J. Corey (Nobel Prize in Chemistry in 1990) in a series of articles in the late 1970s (Scheme 4.2.2) (Corey *et al.*, 1979 and references cited therein), and the total synthesis of erythromycin (known then as erythromycin A) by R. B. Woodward (Nobel Prize in Chemistry in 1965) in a series of articles in 1981, after his death (Scheme 4.2.3) (Woodward *et al.*, 1981 and references cited therein). The Woodward synthesis is particularly elegant, as the dithiadecalin intermediate supplies **both** the C3-C8 and C9-C13 fragments (Scheme 4.2.3).

Once again, erythromycin is such a complex antibiotic that its commercial production by total synthesis will never be feasible, and it is obtained from the submerged culture of free or immobilised *Saccharopolyspora erythraea* (El-Enshasy *et al.*, 2008).

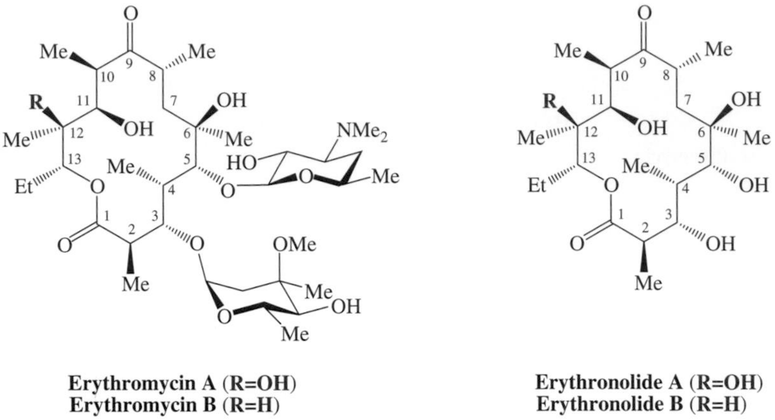

Erythromycin A (R=OH)
Erythromycin B (R=H)

Erythronolide A (R=OH)
Erythronolide B (R=H)

Figure 4.2.2   *Erythromycins A and B and their aglycones, erythronolides A and B*

We have now seen a number of examples of how very complex semi-synthetic antibiotics can be prepared through the combination of fermentation (to give the complex natural product) and chemical modification, so you will no doubt already have spotted that both clarithromycin and roxithromycin are semi-synthetic macrolide antibiotics. Clarithromycin can be obtained in a five-step synthetic procedure, from erythromycin oxime (Brunet *et al.*, 2007), while roxithromycin can also be prepared from this oxime (Massey *et al.*, 1970) in a single step (Scheme 4.2.4) (Gouin d'Ambrieres *et al.*, 1982). What is not so obvious is that azithromycin is also a semi-synthetic macrolide, having originally been produced by PLIVA Pharmaceuticals from erythromycin oxime via a sequence of reactions which included the well-known Beckmann rearrangement (Djokić *et al.*, 1986). For more on the synthesis of the erythromycins, see Paterson and Mansuri (1985).

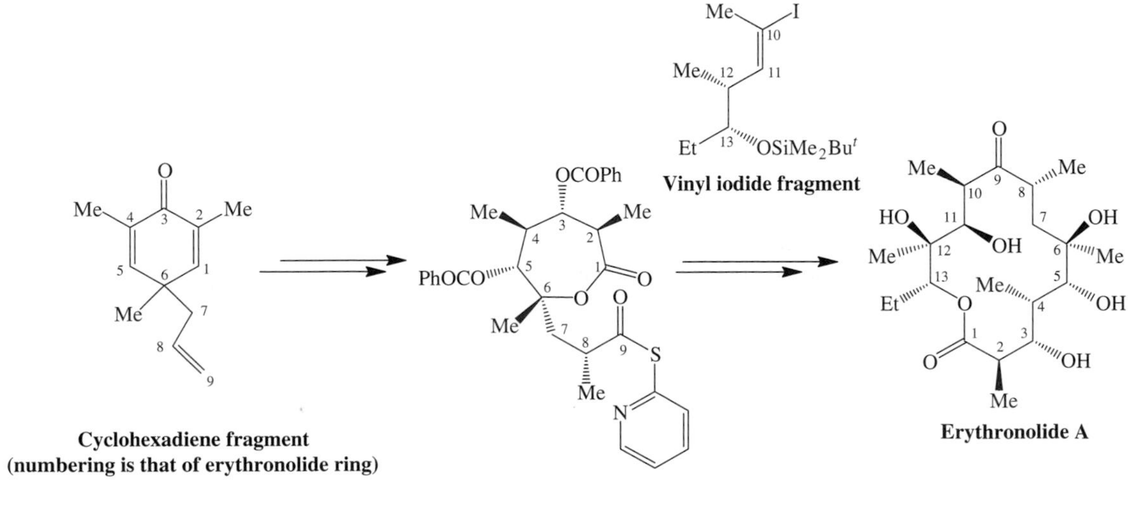

Cyclohexadiene fragment
(numbering is that of erythronolide ring)

Vinyl iodide fragment

Erythronolide A

Scheme 4.2.2   *Corey's total synthesis of erythronolide A (38 steps from the cyclohexadiene fragment; 0.04% overall yield)*

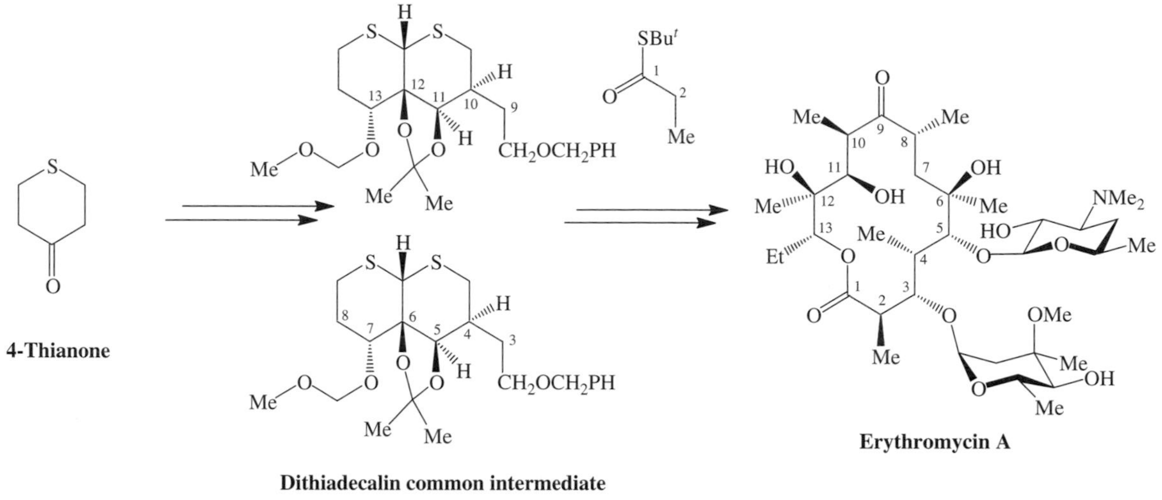

Dithiadecalin common intermediate

Scheme 4.2.3 *Woodward's total synthesis of erythromycin (56 steps from 4-thianone; 0.01% overall yield)*

4.2.3 Bioavailability

Erythromycin is sensitive to acid hydrolysis, undergoing degradation below pH 6.9, both *in vitro* and *in vivo*. Low pH also adversely affects its bioavailability, which is unpredictable (at around 15–45%) (Schentag and Ballow, 1991; Kanatani and Guglielmo, 1994) and dependent upon the exact erythromycin salt used (Billow *et al.*, 1964; Butzler *et al.*, 1979). Even the use of an enteric coating (which is designed to degrade and release the drug only at the higher pH of the small intestine, thereby protecting the drug from the gastric acid, pH 1–3) does not completely solve this problem, as the pH of the small intestine is 6.5–8.0, which still leads to appreciable degradation. The principal product of the acid-catalysed degradation is anhydroerythromycin ('anhydro' indicating the loss of water), which is formed from the hemi-acetal (Scheme 4.2.5) (Hassanzadeh *et al.*, 2006a, 2006b). Anhydroerthryomycin has very low antibacterial activity, but its formation has further serious implications as it is a steroid 6β-hydroxylase inhibitor, as a result of its formation of a complex with cytochrome P450, and so reduces the oxidative metabolism of a range of drugs (Stupans and Sansom, 1991).

To overcome the significant problem of acid instability, and the equally important problem of extreme bitter taste, two main approaches have been taken:

- Different organic acids have been used to prepare erythromycin salts with varying aqueous solubilities, on the principle that an insoluble salt does not dissolve appreciably in saliva and so interacts less with the taste receptors, and provides a range of pharmacokinetic properties. Erythromycin stearate (Figure 4.2.3) is very insoluble in aqueous solution at < 0.5 mg/mL; its low solubility at acidic pH also provides some protection from degradation, leading to slightly enhanced acid stability. Erythromycin is released from the salt in the lower GI tract when the ionised form is in equilibrium with the free amine, allowing absorption of the unionised (amine) form.
- A prodrug strategy has been used: erythromycin ethyl succinate (previously marketed as Erythroped or Eryped) consists of a double succinate ester – the carboxyl group at one end of the succinate is linked to erythromycin through its 2′-OH group and the other end is a simple ethyl ester (Figure 4.2.3) (Martin *et al.*, 1972; Sinkula, 1974).

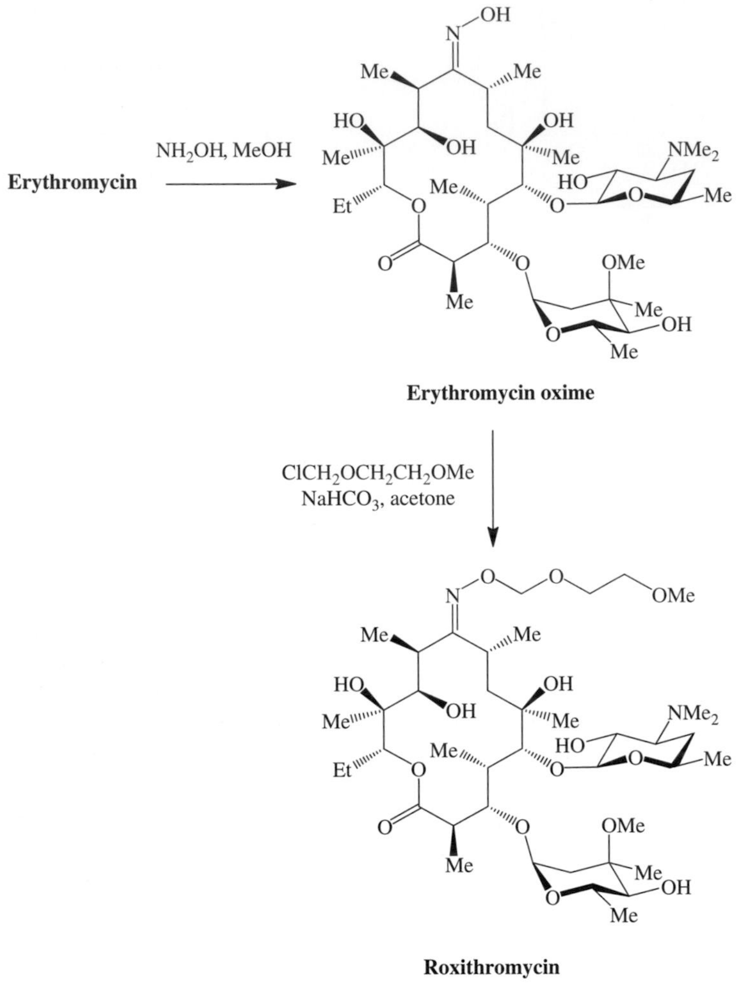

Erythromycin $\xrightarrow{\text{NH}_2\text{OH, MeOH}}$

Erythromycin oxime

ClCH$_2$OCH$_2$CH$_2$OMe
NaHCO$_3$, acetone

Roxithromycin

Scheme 4.2.4   *Preparation of the semi-synthetic macrolide antibiotic roxithromycin*

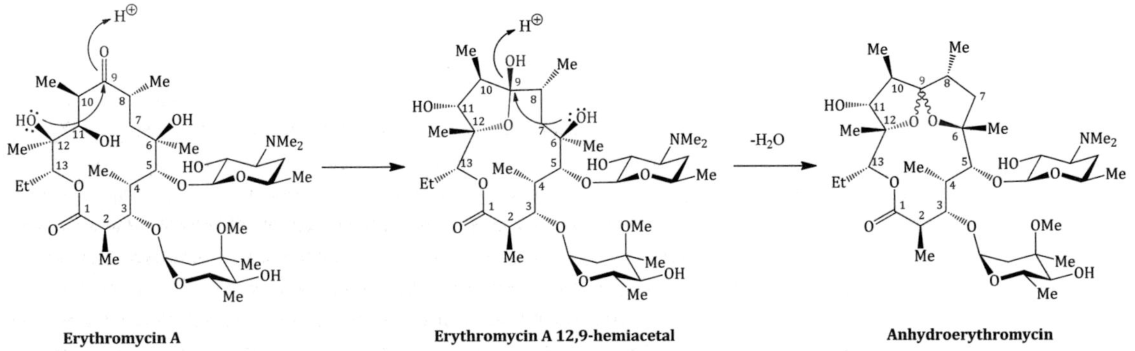

Erythromycin A   **Erythromycin A 12,9-hemiacetal**   **Anhydroerythromycin**

Scheme 4.2.5   *Acid-catalysed degradation of erythromycin*

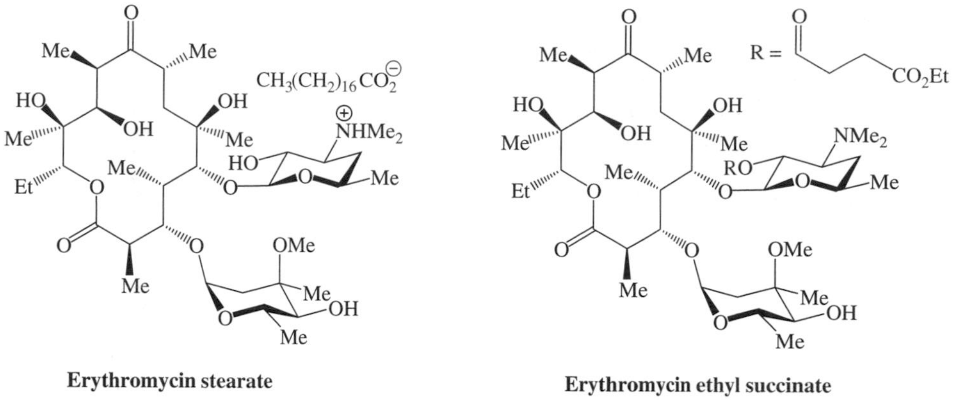

Erythromycin stearate **Erythromycin ethyl succinate**

Figure 4.2.3 *Structures of the salt erythromycin stearate and the prodrug erythromycin ethyl succinate (ethyl succinate prodrug moiety highlighted in blue)*

New salt forms of erythromycin A (Manna *et al.*, 2004) and prodrug forms of erythromycin B, which has greater acid stability (see below), remain of interest (Hassanzadeh *et al.*, 2007).

Pharmacokinetic evaluations of orally administered erythromycin stearate and erythromycin ethyl succinate showed that they were best absorbed, resulting in optimal bioavailability, if taken immediately before food, achieving similar maximal plasma concentrations (C_{max}) of around 2.7–2.8 µg/mL at t_{max} 1–1.5 hours (Malmborg, 1978; Thompson *et al.*, 1980). Taking food before administration leads to a reduction in bioavailability, and a decrease in C_{max} and increase in t_{max}, although absorption after food was still considered to be adequate for achieving MICs and the optimal therapeutic effect. Release of erythromycin from both forms is efficient and bioequivalent to erythromycin base. The acid stability of the other macrolides is considerably better, particularly of azithromycin, which is about 10% degraded after 20 minutes at pH 2 (and 37 °C), while erythromycin will reach the same level of degradation under the same conditions in less than 4 seconds (Lode *et al.*, 1996). The other macrolides, such as azithromycin, clarithyromycin, and the newest relative, telithromycin (which you will meet in Subsection 4.2.9), are relatively well absorbed after oral administration, achieving 37, 52, and 57% bioavailability, respectively, and their absorption is less affected by food than that of erythromycin (Zeitlinger *et al.*, 2009). It is perhaps surprising that these molecules are so well absorbed, considering their high molecular weight, number of groups with hydrogen bonding potential, and interaction with the efflux transporter P-glycoprotein; all of which would be predicted to lead to a decrease in bioavailability. It has been suggested that their unexpectedly high absorption from the gastrointestinal (GI) tract may be the result of facilitated uptake via an as-yet-unknown mechanism (Lan *et al.*, 2009).

The volume of distribution of the macrolide antibiotics, alongside their low serum concentration, indicates high tissue distribution, which is supported by their intracellular accumulation and excellent activity against infection in a wide variety of tissues (Table 4.2.2). By fortunate coincidence, the macrolides accumulate in phagocytic cells, such as polymorphonuclear leukocytes (PMNLs) and macrophages, and are transported to, and subsequently released at, the site of infection, which enhances the macrolide concentration in the presence of the bacteria. Although all macrolides are taken into tissues and cells, including phagocytes, azithromycin, in particular, displays extensive selective tissue accumulation into PMNLs and macrophages and a very long tissue half-life (Amsden, 2001).

Clarithromycin is metabolised in the liver, primarily by oxidative *N*-demethylation and hydroxylation. The latter results in a major metabolite, 14(*R*)-hydroxyclarithromycin, which has comparable antibacterial activity to clarithromycin, is concentrated in tissue, has a longer serum half-life, and undoubtedly makes a contribution to the antibacterial activity of clarithromycin (Martin *et al.*, 2001). Although other macrolides are

Table 4.2.2 *Pharmacokinetic parameters for selected macrolide antibiotics (Amsden, 1996; Rodvold, 1999; Zhanel et al., 2001)*

Macrolide	Volume of distribution, V_d (L/kg)	Peak serum level, C_{max} (μg/mL)	Serum half life, $t_{1/2}$ (h)	I:E ratio[a] (neutrophils)	Dose frequency per day
Azithromycin (500 mg)	100	0.4	68–90	79–220	1
Clarithromycin (500 mg)	2.5	2.4–2.8	3–5	5–20	2
Erythromycin stearate (500 mg)	1.5	0.5–2.9	2–3	1–8	4
Roxithromycin (300 mg)	0.44	8.7	7–10	23.7	1

[a] I:E ratio = intracellular:extracellular ratio.

subject to hepatic metabolism, the metabolites do not exhibit significant antibacterial activity (Kanatani and Guglielmo, 1994). The majority of the macrolide antibiotics are mostly eliminated through the intestinal tract, with a variable, but minor, amount excreted renally. The exception is clarithromycin and its active metabolite, 14(*R*)-hydroxyclarithromycin, which are primarily excreted renally in the urine.

A lot of research has been directed at improving the bioavailability and pharmacokinetic and pharmaco-dynamic properties of the macrolides (Zuckerman *et al.*, 2009; Ma and Ma, 2010), through structural modifications, and this field remains of great interest (Liang *et al.*, 2010; Sugimoto and Tanikawa, 2011).

As already mentioned, erythromycin B has greater acid stability than its better-known A form; these two molecules differ only in the substituent at C12, which is OH in erythromycin A and H in erythromycin B (Figure 4.2.2), removing the possibility of acid-catalysed cyclisation of C12-OH on to the carbonyl group at C9. Despite its similar antibacterial activity and superior acid stability, erythromycin B is not currently licensed (Tyson *et al.*, 2011). For more on the bioavailability of the erythromycins, see Jain and Danziger (2004).

4.2.4 Mode of action and selectivity

During protein synthesis, the ribosome moves along the mRNA from the 5′- to the 3′-end and, once the peptide bond has formed, the non-acylated tRNA leaves the P site and the peptide-tRNA moves from the A to the P site. A new tRNA-amino acid (as specified by the mRNA codon) then enters the A site and the peptide chain grows as amino acids are added, until a stop codon is reached, when it leaves the ribosome through the nascent protein exit tunnel.

Macrolides inhibit protein synthesis by blocking the exit of this nascent protein tunnel (in the 23S ribosomal RNA component of the large ribosome subunit) by binding to a high-affinity site inside the tunnel and close to the peptidyl transferase centre, leading to the arrest of protein elongation and the dissociation of shortened peptidyl-transfer RNAs from the ribosome (Figure 4.2.4) (Schlünzen *et al.*, 2001; Gaynor and Mankin, 2005). Although this tunnel is relatively wide, it narrows near to the peptidyl transferase centre and, as the macrolides bind at this constriction, they block the tunnel. This blockage does not affect the joining together of the first few amino acids and the polymerisation process is only halted once the growing peptide chain becomes large enough to reach the blockage.

The main component of this binding pocket is nucleotide 2058[10] of the ribosomal RNA; in bacteria this nucleotide is adenine (A) and macrolides bind strongly to it, with the desosamine sugar (formation of hydrogen bonds between 2′-OH and A2058 and A2509; electrostatic interaction between the -N$^+$HMe$_2$ cation and the G2505 phosphate anion), the ring hydroxyls (hydrogen bonds between the 6, 11, and 12-OH and nucleotides), and the lactone (hydrophobic interactions) all playing key roles in the macrolide binding to this site

---

[10] Nucleotide 2058 of *E. coli*. The *E. coli* numbering system is used to keep nomenclature uniform.

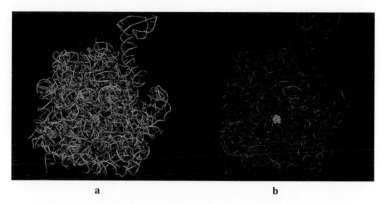

a   b

Figure 4.2.4 *Erythromycin bound inside the nascent protein exit tunnel of the ribosome from* Deinococcus radiodurans; *(a) showing RNA bases, and (b) with bases omitted for clarity and showing only the **RNA backbone** (PDB ID: 1JZY)*

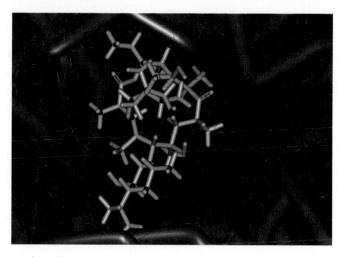

Figure 4.2.5 *Interaction of erythromycin with the nucleotides in its binding site in the protein exit tunnel (PDB ID: 1JZY)*

(Figure 4.2.5) (Schlünzen *et al.*, 2001). In eukaryotic cells, this nucleotide is guanine (G), which is larger and does not have the same favourable interactions with the 14-membered macrolides, so that **the binding affinity of these antibiotics at this site is reduced, the protein tunnel in the eukaryotic ribosome is not blocked by the macrolide, and protein synthesis is unaffected**. This single-nucleotide difference between the prokaryotic and eukaryotic ribosome structures is thus the basis of the selective antibacterial action of the macrolide antibiotics (Böttger *et al.*, 2001).

4.2.5   Bacterial resistance

As we have seen many times already, an understanding of the mode of action of an antibiotic can also help us to understand how bacterial resistance arises. As the macrolides bind to a specific site in the ribosomal protein

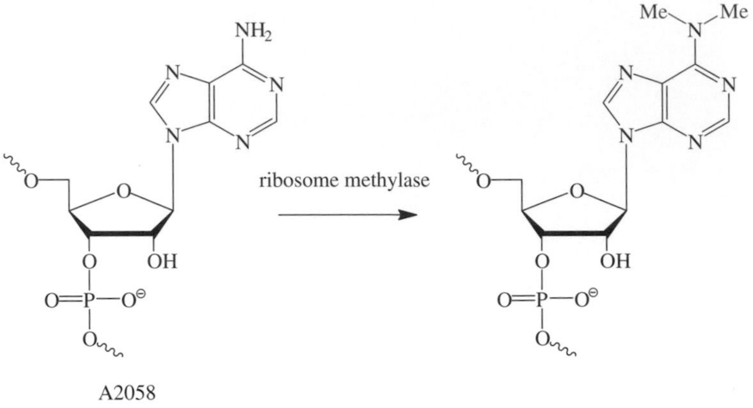

Scheme 4.2.6 *Dimethylation of A2058 by ribosome methylase*

tunnel, it will be no surprise that bacterial resistance can arise due to alterations to this site, resulting in reduced macrolide binding affinity. For example, in *streptococci*, an inducible or constitutive[11] *erm* (erythromycin ribosome methylase) gene of the macrolide-lincosamide-streptogramin B (MLS$_B$)-resistant phenotype gives rise to resistance (Roberts, 2008); in *Streptococcus pneumoniae*, this ribosomal methylase dimethylates a single site, A2058 (on *N*-6) (Scheme 4.2.6), resulting in a decreased binding affinity for erythromycin due to the reduced hydrogen bonding ability and increased size of this nucleotide. This alteration can only take place in a very small time window – it must happen during ribosome assembly, as nucleotide A2058 is buried deep within the ribosome, out of reach of the methylase once ribosome assembly is complete (Leclercq, 2002; Leclercq and Courvalin, 2002).

In *Campylobacter jejuni* and *Campylobacter coli*, which are major causes of gastroenteritis, more than one type of macrolide resistance is involved – modification of the ribosomal target for the macrolides through mutation **and** overexpression of a macrolide efflux pump, CmeABC (which is also involved in the intrinsic and acquired resistance of these bacteria to the quinolones and is the product of the *msr* gene), act in synergy to confer a high level of resistance (Payot *et al.*, 2006). In these cases, alteration of the ribosome occurs as a result of mutations of the 2058 and 2059 nucleotides, again weakening the interaction between the macrolide and its binding site (Matsuoka *et al.*, 1998; Vacher *et al.*, 2005).

Finally, bacteria can develop resistance to the macrolides through the chemical inactivation of the macrolide itself. As the desosamine hydroxyl group (2'-OH) is important for antibacterial activity, deactivation of the macrolide can occur through phosphorylation (addition of phosphate) or glycosylation (addition of a sugar unit) at this position. Another chemical modification occurs in *Enterobacteriaceae*, which express an esterase, encoded by the *ere* gene, which breaks the lactone (ester) bond of the ring to give an inactive, acyclic structure (Wondrack *et al.*, 1996).

4.2.6   Clinical applications

4.2.6.1   Spectrum of activity

The macrolides are broad-spectrum antibiotics and have activity against many Gram positive and some Gram negative bacteria. Erythromycin, the first commericially available macrolide, has an antibacterial spectrum

---

[11] Inducible – only expressed in the presence of antibiotic; constitutive – expressed without any regulation.

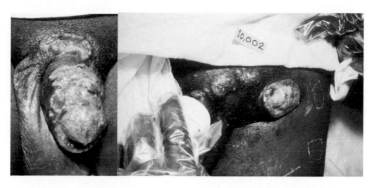

Figure 4.2.6 *Lesions associated with donovanosis (Image courtesy of the Public Health Image Library, ID 5362, 5366. [http://phil.cdc.gov/phil/home.asp, last accessed 29th April 2011].)*

similar to penicillin and is therefore often used as an alternative in patients who are allergic to penicillin. The macrolides also have the added advantage that they are active against bacteria that do not possess a cell wall (e.g. mycoplasma) – such bacteria are naturally resistant to penicillins. Macrolides are generally effective in *pneumococcal*, *streptococcal*, and *Legionella* infections, making them useful agents for the treatment of community-acquired pneumonia. They all have similar *in vitro* activity, although azithromycin appears to show superior activity against *Haemophilus influenzae* (Williamson and Sefton, 1993). In addition, azithromycin has activity against the opportunistic pathogens *Toxoplasma gondii* and *Pneumocystis carinii* (Araujo *et al.*, 1988; Dunne *et al.*, 1999).

4.2.6.2 Donovanosis

Donovanosis is a disease that can affect both men and woman and is a chronic cause of genital ulceration. Initially, the condition presents as a papule or nodule on the genital region, which then begins to ulcerate. The resulting ulcers are described as 'beefy red' and can bleed to the touch. If the ulcers are left untreated, they become very unsightly, causing significant psychological distress to the patient (Figure 4.2.6), and as a result patients with this condition may not seek medical advice due to embarrassment or shame. Once the diagnosis of donovanosis is made and treatment has begun, it is essential to reassure the patient about the favourable treatment outcomes (in other words, they will get better).

Donovanosis is generally considered to be a sexually transmitted infection (STI) and is caused by the Gram negative bacterium *Calymmatobacterium granulomatis,* although there has been recent debate about reclassifying the organism as *Klebsiella granulomatis comb nov*, based on its similarity with *Klebsiella* species (Carter *et al.*, 1999). The condition is relatively rare, but is more prevalent in certain parts of the world, for example, Papua New Guinea, KwaZulu Natal (part of South Africa), and parts of India and Brazil (O'Farrell, 2002).

Diagnosis of donovanosis is made by the presence of Donovan bodies (encapsulated forms of the causative organism, *C. granulomatis*) in tissue or biopsy samples. Once a positive diagnosis for donovanosis has been made, the current guidelines recommend azithromycin (either 1 g weekly or 500 mg daily) as first-line therapy (O'Farrell and Moi, 2010). The weekly dosing allows the azithromycin to be given by DOT,[12] when deemed appropriate, thereby increasing patient compliance, and thus improving treatment outcomes (in theory, at least – as this is a rare disorder there have been no large-scale clinical trials confirming this). In cases

---

[12] DOT stands for **d**irectly **o**bserved **t**herapy and is where an appointed health worker directly supervises patients taking their medication. This technique is also used to manage tuberculosis and was discussed in Subsection 2.2.6.1.

where the patient is pregnant, erythromycin, at a dose of 500 mg four times daily, can be used, although the recent guidelines now suggest azithromycin as an alternative (O'Farrell and Moi, 2010). Treatment should generally continue until the lesions have healed, and if they are not resolved within 6 weeks further tests should be performed to exclude the possibility of cancer.

4.2.6.3   Community-acquired pneumonia

Community-acquired pneumonia (often just abbreviated to CAP) is a relatively common condition which is associated with significant patient morbidity and mortality. Clinical features of the condition include fever, cough, chest pain, and shortness of breath, although patients can also present with non-specific symptoms, such as confusion. The condition may be self-limiting, but antibiotics are still considered essential in the management of CAP, as their use shortens the duration of illness, reduces the risk of complications, and lowers the mortality rate.

A single pathogen is usually identified in the majority of cases (when the aetiology is confirmed – in most cases it is not), but the condition can also be caused by multiple pathogens (when this occurs the condition is known as polymicrobial CAP). *S. pneumoniae* is the commonest cause, but other pathogens, such as *H. influenzae*, *Mycoplasma pneumoniae*, and *Legionella pneumophila*, have been implicated.

When a patient presents with CAP, it is crucial to assess the severity of the condition, as this will determine what treatment they will receive and whether the condition can be managed in the community or should be referred to hospital. To assess the severity of CAP, a tool known as the CURB65 score is used. CURB65 stands for **C**onfusion (does the patient show signs of new mental confusion?), **U**rea (a level raised above 7 mmol/L), **R**espiratory rate (raised above 30 breaths/min), **B**lood pressure (does the patient have a low blood pressure – either systolic (less than 90 mm Hg) or diastolic (less than 60 mm Hg?)), and finally, age (is the patient aged **65** years or more?). In cases where the patient is in hospital, the CURB65 score is used; however, in cases where the patient presents in the community (and the urea level is not easily obtainable), the simplified CRB65 score is used to assess the severity of their condition:

- If the condition is low-severity (e.g. a CURB65 score of 0 or 1, or a CRB65 score of 0), amoxicillin is recommended as monotherapy, but in cases of penicillin allergy, clarithromycin is used.
- When the condition is of moderate severity (e.g. a CURB65 score of 2), amoxicillin should be combined with clarithromycin.
- In cases of high severity (e.g. a CURB65 score of 3–5), co-amoxiclav should be combined with clarithromycin. Clarithromycin is generally preferred over erythromycin, as it causes fewer GI side effects and has an easier dosage regimen (it is given twice daily rather than four times daily).

In all of the scenarios described above, the antibiotics are given empirically (this should not be a new concept to you as we covered it in Subsection 3.2.6, when we discussed the use of trimethoprim to treat urinary tract infections in women) as they have a good spectrum of activity against a variety of organisms that cause CAP (and in particular the most common organism, *S. pneumoniae*). Occasionally, however, patients fail to respond to empirical antibiotic therapy and, if this does occur, cultures should be taken from the patient with a view to determining the causative organism. In cases where the organism has been identified as *M. pneumoniae*, clarithromycin (given either IV or orally) is recommended as first-line therapy.

One form of pneumonia that deserves a special mention is that caused by *Legionella* species (e.g. *Legionella pneumophila*) – legionnaires' disease. This is not exclusively a form of CAP, as it can be contracted in both a hospital and the community and is rather unique as, unlike the other forms of pneumonia, it is not spread person to person – it is actually contracted by inhaling water vapour containing *Legionella* bacteria. These bacteria are found in natural water (e.g. rivers and lakes) and man-made water systems (e.g. water fountains), and are usually present in low numbers. Under certain conditions (such as a water temperature between 20 and 45 °C), however, the bacteria will flourish and, if ingested, can ultimately lead to infection.

There have been many sources of contamination that have led to legionnaires' disease, with examples including air conditioning units, aerated hot tubs, water fountains, sprinkler systems, and possibly even windscreen wiper fluid in cars (Wallensten *et al.*, 2010). Indeed, the condition gets its rather unusual name as it was first recognised in 1976 during an outbreak of pneumonia among American legionnaires attending a conference – 29 out of the reported 182 cases were fatal, illustrating how serious the condition can be (Fraser *et al.*, 1977).

If legionnaires' disease is suspected (clinical features which point to *Legionella* infection include encephalopathy and other neurological symptoms), or the patient presents with high-severity CAP, a *Legionella* urinary antigen test should be carried out to confirm a *Legionella* infection.

Erythromycin was once considered the agent of choice to treat legionnaires' disease, but nowadays a quinolone antibiotic is recommended as first-line therapy (British Thoracic Society, 2009). In cases where a quinolone cannot be used, or is not tolerated, a macrolide (such as clarithromycin) is recommended. There have been no large-scale randomised controlled trials comparing the efficacy of macrolides and quinolones in the management of legionnaires' disease, but results from a recent retrospective study measuring the time to clinical stability and length of hospital stay suggest that the use of levofloxacin is more favourable than macrolides – although it should be pointed out that because of low patient numbers, the results were not statistically significant (Griffin *et al.*, 2010).

4.2.6.4 Eradication of Helicobacter pylori

The macrolides are also used (alongside other antibiotics) to eradicate *H. pylori*. We have already discussed *H. pylori* and the role it has in gastritis, peptic ulcer disease, and gastric cancer in Subsection 2.3.6, when we covered the clinical uses of the nitroimidazoles. Indeed, the macrolides (most commonly, clarithromycin) are used in combination with the nitroimidazoles and a proton pump inhibitor to eradicate *H. pylori*. In general, there are two common combinations used:

- PAC, which consists of a proton pump inhibitor, clarithromycin, and amoxicillin.
- PCM, which consists of a proton pump inhibitor, clarithromycin, and metronidazole.

Both regimens are thought to successfully eradicate *H. pylori* in around 80% of cases (Huang and Hunt, 1999; Felga *et al.*, 2010). The optimum dose of clarithromycin used in these regimens has been open to debate, with some clinicians using clarithromycin at a dose of 250 mg twice daily and others using 500 mg twice daily. A meta-analysis which examined the doses of clarithromycin found that when the PAC regimen was used, pooled data from clinical trials showed eradication rates of 79.8% with clarithromycin 250 mg twice daily compared to 89.6% with clarithromycin 500 mg twice daily (Huang and Hunt, 1999). In this case, the 500 mg twice-daily dose of clarithromycin was statistically more effective at eradicating *H. pylori*. When the PCM regimen was used, however, the analysis found that doubling the dose of clarithromycin from 250 to 500 mg twice daily was not statistically more effective at eradicating *H. pylori*, with eradication rates of 87.4 and 88.9% reported, respectively (Huang and Hunt, 1999). The dose of clarithromycin used to eradicate *H. pylori* is therefore dependent on which regimen is being prescribed; if the **PAC** regimen is used, clarithromycin **500 mg** twice daily is recommended, but if the **PCM** regimen is used, the lower dose of clarithromycin **250 mg** twice daily is recommended.

Other macrolides can also be used to eradicate *H. pylori* infection, although erythromycin tends not to be used, as its use is associated with high failure rates (Beales, 2001). The newer macrolide, azithromycin, on the other hand, appears to be effective at eradicating *H. pylori* when used in combination with other antibiotics. One meta-analysis found that when azithromycin was used, the eradication rate was 72% (compared to 70% when it was not used) (Dong *et al.*, 2009). Azithromycin has the added advantage that, due to its favourable pharmacokinetics, it can reach high gastric tissue concentrations, which persist for several days, meaning that it can be taken once daily for only 3 days during a 7-day treatment course. If you look again at Table 2.3.2 (which shows the common treatment regimens for eradicating *H. pylori*), this is clearly advantageous given the complexity and multiple dosing of the other regimens (that is, patients will find azithromycin-containing regimens easier to take, which should improve compliance with therapy). One study, however, does not share the optimism associated with the use of azithromycin for *H. pylori* eradication; this study found low eradication rates (at around 40%) when using azithromycin and concluded that it should not be recommended for *H. pylori* infection (Silva *et al.*, 2008). This, however, serves as a good example of the importance of critically reviewing scientific literature: if we look at the doses of the agents used in the study (azithromycin 500 mg daily, omeprazole 20 mg daily, and amoxicillin 500 mg three times daily) and compare them to Table 2.3.2, you will notice that the doses and dose frequency are surprisingly different. The doses used in the study by Silva *et al.* are rather low compared to what is recommended, and could even be considered as 'sub-therapeutic'. This is a possible explanation for the low eradication rates (which are not necessarily due to the fact that azithromycin has poor activity against *H. pylori*). The use of azithromycin for *H. pylori* eradication is currently not commonplace but it may have a role in the future – particularly if patient compliance becomes a problem.

4.2.6.5   Pertussis

Pertussis is an acute respiratory tract infection caused by the Gram negative bacterium *Bordetella pertussis*, which is highly contagious and is spread by inhaling contaminated aerosol droplets in the air (which are released through coughing or sneezing). It is characterised by bouts of severe coughing, which are often followed by a characteristic 'whoop' sound as the patient inhales to get their breath (yes, it really does sound like a whoop – there are plenty of examples on YouTube for you to listen to). The condition, also referred to as whooping cough, has three phases: a catarrhal phase (a dry cough may develop with signs of an upper RTI), a paroxysmal phase (the severe coughing stage often accompanied by the whoop), and a convalescent phase (when the patient starts to get better). Most people who develop pertussis make a full recovery, but complications can occasionally occur, and these include pneumonia, seizures, and, rarely, encephalopathy.

Macrolides are used first-line to treat pertussis and should be taken within 21 days of the onset of the cough (at this point, the patient is classed as still being infectious) (Health Protection Agency, 2011). They show good activity against *B. pertussis* and, encouragingly, there have been few reports of macrolide-resistant cases. Generally, erythromycin is the agent of choice for treatment of the infection, but as it can cause GI disturbance through interaction with motilin receptors (see Section 4.2.7), patient compliance may be an issue. When this is a problem, the newer macrolides, such as azithromycin or clarithromycin, can be used as they offer the advantage of a better side-effect profile. In addition, in infants below the age of 1 month, azithromycin is considered to be the agent of choice, as erythromycin has been associated with hypertrophic pyloric stenosis (a narrowing of the pylorus, the lower part of the stomach) in this patient group (Cooper *et al.*, 2002).

A Cochrane systematic review found that macrolides are effective in eradicating *B. pertussis*, and short courses (e.g. azithromycin for 3–5 days, or clarithromycin or erythromycin for 7 days) were as efficient as longer courses of treatment (e.g. for 14 days) (Altunaiji *et al.*, 2007). Although macrolides appear to be effective at eradicating *B. pertussis*, the review found they tend not to alter the clinical course of the illness, so the main reason for prescribing antibiotics in this condition is to eradicate *B. pertussis* and thereby prevent secondary transmission of the infection.

So far we have only considered the use of macrolides for the treatment of acute infection – they can also be used for preventative therapy when a patient has been exposed to someone with pertussis, and, interestingly, the doses for both indications are the same. Prophylaxis is recommended for all close contacts (meaning a family member, or someone living in the same household) who have been exposed to a patient who is still infectious (i.e. within 21 days of onset) **and** if there is a vulnerable close contact present (e.g. a newborn infant or an immunocompromised person) (Health Protection Agency, 2011).

4.2.6.6   Miscellaneous

As you will have probably gathered, there are many clinical uses of the macrolides, and we obviously cannot discuss all of them. Some of the disease states for which the macrolides are used have already been discussed, and these include:

- The prophylaxis and treatment of mycobacterium avium complex (MAC) infection when clarithromycin or azithromycin is used (see Subsection 2.2.6 for more details).
- Treatment of chlamydia with a one-off single dose of azithromycin (see Subsection 4.3.6).
- Primary prophylaxis of rheumatic fever in penicillin allergy (see Subsection 3.1.6).
- Treatment of typhoid fever caused by multiple antibacterial resistant organisms (see Subsection 2.1.6) with azithromycin.

4.2.7   Adverse drug reactions

There are several important adverse drug reactions associated with the macrolide antibiotics: GI disturbances and their pro-arrhythmic action, hepatotoxicity, effects on respiratory mucus transport, skin rashes, and, rarely, hearing loss.

4.2.7.1   Effects upon GI motility

The macrolide antibiotics are associated with an increased incidence of GI disturbances: nausea, vomiting, abdominal discomfort, and diarrhoea. The incidence of these disturbances ranges from approximately 7% in patients treated with erythromycin to approximately 2.5% in patients treated with azithromycin (Hopkins, 1991). The mechanism of this adverse reaction was highlighted in the mid-1980s as being associated with motilin (Itoh *et al.*, 1984; Zara *et al.*, 1985), a polypeptide consisting of 22 amino acid units which stimulates the secretion of peptin. In effect, it stimulates peristalsis and accelerates gastric emptying. Motilin receptor stimulation occurs at concentrations much lower than the MICs and the use of macrolide antibiotics may lead to over-stimulation and resulting GI disturbances (Curry *et al.*, 2001; Zatman *et al.*, 2001).

The greatest adverse effects are seen with macrolides composed of a 14-membered ring, compared to those with a 16-membered ring (Sifrim *et al.*, 1992), and several structure–activity relationships have been conducted in an attempt to identify structural components responsible for motilin receptor activation. Several non-antibiotic, prokinetic agents (motilides) have been developed as a consequence of some of these studies, such as mitemcinal (GM-611) (Figure 4.2.7), commonly known as motilides (Stanghellini *et al.*, 2003). Other structural activity observations are:

- An enol configuration in the macrolide ring has a marked effect upon potency (Omura *et al.*, 1987).
- 3′-Amino substitution also increases contractile potency of the motilides (Sunazuka *et al.*, 1989; Tsuzuki *et al.*, 1989).
- Sugar attachment to the macrolide structure maintains potency.

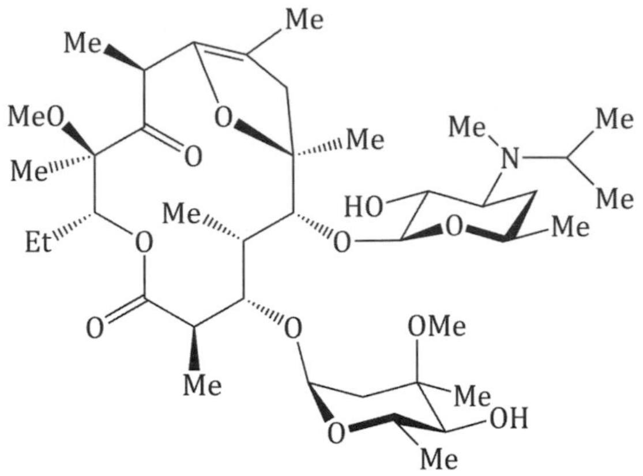

Figure 4.2.7   *Mitemcinal (GM-611)*

It has been deduced that macrolides with a low affinity for the motilin receptor have fewer GI effects and that the synthesised motilides with high motilin receptor activity have low antibacterial activity (Peeters and Depoorte, 1994).

The development of motilides has not produced clinically important results and the use of macrolide antibiotics for GI tract motility problems remains an off-licence use. Newer macrolide antibiotics have not demonstrated improved activity with respect to adverse events; for example, the recent macrolide telithromycin has been shown to induce diarrhoea and nausea in 13.3 and 8.1% of cases, respectively (Zhanel *et al.*, 2002).

4.2.7.2   Pro-arrhythmic potential

The pro-arrhythmic potential of several of the macrolide antibiotics is now well established (Stramba-Badiale *et al.*, 1997; Mishra *et al.*, 1999; Piquette, 1999; Woywodt *et al.*, 2000). Studies have shown that the macrolides prolong the QT-wave interval (represented on the ECG), thus increasing the likelihood of cardiac arrhythmias, by inhibiting the rapid component of the delayed rectifier potassium current (K_{ir}) through the block of the potassium channel product of the human ether-a-go-go gene (HERG) (Volberg *et al.*, 2002).

Potassium currents that regulate cardiac action depend upon specific cell types. The inwardly rectifying potassium channels (K_{ir} or IRK) pass potassium ions back into cells, returning the membrane back to the resting potential. This is in contrast to typical potassium channels, which carry outwards potassium currents and are considered 'outward-rectifying'. The inward-rectifying currents or channels are classified as rapid-acting and slow-acting, and the pharmacological block of the rapid channels may explain a prolonged cardiac repolarisation and consequently the potential for arrhythmias to develop, namely torsade des pointes. This action may be affected by macrolide antibiotics, especially when they are administered in combination with other agents, such as cisapride (Daleau *et al.*, 1995; Volberg *et al.*, 2002).

4.2.7.3   Skin rash

Very little information exists on the possible mechanism of skin allergy associated with the macrolide antibiotics, but it has been suggested that the skin reaction to these compounds is a general hypersensitivity reaction and constitutes one of the many hypersensitivity reactions (up to 5%) associated with a number of drugs and several body systems (Demoly and Bousquet, 2001). One study demonstrated that 13.7% of patients who had previously reported hypersensitivity to one or more drugs were hypersensitive to macrolide antibiotics (Messaad *et al.*, 2004). The symptoms associated with macrolide hypersensitivity are urticaria and rash, which is described as an immediate IgE-type reaction.

4.2.7.4 Hearing loss

Although extremely rare, several reports have described hearing loss associated with the use of macrolide antibiotics (Karmody and Weinstein, 1977; Demaldent *et al.*, 1984; Haydon *et al.*, 1984; Sacristan *et al.*, 1993), although hearing was restored in all cases once drug administration ceased.

4.2.8 Drug interactions

The macrolide antibiotics are responsible for many significant clinical drug interactions. They are metabolised by cytochrome CYP 3A4, which is responsible for up to 50% of the biotransformations of all therapeutic agents (Gibbs and Hosea, 2003). Recent data have shown that they may also interact with the efflux carrier system P-glycoprotein (see Subsection 4.2.8.1). Macrolide antibiotics thus have the potential to be involved in many drug–drug interactions.

The macrolide antibiotics are metabolised by CYP 3A4 into nitrosoalkanes (Periti and Mazzei, 1992). CYP 3A4 comprises a reactive iron (III) haem centre and a reactive apoprotein component (other isoenzymes differ in their apoprotein component, but all contain iron in the ferric (3^+) state). With the assistance of NADPH and molecular oxygen, these isoenzymes actively demethylate and oxidise macrolide antibiotics into nitrososalkanes, which then complex with the enzyme (Figure 4.2.8 and Scheme 4.2.7).

The formation of metabolic intermediates (MIs) from the macrolides may thus result in the destruction of the isoenzyme or time-dependent reversibility, temporarily depleting the intestinal and liver concentrations of CYP 3A4 and subsequently affecting the metabolism of all drugs metabolised by CYP 3A4. The *de novo* synthesis and consequent replacement of CYP 3A4 is often the rate-limiting step with drug–drug interactions involving CYP 3A4 (and therefore the macrolide antibiotics). Erythromycin and the other macrolide antibiotics contain aliphatic amine(s), which are oxidised to reactive nitrosoalkane intermediates that coordinate tightly with the ferrous heme moiety of CYP 3A4. This complex formation requires the presence of a primary amine, but demethylation can result in secondary and tertiary amines becoming involved (Scheme 4.2.7).

This *N*-demethylation has been observed with erythromycin, to form nitrosoalkanes and subsequently nitrosoalkane-MI complexes (Periti and Mazzei, 1992). An additional structural requirement for the formation

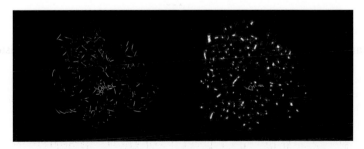

Figure 4.2.8 *Erythromycin bound to CYP 3A4, with ribbon (left) and surface (right); the catalytic iron residue is shown as a red sphere (PDB ID: 2J0D) [Ekroos and Sjögren, 2006]*

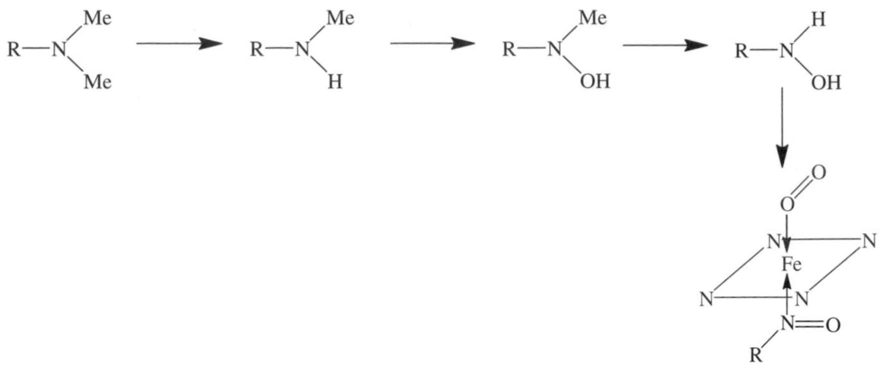

Scheme 4.2.7 *Metabolism of macrolide (RNMe$_2$) and binding of nitrosoalkane metabolic intermediate (MI) to CYP 3A4*

of the MI complex is that the macrolide be a 14-membered ring and not 16-membered, as the latter do not form MI complexes (Periti *et al.*, 1992). Clarithromycin forms complexes to a lesser extent (Periti *et al.*, 1992), and this may be due to other factors, such as pKa, steric hindrance, and lipophilicity, affecting the degree of complex formation.

The macrolide antibiotics are categorised into three groups according to their ability to inhibit the CYP 3A4 isoform: I, II, and III (Pessayre *et al.*, 1985; Gascon and Dayer, 1991; Lindstrom and Hanssen, 1993; von Rosentstiel and Adam, 1995). Group I has strong ligand affinity and is represented by erythromycin; Group II has intermediate affinity and is represented by clarithromycin; and the low-affinity group III is represented by azithromycin.

4.2.8.1   Macrolide drug interactions

Benzodiazepines

The benzodiazepines – alprazolam, midazolam, and temazepam – are all metabolised by CYP 3A4 (von Moltke *et al.*, 1995), and the plasma concentrations, area under the curve (AUC), and elimination half-life of the benzodiazepines are raised with the concomitant use of erythromycin and clarithromycin.

Neuroleptics

Several reports have demonstrated a clinical interaction between erythromycin and clozapine, which is most likely due to reduced hepatic clearance via CYP 3A4, and it is recommended that the use of macrolides with clozapine is avoided. Symptoms encountered with concomitant use were seizures, somnolence, difficulty in ambulation and coordination, and disorientation (Funderberg *et al.*, 1994; Yasui *et al.*, 1996).

Two fatal reports of ventricular dysrhythmia have been described with the simultaneous use of the neuroleptic pimozide and clarithromycin, and these were most probably due to QT-wave prolongation (Flockhart *et al.*, 1996). Clarithromycin has been shown to reduce the hepatic clearance of pimozide, with a resulting increase in QT-wave prolongation, and the use of pimozide with macrolide antibiotics is thus contraindicated (Desta *et al.*, 1999). Pimozide is primarily metabolised (*N*-dealkylation) via CYP 3A4, with contributions from CYP 1A2, and may inhibit the metabolism of drugs that are substrates for CYP 2D6 (Desta *et al.*, 1998).

Hydroxymethylglutaryl Co-Enzyme A Reductase Inhibitors (HMG CoA Inhibitors, Statins)

The hydroxymethylglutaryl co-enzyme A reductase inhibitors, with the exception of fluvastatin (which is metabolised by CYP 2C9) and pravastatin (which is not metabolised by the P450 enzymes), are metabolised by CYP 3A4. The statins exhibit toxicity against mammalian skeletal muscle, rhabdomyolysis, myopathy, and myalgia, and there is an increased likelihood of toxicity when they are used in combination with macrolide antibiotics (Illingworth and Tobert, 1994). One study has demonstrated a six-fold increase in the AUC of simvastatin with erythromycin (Kantola *et al.*, 1998). Rhabdomyolysis has been demonstrated with lovastatin and erythromycin, clarithromycin and azithromycin, so that the concomitant use of statins and macrolide antibiotics should always be done with caution (von Rosentstiel and Adam, 1995; Grunden and Fisher, 1997).

Theophylline

Erythromycin and clarithromycin have been shown to decrease the clearance of theophylline by 20–25% (Amsden, 1995; von Rosentstiel and Adam, 1995). Theophylline is metabolised by *N*-demethylation (primarily by CYP 1A2) and 8-hydroxylation (by CYP 1A2, CYP 2E1, and CYP 3A4) (Fuhr *et al.*, 1992; Sarkar *et al.*, 1992; Tjia *et al.*, 1996). It would seem, however, that inhibitors of CYP 3A4 inhibit both *N*-demethylation and 8-hydroxylation *in vitro*, which is not wholly reflected in *in vivo* studies. The interaction of macrolide antibiotics with theophylline is unpredictable, and careful monitoring of dual use is recommended.

Carbamazepine

Several published case reports indicate that toxic levels of carbamazepine are produced by inhibition of CYP 3A4 (Barzaghi *et al.*, 1987). Careful monitoring of the co-administration of carbamazepine and erythromycin/clarithromycin should be performed. Azithromycin, on the other hand, appears to be free from any interaction with carbamazepine (Watkins *et al.*, 1997; Garey and Amsden, 1999).

Class IA Anti-Arrhythmic Agents

The metabolism of quinidine is significantly affected by the co-administration of erythromycin, with a 50% decrease in the clearance of quinidine reported (Guengerich *et al.*, 1986).

P-Glycoprotein (Pgp) Efflux System

The primary role of Pgp as an efflux mechanism is to excrete metabolites from the epithelial cells of the kidney into the urine, from the jejunum into the GI tract lumen, and from the pancreas and liver into the bile (Thiebut and Tsuruo, 1987). Erythromycin and clarithromycin have been found to inhibit Pgp in a variety of tissues, thus effectively increasing serum concentrations of drugs and their metabolites normally excreted via this efflux pump mechanism (Rodriguez and Abernethy, 1999; Siedlik and Olson, 1999; Yu, 1999). For more on the P-glycoprotein efflux system, see Aller and Yu (2009).

4.2.9 Recent developments

One set of compounds that has had a great deal of interest over recent years is the ketolides. These compounds share similar structural characteristics with the conventional macrolides, but differ by possessing a keto group at the C3 position rather than a cladinose group (sugar unit); hence the name '**keto**lides' (Figure 4.2.9). The ketolides were primarily designed to overcome the problems associated with macrolide-resistant streptococci (e.g. *Streptoccus pneumoniae*), making them potentially useful agents in the management of respiratory infections, such as CAP. To date, only one ketolide, telithromycin, has made it to market, although several others are in the later stages of development. Telithromycin has a similar spectrum to other conventional macrolides, but also shows activity against penicillin and erythromycin-resistant *S. pneumoniae*. Unfortunately, however, the use of telithromycin has been associated with severe hepatotoxicity and several fatal cases of liver failure have been reported (limiting its clinical use) (Clay *et al.*, 2006). It has been suggested that the mechanism of toxicity is due to telithromycin inhibiting nicotinic acetylcholine receptors at the vagus nerve, innervating the liver, with the pyridine ring of telithromycin playing a significant part in this interaction and the associated toxicity problems (Bertrand *et al.*, 2010).

Examples of ketolides that have not yet made it to market include solithromycin (Putman *et al.*, 2011) and cethromycin (Rafie *et al.*, 2010); as you can see from Figure 4.2.9, these agents lack a pyridine ring, so should, in theory, be safer alternatives to telithromycin. The initial safety data on both compounds appear to support

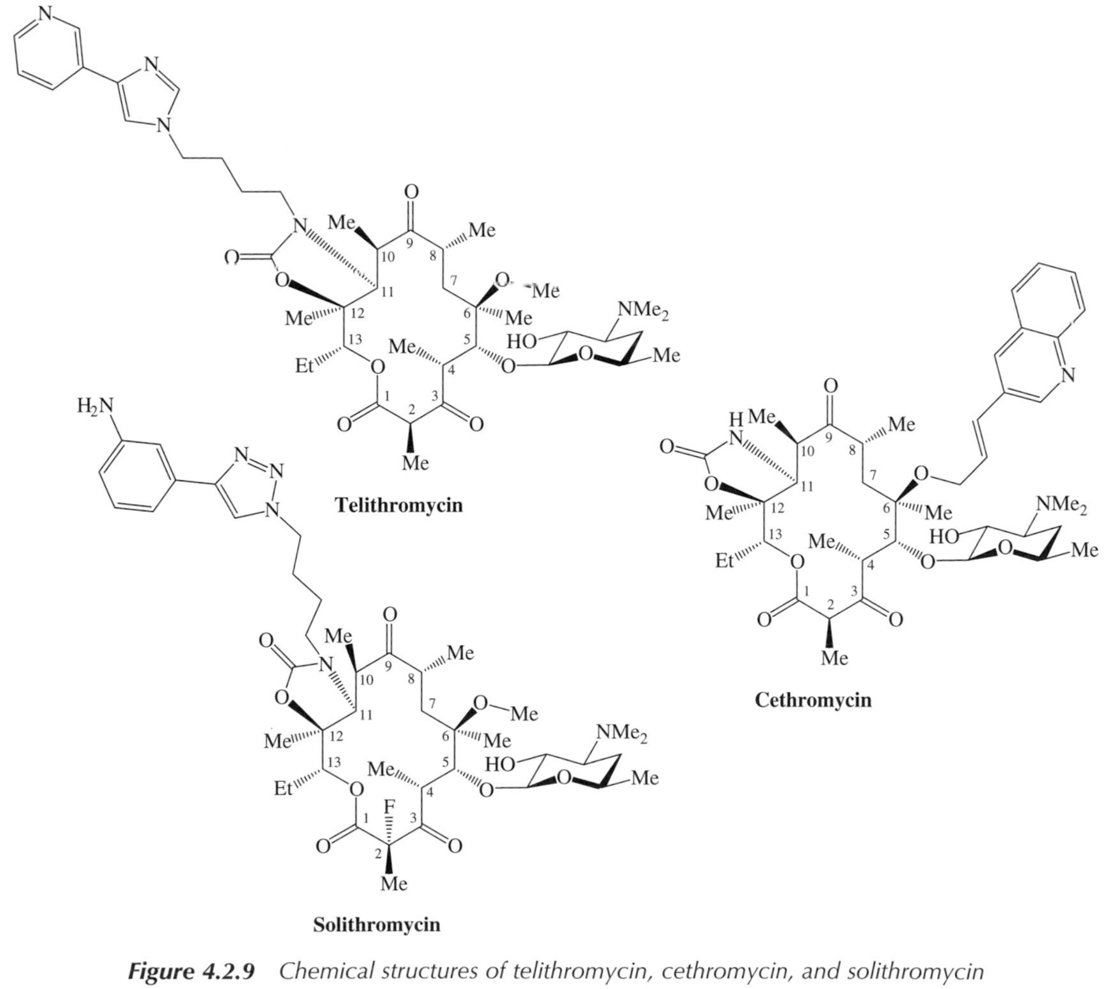

Figure 4.2.9   *Chemical structures of telithromycin, cethromycin, and solithromycin*

this theory, but it will be several years (and will require post-marketing surveillance[13] – should they be licensed) before any firm conclusions can be made regarding the long-term safety profiles of these agents.

References

S. M. Altunaiji, R. H. Kukuruzovic, N. C. Curtis, and J. Massie, *Cochrane Database of Systematic Reviews*, 2007, Issue 3.

G. W. Amsden, *Ann. Pharmacother.*, 1995, **29**, 906–917.

G. W. Amsden, *Clin. Ther.*, 1996, **18**, 56–72.

G. W. Amsden, *Int. J. Antimicrob. Agents*, 2001, **18**, S11–S15.

F. G. Araujo, D. R. Guptill, and J. S. Remington, *Antimicrob. Agents Chemother.*, 1988, **32**, 755–757.

I. L. P. Beales, *BMC Gastroenterol.*, 2001, **1**, 7.

D. Bertrand, S. Bertrand, E. Neveu, and P. Fernandes, *Antimicrob. Agents Chemother.*, 2010, **54**, 5399–5402.

B. W. Billow, E. A. Thompson, E. A. Stern, and A. Florio, *Curr. Ther. Res. Clin. Exp.*, 1964, **6**, 381–384.

E. K. Böttger, B. Springer, T. Prammananan, Y. Kidan, and P. Sander, *EMBO Reports*, 2001, **2**, 318–323.

British Thoracic Society Community Acquired Pneumonia in Adults Guideline Group, Guidelines for the Management of Community Acquired Pneumonia in Adults: Update 2009, 2009 (http://www.brit-thoracic.org.uk/Portals/0/Clinical%20Information/Pneumonia/Guidelines/CAPGuideline-full.pdf, last accessed 10 March 2012).

E. Brunet, D. M. Muñoz, F. Parra, S. Mantecón, O. Juanes, J. C. Rodríguez-Ubis, M. C. Cruzado, and R. Asensio, *Tetrahedron Lett.*, 2007, **48**, 1321–1324.

N. Barzaghi, G Gatti, and F. Crema, *Br. J. Clin. Pharmacol.*, 1987, **17**, 827–829.

R. L. Bunch and J. M. McGuire, US Patent 2653899, 29 September 1953.

J. P. Butzler, R. Vanhoof, N. Clumeck, P. D. Mol, M. P. Vanderlinden, and E. Yourassowsky, *Chemother.*, 1979, **25**, 367–372.

J. S. Carter, F. J. Bowden, I. Bastian, G. M. Myers, K. S. Sriprakash, and D. J. Kemp, *Int. J. Syst. Bacteriol.*, 1999, **49**, 1695–1700.

K. D. Clay, J. S. Hanson, S. D. Pope, R. W. Rissmiller, P. P. Purdum, and P. M. Banks, *Ann. Intern. Med.*, 2006, **144**, 415–420.

W. O. Cooper, M. R. Griffin, P. Arbogast, G. B. Hickson, S. Gautam, and W. A. Ray, *Arch. Pediatr. Adolesc. Med.*, 2002, **156**, 647–650.

E. J. Corey, P. B. Hopkins, S. Kim, S. Yoo, K. P. Nambiar, and J. R. Falck, *J. Amer. Chem. Soc.*, 1979, **101**, 7131–7134.

S. A. Crann and J. Schacht, *Audiol. Neuro-Otol.*, 1996, **1**, 80–85.

S. A. Crann, M. Y. Huang, J. D. McLaren, and J. Schacht, *Biochem. Pharmacol.*, 1992, **43**, 1835–1839.

J. I. Curry, T. D. Lander, and M. D. Stringer, *Aliment. Pharmacol. Ther.*, 2001, **15**, 595–603.

P. Daleau, E. Lessard, and M. F. Groleau, *Circulation*, 1995, **91**, 3010–3016.

J. E. Demaldent, A. Rolland, and Y. Mongrolle, *Ann. Otolaryngol. Chir. Cervicofac.*, 1984, **101**, 643–647.

P. Demoly and J. Bousquet, *Curr. Opin. Allergy Clin. Immunol.*, 2001, **1**, 305–310.

Z. Desta, T. Kerbusch, N. Soukhova, E. Richard, J.W. Ko, and D.A. Flockhart, *J. Pharmacol. Exp. Ther.*, 1998, **285**, 428–437.

Z. Desta, T. Kerbusch, and D. A. Flockhart, *Clin. Pharmacol. Ther.*, 1999, **65**, 10–20.

J. Dong, X. F. Yu, and J. Zou, *World J. Gastroenterol.*, 2009, **15**, 6102–6110.

S. Djokić, G. Kobrehel, G. Lazaravski, N. Lopotar, Z. Tamburagev, B. Kamenar, A. Nagl, and I. Vicković, *J. Chem. Soc., Perkin Trans I*, 1986, 1881–1890.

M. W. Dunne, S. Bozzette, J. A. McCutchan, M. P. Dubé, F. R. Sattler, D. Forthal, C. A. Kemper, and D. Havlir, *Lancet*, 1999, **354**, 891–895.

---

[13] When a drug is newly licensed, it is intensively monitored to ensure that any rare or long-term adverse effects are identified. This is known as post-marketing surveillance and in the UK such drugs are classed as 'Black Triangle' drugs, as the ▼ symbol is used to show they are being intensively monitored.

M. Ekroos and T. Sjögren, *Proc. Natl. Acad. Sci. USA*, 2006, **103**, 13 682–13 687.

H. A. El-Enshasy, N. A. Mohamed, M. A. Farid, and A. I. El-Diwany, *Bioresource Technol.*, 2008, **99**, 4263–4268.

N. O'Farrell, *Sex. Tranm. Infect.*, 2002, **78**, 452–457.

N. O'Farrell and H. Moi, *Int. J. STD AIDS*, 2010, **21**, 609–610.

G. Felga F. M. Silva, R. C. Barbuti, T. Navarro-Rodriguez, S. Zaterka, and J. N. Eisig, *J. Infect. Dev. Ctries.*, 2010, **4**, 712–716.

D. A. Flockhart, E. Richard, R. L. Woosley, P. L. Pearle, and M. D. A. Drici, *Clin. Pharmacol. Ther.*, 1996, **59**, 189.

E. H. Flynn, M. V. Sigal, P. W. Wiley, and K. Gerzon, *J. Amer. Chem. Soc.*, 1954, **76**, 3121–3131.

D. W. Fraser, T. R. Tsai, W. Orenstein, W. E. Parkin, H. J. Beecham, R. G. Sharrar, J. Harris, G. F. Mallison, S. M. Martin, J. E. McDade, C. C. Shepard, and P. S. Brachman, *N. Engl. J. Med.*, 1977, **297**, 1189–1997.

E. Fuhr, J. Doehmer, and N. Battula, *Biochem. Pharmacol.*, 1992, **43**, 225–235.

L. G. Funderberg, J. E. Vertrees, and J. E. True, *Am. J. Psychiatry*, 1994, **151**, 1840–1841.

K. W. Garey and G. W. Amsden, *Ann. Pharmacother.*, 1999, **33**, 218–228.

P. M. Gascon and P. Dayer, *Clin. Pharmacol. Ther.*, 1991, **49**, 158.

M. Gaynor and A. S. Mankin, *Frontiers in Med. Chem.*, 2005, **2**, 21–35.

M. A. Gibbs and N. A. Hosea, *Clin. Pharmacokinet*, 2003, **42**, 969–984.

S. Gouin d'Ambrieres, A. Lutz, and J.-C. Gasc, US Patent 4349545, 14 September 1982.

A. T. Griffin, P. Peyrani, T. Wiemken, and F. Arnold, *Int. J. Tuberc. Lung Dis.*, 2010, **14**, 495–499.

J. W. Grunden and K. A. Fisher, *Ann. Pharmacother.*, 1997, **31**, 859–863.

F. P. Guengerich, D. Muller Enoch, and I. A. Blair, *Mol. Pharmacol.*, 1986, **30**, 287–295.

D. R. Harris, S. G. McGeachin, and H. H. Mills, *Tetrahedron Lett.*, 1965, **6**, 679–685.

A. Hassanzadeh, P. A. Gorry, G. A. Morris, and J. Barber, *J. Med. Chem.*, 2006a, **49**, 6334–6342.

A. Hassanzadeh, M. Helliwell, and J. Barber, *Org. Biomol. Chem.*, 2006b, **4**, 1014–1019.

A. Hassanzadeh, J. Barber, G. A. Morris, and P. A. Gorry, *J. Phys. Chem. A*, 2007, **111**, 10 098–10 104.

R. C. Haydon, J. W. Thelin, and W. E. Davis, *Otolaryngol. Head Neck Surg.*, 1984, **92**, 678–684.

Health Protection Agency, HPA Guidelines for the Public Health Management of Pertussis, 2011 (http://www.hpa.org.uk/web/HPAwebFile/HPAweb_C/1287142671506, last accessed 10 March 2012).

S. Hopkins, *Am. J. Med.*, 1991, **91**, 40S–45S.

J. Huang and R. H. Hunt, *Aliment. Pharmacol. Ther.*, 1999, **13**, 719–729.

D. R. Illingworth and J. A. Tobert, *Clin. Ther.*, 1994, **16**, 366–385.

Z. Itoh, M. Nakaya, T. Suzuki, H. Arai, and K. Wakabayashi, *Am. J. Physiol.*, 1984, **247**, G688–G694.

R. Jain and L. H. Danziger, *Curr. Pharm. Design*, 2004, **10**, 3045–3053.

M. S. Kanatani and B. J. Guglielmo, *West. J. Med.*, 1994, **160**, 31–37.

T. Kantola, K. T. Kivisto, and P. J. Neuvonen, *Clin. Pharmacol. Ther.*, 1998, **64**, 177–182.

C. S. Karmody and L. Weinstein, *Annotol. Rhinol. Laryngol.*, 1977, **86**, 9–11.

T. Lan, A. Rao, J. Haywood, C. B. Davis, C. Han, E. Garver, and P. A. Dawson, *Drug Metab. Dispos.*, 2009, **37**, 2375–2382.

R. Leclercq, *Clin. Infect. Diseases*, 2002, **34**, 482–492.

R. Leclercq and P. Courvalin, *Antimicrob. Agents and Chem.*, 2002, **46**, 2727–2734.

J.-H. Liang, L.-J. Dong, H. Wang, K. An, X.-L. Li, L. Yang, G.-W. Yao, and Y.-C. Xu, *Eur. J. Med. Chem.*, 2010, **45**, 3627–3635.

T. D. Lindstrom and B. R. Hanssen, *Antimicrob. Agents Chemother.*, 1993, **37**, 265–269.

H. Lode, K. Borner, P. Koeppe, and T. Schaberg, *J. Antimicrob. Chemother.*, 1996, **37** (Suppl. C), 1–8.

C. Ma and S. Ma, *Mini-Rev. Med. Chem.*, 2010, **10**, 272–286.

A.-S. Malmborg, *Curr. Med. Res. Opin.*, 1978, **5** (Suppl. 2), 15–18.

P. K. Manna, V. Kumaran, G. P. Mohanta, and R. Manavalan, *Acta Pharm.*, 2004, **54**, 231–242.

S. J. Martin, C. G. Garvin, C. R. McBurney, and E. G. Sahloff, *J. Antimicrob. Chemother.*, 2001, **47**, 581–587.

Y. C. Martin, P. H. Jones, T. J. Perun, W. E. Grudy, S. Bell, R. R. Bower, and N. L. Shipkowitz, *J. Med. Chem.*, 1972, **15**, 635–638.

E. H. Massey, B. Kitchell, L. D. Martin, K. Gerzon, and H. W. Murphy, *Tetrahedron Lett.*, 1970, **11**, 157–160.

M. Matsuoka, K. Endou, H. Kobayashi, M. Inoue, and Y. Nakajima, *FEMS Microbiol. Letters*, 1998, **167**, 221–227.

D. Messaad, H. Sahia, S. Benahmed, P. Godard, J. Bousquet, and P. Demoly, *Ann. Intern. Med.*, 2004, **140**, 1001–1006.

A. Mishra, H. S. Friedman, and A. K. Sinha, *Chest*, 1999, **115**, 983–986.

S. Omura, K. Tsuzuki, T. Sunazuka, S. Marui, H. Toyoda, N. Inatomi, and Z. Itoh, *J. Med. Chem.*, 1987, **30**, 1941–1943.

I. Paterson and M. M. Mansuri, *Tetrahedron*, 1985, **41**, 3569–3624.

S. Payot, J.-M. Bolla, D. Corcoran, S. Fanning, F. Mégraud, and Q. Zhang, *Microbes and Infection*, 2006, **8**, 1967–1971.

P. Periti, T. Mazzei, E. Mini, A. Novelli, *Clin. Pharmacokin.*, 1992, **23**, 106–131.

D. Pessayre, D. Larrey, C. Funck-Brentano, and J. P. Benhamou, *J. Antimicrob. Chemother.*, 1985, **16** (Suppl. A), 181–194.

R. K. Piquette, *Ann. Pharmacother.*, 1999, **33**, 22–26.

S. Rafie, C. MacDougall, and C. L. James, *Pharmacotherapy*, 2010, **30**, 290–303.

M. C. Roberts, *FEMS Microbiol. Lett.*, 2008, **282**, 147–159.

I. Rodriguez and D. R. Abernethy, *Circulation*, 1999, **99**, 472–474.

K. A. Rodvold, *Clin. Pharmacokinet.*, 1999, **37**, 385–398.

J. A. Sacristan, M. Angeles de Cos, and J. Soto, *Am. J. Otol.*, 1993, **14**, 186–188.

M. A. Sarkar, C. Hunt, P. S. Guzelian, and H. T. Karnes, *Drug Metab. Dispos.*, 1992, **20**, 31–37.

J. J. Schentag and C. H. Ballow, *Am. J. Med.*, 1991, **91** (Suppl. A), 5–11.

F. Schlünzen, R. Zarivach, J. Harms, A. Bashan, A. Tocilj, R. Albrecht, A. Yonath, and F. Franceschi, *Nature*, 2001, **413**, 814–821.

P. H. Siedlik and S. C. Olson, *J. Clin. Pharmacol.*, 1999, **39**, 501–504.

D. Sifrim, J. Janssens, and G. Vantrappen, *Int. J. Clin. Pharmacol. Res.*, 1992, **12**, 71–79.

F. M. Silva, J. N. Eisig, A. C. Teixeira, R. C. Barbuti, T. Navarro-Rodriguez, and R. Mattar, *BMC Gastroenterol.*, 2008, **8**, 20.

A. A. Sinkula, *J. Pharm. Sci.*, 1974, **63**, 842–848.

V. Stanghellini, F. De Ponti, R. De Giorgio, G. Barbara, C. Tosetti, and R. Corinaldesi, *Drugs*, 2003, **63**, 869–872.

M. Stramba-Badiale, F. Nador, N. Porta, S. Guffanti, M. Frediani, C. Colnaghi, F. Grancini, G. Motta, V. Carnelli, and P. J. Schwartz, *Am. Heart J.*, 1997, **133**, 108–111.

I. Stupans and L. N. Sansom, *Biochem. Pharmacol.*, 1991, **42**, 2085–2090.

T. Sugimoto and T. Tanikawa, *ACS Med. Chem. Lett.*, 2011, **2**, 234–237.

T. Sunazuka, K. Tsuzuki, S. Marui, H. Toyoda, S. Omura, N. Inatomi, and Z. Itoh, *Chem. Pharm. Bull. (Tokyo)*, 1989, **37**, 2701–2709.

F. Thiebut and T. Tsuruo, *Proc. Natl. Acad. Sci. USA.*, 1987, **84**, 7735–7738.

P. J. Thompson, K. R. Burgess, and G. E. Marlin, *Antimicrob. Agents Chemother.*, 1980, **18**, 829–831.

J. F. Tjia, J. Colbert, and D. J. Back, *J. Pharmacol. Exp. Ther.*, 1996, **276**, 912–917.

K. Tsuzuki, T. Sunazuka, S. Marui, H. Toyoda, S. Omura, N. Inatomi, and Z. Itoh, *Chem. Pharm. Bull. (Tokyo)*, 1989, **37**, 2687–2700.

P. Tyson, A. Hassanzadeh, M. N. Mordi, D. G. Allison, V. Marquez, and J. Barber, *Med. Chem. Commun.*, 2011, **2**, 331–336.

S. Vacher, A. Menard, E. Bernard, and F. Megraud, *Microb. Drug Resist.*, 2005, **11**, 40–47.

L. L. von Moltke, D. J. Greenblatt, J. Schmider, J. S. Harmatz, and R. I. Shader, *Clin. Pharmacokinet.*, 1995, **29**, 33–44.

N. A. von Rosentstiel and D. Adam, *Drug Safety*, 1995, **13**, 105–122.

A. Wallensten, I. Oliver, K. Ricketts, G. Kafatos, J. M. Stuart, and C. Joseph, *Eur. J. Epidemiol.*, 2010, **25**, 661–665.

V. S. Watkins, R. E. Polk, and J. L. Stotka, *Ann. Pharmacother.*, 1997, **31**, 349–356.

J. D. Williams and A. M. Sefton, *J. Antimicrob. Chemother.*, 1993, **31** (Suppl. C), 11–26.

L. Wondrack, M. Massa, B. V. Yang, and J. Sutcliffe, *Antimicrob. Agents Chemother.*, 1996, **40**, 992–998.

R. B. Woodward, E. Logusch, K. P. Nambiar, K. Sakan. D. E. Ward, B.-W. Au-Yeung, P. Balaram, L. J. Browne, P. J. Card, C. H. Chen, R. B. Chenevert, A. Fliri, K. Frobel, H.-J. Gais, D. G. Garratt, K. Hayakawa, W. Heggie, D. P. Hesson, D. Hoppe, I. Hoppe, J. A. Hvatt. D. Ikeda, P. A. Jacobi, K. S. Kim, Y. Kobuke, K. Kojima, K. Krowicki, V. J. Lee, T. Leutert, S. Malchenko, J. Martens, R. S. Matthews, B. S. Ong, J. B. Press, T. V. Rajan Babu, G. Rousseau, H. M. Sauter, M. Suzuki, K. Tatsuta, L. M. Tolbert, E. A. Truesdale, I. Uchida, Y. Ueda, T. Uyehara, A. T. Vasella, W. C. Vladuchick, P. A. Wade, R. M. Williams, and H. N. C. Wong, *J. Am. Chem. Soc.*, 1981, **103**, 3210–3213, 3213–3215, 3215–3217.

A. Woywodt, U. Grommas, W. Buth, and W. Rafflenbeul, *Postgrad. Med. J.*, 2000, **76**, 651–653.

N. Y. Yasui, K. Otani, and S. Kaneko, *Clin. Pharmacol. Ther.*, 1996, **59**, 514–519.

D. K. Yu, *J. Clin. Pharmacol.*, 1999, **39**, 1203–1211.

G. P. Zara, H. H. Thompson, M. A. Pilot, and H. D. Ritchie, *J. Antimicrob. Chemother.*, 1985, **16** (Suppl. A), 175–179.

T. F. Zatman, J. E. Hall, and M. Harmer, *Br. J. Anaesth.*, 2001, **86**, 869–871.

M. Zeitlinger, C. C. Wagner, and B. Heinisch, *Clin. Pharmacokinet.*, 2009, **48**, 23–38.

G. G. Zhanel, M. Dueck, D. J. Hoban, L. M. Vercaigne, J. M. Embil, A. S. Gin, and J. A. Karlowsky, *Drugs*, 2001, **61**, 443–498.

G. G. Zhanel, M. Walters, A. Noreddin, L. M. Vercaigne, A. Wierzbowski, J. M. Embil, A. S. Gin, S. Douthwaite, and D. J. Hoban, *Drugs*, 2002, **62**, 1771–1804.

J. M. Zuckerman, F. Qamar, and B. R. Bono, *Infect. Dis. Clinics N. Amer.*, 2009, **23**, 997–1026.

4.3

Tetracycline antibiotics

Key Points

- Tetracyclines are broad-spectrum agents with excellent bioavailability and activity against both Gram positive and Gram negative bacteria; resistance and issues of toxicity restrict their regular clinical use.
- New classes of tetracycline are being developed; the first clinical example, tigecycline, a glycylcycline, shows potency across a wide range of bacteria and low susceptibility to resistance.
- For further information, see Chopra and Roberts (2001) and Zhanel *et al.* (2004).

Tetracycline antibiotics used clinically include doxycycline, minocycline, and tetracycline, while tigecycline, which has the same four-ringed structure and is a derivative of minocycline, is the first member of the glycylcyclines to be approved; their structures are shown in Figure 4.3.1 and their therapeutic indications are listed in Table 4.3.1.

4.3.1 Discovery

The tetracycline antibiotics are the third example so far in this section, of naturally occurring molecules from a microbial source that interfere with bacterial protein synthesis. Once again, the first in this series, chlortetracycline (originally named aureomycin), resulted from a programme of screening soil microorganisms for potential new antibiotics. This discovery is attributed to Benjamin M. Duggar, a retired botanist, with expertise across a wide range of plant and microorganism physiology. He retired from Wisconsin University in 1943, when he was 71, but was approached to act as consultant for Lederle Laboratories in New York (part of American Cyanamid Company, now part of Wyeth Pharmaceuticals), who were supporting the war effort by searching for new antibiotic and antimalarial agents (Walker, 1982).

Although he is solely credited with the discovery of chlortetracycline in 1948, Duggar was part of a larger team, under the direction of Yellapragada SubbaRow, which was systematically investigating soil microorganisms for natural products with desirable pharmaceutical activities. Duggar requested some local samples

Antibacterial Agents: Chemistry, Mode of Action, Mechanisms of Resistance and Clinical Applications, First Edition.
Rosaleen J. Anderson, Paul W. Groundwater, Adam Todd and Alan J. Worsley.
© 2012 John Wiley & Sons, Ltd. Published 2012 by John Wiley & Sons, Ltd.

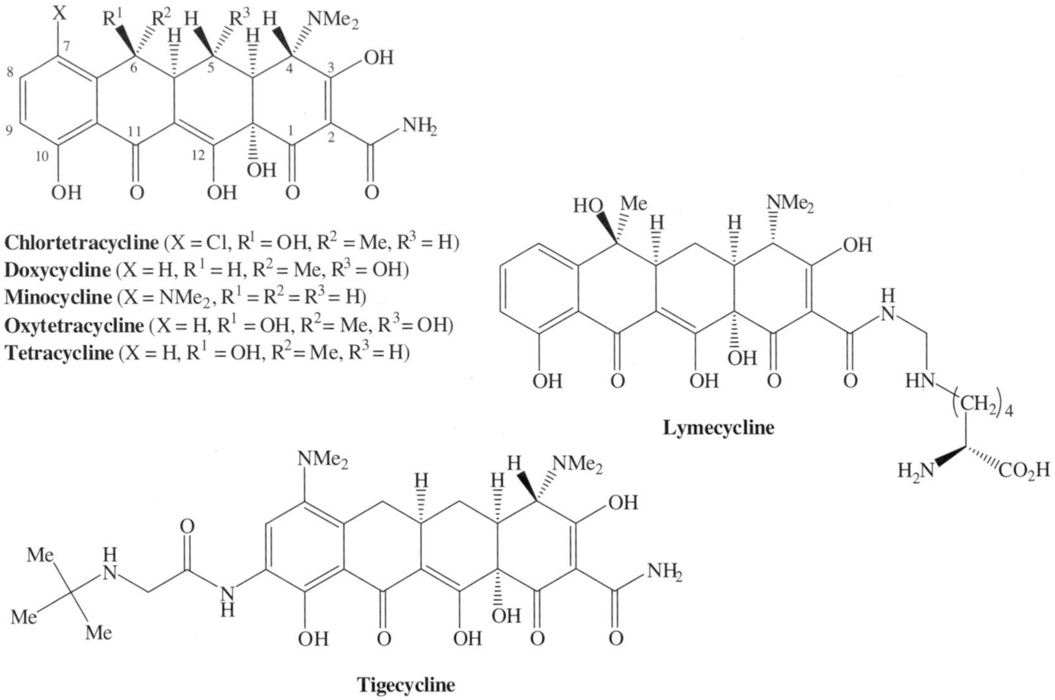

Chlortetracycline (X = Cl, R^1 = OH, R^2 = Me, R^3 = H)
Doxycycline (X = H, R^1 = H, R^2 = Me, R^3 = OH)
Minocycline (X = NMe$_2$, R^1 = R^2 = R^3 = H)
Oxytetracycline (X = H, R^1 = OH, R^2= Me, R^3= OH)
Tetracycline (X = H, R^1 = OH, R^2= Me, R^3 = H)

Lymecycline

Tigecycline

Figure 4.3.1   *Tetracycline antibiotics*

from the soil microbiologist at the University of Missouri, William Albrecht, from which he cultured a golden mould, which produced a yellow pigment that displayed growth inhibitory properties against bacteria, such as streptococci. He identified the mould as a *Streptomyces* species that had not previously been catalogued; to reflect its colour, he named it *Streptomyces aureofaciens* and the antibiotic it produced aureomycin (Duggar, 1948). Note that the 'mycin' part of the name corresponds with its isolation from a Streptomyces species, like the aminoglycoside streptomycin and the macrolide erythromycin, which we have just met.

Aureomycin was quickly released to clinicians and other researchers to obtain evaluative data of its activity and efficacy; it gathered support with glowing testimonials of its broad spectrum of activity, including against

Table 4.3.1   *Therapeutic indications for the tetracycline antibiotics*

Tetracycline antibiotic	Indications
Doxycycline	Chronic prostatitis, sinusitis, syphilis, uncomplicated genital chlamydial infection, pelvic inflammatory disease, acne vulgaris, rosacea, Lyme disease, community-acquired pneumonia
Lymecycline	Acne vulgaris
Minocycline	Acne vulgaris, prophylaxis of asymptomatic meningococcal carrier state (no longer recommended)
Oxytetracycline	Acne vulgaris, rosacea
Tetracycline	Acne vulgaris, rosacea, non-gonococcal urethritis, chronic bronchitis
Tigecycline	Complicated intra-abdominal or skin/soft tissue infections

streptomycin- and penicillin-resistant organisms (Wright and Schreiber, 1949; Cantor, 1950; Kiser *et al.*, 1952). It was found to be as effective as penicillin and streptomycin and had the significant advantage of being the first antibiotic that was effective when administered orally.

Following the discovery of aureomycin, other tetracyclines were soon discovered:

- Oxytetracycline (originally called terramycin) in 1949 from *Streptomyces rimosus* by Pfizer (Finlay *et al.*, 1950).
- Tetracycline (marketed originally as achromycin (Darken *et al.*, 1960)) in 1953 from *S. aureofaciens* when cultured with a chlorination inhibitor (Goodman and Matrishin, 1968).
- Demethylchlortetracycline in 1957 from *S. aureofaciens* (McCormick *et al.*, 1957; Wilson, 1961). This was the last natural tetracycline to be identified and was originally called declomycin or ledermycin. It is still marketed as the latter (or under its generic name, demeclocycline) by Lederle Laboratories.

The discovery of this new class of antibiotics is not without controversy, though: Pfizer, American Cyanamid, and Bristol-Myers formed a monopoly that maintained artificially high prices for tetracycline over several years before the US Federal Trade Commission halted the violations after a series of high-profile investigations, charges, and appeals heard in the high court (Anon, 1964; US Court of Appeals, 1968).

When it was discovered that the hydrogenation of chlortetracycline resulted in dechlorination and conversion into tetracycline (which was as active as chlortetracycline) (Stephens *et al.*, 1952; Conover, 1955), the possibility that synthetic modification of tetracyclines might provide alternative agents with antibacterial activity was realised. During the next 15–20 years, many semi-synthetic analogues were prepared, including lymecycline, doxycycline, and minocycline; some of these second-generation semi-synthetic tetracyclines were even more potent than chlortetracycline and are still marketed today. Structural modification has continued and has resulted in the discovery of a third-generation tetracycline, *t*-butylglycylamidominocycline (tigecycline, originally labelled GAR-936) (Petersen *et al.*, 1999), with more in development and in clinical trials (Sun *et al.*, 2008; Brötz-Oesterhelt and Sass, 2010).

Structurally, the tetracyclines are based on a four-ring (tetracyclic or octahydronaphthacene) system, hence the name; the rings are labelled A, B, C, and D (Figure 4.3.2). One face consists of carbonyl, phenol, alcohol, and enol oxygen atoms, with high polarity and metal ion binding ability, while the other face is substantially less polar. There are a number of substitution patterns commonly found in the antibiotic tetracyclines and significant deviation from these leads to greatly reduced antibacterial activity (Chopra and Roberts, 2001; Zhanel *et al.*, 2004).

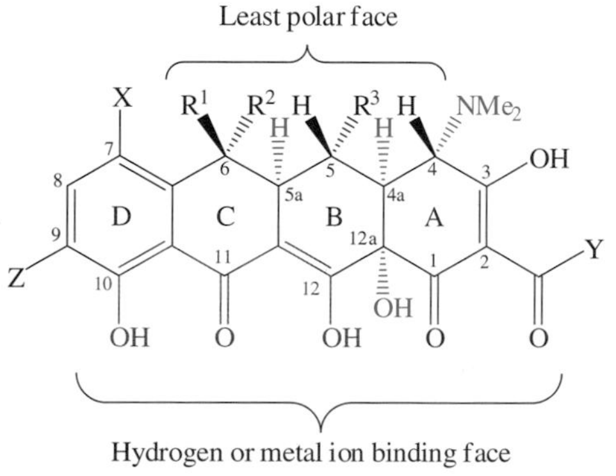

Ring requirements
Rings A,B,C,D linearly fused
Rings D,C,B: C-10, C-11, C-12 phenolic keto-enol system
Ring A: C-1, C-2, C-3 tricarbonyl-type keto-enol system

R^1 = H or OH; R^2 = H or CH_3; R^3 = H or OH
C-4: α-NMe_2; C-4a and C-5a: α-H; C-12a: α-OH

X = H (tetracyclines); Cl (chlortetracyclines); NR_2 (minocycline and analogues, including tigecycline) or NO_2
Y = NH_2 or NHR (R must contain amino group)
Z = H or glycylamido

Figure 4.3.2   *Requirements for tetracycline antibiotic activity (Chopra and Roberts, 2001; Zhanel et al., 2004)*

Much of the development of new tetracycline antibiotics has been driven by the instability of the first-generation tetracyclines, particularly chlortetracycline, oxytetracycline, and tetracycline, which can lead to degradation during storage and even production of a toxic product. We will look at some of the reactions of tetracyclines and the resultant effects upon bioavailability in Subsection 4.3.3.

4.3.2   Synthesis

Looking back at Figure 4.3.1, you can see that the tetracyclines have a number of chiral centres and functional groups, so you will not be surprised to learn that fermentation methods are considered to be the most cost-effective for their production, and for the production of the base structures for semi-synthetic analogues, such as lymecycline and tigecycline (Khosla and Tang, 2005). As you will see further on in this subsection, there is now an efficient chemical synthetic method for multigram quantities of a key intermediate in tetracycline synthesis, which offers the possibility of analogue synthesis (Brubaker and Myers, 2007). The first patented fermentations of *S. aureofaciens* were for the production of chlortetracycline (Duggar, 1948; Neidercorn, 1952) and tetracycline (Goodman *et al.*, 1959); much work since then has focussed on optimising the selectivity for, and the yields of, the desired tetracyclines, especially since some *Streptomyces* species can produce more than one tetracycline, depending upon the fermentation conditions (Bêhal, 1987, 2000). In early 2011, there were almost 3000 patents relating to the biosynthesis and synthesis of tetracycline and its analogues (worldwide.espacenet.com), including some filed in 2010 and 2011 – proof that there is still interest in the production and use of tetracyclines. It should be noted, however, that some of these were for non-antibiotic uses of tetracyclines, briefly mentioned in Subsection 4.3.4.

4.3.2.1   Biosynthesis of tetracyclines

The biosynthesis of tetracycline antibiotics is related to the bacterial synthesis of fatty acids through the bacterial type II polyketide synthase pathway, consisting of a well-studied set of enzymes (for examples, see Khosla, 2009; Zhang and Tang, 2009), although the synthesis of tetracyclines is unique to certain bacteria (Clardy *et al.*, 2009). The biosynthetic pathways to tetracycline and oxytetracycline are the most studied (for example, Petković *et al.*, 2006; Pickens and Tang, 2009). The biosynthetic pathway to natural tetracyclines is available from the KEGG database (http://www.genome.jp/kegg/pathway/map/map00253.html, last accessed 10 March 2012); in summary, it uses the precursor, malonamyl coenzyme A (CoA), which is obtained from acetyl CoA via malonyl CoA and glutamine (Wang *et al.*, 1986), and proceeds through two common key intermediates, 6-methylpretetramide and 4-ketoanhydrotetracycline (Scheme 4.3.1 and Table 4.3.2) (Clardy *et al.*, 2009; Pickens and Tang, 2009).

For clarity, in Scheme 4.3.1 the precursor unit, malonamyl CoA, is coloured pink throughout; the consecutive acetyl units added to malonamyl CoA from acetyl CoA are coloured alternately black and red – you can see that the cycle of acetyl addition occurs eight times until the linear nonaketamide is formed. A series of enzymic reactions involving OxyJ, OxyK, and OxyN leads to pretetramide (not shown in Scheme 4.3.1), which is converted into 6-methylpretetramide by OxyF. This series of reactions results in cyclisation of the nonaketamide to the tetracyclic structure of 6-methylpretetramide, followed by methylation at C6, with the new bonds formed in this sequence shown in blue. If you trace the sequentially added acetyl units, you can see how the two-carbon units form the backbone of the structure. Oxidation at C4 (by OxyE), and hydroxylation at C12a by OxyL, provides the second key intermediate, 4-ketoanhydrotetracycline, at which the tetracycline biosynthetic paths diverge. Chlortetracycline results from Cts4 halogenase action at C7 (Dairi *et al.*, 1995), followed by amination at C4 (OxyQ) and OxyT *N,N*-dimethylation (the methyl groups are provided by *S*-adenosyl methionine); OxyS catalyses the stereospecific hydroxylation at C6, leaving a stereospecific

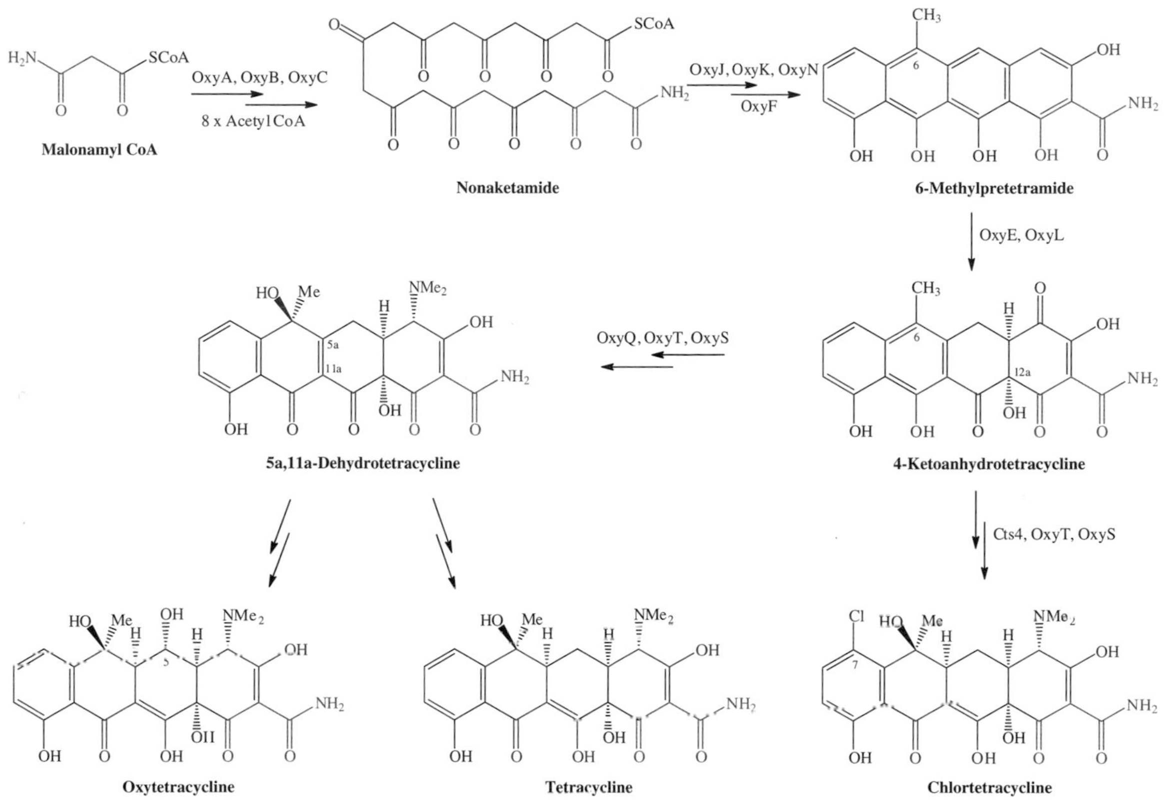

Scheme 4.3.1 *Biosynthetic pathway to tetracycline antibiotics (Clardy et al., 2009; http://www.genome.jp/kegg/pathway/map/map00253.html, last accessed 10 March 2012)*

reduction at C5a required to produce chlortetracycline. In the other pathway, amination at C4, dimethylation, and hydroxylation at C6 provide 5a,11a-dehydrotetracycline, from which oxytetracycline and tetracycline are obtained.

Table 4.3.2 *Enzymes involved in tetracycline biosynthesis*

Enzyme	Function
OxyA	Ketosynthase
OxyB	Chain length factor
OxyC	Acyl carrier protein
OxyJ	Ketoreductase
OxyK	Aromatase
OxyN	Cyclase
OxyF	C-methyltransferase
OxyE	Flavin-dependent monooxygenase
OxyL	NADPH-dependent dioxygenase
OxyQ	Aminotransferase
OxyT	*N,N*-dimethyltransferase
OxyS	Monooxygenase that hydroxylates stereospecifically at C6
Cts4	Halogenase

Much detailed research has been directed at elucidating this pathway; most was carried out on the enzymes of the oxytetracycline-producing species *Streptomyces rimosus*, hence the Oxy names, but the enzymes are the same or similar for the other tetracycline-producing species (Zhang *et al.*, 2007; Petković *et al.*, 2010; Pickens and Tang, 2010).

4.3.2.2   Chemical synthesis of tetracyclines

The literature related to the chemical synthesis of tetracyclines resembles a 'Who's Who' of synthetic organic chemistry:

- R. B. Woodward solved the structure, complete with stereochemistry, in 1952 (this was revised slightly in the 1960s with the help of X-ray crystallography) (Hochstein *et al.*, 1953; Donohoe *et al.*, 1963; von Wittenau *et al.*, 1965).
- Woodward and Conover (who first synthesised tetracycline by hydrogenation of chlortetracycline) synthesised a biologically active tetracycline, named sancycline (Korst *et al.*, 1968), albeit in 25 steps and 0.002% overall yield.
- Shemyakin synthesised a tetracycline natural product, (±)-12a-deoxy-5a,6-anhydrotetracycline (Gurevich *et al.*, 1967).
- Muxfeldt identified the major problems with the synthesis of tetracyclines: the complexity of the required stereochemistry and the sensitivity of the tetracycline functional groups to both mild acid and base during initial studies (Muxfeldt and Rogalski, 1965), then later achieved the total synthesis of (±)-5-oxytetracycline in 22 steps and 0.06% (Muxfeldt *et al.*, 1968, 1979).
- Stork concentrated on achieving the correct stereochemistry at each centre in the basic tetracycline structure, producing (±)-12a-deoxytetracycline in 16 steps and an impressive 18–25% yield, although this structure has little antimicrobial activity (Stork *et al.*, 1996).
- Tatsuta and co-workers exploited the natural stereochemical definition of carbohydrates for their starting materials and achieved the total synthesis of natural (−)-tetracycline from D-glucosamine in 34 steps and 0.002% yield (Tatsuta *et al.*, 2000; Tatsuka and Hosokawa, 2005), including a solution to the difficult stereospecific hydroxylation of C12a.
- More recently, Myers and co-workers developed a highly effective synthetic approach to natural tetracyclines, their analogues, and their precursors (Charest *et al.*, 2005a, 2005b; Brubaker and Myers, 2007; Myers *et al.*, 2007, 2011; Sun *et al.*, 2008), which takes account of the considerable challenge in achieving the correct stereochemistry, particularly at C12a, and provides versatility for the synthesis of many new analogues for microbiological evaluation (Myers *et al.*, 2007; Sun *et al.*, 2008).

A discussion of the chemical strategies for synthesising tetracyclines, the reactions required, and their stereochemical complexities would be a section in itself, so we will restrict ourselves here to consideration of the most recent syntheses, which have enabled multigram quantities of optically pure tetracyclines to be achieved and thus offer potential commercial synthetic routes to these agents. Myers and colleagues recognised that one key intermediate **1**, providing the A and B rings of tetracyclines, allows the synthesis of a wide range of tetracycline antibiotics and their analogues (Scheme 4.3.2).

Initially, they developed a synthesis to this important intermediate from benzoic acid and achieved intermediate **1** in 21% overall yield after 7 steps (Charest *et al.*, 2005a). Using this route, (−)-tetracycline could be synthesised in 17 steps and 1.1% yield from benzoic acid. Conversion of **1** into **2** and **3** provided routes to tetracyclines with no hydroxyl group at C5 (including an alternative synthesis of tetracycline) and 5-hydroxytetracyclines, respectively. Using this approach, the synthesis of (−)-6-deoxytetracycline was achieved in 14 steps and 7% yield via intermediate **2**, while (−)-doxycycline was isolated in 8.3% yield

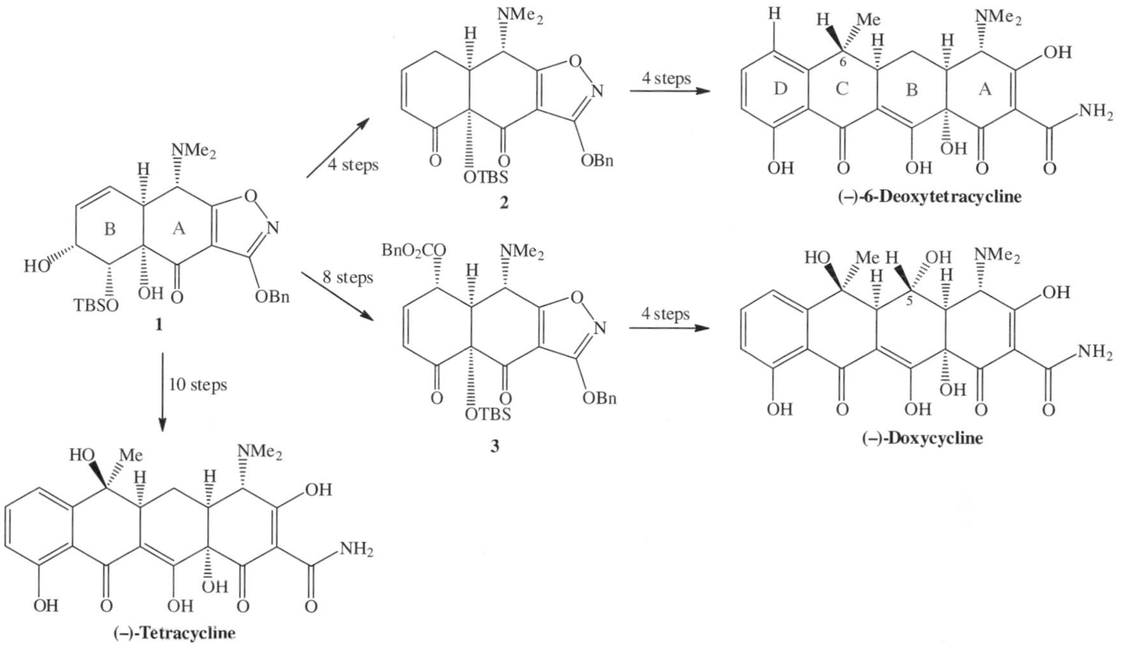

Scheme 4.3.2 *Key intermediates in the synthesis of tetracycline antibiotics (Myers et al., 2007, 2011)*

after 18 steps via intermediate **3**; the yields and number of steps relate to the total synthesis from benzoic acid (Charest *et al.*, 2005a, 2005b; Myers *et al.*, 2007, 2011). Not content with these impressive achievements, Myers and Brubaker re-designed and improved the synthesis of intermediate **2** (Scheme 4.3.3), from which most tetracyclines can be accessed; they identified an alternative cheap and readily available commercial starting material, methyl 3-hydroxy-5-isoxazolecarboxylate **4**, along with an improved synthetic strategy to improve stereoselectivity and yields, and obtained intermediate **2** in 9 steps and 21% overall yield from starting material **4** (Brubaker and Myers, 2007).

There are several noteworthy features in this revised synthesis of intermediate **2** (Scheme 4.3.3):

- First, the introduction of a stereogenic (chiral) centre into structure **5** is carried out in high yield and with a high enantiomeric excess, and this enantiomeric ratio is maintained during the S_N2 replacement of the hydroxyl group (as a mesylate) with a dimethylamino group in the synthesis of **6**.
- The resulting stereogenic centre at C6 becomes C4 in the tetracycline and already has the correct stereochemistry, which is retained throughout the remainder of the synthesis.
- Intermediate **7** has a new stereogenic centre bearing a hydroxyl group; you may be concerned that the stereochemistry is not defined at this new centre, but this group is oxidised to a carbonyl during the synthetic sequence that results in intermediate **8**, so the mixed stereochemistry does not matter at this stage.

So far, we have not considered how rings C and D can be constructed, yet this is just as important for the tetracycline structure. The synthetic strategy adopted by the Myers group uses intermediate **1**, **2**, or **3** as appropriate to provide rings A and B of the tetracycline with the correct stereochemistry and functionality, then elaborates this basic structure by construction of the C-ring, while adding the D-ring through a generalised Michael–Dieckmann reaction sequence (using a carbanion formed from a variety of D-ring precursors), followed by deprotection of all the functional groups (Scheme 4.3.4) (Charest *et al.*, 2005a).

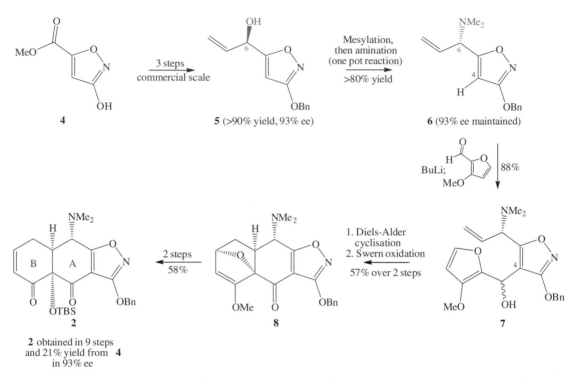

Scheme 4.3.3 *Improved synthesis of key intermediate* **2** *for large-scale tetracycline synthesis (Brubaker and Myers, 2007)*

One great advantage of this approach is the ease with which varying functionality can be added throughout the structure, particularly substituents at C5 (X), C6 (R), C7 (Y), C9 (Z), and even an extra (E) ring; a great many analogues have been made, and some of these combine strong antibacterial activity with activity against strains resistant to first- and second-generation tetracycline antibiotics (Myers *et al.*, 2011; Sun *et al.*, 2011). The synthetic routes to the tetracyclines developed by the Myers group have been sufficiently successful to support a spin-out company, Tetraphase, which has several tetracyclines in early clinical trials. Other groups have also pursued C9-substituted tetracyclines (for example, Koza and Nsiah, 2002; Sum *et al.*, 2006); the clinical success of tigecycline (see Subsections 4.3.4 and 4.3.5) and the development of amadacycline (which is in clinical trials and is discussed in Subsection 4.3.9) provided the rationale for their evaluation.

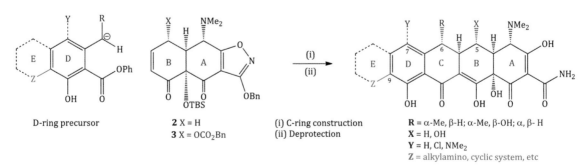

Scheme 4.3.4 *Elaboration of intermediates* **2** *and* **3** *into tetracyclines and their analogues (Charest et al., 2005a)*

4.3.3 Bioavailability

The pharmacokinetics and pharmacodynamics of the tetracycline antibiotics were reviewed recently (Agwuh and MacGowan, 2006; Barbour *et al.*, 2010), but they are not fully understood, with several seemingly contradictory observations. The tetracyclines display time-dependent effects, yet the general concentration-dependent parameters (of exposure time at a concentration above the MIC) provide good clinical results, with a strong post-antibiotic effect. Although generally considered to be bacteriostatic, there is evidence of bactericidal activity with certain tetracycline antibiotics against specific bacteria, when used at an appropriate concentration (Zhanel *et al.*, 2004; Barbour *et al.*, 2010).

The instability of the first-generation tetracyclines during storage has already been mentioned; we will consider the reactions of tetracyclines more carefully here, as they have an effect upon bioavailability and even upon the safety of the products.

The first-generation tetracyclines, although clinically successful, were found to be unstable to acidic, basic, and neutral pH during storage and in solution, including in the gastrointestinal (GI) tract after administration, decreasing their bioavailability (Walton *et al.*, 1970; Ali and Strittmatter, 1978; Wu and Fassihi, 2005). Two main reactions occur in the presence of acid: epimerisation at C4, to produce epitetracycline (Hussar *et al.*, 1968) (Scheme 4.3.5), and dehydration of 6-hydroxytetracyclines across C5a-C6 (with loss of the OH group at C6), to give the anhydrotetracycline derivative, as demonstrated for tetracycline in Scheme 4.3.6.

The dehydration (shown in the upper part of Scheme 4.3.6 for chlortetracycline) proceeds through an E1 mechanism, involving protonation of the C6-OH and loss of a good leaving group (H_2O), to produce a stabilised tertiary carbocation at C6, followed by loss of a proton from C5a to form the alkene group of anhydrotetracycline. The slightly improved acid stability of demeclocycline and slower dehydration is due to fact that the secondary carbocation that has to be formed at C6 is less stable, and this process thus has a greater activation energy. In this latter case, reversible protonation of the C6-OH favours demeclocycline, instead of proceeding to the carbocation (Scheme 4.3.6).

Epimerisation at C4 of the anhydrotetracycline derivatives can also occur, producing the corresponding epianhydrotetracycline derivative (Sokoloski *et al.*, 1977). Some of the products formed by acid- or base-catalysed degradation are themselves active as antibiotics, while others, for example anhydrotetracycline and epianhydrotetracycline, are toxic (Mull, 1966; Kunin, 1967). The adverse effects (which can manifest as Fanconi syndrome; see Subsection 4.3.7) of using tetracyclines that have degraded during storage were observed soon after these agents had been adopted into regular use, but it was several more years before reliable analytical methods for their analysis and quality control were developed (Frimpter *et al.*, 1963; Gross, 1963; Butterfield *et al.*, 1973). We will discuss the adverse effects of tetracyclines in Subsection 4.3.7; our main concern here is that the bioavailability of some tetracyclines (particularly the first-generation members) can be reduced by their degradation during storage, particularly in solution (Wu and Fassihi, 2005), or as a result of GI-induced reactions (Okeke and Lamikanra, 1995; Dos Santos *et al.*, 1998). The second-generation

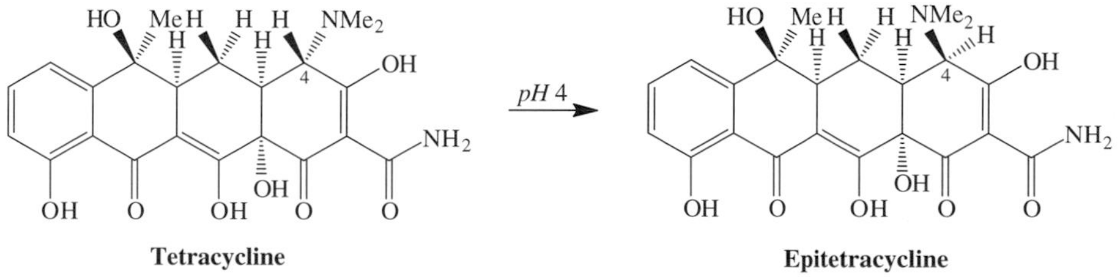

Scheme 4.3.5 *Acid-catalysed epimerisation at C4 of tetracycline*

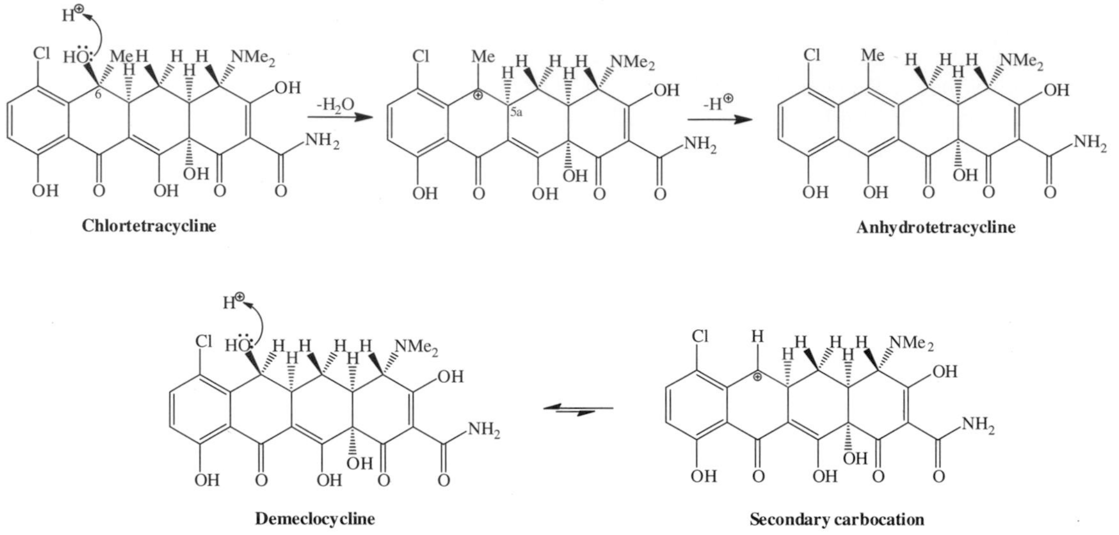

Scheme 4.3.6 *Acid-catalysed dehydration of chlortetracycline and improved stability of demethylchlortetracycline (demeclocycline)*

tetracyclines, doxycycline and minocycline, and the third-generation, tigecycline, are more stable to acidic pH, as they do not have a C6-OH substituent to be protonated and eliminated as water.

Lymecycline is a prodrug form of tetracycline that is hydrolysed at acidic and neutral pH *in vivo* to tetracycline, formaldehyde (methanal), and the amino acid lysine after oral and parenteral administration (Scheme 4.3.7). Interestingly, it has lower oral bioavailability than the parent compound, tetracycline (Sjölin-Forsberg and Hermansson, 1984).

The balance of lipophilicity to hydrophilicity in tetracyclines is affected by pH (Chen and Lin, 1998); such a property is usually a clue that there are functional groups in the molecule under investigation which are ionisable at physiological pH values. Each tetracycline has at least one readily ionisable amine group (at C4) and two enol groups (at C3 and C12) that are also relatively easily ionised. First- and second-generation tetracyclines have three pK_a values in the physiological pH range (generally around 3.2, 7.6, and 9.6) – the two enol groups, at C3 and C12, are the most acidic (and so have the lower pK_a values). The pK_a of the phenol group at C10 ($\sim$12) means that it is not ionised at physiological pH values. The acid–base equilibria for tetracycline, which are typical of the tetracycline antibiotics, are demonstrated in Scheme 4.3.8 (Jin *et al.*, 2007).

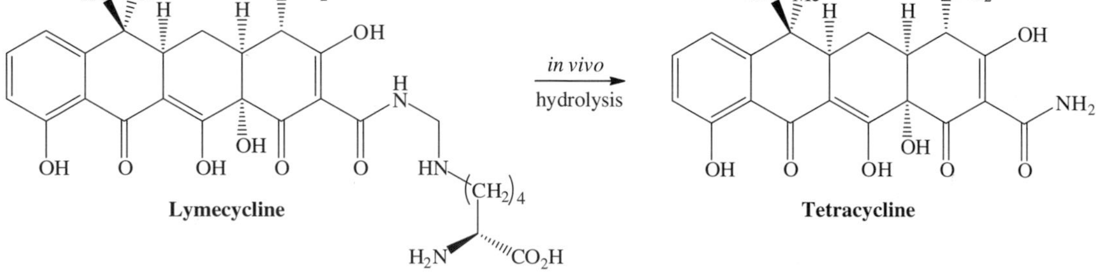

Scheme 4.3.7 In vivo *hydrolysis of lymecycline to form tetracycline*

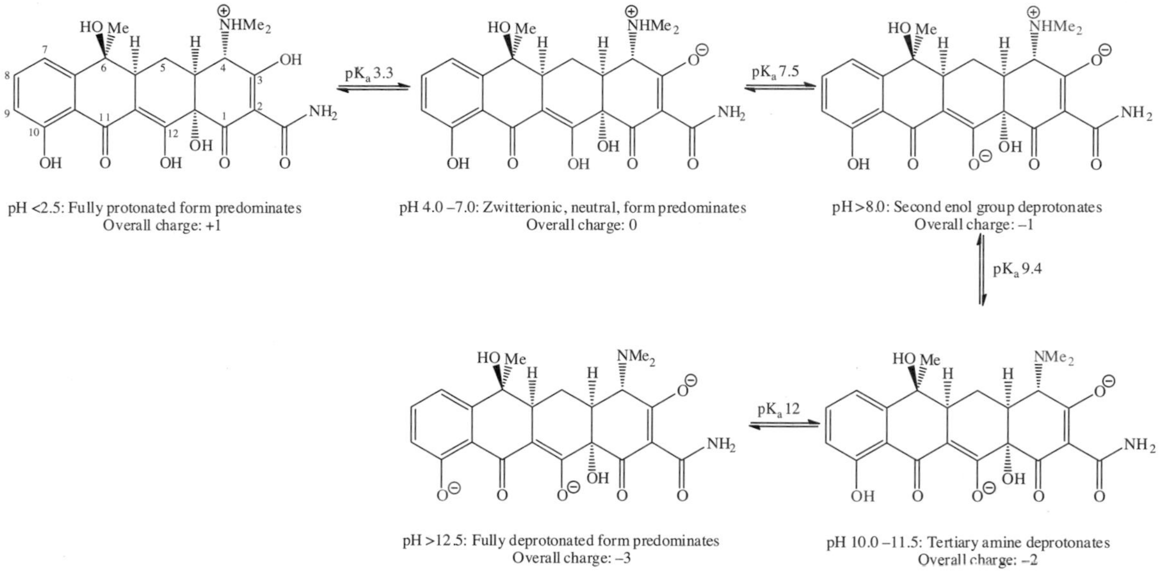

Scheme 4.3.8   *Acid–base equilibria for tetracycline (Jin et al., 2007)*

The fully protonated form of the tetracyclines (with an overall charge of $+1$) is likely to predominate in the acidic medium of the stomach, minimising absorption from this compartment (as neutral species are absorbed best). It is not surprising to find that tetracycline antibiotics are absorbed chiefly from the duodenum, where the pH is around 6–6.5 and the overall neutral form of the tetracycline predominates (Colaizzi and Klink, 1969). By modifying the pH, greater aqueous solubility of the tetracycline antibiotics can be achieved, making them suitable for parenteral administration, and oxytetracycline, lymecycline, doxycycline, and minocycline have been formulated in this way (Chopra and Roberts, 2001).

The third-generation antibiotic tigecycline has limited oral bioavailability and is administered by IV infusion, due to its greater hydrophilicity and reduced lipophilicity as a result of its extra ionisable groups (Meagher *et al.*, 2005). Tigecycline has two extra ionisable groups (Figure 4.3.3):

- The *t*-butylamine on the side chain at C9 (a secondary aliphatic amine, so expected to be basic, with a pK_a (of the conjugate acid, R_3NH^+) of 8.5–10).
- The dimethylamino group at C7 (an aromatic amine, so expected to be a weak base, due to resonance of the nitrogen lone pair with the π-system of the aromatic ring, with a pK_a of 3.5–5).

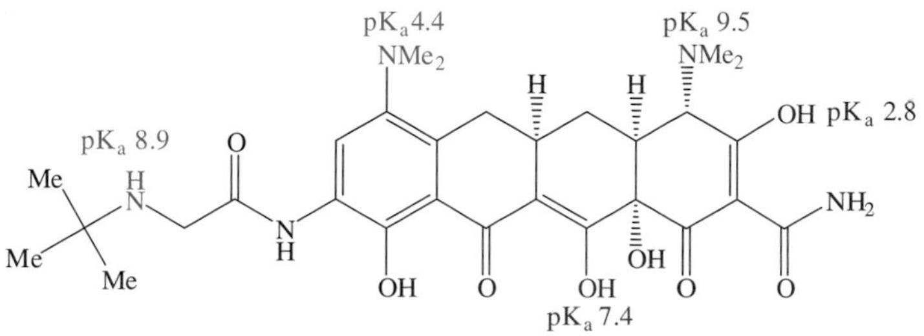

Figure 4.3.3   *Tigecycline pKa values (Tygacil_BioPharmr)*

Besides being affected by pH, the absorption of the first-generation tetracyclines in particular, and some of the second-generation analogues, is adversely affected by their concurrent administration with food (Schimdt and Dalhoff, 2002; Agwuh and MacGowan, 2006), dairy products, and other metal-ion-containing preparations (such as antacid treatments), although there is some evidence that the absorption of lymecycline is less affected by milk (Ericson and Gnarpe, 1979). Tetracycline has a logP value of 0.09 and a bioavailability of around 75%, which is reduced by about 50% when co-administered with food (Miller *et al.*, 1977; Zhanel *et al.*, 2004). By comparison, doxycycline and minocycline have a greater bioavailability (90–100%) (Zhanel *et al.*, 2004), their absorption is not affected by food, and they are well absorbed after oral administration. The greater lipophilicity of these agents (logP values of 0.95 and 1.12, respectively) (Colaizzi and Klink, 1969) undoubtedly make a major contribution to these properties. Although minocycline offers improved oral bioavailability over the majority of other tetracyclines, the risk of side effects when used for a protracted period and of the development of resistance has limited it largely to the treatment of acne vulgaris. A recent review suggests that it may offer potential for the systemic treatment of community-associated MRSA and of the significant nosocomial threat of *Acinetobacter baumanii* (Bishburg and Bishburg, 2009).

Tetracyclines are known to bind strongly to a range of metal ions, with the strongest and most significant binding, in terms of bioavailability, mode of action, mechanisms of resistance, and adverse events, being to magnesium, calcium, iron, and copper (Agwuh and MacGowan, 2006). Chelation of tetracyclines to metal ions in the GI tract adversely affects the absorption of both the tetracycline and the metal ion through the precipitation of insoluble metal-tetracycline complexes (Agwuh and MacGowan, 2006), which provides a scientific rationale for avoiding the administration of tetracyclines alongside metal ion preparations and dairy products. The ability of tetracyclines to coordinate metal ions means that these antibiotics should not be administered to children, due to their ability to sequester calcium and other metal ions at a crucial time in bone and teeth development. In the USA and Australia, children under the age of eight are contraindicated, while in the UK tetracyclines should not be given to children under the age of 12. European guidelines on the third-generation tetracycline tigecycline recommend that it is not used for children and adolescents under the age of 18 years, due to lack of data on its safety and efficacy in these patient groups. The reason for the strong metal ion binding can be seen by considering the tetracycline structure – the lower face of the molecule (as can be seen in Scheme 4.3.9) has several oxygen atoms and so is ideal for binding to a metal ion; the C12 (as the enolate) and C11 oxygen atoms are accepted to be the major binding site (Jin *et al.*, 2007; Palm *et al.*, 2008).

We will return to tetracycline-magnesium complexes again later in this section, when we consider the uptake of tetracyclines into bacterial cells and also when we look at the mode of action (Subsection 4.3.4) and mechanisms of resistance (Subsection 4.3.5).

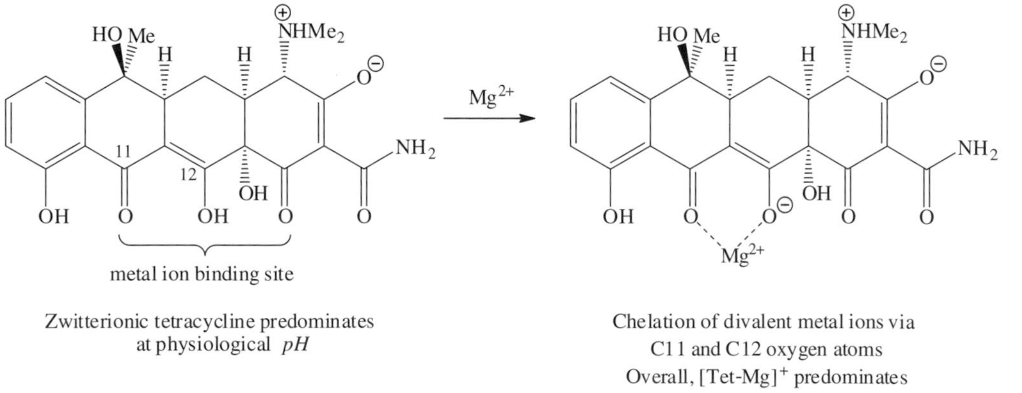

Zwitterionic tetracycline predominates at physiological *pH*

Chelation of divalent metal ions via C11 and C12 oxygen atoms
Overall, $[\text{Tet-Mg}]^+$ predominates

Scheme 4.3.9　*Tetracycline-magnesium ion chelation (Jin et al., 2007; Palm et al., 2008)*

Table 4.3.3 *Pharmacokinetic parameters for selected tetracycline antibiotics (Zhanel et al., 2004; Agwuh and MacGowan, 2006; Hoffmann et al., 2007; Barbour et al., 2010)*

Tetracycline antibiotic (dose)[a]	Half-life in serum, $t_{1/2}$ (hrs)	Volume of distribution (L/kg)	C_{max} (µg/mL)	Daily dose regimen related to $t_{1/2}$
Doxycycline (200 mg oral)	12	0.7	2.5	2/day
Lymecycline (400 mg)	8–10	-	2.1	1–2/day
Minocycline (200 mg oral)	11	0.14–0.7	2.5	2/day
Oxytetracycline (250 mg)	6–10	1.8–1.9	2	4/day
Tetracycline (500 mg oral)	6–10	1.3–1.6	3–4	4/day
Tigecycline (12.5 mg IV)	36	>10	0.11	2/day

[a] The dose relates to the measurement of the pharmacokinetic parameters and not necessarily to the usual clinical dose.

In general, the tetracycline antibiotics are not metabolised, except for tetracycline, of which about 5% is metabolised, and tigecycline, of which 5–20% is metabolised (Meagher *et al.*, 2005); they are excreted by both the urinary (<50%) and the faecal (>40%) routes (Agwuh and MacGowan, 2006). The urinary excretion of tetracyclines has been found to be affected by pH; as expected, it is significantly increased at pH values above 8, at which values greater ionisation and hydrophilicity are also to be expected (Jaffe *et al.*, 1973).

The tetracyclines exhibit a high volume of distribution and generally good tissue penetration; alongside their long half-lives and significant post-antibiotic effect, most can be administered once or twice daily (Table 4.3.3).

We discussed above how metal ion binding adversely affects the absorption of tetracyclines from the GI tract, but it is an important and crucial part of the uptake of tetracyclines into bacterial cells. The chelation to a magnesium ion forms a tetracycline-magnesium cationic complex (with an overall charge of +1), which is transported through the outer membrane into the periplasm by porins, such as the well-characterised *OmpF* and *OmpC* examples from Gram negative bacteria (Nikaido, 1994, 2003). You will remember from Section 1.1.2 that porins are protein pores which transport a range of molecules through the outer membrane into the periplasm; they are known to be responsible for the transport of other antibacterial agents, such as quinolones and β-lactams, besides tetracyclines (Jaffe *et al.*, 1982; Mortimer and Piddock, 1993). The mechanism for porin-mediated transport relies upon the Donnan potential[14] across the outer membrane and leads to accumulation of the tetracycline-magnesium cationic complex in the periplasm (Zhanel *et al.*, 2004). It is probable that the tetracycline-magnesium complex dissociates in the periplasm, perhaps due to the lower pH in this compartment, which is sufficiently acidic to drive the reprotonation of the enol oxygen at C12. The released zwitterionic tetracycline (overall neutral charge) is in equilibrium with a small proportion of the uncharged form, which is weakly lipophilic and able to diffuse through the cytoplasmic (inner) membrane in an energy-requiring process (Scheme 4.3.10) (Nikaido and Thanassi, 1993). Tetracyclines are presumed to adopt a similar uncharged tetracycline diffusion entry route through the simpler cell membrane of Gram positive bacteria.

You may be wondering why the diffusion of the neutral tetracycline across the bacterial cytoplasmic membrane is energy-requiring. Live bacteria have a difference in pH between the cytoplasm and periplasm of about 1.7 pH units (Nikaido and Thanassi, 1993), so that, after the neutral tetracycline passes through the cytoplasmic membrane, it becomes trapped at the higher pH of the cytoplasm, releasing a proton to form a greater proportion of tetracycline in a more hydrophilic form (overall molecular charge of −1)

---

[14] The Donnan potential is an electrical potential caused by the small difference in charge on either side of a semi-permeable membrane due to inequivalent concentrations of an ionised substance on each side.

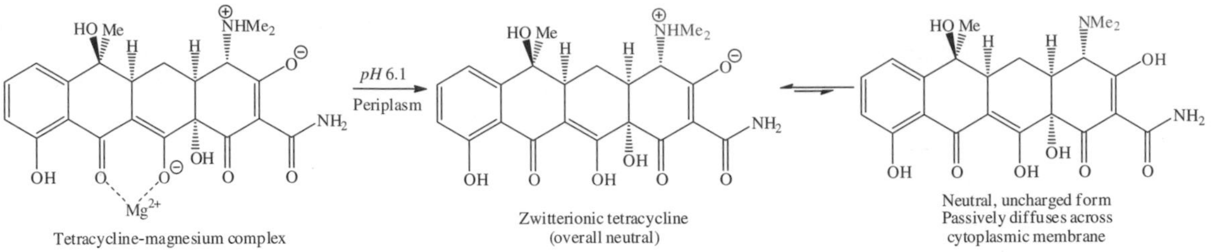

Scheme 4.3.10 *A tetracycline-magnesium complex dissociates to the zwitterion in the periplasm and equilibrates with the neutral form, which enters bacterial cells by passive diffusion (Nikaido and Thanassi, 1993)*

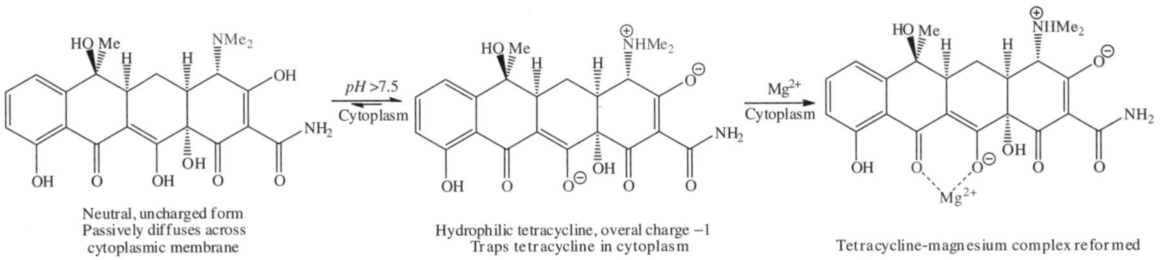

Scheme 4.3.11 *In the cytoplasm, the hydrophilic form of the tetracyclines predominates and reforms the tetracycline-magnesium complex (Nikaido and Thanassi, 1993)*

(Scheme 4.3.11). To maintain the pH of the cell, and the difference with the external pH, the bacteria must continue to transport protons out of the cytoplasm, which is an energy-dependent process (Nikaido and Thanassi, 1993), so it is not actually the diffusion of tetracycline into the cell that is energy-requiring, but the maintenance of pH that is required as a result.

The cytoplasmic pH is generally around 7.1–7.8 (Wilks and Slonczewski, 2007) and the magnesium ion concentration is about 1 mM, allowing the reformation of the tetracycline-magnesium complex. What happens next is part of the mode of action, which we will investigate in the next subsection.

4.3.4   Mode of action and selectivity

Tetracyclines are known to bind to a high-affinity binding site in the 30S ribosomal subunit, near to the 16S subunit, and to cause inhibition of bacterial protein synthesis, through the competitive binding of the tetracycline antibiotic instead of the charged tRNA at the peptidyl transferase centre A site (Spahn and Prescott, 1996). Considering that the aminoglycosides bind more or less to the same site, it is perhaps surprising to find that the antibiotic mode of action of tetracyclines is much less well defined and understood.

In Section 1.1.3, we looked at the basic process of translation; earlier in this section, when we considered the mode of action of aminoglycosides (Subsection 4.1.4), we also looked at the recognition of cognate tRNA and its differentiation from non-cognate tRNA. The involvement of other processes and species was only mentioned briefly, but we must also consider the role of the elongation factor EF-Tu in protein synthesis. Elongation factors are involved in exactly what their name suggests – they play a key role in the elongation of the growing peptide/protein chain. There are two elongation factors involved in prokaryotic protein synthesis: **EF-Tu** is involved in the initial binding of charged tRNA (amino acid linked to tRNA) to rRNA in the A site and is part of the proof-reading process, while, after transfer of the amino acid from charged tRNA to the growing peptide chain, **EF-G** assists in moving the now-uncharged tRNA to the P site prior to its leaving the ribosome.

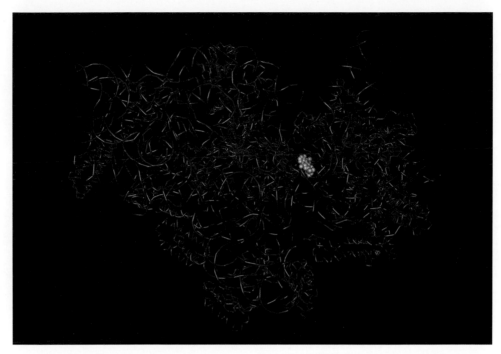

Figure 4.3.4 *Tetracycline bound to rRNA in the A site (or Tet-1 site), which lies between the head (lower) and the body (upper) of the 30S subunit; the dsRNA backbone is shown as a blue ribbon ribosomal proteins are shown in red; and the magnesium ion (green sphere) is clearly visible chelated to tetracycline (yellow) (base pairs have been removed for clarity) (PDB 1HNW) (Brodersen et al., 2000) (Reprinted from D. E. Brodersen, W. M. Clemons Jr., A. P. Carter, et al., Cell, 103, 1143–1154, 2000, with permission of Elsevier.)*

In common with EF-Tu, EF-G uses the hydrolysis of GTP[15] to drive a conformational change, which causes the tRNA to move from the A site to the P site. It is EF-Tu that is involved in the mode of action of tetracyclines, so we need to look at its role a little more closely:

- After initiation of translation, EF-Tu binds to a charged tRNA and this complex enters the A site, where the charged tRNA presents its anti-codon to the mRNA codon.
- If the codon/anti-codon triplets match, EF-Tu hydrolyses a GTP molecule (to form GDP and phosphate), causing a conformational change in EF-Tu that releases the cognate charged tRNA.
- The cognate charged tRNA then fully enters the A site so that protein synthesis can take place involving the amino acid it carries.
- If the codon/anti-codon triplets do not match, GTP is not hydrolysed and the charged tRNA is not released for protein synthesis to occur.

Tetracyclines do not interfere with the process of EF-Tu binding to charged tRNA, nor with the initial binding of the EF-Tu/charged tRNA complex or with the hydrolysis of GTP; but somehow they prevent the full binding of the cognate charged tRNA to the A site, which leads to termination of peptide chain growth and inhibition of protein synthesis. They also prevent binding of release factors to the A site at the end of protein synthesis (Brodersen *et al.*, 2000; Pioletti *et al.*, 2001).

---

[15] GTP = guanosine triphosphate; GDP = guanosine diphosphate. There are no prizes for guessing what GMP stands for.

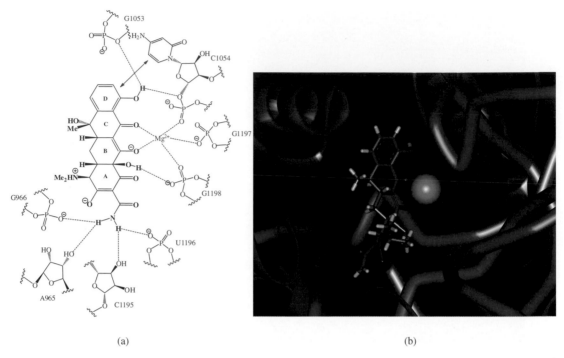

Figure 4.3.5 *(a) Possible interactions between tetracycline and bases of helices H34 and H31 of the 16S part of the 30S subunit of rRNA, including magnesium ion chelation, hydrogen bonding, and aromatic stacking (Reprinted from D. E. Brodersen, W. M. Clemons Jr., A. P. Carter, et al., Cell, 103, 1143–1154, 2000, with permission of Elsevier.). (b) Chelation of tetracycline to magnesium ion (green) at the A site of rRNA (Reprinted from M. Pioletti, F. Schlunzen, J. Harms, et al., EMBO J., 20, 1829–1839, 2001, with permission of Nature Publishing Group.). The relative planarity of the B, C, and D rings is visible, to which the A ring (at the bottom of the molecule) is almost perpendicular*

Several tetracycline binding sites have been observed, including a primary site close to the A site and five others distributed around the upper part of the 30S ribosomal subunit (Pioletti *et al.*, 2001). The primary site (the A site or Tet-1 site) lies in the 30S subunit, close to the 16S section; the tetracycline binding pocket is formed by a distortion in the minor groove of helix-34 (H34), involving RNA residues 1196–1200 and 1053–1056, along with residues 964–967 of H31 (Figure 4.3.4).

Tetracycline interacts with the sugar-phosphate backbone of H34 through coordination with a magnesium ion (Figure 4.3.5) (Brodersen *et al.*, 2000; Pioletti *et al.*, 2001). Its binding to the constant sugar-phosphate backbone offers a rationale for its broad-spectrum activity, which may derive from the lack of requirement for specific bases in the binding site (Connell *et al.*, 2003).

Tigecycline also binds to the ribosomal A site and sterically prevents the binding of charged tRNA, resulting in the same outcome: inhibition of protein synthesis, although tigecycline binds to the ribosome more strongly than does tetracycline (Connell *et al.*, 2003; Olson *et al.*, 2006). Tigecycline associates with the magnesium ion and interacts with the same bases of H34 and H31, and extra intermolecular interactions may account for the increased affinity of tigecycline, such as additional hydrogen bonding through the *t*-butylamide side chain at C9 (Olson *et al.*, 2006).

The ribosomal binding of the tetracycline antibiotics is reversible, which may be the basis of the bacteriostatic action, allowing bacterial cell growth to resume once the agent is removed (Chopra and Roberts, 2001; Zhanel *et al.*, 2004).

We have always concentrated on the selectivity of antibacterial agents, and the selective interaction of the tetracyclines with prokaryotic over eukaryotic ribosomes is poorly understood but is assumed to be related to their differences in structure. The tetracyclines do not have any notable affinity for the 80S eukaryotic ribosomes (Roberts, 2003) or the 50S subunit (Epe and Woolley, 1984).

In the search for other tetracyclines which are better able to evade resistance mechanisms, agents with greater lipophilicity have been developed, several of which have been found to be bactericidal, rather than bacteriostatic. These 'atypical tetracycline' agents, such as the anhydrotetracyclines, do not target the ribosomes – their mode of action relies upon disruption of the bacterial cell membrane (Chopra, 1994). As we saw in Figure 4.3.5, tetracyclines are not totally planar, but are almost 'L' shaped, with three rings, B, C, and D, almost planar and the A ring at an angle to that plane; along with their enhanced lipophilicity, their shape enables them to accumulate in the membrane, which disturbs its function and leads to cell death. The atypical tetracyclines are not currently in clinical use as this membrane-disturbing ability extends to eukaryotic cell plasma membranes, conferring significant toxicity upon these molecules (Chopra, 2001).

It is interesting to note that some of the first- and second-generation tetracyclines were found to act upon other macromolecules besides the ribosome (Sapadin and Fleischmajer, 2006; Griffin *et al.*, 2010, 2011). The medicinal potential of these properties was quickly realised and many of the recent patents and tetracyclines in clinical trials relate to the discovery, biosynthesis, or synthesis of agents with non-antibacterial activities. For more information on the mode of action and selectivity of the tetracyclines, see Chopra and Roberts (2001), Zhanel *et al.* (2004) and Zakeri and Wright (2008).

4.3.5   Bacterial resistance

The broad-spectrum activity and low incidence of side effects of the tetracyclines resulted in their widespread clinical use, and also their use in veterinary and agricultural applications, for several decades after their discovery. For example, tetracyclines were found to enhance the growth of cattle and, as a result, cattle feed was supplemented with tetracycline at a low level (< 200 g/ton), which introduced tetracycline into the environment at sub-lethal concentrations for bacteria, encouraging the development of resistance (Chopra and Roberts, 2001). Concerns over this practice were raised nearly 50 years ago (Manten, 1963) and continue to be raised, with observations of increased risk of antibiotic-resistant strains of bacteria in farm workers (Levy, 2001; Roberts, 2002), especially *Escherichia coli*, *Enterococcus faecium*, and *Salmonella typhimurium* (Gyles, 2008; Anjum *et al.*, 2011). There is some disagreement over whether or not this resistance, be it antibiotic-specific or multidrug resistance, represents a health threat (Cox and Popken, 2010; Mathers *et al.*, 2011). However it has arisen, it now limits the clinical use of the first- and second-generation tetracyclines (Chopra and Roberts, 2001; Zhanel, 2004; Zakeri and Wright, 2008). Continued research into novel antibiotic tetracyclines indicates our reluctance to relegate these valuable antibiotics to the past, as the number of clinical antibacterial choices with strong activity across both Gram positive and Gram negative pathogens is decreasing at an alarming rate and the development of new tetracyclines, such as tigecycline and amadacycline (which you will meet in Subsection 4.3.9), with a low incidence of resistance provides hope for the future of antibacterial chemotherapy.

Bacterial resistance to antibiotic tetracyclines has been comprehensively reviewed (Speer *et al.*, 1992; Chopra and Roberts, 2001), and a Web site, updated twice annually, provides information on the different tetracycline resistance mechanisms and the genes that encode the various proteins conferring resistance (http://faculty.washington.edu/marilynr/, last accessed 10 March 2012).

There are three main mechanisms of tetracycline resistance:

- Efflux from the cell by specific tetracycline efflux pumps and multidrug resistance efflux pumps.
- Rescue of inhibited ribosomes by ribosomal protection proteins.
- Enzymatic inactivation of the tetracyclines.

The first two are the major mechanisms seen in clinically relevant bacteria (Zhanel *et al.*, 2004) and each results from the production of a protein, or set of proteins, which mediates the specific activity leading to tetracycline resistance. In the majority of Gram positive bacteria, a single gene encodes a protein that confers tetracycline resistance, usually an efflux pump or a ribosomal protection protein. In a few Gram positive bacteria and most Gram negative bacteria, several proteins may be required for resistance, encoded by a single unit called a **resistance determinant**, which includes all the necessary genes for that mechanism of resistance (Roberts, 1996). For example, a typical resistance determinant in a Gram negative bacterium includes genes for both an efflux pump and a repressor, along with overlapping promoters and operators (Hillen and Berens, 1994). The genes and resistance determinants encoding for these proteins are largely found on plasmids or transposons, both mobile elements that can be transferred from one bacterial strain to another (Speer *et al.*, 1992), accounting for the increased observation of clinical resistance cases. In fact, tetracycline-resistance-gene-containing transposons are activated towards transfer by the presence of tetracycline, even between Gram positive and Gram negative bacterial species (Nikaido, 2009).

As of mid-2011, there were 43 different genes encoding for tetracycline resistance: 27 of these code for energy-dependent efflux pumps, 12 for ribosomal protection proteins, 3 for inactivating enzymes, and 1 remains unclassified. A system of nomenclature has been agreed upon for acquired tetracycline resistance genes, which uses *tet* for genes conferring tetracycline resistance and Tet for the consequent proteins, with the class of resistance determinant in parentheses directly afterwards (Levy *et al.*, 1989, 1999). Another letter (or, recently, a number) is placed before the parentheses when there is more than one family member in a class, indicating the position of that particular gene in the class; the A gene is always the first in any class. Each class of efflux resistance determinants in Gram negative bacteria starts with a *tetA* gene and a *tetR* gene, which code for an efflux pump, TetA, and a repressor protein, TetR, respectively (Hillen and Berens, 1994). For example, TetA(B) refers to an efflux pump protein of the Tet(B) class and TetR(B) refers to the repressor protein in the same class, encoded by the *tetA(B)* and *tetR(B)* genes. We will look at each type of resistance protein, including the repressor protein, next (Table 4.3.4). The equivalent nomenclature of *otr* and Otr is used for those genes and proteins, respectively, relating to oxytetracycline, which are presumed to be equivalent to their tetracycline analogues (Bishburg and Bishburg, 2009). For more general information on bacterial resistance to the tetracyclines, see Speer *et al.* (1992); Chopra and Roberts (2001).

4.3.5.1   Efflux pumps

Tetracycline efflux pumps belong to the class of proton-dependent efflux systems; some of these systems are specific to tetracyclines (Roberts, 1997; Chopra, 2002), whereas others have a wide range of substrates they can expel from the cell, including various drugs and metal ions (Bay and Turner, 2009; Ma *et al.*, 2009). They are membrane-bound, energy-requiring tetracycline transporters, expelling the monocationic tetracycline-magnesium ion chelate $(Tet-Mg)^+$ from the bacterial cell in exchange for a proton (H^+); the term 'antiporter' is often used to describe them. The genes coding for efflux pumps belong to the well-studied major facilitator superfamily (MFS) (Paulsen *et al.*, 1996; Mazurkiewicz *et al.*, 2005).

Of the (currently) 27 classes of bacterial efflux pump, the vast majority are found exclusively in Gram negative bacteria and are usually associated with large conjugative plasmids, which commonly also carry genes for resistance to several other agents, such as heavy metals and toxins. Exposure of bacteria to any of these exogenous agents selects for this multiple-resistance plasmid and confers cross-resistance, which is believed to have been a major contributor to the recent exponential growth in the number of multidrug-resistant bacteria (Chopra and Roberts, 2001; Schweizer, 2003; Zakeri and Wright, 2008). The most common efflux proteins found in Gram negative bacteria include Tet(A), Tet(B), Tet(C), Tet(D), Tet(E), and Tet(G); in particular, Tet(B) has been observed in more than 40 Gram negative genera (Chopra and Roberts, 2001;

Table 4.3.4   *Selected tetracycline resistance proteins (Chopra and Roberts, 2001; Zhanel et al., 2004; Poole, 2007)*

Protein conferring tetracycline resistance	Class of resistance mechanism	Associated bacterial species	Location of gene	Notes
Tet(B)	Efflux pump	Gram negative species	Plasmid	Most common efflux pump in Gram negative bacteria
Tet(K)	Efflux pump	Gram positive species *Staphylococcus aureus* (G+)	Plasmid	Confers resistance to most tetracyclines, but not tigecycline
Tet(L)	Efflux pump	Gram positive species *Bacillus subtilis* (G+)	Chromosomal	Confers resistance to tetracycline and doxycycline, not to minocycline or tigecycline
Tet(M)	Ribosomal protection protein	*Neisseria gonorrhoeae* (G−) Also common in Gram positive bacteria	Chromosomal	High-level resistance *tet(M)* is widely distributed across bacteria
Tet(O)	Ribosomal protection protein	*Campylobacter jejuni*	Plasmid	Widespread across Gram positive bacteria
TetA(P)	Efflux pump	Clostridia species	Plasmid	Found in some Gram positive species; commonly found with TetB(P)
TetB(P)	Ribosomal protection protein	Clostridia species	Plasmid	Found in some Gram positive species; commonly found with TetA(P)

Poole, 2007). Efflux proteins found in Gram positive bacteria often belong to the classes Tet(K) and Tet(L) and are usually associated with small plasmids with ready transmission between species, although they have been incorporated into the chromosomes of certain bacteria, such as staphylococci and *B. subtilis* (Chopra and Roberts, 2001). Gram positive efflux pumps are active against tetracycline, chlortetracycline, and occasionally minocycline, but not tigecycline. The greater lipophilicity of the second-generation tetracyclines, such as doxycycline and minocycline, confers greater ability to evade efflux systems and overcome resistance due to this mechanism (McMurry *et al.*, 1982), but even these tetracyclines succumb to certain Gram positive and Gram negative efflux pumps.

By now, you may be wondering why all bacteria do not constantly express an efflux pump to protect themselves from tetracycline, or from a range of threatening agents with a multidrug-resistance efflux pump. Fortunately for us, it's not that simple! Although the genes encoding efflux pumps have become relatively common, they cannot be expressed by bacteria all of the time, as they are detrimental to the cell. Unregulated efflux pump expression can result in the non-selective export of multiple cations, loss of membrane proton potential, and consequent cell death (Grkovic *et al.*, 2002). The co-existence of the repressor protein TetR ensures that Gram negative efflux pumps are only produced when needed (when the bacteria are challenged by a tetracycline antibiotic). TetR acts as an exquisitely sensitive molecular detector for tetracycline, at concentrations well below the MIC, and is activated to switch on expression of both TetA and TetR when tetracycline binds to it, which results in active efflux of the tetracycline-metal ion complex from the cell. This resistance mechanism is very finely tuned, as TetR responds immediately tetracycline is detected in the cell in order to maintain the cellular concentration of tetracycline below its pharmacologically active level (Zakeri and Wright, 2008). There are more than 2000 members of the TetR family (Ramos *et al.*, 2005); although TetR is frequently the nomenclature used, there are many examples with different names across a wide range of bacteria. Gram positive bacterial efflux pumps, such as Tet(K) and Tet(L), are not regulated by a repressor protein, but are instead suppressed, by a currently unknown mechanism.

Inhibition of tetracycline efflux pumps offers a possible method for the restoration of tetracycline sensitivity in bacteria and has attracted some attention for the potential development of novel compounds for synergistic use with tetracyclines (Alibert-Franco *et al.*, 2009; Pagès and Amaral, 2009; Zechini and Versace, 2009; Piddock *et al.*, 2010; Garvey *et al.*, 2011)

4.3.5.2   Ribosomal protection proteins

These cytoplasmic proteins, as the name suggests, protect ribosomes from tetracycline inhibition and confer resistance to those species that express them. Ribosomal protection proteins confer a wider spectrum of tetracycline resistance than is offered by efflux pumps (except for the Tet(B) efflux pump system) (Chopra and Roberts, 2001). When a tetracycline binds to a ribosome, the ribosomal protection proteins operate as a salvage mechanism; they are thought to bind to the H34 helix at a site distant from the A site and cause allosteric changes to the the Tet-1 binding site, which decreases the binding affinity of the tetracycline, causing it to be released from the ribosome (Spahn *et al.*, 2001). The *N*-terminal sequences of the ribosomal protection proteins Tet(M) and Tet(O) show a high degree of homology to the elongation factors EF-Tu and EF-G and, like these proteins, have GTP-ase activity, upon which their action is dependent (Taylor and Chau, 1996; Connell *et al.*, 2003). Tet(M) is expressed by a wide range of bacteria and is the most common ribosomal protection protein.

4.3.5.3   Enzymatic degradation of tetracyclines

Three genes that confer resistance to tetracycline through enzymatic modification and degradation have been identified as *tetX*, *tet34*, and *tet37* (Wright, 2005), and the Tet(X) enzyme is the best-studied (Yang *et al.*, 2004). Tet(X) is an NADP-dependent oxidoreductase (redox) enzyme that catalyses hydroxylation at C11a, which disturbs chelation of the magnesium ion and prevents binding of the tetracycline to the ribosome (remember, Mg^{2+} chelation by a tetracycline is essential for its activity). The hydroxylation at C11a results in formation of a carbonyl group at C12, which is attacked by C6-OH to form a hemiketal, which then spontaneously decomposes to inactive products, as illustrated for oxytetracycline in Scheme 4.3.12 (Yang *et al.*, 2004).

Tigecycline is also susceptible to this type of resistance, even though it lacks the hydroxyl group at C6. The result of C11a hydroxylation of tigecycline by Tet(X) is weaker binding to the magnesium ion, which is crucial for binding to the ribosome, and decreased inhibition of protein translation – the essential mode of action of tetracyclines (Moore *et al.*, 2005).

Fortunately, the genes encoding these tetracycline deactivation enzymes have so far been found only in a very few bacteria, originally transposon-borne in the strict anaerobe *Bacteroides fragilis*, which cannot grow in an oxygen atmosphere and therefore has no oxygen source to carry out this hydroxylation (Yang *et al.*, 2004). The *tet(X)* gene has since been observed in one *Sphingobacterium* species, which is an aerobic bacterium with the potential for tetracycline resistance by this mechanism; although the *tet(X)* gene does not appear to be common, it does represent a possible future threat and current research aims to develop Tet(X) inhibitors, before clinical resistance becomes a problem, using the crystal structure for guidance (Volkers *et al.*, 2011).

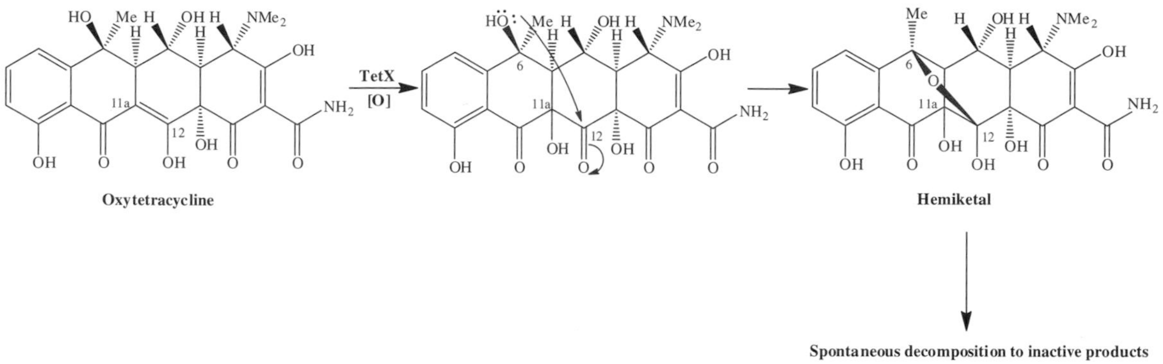

Scheme 4.3.12 *Enzymatic inactivation of oxytetracycline by Tet(X) (Yang et al., 2004)*

4.3.5.4 Glycylcyclines

The development of tigecycline and similar tetracycline derivatives offers hope to clinicians for new highly active antibacterial agents, with a broad spectrum of activity across both Gram positive and Gram negative species due to the limited resistance to their action (Zhanel *et al.*, 2004; Hawkey and Finch, 2007). So far, tigecycline has **not** been found to be acted on by ribosomal protection proteins across a range of bacteria, which appears to be due to the tighter binding of tigecycline to the ribosome, so that the ribosomal protection proteins, such as Tet(M), are unable to displace it (Zhanel *et al.*, 2004; Šeputienė *et al.*, 2010). As we have discovered, efflux pumps are widespread across bacteria and are a major source of resistance. Part of the excitement around the development of tigecycline is its ability, so far, to evade the activity of the majority of efflux pumps, both *in vitro* and in the clinic (Tuckman *et al.*, 2007; Kelesidis *et al.*, 2008). Sadly, however, a celebration would be premature; read on to the end of the next subsection for the latest setback to infection control!

4.3.6 Clinical applications

4.3.6.1 Spectrum of activity

As we have seen, the tetracyclines are considered to be broad-spectrum antibiotics as they have a wide range of activity against Gram negative bacteria, Gram positive bacteria, and atypical organisms such as chlamydiae, mycoplasmas, and rickettsiae (Chopra and Roberts, 2001). In general, there is little difference in the antimicrobial spectrum of the tetracyclines, with the exception of minocycline, which is considered to have a broader spectrum and which also has activity against *Neisseria meningitidis* (Brown and Fallon, 1980).

Unfortunately, despite this broad spectrum of activity, the clinical value of the tetracyclines has decreased over the years due to increased bacterial resistance; tigecycline was developed to overcome the problems associated with tetracycline resistance and has activity against tetracycline-resistant organisms, including meticillin resistant *Staphylococcus aureus* (MRSA) and vancomycin resistant enterococci (VRE) (Zhanel *et al.*, 2004).

4.3.6.2 Acne vulgaris and rosacea

Acne vulgaris (sometimes shortened to 'acne') is a disorder of the pilosebaceous glands that frequently occurs in teenagers. It is characterised by the presence of non-inflammatory or inflammatory lesions (or sometimes a mixture of the two), which principally affect the face, back, and chest. The non-inflammatory lesions consist of open or closed comedones (more commonly known as blackheads and whiteheads), while the inflammatory lesions typically present as papules and pustules, but in more severe cases can present as nodules and cysts, which may lead to scarring. The cause of acne is complex and not completely understood, but there are several important factors that appear to contribute to its pathophysiology, including: androgen-mediated increased sebum secretion, abnormal keratinisation, and colonisation and proliferation of the bacterium *Propionibacterium acnes*, which causes (through enzyme and cytokine release) dermal inflammation. It is these last two events (i.e. the *P. acnes* colonisation and subsequent inflammation) that the tetracyclines appear to target. Indeed, tetracylines reduce the number of *P. acnes* colonies within the follicles, and also inhibit the production of bacterial-induced pro-inflammatory cytokines. The tetracylines thus have antibacterial and anti-inflammatory effects, both of which contribute to their efficacy in the management of acne (Jain *et al.*, 2002).

There are many pharmacological options available for the treatment of acne; some target non-inflammatory lesions (such as open and closed comedones) and some inflammatory lesions (such as papules and pustules). The tetracyclines are used to treat inflammatory acne and are available as either topical or systemic formulations. Topical antibiotics are generally used to treat mild to moderate inflammatory acne, while systemic antibiotics are used to manage moderate to severe inflammatory acne. If systemic antibacterial therapy is indicated, tetracyclines are considered the agents of choice (Dréno *et al.*, 2004), but there is no evidence to suggest that one tetracycline is more effective than another (Simonart *et al.*, 2008), although the use of minocycline is not recommended first-line (Garner *et al.*, 2003), as it is more expensive and is associated with greater risk of adverse effects.

The choice of tetracycline is generally dependent on cost, patient preference, side-effect profile, and dosage frequency. The older tetracylines (oxytetracyline and tetracycline) are given twice daily, while the newer agents (doxycycline and lymecycline) offer a more convenient once-daily dosing regimen, which is often preferred by patients.

Unfortunately, tetracycline-resistant strains of *P. acnes* have emerged, which may be associated with treatment failure (Eady *et al.*, 2003), and in order to minimise the probability of resistance occurring, the concomitant use of different topical and systemic antibiotics is not recommended. In addition, if no improvement occurs after three months of systemic therapy, it is recommended that the antibacterial drug is changed.

Acne vulgaris can also be mistaken for a skin condition known as rosacea (which is confusingly sometimes called 'acne rosacea') (Figure 4.3.6). This is a condition that affects adults and is characterised by facial flushing, persistent erythema, papules, and pustules, but in contrast to acne vulgaris, is not associated with

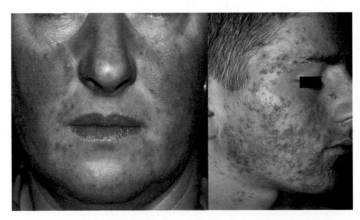

Figure 4.3.6   *The presentation of rosacea (left) and acne vulgaris (right) (Reprinted from 'How to Get Rid Of Acne - Acne Rosacea, What Is It?,' Online, [http://howtogetridofacnex.org/how-to-get-rid-of-acne-acne-rosacea-what-is-it/] 2011.)*

comedones. Rosacea may coexist with acne vulgaris, but is classed as a separate condition. Mild or moderate rosacea is managed using topical metronidazole (see Subsection 2.3.6), but if the condition is moderate or severe, a tetracycline can be used (although only some are licensed for this indication). Surprisingly, tetracyclines are effective at sub-therapeutic antibacterial doses (e.g. doxycycline can be used at a dose of 40 mg daily to manage rosacea, while the conventional antibacterial dose is 100 mg daily), indicating that it is the anti-inflammatory properties of the tetracylines that elicit the therapeutic effect in rosacea.

4.3.6.3   Syphilis

Syphilis is a complex systemic disease caused by the coiled spirochaete bacterium *Treponema pallidum* (Figure 4.3.7). It can present with a number of possible clinical manifestations, many of which are severe, and consequently has been described as the 'great imitator' or the 'great imposter' of medicine.

Infection with syphilis occurs in several stages. The first stage, primary syphilis, is characterised by the development of a lesion (known as a chancre) at the site of initial infection. This usually occurs at the anogenital region, but can also occur in other areas, such as the lips, mouth, and fingers (Figure 4.3.8). The lesions tend to be singular and painless, and, if left untreated, will spontaneously heal after 4–5 weeks (French, 2007). After this time, the *T. pallidum* infection becomes systemic, leading to the development of secondary syphilis, which is characterised by fever, malaise, generalised lymphadenopathy, and a widespread rash. Once again, after around 3–6 weeks the manifestations of secondary syphilis will resolve spontaneously. This is then followed by a latent period of infection, where the patient has no signs or symptoms.

The latent infection can persist for life, but in around 35% of cases, tertiary manifestations of the disease can develop (French, 2007). Tertiary syphilis can affect (and damage) many different organ systems, including the brain, heart, and bones, causing significant morbidity, and in some cases mortality. The three main clinical manifestations of tertiary syphilis are neurosyphilis, cardiovascular syphilis, and gummatous syphilis.

Intramuscular benzathine penicillin or procaine penicillin (see Section 5.1) are the agents of choice for the management of syphilis, but when a patient is allergic to penicillin, or refuses parenteral therapy, doxycyline can be used as an alternative. In the case of primary and secondary syphilis, it is recommended that

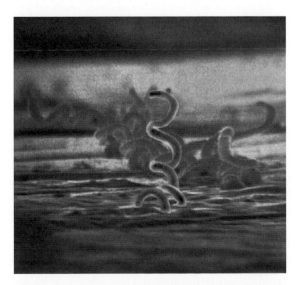

Figure 4.3.7   *Electron micrograph of Treponema pallidum, the causative agent of syphilis (Image courtesy of the Public Health Image Library, ID1977. [http://phil.cdc.gov/phil/home.asp, last accessed 29th April 2011].)*

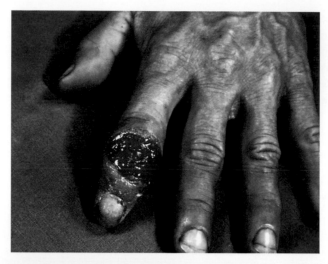

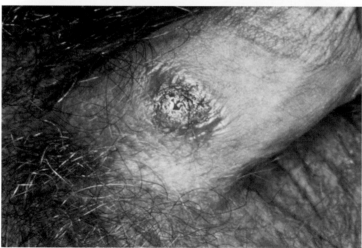

Figure 4.3.8   *A primary syphilitic lesion (or chancre) on the penis (lower) and finger (upper) (Image courtesy of the Public Health Image Library, ID6758. [http://phil.cdc.gov/phil/home.asp, last accessed 29th April 2011].) (Image courtesy of the Public Health Image Library, ID4147. [http://phil.cdc.gov/phil/home.asp, last accessed 29th April 2011].)*

doxycyline 100 mg is taken twice daily for 14 days, but where cardiovascular syphilis or gummatous syphilis has developed, it is recommended that the duration of treatment is extended to 28 days (French *et al.*, 2009). Finally, in the case of neurosyphilis, the recommended dose of doxycycline is increased to 200 mg twice daily for 28 days (French *et al.*, 2009). To date, there have been no large randomised controlled trials comparing the efficacy of pencillins and tetracyclines, but analysis of retrospective data suggests that doxycycline is a suitable alternative to benzathine penicillin (Ghanem *et al.*, 2006).

4.3.6.4   Chlamydia

Chlamydia is a common sexually transmitted infection (STI) caused by the bacterium *Chlamydia trachomatis* (Bébéar and de Barbeyrac, 2009). The infection can affect both men and women and, as it often presents without any symptoms, can remain undetected. Symptoms of infection (if present at all) typically include urethral discharge in men and an abnormal vaginal discharge in women. If the infection spreads to the upper genital tract, it is considered complicated (as opposed to uncomplicated, when the infection has not ascended to the upper genital tract), which, in women can, in the absence of treatment, cause pelvic inflammatory disease (PID), resulting in an increased risk of infertility, ectopic pregnancy, and chronic pelvic pain.

In the case of an uncomplicated chlamydial infection, either azithromycin (as a 1 g single dose) or doxycycline (at a dose of 100 mg twice daily for 7 days) can be used to manage the infection (Scottish Intercollegiate Guidelines Network, 2009). Both agents appear equally effective and have similar tolerability (Lau and Qureshi, 2002), but when patient compliance factors are taken into account, azithromycin is often preferred.

On occasions when PID is suspected, a slightly more complicated antibiotic regimen is recommended; for example, oral doxycycline can be used in combination with oral metronidazole and intramuscular ceftriaxone (Royal College of Obstetricians and Gynaecologists, 2008). In this case, intramuscular ceftriaxone is given as a one-off single dose, while the doxycycline and metronidazole are taken for 14 days.

In cases where a patient has signs or symptoms that strongly suggest a chlamydial infection, treatment should be started without waiting for laboratory confirmation. In addition, as chlamydia is an STI, tracing of all sexual contacts in the last 6 months is recommended.

4.3.6.5   Lyme disease

Lyme disease is an illness caused by the tick-borne spirochete *Borrelia burgdorferi* (Marques, 2010), which was first described in Lyme, Connecticut (hence the name). The original publication reported that 51 patients from three communities in Connecticut developed an unusual illness that was characterised by recurrent attacks of swelling and pain in the large joints, and was occasionally preceded by a skin lesion (Steere *et al.*, 1977). The authors concluded that the condition was, at the time, an unrecognised clinical entity, and named it 'Lyme arthritis'. Because of the rather unusual epidemiology, it was also suggested that an arthropod vector of some description could transmit the condition. Later research showed this to be the case, when a new spirochete was isolated from *Ixodes* ticks and it was suggested that it might have a role in the aetiology of Lyme disease (Burgdorfer *et al.*, 1982). The newly discovered spirochete, later called *Borrelia burgdorferi* (yes, you've guessed it, it was named after the author of the original publication), is transmitted to humans by the bite of an *Ixodes* tick, possibly causing Lyme disease. Other pathogenic species that cause Lyme disease include *Borrelia afzelii* and *Borrelia garinii* (in Europe), while Lyme disease in the USA is caused by *Borrelia burgdorferi*.

Lyme disease is, for clinical purposes, divided into two stages – an early stage and a late stage – with the early stage characterised by the development of a skin lesion at the site of the tick bite. The lesion, known as erythema migrans, starts as a red papule, which gradually expands, and can eventually end up looking like a bullseye from a dartboard (Figure 4.3.9). The skin lesion can also be accompanied by flu-like symptoms, including fatigue, headache, and arthralgia (Steere, 2001). After several weeks, the patient may develop cardiac (e.g. atrioventricular block), neurological (e.g. facial palsy), and rheumatological (e.g. intermittent joint swelling) problems, indicating disseminated infection. Untreated Lyme disease (or Lyme disease that has not been treated correctly) can lead to late-stage disease, manifestations of which include encephalopathy, encephalomyelitis (inflammation of the brain and spinal cord), chronic arthritis (typically involving the knees), and peripheral neuropathy.

Antibiotics can be used in the management of Lyme disease, although the choice depends on the severity of the condition and other patient factors. For early-stage Lyme disease (such as the presentation of erythema migrans), oral doxycycline (at a dose of 100 mg twice daily for 14 days) is recommended. If, however, the patient is under 12 years of age, or pregnant, amoxicillin can be used as an alternative

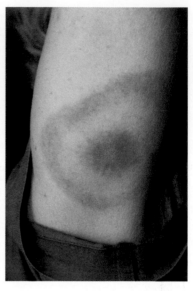

Figure 4.3.9 *Left: the Ixodes pacificus tick, which has been shown to transmit Borrelia burgdorferi, the causative agent of Lyme disease. Right: the characteristic presentation of erythema migrans (in the pattern of a bullseye) (Image courtesy of the Public Health Image Library, ID8685. [http://phil.cdc.gov/phil/home.asp, last accessed 29th April 2011].) (Image courtesy of the Public Health Image Library, ID9875. [http://phil.cdc.gov/phil/home.asp, last accessed 29th April 2011].)*

(Wormser *et al.*, 2006). For late manifestations of Lyme disease (e.g. Lyme arthritis) without neurological involvement, oral doxycycline can be used (at a dose of 100 mg twice daily for 28 days), while patients with arthritis and with evidence of neurological disease should be given parenteral ceftriaxone (Wormser *et al.*, 2006). Single-dose doxycycline may also have a role to play in the prevention of Lyme disease. Indeed, a recent systematic review examining various antibiotics (including doxycycline) suggests that, following an *Ixodes* tick bite, one case of Lyme disease is prevented for every 50 patients taking prophylactic therapy (Warshafsky *et al.*, 2010).

4.3.6.6   Miscellaneous

Doxycycline is occasionally used to manage acute sinusitis – a condition characterised by inflammation of the mucosal lining of the paranasal sinuses. The cause of sinusitis is usually viral in origin, but it can, on occasion, be caused by bacteria such as *Haemophilus influenzae* and *Streptococcus pneumoniae* (Ah-See and Evans, 2007). The classic symptom of acute sinusitis is pain and tenderness over the sinus, which is amplified when the patient bends their head forward. Management of the symptoms of sinusitis usually includes the use of analgesics and decongestants, but if the symptoms are severe or there is purulent discharge lasting at least 7 days, doxycycline may be used.

One agent that has yet to be mentioned in this section is tigecycline (a new tetracycline analogue). The therapeutic indications for this agent are significantly fewer than for the other tetracyclines (e.g. doxycycline), as it is kept in reserve for the treatment of serious infections caused by bacteria that are resistant to other antibiotics. Currently, tigecycline is used to treat complicated skin and soft-tissue infections (cSSTIs) and complicated abdominal infections caused by multiple-drug-resistant bacteria. The FDA has, however, recently released a safety alert which states that tigecycline is associated with an increased mortality rate (in comparison to that of other antibiotics) when used to treat a variety of infections, including complicated skin and skin-structure infections (cSSSIs), ventilator-associated pneumonia, and diabetic foot infection (FDA, 2010). As a result of this information, the FDA now recommends that when a patient has a severe infection, an alternative to tigecycline be considered. This is clearly a significant setback for tigecycline and it is not yet known how this information will affect the long-term future of this drug.

4.3.7   Adverse drug reactions

A series of adverse events has been associated with the tetracyclines, the most serious of which are hepatic and, rarely, renal failure (Miller and McGarity, 2009). Renal damage is caused by inhibition of protein synthesis (which is, of course, the mechanism of action of these agents), which ultimately provokes catabolism (Shils, 1963; George and Evans, 1971; Phillips *et al.*, 1974; Bihorac *et al.*, 1999). The isolated cases of tetracycline-induced acute interstitial nephritis are reported to be caused by an allergic-type reaction, with the presence of rash, eosinophilia, and nephritis (Bihorac *et al.*, 1999).

4.3.7.1   Hepatotoxicity

Tetracyclines have been reported to cause hepatotoxicity, primarily in the form of hepatitis with and without lupus-type symptoms (Carson *et al.*, 1993; Gough *et al.*, 1996; Bhat *et al.*, 1998) One particular form of hepatitis associated with the tetracyclines is non-alcoholic steatohepatitis (NASH) (Mukherjee and Mukherjee, 1969; Mikhail *et al.*, 1980; Yin *et al.*, 2006), through the inhibition of the mitochondrial beta-oxidation of fatty acids[16] (Freneaux *et al.*, 1988). This is associated with triglyceride accumulation and lipid peroxidation, followed by elevation of serum alanine and aspartate aminotransferases (Freneaux *et al.*, 1988). The hepatocytes with accumulated fatty acids have greatly elevated CYP 450 enzymes, which result in drug-mediated hepatotoxicity (Shen *et al.*, 2007). Tetracycline hepatotoxicity can be summarised as induced mitochondrial damage and lipid accumulation, followed by oxidative stress, and eventually cell death. The accumulation of lipids was once believed to be the main cause of cell death, but this is now considered to be due to oxidative damage (Gomez-Lechon *et al.*, 2007). Consequently, autopsies have identified fatty liver with tetracycline use (Robinson and Rywlin, 1970). It has been suggested that tetracyclines have an effect upon peroxisome proliferator-activated receptor alpha (PPARα): a ligand-activated transcription factor which controls the rate of beta-oxidation in human hepatic cells (Tachibana *et al.*, 2009). In addition, the use of tetracyclines results in reduced secretion of triglycerides from the hepatocytes. The exact mechanism for this is still unknown but one theory is that, as the tetracyclines are protein synthesis inhibitors, they may also inhibit apoproteins, reducing the formation of some lipoproteins, such as very low-density lipoproteins (VLDLs), thereby affecting the movement of triglycerides out of the cells (Zimmerman, 1978). This disturbance of the relationship between triglycerides and apoproteins most likely occurs in the Golgi body, and is probably due to the inhibition of the microsomal triglyceride transfer protein (MTP) (Deboyser *et al.*, 1989; Letteron *et al.*, 2003).

4.3.7.2   Fanconi syndrome

Fanconi syndrome is a disease of the kidneys, whereby glucose, amino acids, phosphate, uric acid, and bicarbonate are excreted without significant reabsorption. The area of the kidney affected is the proximal tubule and, as already noted in Subsection 4.3.3, some degradation products of tetracycline may cause a drug-induced Fanconi syndrome (Montoliu *et al.*, 1981).

---

[16] Mitochondrial beta-oxidation is the process whereby fatty acids are broken down into acetyl-coenzyme A units.

Some mechanisms of drug-induced Fanconi syndrome have been suggested; P-glycoprotein (Pgp), human organic anion transporters (hOAT), and the multidrug-resistance-associated protein 2 (Mrp2) superfamily of transporters have been found to be localised on the luminal membrane of proximal tubule cells, and on other cells with excretory and barrier function (Chaub *et al.*, 1997; Sweet *et al.*, 1999; Babu *et al.*, 2002).

These Mrp2 and P-glycoprotein receptors are important determinants of drug absorption, excretion, and thus, in this case, toxicity. In addition, hOAT4 is responsible for the absorption, as well as efflux, of tetracyclines in the renal proximal tubule, which may be connected to the degradation product nephrotoxicity.

4.3.8   Drug interactions

The majority of drug interactions involving the tetracyclines centre on the absorption of this class of drugs. For example, the use of antacids containing aluminum, calcium, bismuth, and magnesium ions will reduce the absorption of tetracyclines. We discussed the binding of magnesium ions to tetracyclines in Subsection 4.3.3 and you should refer back to that subsection to remind yourself how, and where, a metal ion can bind. The co-administration of tetracycline and antacids will thus result in the diminished absorption of the tetracycline and a reduced therapeutic effect. Patients should be advised to take these preparations at least 2 hours apart from one other in order to reduce the likelihood of this interaction.

The other major drug interaction involving the tetracyclines is that of doxycycline and rifampicin. Although the mechanism has not been fully established, it is believed that rifampicin (which induces CYP 3A4) increases the metabolism of doxycycline, thus reducing its overall plasma levels and its therapeutic efficacy. Other agents, such as phenytoin, carbamazepine, primidone, and phenobarbital, which induce liver enzymes, may also increase the metabolism of doxycycline and thus reduce therapeutic plasma levels (Alestig, 1974; Neuvonen and Pentilla, 1974)

4.3.9   Recent developments

Amadacycline (formally known as PTK-0796) is a first-in-class aminomethylcycline antibiotic that is structurally related to the tetracyclines (Figure 4.3.10). This agent shows potent *in vitro* activity against various multidrug-resistant pathogens, including MRSA and VRE, and, perhaps most encouragingly, it shows comparable efficacy and safety profiles to linezolid when used to treat cSSSIs (Wang *et al.*, 2009). The development of amadacycline is in the advanced stages, as it has just completed phase III clinical trials; the marketing rights have been recently bought by Novartis, who claim that the agent can be given once daily (either by an oral tablet or by an IV infusion) to treat infections caused by MRSA.

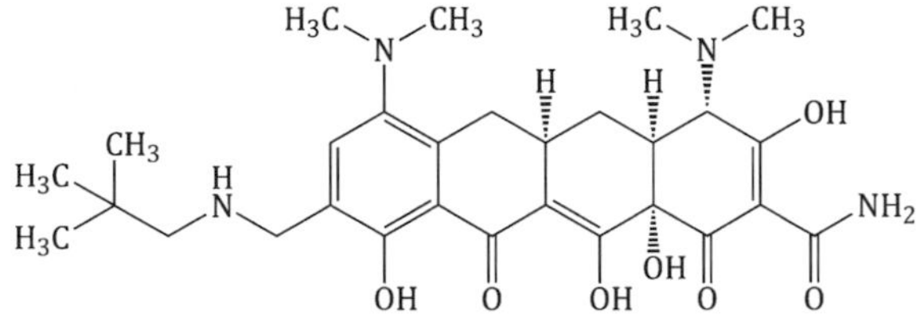

PTK 0796

Figure 4.3.10   *Amadacycline (PTK-0796)*

References

K. W. Ah-See and A. S. Evans, *Brit. Med. J.*, 2007, **334**, 358–361.

K. K. Alestig, *Scand. J. Infect. Dis.*, 1974, **6**, 265–271.

S. L. Ali and T. Strittmatter, *Int. J. Pharmaceut.*, 1978, **1**, 185–188.

S. Alibert-Franco, B. Pradines, A. Mahmoud, A. Davin-Regli, and J.-M. Pagès, *Curr. Med. Chem.*, 2009, **16**, 301–317.

M. F. Anjum, S. F. Choudhary, V. Morrison, L. C. Snow, M. Mafura, P. Slickers, R. Ehricht, and M. J. Woodward, *J. Antimicrob. Chemother.*, 2011, **66**, 550–559.

Anon, *Chem. Eng. News*, 1964, **42**, 21–23.

E. Babu, M. Takeda, S. Narikawa, Y. Kobayashi, T. Yamamoto, S. H. Cha, T. Sekine, D. Sakthiskaran, and H. Endou, *Jap. J. Pharmacol.*, 2002, **88**, 69–76.

A. Barbour, F. Scaglione, and H. Derendorf, *Int. J. Antimicrob. Agents*, 2010, **35**, 431–438.

D. C. Bay and R. J. Turner, *BMC Evolut. Biol.*, 2009, **9**, 140–168.

C. Bébéar and B. de Barbeyrac, *Clin. Microbiol. Infect.*, 2009, **15**, 4–10.

V. Bêhal, *Crit. Rev. Biotechnol.*, 1987, **5**, 275–318.

V. Bêhal, *Adv. Appl. Microbiol.*, 2000, **47**, 113–156.

G. Bhat, J. Jordan, Jr, and S. D. O. Sokalski, *J. Clin. Gastroenterol.*, 1998, **27**, 74–75.

A. Bihorac, C. Ozner, E. Akoglu, and S. Kullu, *Nephron*, 1999, **81**, 72–75.

E. Bishburg and K. Bishburg, *Int. J. Antimicrob. Agents*, 2009, **34**, 395–401.

D. E. Brodersen, W. M. Clemons, Jr, A. P. Carter, R. J. Morgan-Warren, B. T. Wimberley, and V. Ramakrishnan, *Cell*, 2000, **103**, 1143–1154.

H. Brötz-Oesterhelt and P. Sass, *Future Microbiol.*, 2010, **5**, 1553–1579.

W. Brown and R. J. Fallon, *J. Antimicrob. Chemother.*, 1980, **6**, 91–95.

J. D. Brubaker and A. G. Myers, *Org. Lett.*, 2007, **9**, 3523–3525.

W. Burgdorfer, A. G. Barbour, S. F. Hayes, J. L. Benach, E. Grunwaldt, and J. P. Davis, *Science*, 1982, **216**, 1317–1319.

A. G. Butterfield, D. W. Hughes, N. J. Pound, and W. L. Wilson, *Antimicrob. Agents Chemother.*, 1973, **4**, 11–15.

J. Cantor, *Amer. J. Dig. Dis.*, 1950, **17**, 340–342.

J.L. Carson, B.L. Strom, A. Duff, A. Gupta, M. Shaw, F.E. Lundin, and K. Sa, *Arch. Int. Med.*, 1993, **119**, 576–583.

M. G. Charest, C. D. Lerner, J. D. Brubaker, D. R. Siegel, and A. G. Myers, *Science*, 2005a, **308**, 395–398.

M. G. Charest, D. R. Siegel, and A. G. Myers, *J. Am. Chem. Soc.*, 2005b, **127**, 8292–8293.

P. Chaub, J. Kartenbeck, and J. König, *J. Am. Soc. Nephrol.*, 1997, **8**, 1213–1221.

Y. Chen and C. Lin, *J. Chromatogr. A*, 1998, **802**, 95–105.

I. Chopra, *Antimicrob. Agents Chemother.*, 1994, **38**, 637–640.

I. Chopra, *Curr. Opin. Pharmacol.*, 2001, **1**, 464–469.

I. Chopra, *Drug Resis. Updates*, 2002, **5**, 119–125.

I. Chopra and M. Roberts, *Microbiol. Mol. Biol. Rev.*, 2001, **65**, 232–260.

J. Clardy, M. Fischback, and C. Currie, *Curr. Biol.*, 2009, **19**, R437–R441.

J. L. Colaizzi and P. R. Klink, *J. Pharm. Sci.*, 1969, **58**, 1184–1189.

S. R. Connell, D. M. Tracz, K. H. Nierhaus, and D. E. Taylor, *Antimicrob. Agents Chemother.*, 2003, **47**, 3675–3681.

L. H. Conover, US Patent 2699054, 11 January 1955.

L. A. Cox, Jr., and D. A. Popken, *Risk Anal.*, 2010, **30**, 432–457.

T. Dairi, T. Nakano, K. Aisaka, R. Katsumata, and M. Hasegawa, *Biosci. Biotech. Biochem.*, 1995, **59**, 1099–1106.

M. A. Darken, H. Berenson, R. J. Shirk, and N. O. Sjolander, *Appl. Microbiol.*, 1960, **8**, 46–51.

D. Deboyser, F. Goethals, G. Krack, and M. Roberfroid, *Toxicol. Appl. Pharmacol.*, 1989, **97**, 473–479.

J. Donohoe, J. D. Dunitz, K. N. Trueblood, and M. S. Webster, *J. Am. Chem. Soc.*, 1963, **85**, 851–856.

H. F. Dos Santos, W. B. De Almeida, and M. C. Zerner, *J. Pharm. Sci.*, 1998, **87**, 190–195.

B. Dréno, V. Bettoli, F. Ochsendorf, A. Layton, H. Mobacken, and H. Degreef, *Eur. J. Dermatol.*, 2004, **14**, 391–399.

B. M. Duggar, *Ann. N.Y. Acad. Sci.*, 1948, **51**, 177–181.

E. A. Eady, M. Gloor, and J. J. Leyden, *Dermatology*, 2003, **206**, 54–56.

B. Epe and P. Woolley, *EMBO J.*, 1984, **3**, 121–126.

S. Ericson and H. Gnarpe, *J. Int. Med. Res.*, 1979, **7**, 471–472.

FDA, Drug Saftey Communication: Increased Risk of Death with Tygacil (Tigecycline) Compared to Other Antibiotics Used to Treat Similar Infections, 2010 (http://www.fda.gov/Drugs/DrugSafety/ucm224370.htm, last accessed 10 March 2012).

A. C. Finlay, G. L. Hobby, S. Y. P'an, P. P. Regna, J. B. Routien, D. B. Seeley, G. M. Shull, B. A. Sobin, I. A. Solomons, J. W. Vinson, and J. H. Kane, *Science*, 1950, **111**, 85.

P. French, *Brit. J. Med.*, 2007, **334**, 143–147.

P. French, M. Gomberg, M. Janier, B. Schmidt, P. van Voorst Vader, and H. Young, *Int. J. STD AIDS*, 2009, **20**, 300–309.

E. Freneaux, G. Labbe, P. Letteron, D. le The, C. Degott, J. Geneve, D. Larrey, and D. Pessayre, *Hepatol.*, 1988, **8**, 1056–1062.

G. W. Frimpter, A. E. Timpanelli, W. J. Eisenmenger, H. S. Stein, and L. I. Ehrlich, *JAMA*, 1963, **184**, 111–113.

S. E. Garner, A. Eady, C. M. Popescu, J. Newton, and A. Li Wan Po, *Cochrane Database of Systematic Reviews*, 2003, Issue 1.

M. I. Garvey, M. M. Rahman, S. Gibbons, and L. J. V. Piddock, *Int. J. Antimicrob. Agents*, 2011, **37**, 145–151.

C. R. George and R. A. Evans, *Med. J. Aust.*, 1971, **1**, 1271–1273.

K. G. Ghanem, E. J. Erbelding, W. W. Cheng, and A. M. Rompalo, *Clin. Infect. Dis.*, 2006, **42**, e45–e49.

M. J. Gomez-Lechon, M. T. Donato, A. Martiez-Romero, N. Jimenez, J. V. Castell, and J. E. O'Connor, *Chem. Biol. Interact.*, 2007, **165**, 106–116.

J. J. Goodman and M. Matrishin, *Nature*, 1968, **219**, 291–292.

J. J. Goodman, M. Matrishin, R. W. Young, and J. R. D. McCormick, *J. Bacteriol.*, 1959, **78**, 492–499.

A. Gough, S. Chapman, and K. Wagstaff, *Brit. Med. J.*, 1996, **312**, 169–172.

M. O. Griffin, E. Fricovsky, G. Ceballos, and F. Villarreal, *Am. J. Physiol. Cell Physiol.*, 2010, **299**, C539–C548.

M. O. Griffin, G. Cabellos, and F. J. Villareal, *Pharmacol. Res.*, 2011, **63**, 102–107.

S. Grkovic, M. H. Brown, and R. A. Skurray, *Microbiol. Mol. Biol. Rev.*, 2002, **66**, 671–701.

J. M. Gross, *Ann. Intern. Med.*, 1963, **58**, 523–528.

A. I. Gurevich, M. G. Karapetyan, M. N. Kolosov, V. G. Korobko, V. V. Onoprienko, S. A. Popravko, and M. M. Shemyakin, *Tet. Lett.*, 1967, **8**, 131–134.

C. L. Gyles, *Anim. Health Res. Rev.*, 2008, **9**, 149–158.

P. Hawkey and R. Finch, *Clin. Microbiol. Infect.*, 2007, **13**, 354–362.

W. Hillen and C. Berens, *Annu. Rev. Microbiol.*, 1994, **48**, 345–369.

F. A. Hochstein, C. R. Stephens, L. H. Conover, P. P. Regna, R. Pasternack, P. N. Gordon, F. J. Pilgrim, K. J. Brunings, and R. B. Woodward, *J. Am. Chem. Soc.*, 1953, **75**, 5455–5475.

M. Hoffmann, W. DeMaio, R. A. Jordan, R. Talaat, D. Harper, J. Speth, and J. Scatina, *Drug Metab. Dispos.*, 2007, **35**, 1543–1553.

D. A. Hussar, P. J. Niebergall, E. T. Sugita, and J. T. Doluisio, *J. Pharm. Pharmacol.*, 1968, **20**, 539–546.

A. Jaffe, Y. A. Chabbert, and O. Semonin, *Antimicrob. Agents Chemother.*, 1982, **22**, 942–948.

J. M. Jaffe, J. L. Colaizzi, R. I. Poust, and R. H. McDonald, Jr, *J. Pharmacokin. Biopharm.*, 1973, **1**, 267–282.

A. Jain, L. Sangal, E. Basal, G. P. Kaushal, and S. K. Agarwal, *Dermatol. Online J.*, 2002, **8**, 2.

L. Jin, X. Amaya-Mazo, M. E. Apel, S. S. Sankisa, E. Johnson, M. A. Zbysynska, and A. Han, *Biophys. Chem.*, 2007, **128**, 185–196.

T. Kelesidis, D. E. Karageorgopoulos, I. Kelesidis, and M. E. Falagas, *J. Antimicrob. Chemother.*, 2008, **62**, 895–904.

J. S. Kiser, G. C. De Mello, D. H. Reichard, and J. H. Williams, *J. Infect. Dis.*, 1952, **90**, 76–80.

C. Khosla, *J. Org. Chem.*, 2009, **74**, 6416–6420.

C. Khosla and Y. Tang, *Science*, 2005, **308**, 367–368.

J. K. Korst, J. D. Johnston, K. Butler, E. J. Bianco, L. H. Conover, and R. B. Woodward, *J. Am. Chem. Soc.*, 1968, **90**, 439–457.

D. J. Koza and Y. A. Nsiah, *Bioorg. Med. Chem. Lett.*, 2002, **12**, 2163–2165.

C. M. Kunin, *JAMA*, 1967, **202**, 204–208.

C. Y. Lau and A. K. Qureshi, *Sex Transm. Dis.*, 2002, **29**, 497–502.

P. Letteron, A. Sutton, A. Mansouri, B. Fromenty, and D. Pessayre, *Hepatology*, 2003, **38**, 133–140.

S. B. Levy, *Clin. Infect. Dis.*, 2001, **33** (Suppl. 3), S124–S129.

S. B. Levy, L. M. McMurry, V. Burdett, P. Courvalin, W. Hillen, M. C. Roberts, and D. E. Taylor, *Antimicrob. Agents Chemother.*, 1989, **33**, 1373–1374.

S. B. Levy, L. M. McMurry, T. M. Barbosa, V. Burdett, P. Courvalin, W. Hillen, M. C. Roberts, J. I. Rood, and D. E. Taylor, *Antimicrob. Agents Chemother.*, 1999, **43**, 1523–1524.

Z. Ma, F. E. Jacobsen, and D. P. Gledroc, *Chem. Rev.*, 2009, **109**, 4644–4681.

A. Manten, *Bull. Wld. Hlth. Org.*, 1963, **29**, 387–400.

A. R. Marques, *Curr. Allergy Asthma Rep.*, 2010, **10**, 13–20.

J. J. Mathers, S. C. Flick, and L. A. Cox, Jr, *Environ. Int.*, 2011, **37**, 991–1004.

P. Mazurkiewicz, A. J. M. Driessen, and W. N. Konings, *Curr. Issues Mol. Biol.*, 2005, **7**, 7–21.

J. R. D. McCormick, N. O. Sjolander, U. Hirsch, E. R. Jensen, and A. P. Doerschuk, *J. Am. Chem. Soc.*, 1957, **79**, 4561–4563.

L. M. McMurry, J. C. Cullinane, and S. B. Levy, *Antimicrob. Agents Chemother.*, 1982, **22**, 791–799.

A. K. Meagher, P. G. Ambrose, T. H. Grasela, and E. J. Ellis-Grosse, *Diagn. Micr. Infec. Dis.*, 2005, **52**, 165–171.

T. H. Mikhail, K. M. Ibrahim, R. Awadallah, and E. Z. Mna, *Ernahrungswiss*, 1980, **19**, 173–178.

G. H. Miller, H. L. Smith, W. L. Rock, and S. Hedberg, *J. Pharm. Sci.*, 1977, **66**, 88–92.

G. S. Miller and G. J. McGarity, *JADA*, 2009, **140**, 56–57.

J. Montoliu, M. Carrera, A. Darnell, and L. Revert, *Br. Med. J. (Clin. Res. Ed.)*, 1981, **283**, 1576–1577.

I. F. Moore, D. W. Hughes, and G. D. Wright, *Biochemistry*, 2005, **44**, 11 829–11 835.

P. G. S. Mortimer and L. J. V. Piddock, *J. Antimicrob. Chemother.*, 1993, **32**, 195–213.

D. Mukherjee and S. Mukherjee, *J. Antibiot. (Tokyo)*, 1969, **22**, 45–48.

M. M. Mull, *Am. J. Dis. Child*, 1966, **112**, 483–493.

H. Muxfeldt and W. Rogalski, *J. Am. Chem. Soc*, 1965, **87**, 933–934.

H. Muxfeldt, G. Hardtmann, F. Kathawala, E. Vedejs, and J. B. Mooberry, *J. Am. Chem. Soc.*, 1968, **90**, 6534–6536.

H. Muxfeldt, G. Haas, G. Hardtmann, F. Kathawala, J. B. Mooberry, and E. Vedejs, *J. Am. Chem. Soc.*, 1979, **101**, 689–701.

A. G. Myers, J. D. Brubaker, C. Sun, and Q. Wang, World Patent WO 2007/117639, 18 October 2007.

A. G. Myers, M. G. Charest, C. D. Lerner, J. D. Brubaker, and D. R. Siegel, US Patent US 2011/0009371, 13 January 2011.

P. J. Neuvonen and O. Pentilla, *Br. Med. J.*, 1974, **1**, 535–536.

H. Nikaido, *J. Biol. Chem.*, 1994, **269**, 3905–3908.

H. Nikaido, *Microbiol. Mol. Biol. Rev.*, 2003, **67**, 593–656.

H. Nikaido, *Annu. Rev. Biochem.*, 2009, **78**, 119–146.

H. Nikaido and D. G. Thanassi, *Antimicrob. Agents Chemother.*, 1993, **37**, 1393–1399.

M. W. Olson, A. Ruzin, E. Feyfant, T. S. Rush, III, J. O'Connell, and P. A. Bradford, *Antimicrob. Agents Chemother.*, 2006, **50**, 2156–2166.

J.-M. Pagès and L. Amaral, *Biochim. Biophys. Acta*, 2009, **1794**, 826–833.

G. J. Palm, T. Lederer, P. Orth, W. Saenger, M. Takahashi, W. Hillen, and W. Hinrichs, *J. Biol. Inorg. Chem.*, 2008, **13**, 1097–1110.

I. T. Paulsen, M. H. Brown, and R. A. Skurray, *Microbiol. Rev.*, 1996, **60**, 575–608.

P. J. Petersen, N. V. Jacobus, W. J. Weiss, P. E. Sum, and R. T. Testa, *Antimicrob. Agents Chemother.*, 1999, **43**, 738–744.

H. Petković, J. Cullum, D. Hranueli, I. S. Hunter, N. Perić-Concha, J. Pigac, A. Thamchaipenet, D. Vujaklija, and P. F. Long, *Microbiol. Mol. Biol. Rev.*, 2006, **70**, 704–728.

H. Petković, P. Raspor, and U. Lesnik, US Patent US 2010/0035847, 11 February 2010.

M. E. Phillips, J. B. Eastwood, J. R. Curtis, P. D. Gower, and H. E. De Wardener, *Br. Med. J.*, 1974, **2**, 149–151.

L. B. Pickens and Y. Tang, *Metab. Eng.*, 2009, **11**, 69–75.

L. B. Pickens and Y. Tang, *J. Biol. Chem.*, 2010, **285**, 27 509–27 515.

L. J. V. Piddock, M. I. Garvey, M. M. Rahman, and S. Gibbons, *J. Antimicrob. Chemother.*, 2010, **65**, 1215–1223.

M. Pioletti, F. Schlunzen, J. Harms, R. Zarivach, M. Gluhmann, H. Avila, A. Bashan, H. Bartels, T. Auerbach, C. Jacobi, T. Hartsch, A. Yonath, and F. Franceschi, *EMBO J.*, 2001, **20**, 1829–1839.

K. Poole, *Ann. Med.*, 2007, **39**, 162–176.

J. L. Ramos, M. Martínez-Bueno, A. J. Molina-Henares, W. Terán, K. Watanabe, X. Zhang, M. T. Gallegos, R. Brennan, and R. Tobes, *Microbiol. Mol. Biol. Rev.*, 2005, **69**, 326–356.

M. C. Roberts, *FEMS Microbiol. Rev.*, 1996, **19**, 1–24.

M. C. Roberts, *Ciba Found. Symp.*, 1997, **207**, 206–218.

M. C. Roberts, *Mol. Biotechnol.*, 2002, **20**, 261–283.

M. C. Roberts, *Clin. Infect. Dis.*, 2003, **36**, 462–467.

M. J. Robinson and A.M. Rywlin, *Am. J. Dig. Dis.*, 1970, **15**, 857–862.

Royal College of Obstetricians and Gynaecologists, Management of Acute Pelvic Inflammatory Disease. Green-top Guideline, No. 32, November 2008 (http://www.rcog.org.uk/files/rcog-corp/uploaded-files/T32PelvicInflamatoryDisease2008MinorRevision.pdf, last accessed 10 March 2012).

A. N. Sapadin and R. Fleischmajer, *J. Am. Acad. Dermatol.*, 2006, **54**, 258–265.

Scottish Intercollegiate Guidelines Network, Management of Genital Chlamydia Trachomatis Infection (109). A National Clinical Guideline, March 2009 (http://www.sign.ac.uk/pdf/sign109.pdf, last accessed 10 March 2012).

H. P. Schweizer, *Genet. Mol. Res.*, 2003, **2**, 48–62.

V. Šeputienė, J. Povilonis, J. Armalytė, K. Sužiedėlis, A. Pavilonis, and E. Sužiedėlienė, *Medicina (Kaunas)*, 2010, **46**, 240–248.

C. Shen, G. Zhang, and Q. Meng, *Biochem. Eng. J.*, 2007, **34**, 267–272.

M. E. Shils, *Ann. Int. Med.*, 1963, **58**, 389–408.

T. Simonart, M. Dramaix, and V. De Maertelaer, *Br. J. Dermatol.*, 2008, **158**, 208–216.

G. Sjölin-Forsberg and J. Hermansson, *Br. J. Clin. Pharmac.*, 1984, **18**, 529–533.

T. D. Sokoloski, L. A. Mitscher, P. H. Yuen, J. V. Juvurkar, and B. Hoener, *J. Pharm. Sci.*, 1977, **66**, 1159–1165.

C. M. T. Spahn and C. D. Prescott, *J. Mol. Med.*, 1996, **74**, 423–439.

C. M. T. Spahn, G. Blaha, R. K. Aggarwal, P. Penczek, R. A. Grassucci, C. A. Trieber, S. R. Connell, D. E. Taylor, K. H. Nierhaus, and J. Frank, *Mol. Cell*, 2001, **7**, 1037–1045.

B. S. Speer, N. B. Shoemaker, and A. A. Salyers, *Clin. Microbiol. Rev.*, 1992, **5**, 387–399.

A. C. Steere, *N. Engl. J. Med.*, 2001, **345**, 115–125.

A. C. Steere, S. E. Malawista, D. R. Snydman, R. E. Shope, W. A. Andiman, M. R. Ross, and F. M. Steele, *Arthritis Rheum.*, 1977, **20**, 7–17.

C. R. Stephens, L. H. Conover, F. A. Hochstein, P. P. Regna, F. J. Pilgrim, and K. J. Brunings, *J. Am. Chem. Soc.*, 1952, **74**, 4976–4977.

G. Stork, J. J. La Clair, P. Spargo, R. P. Nargund, and N. Totah, *J. Am. Chem. Soc.*, 1996, **118**, 5304–5305.

P.-E. Sum, A. T. Ross, P. J. Petersen, and R. T. Testa, *Bioorg. Med. Chem. Lett.*, 2006, **16**, 400–403.

C. Sun, Q. Wang, J. D. Brubaker, P. M. Wright, C. D. Lerner, K. Noson, M. Charest, D. R. Siegel, Y.-M. Wang, and A. G. Myers, *J. Am. Chem. Soc.*, 2008, **130**, 17 913–17 927.

C. Sun, D. K. Hunt, R. B. Clark, D. Lofland, W. J. O'Brien, L. Plamondon, and X. Y. Xiao, *J. Med. Chem.*, 2011, **54**, 3704–3731.

D. H. Sweet, D. S. Miller, and J. B. Pritchard, *Am. J. Physiol.*, 1999, **276**, F864–F873.

K. Tachibana, K. Takeuchi, H. Inada, D. Yamasaki, K. Ishimoto, T. Tanaka, T. Hamakubo, J. Sakai, T. Kodama, and T. Doi, *Biochem. Biophys. Res. Commun.*, 2009, **389**, 501–505.

K. Tatsuta and S. Hosokawa, *Chem. Rev.*, 2005, **105**, 4707–4729.

K. Tatsuta, T. Yoshimoto, H. Gunji, Y. Okado, and M. Takahashi, *Chem. Lett.*, 2000, **29**, 646–647.

D. E. Taylor and A. Chau, *Antimicrob. Agents Chemother.*, 1996, **40**, 1–5.

M. Tuckman, P. J. Petersen, A. Y. M. Howe, M. Orlowski, S. Mullen, K. Chan, P. A. Bradford, and C. H. Jones, *Antimicrob. Agents Chemother.*, 2007, **51**, 3205–3211.

US Court of Appeals *The Federal Reporter*, 1968, **401**, F.2d 574.

G. Volkers, G. J. Palm, M. S. Weiss, G. D. Wright, and W. Hinrichs, *FEBS Lett.*, 2011, **585**, 1061–1066.

M. S. von Wittenau, R. K. Blackwood, L. H. Conover, R. H. Glauert, and R. B. Woodward, *J. Am. Chem. Soc.*, 1965, **87**, 134–135.

J. C. Walker, *Ann. Rev. Phytopathol.*, 1982, **20**, 33–39.

V. C. Walton, M. R. Howlett, and G. B. Selzer, *J. Pharm. Sci.*, 1970, **59**, 1160–1164.

I.-K. Wang, L. C. Vining, J. A. Walter, and A. G. McInnes, *J. Antibiot. (Tokyo)*, 1986, **39**, 1281–1287.

Y. Wang, R. Castaner, J. Bolos, and C. Estivill, *Drugs Fut.*, 2009, **34**, 11.

S. Warshafsky, D. H. Lee, L. K. Francois, J. Nowakowski, R. B. Nadelman, and G. P. Wormser, *J. Antimicrob. Chemother.*, 2010, **65**, 1137–1144.

J. C. Wilks and J. L. Slonczewski, *J. Bacteriol.*, 2007, **189**, 5601–5607.

J. R. Wilson, *Med. Press*, 1961, **245**, 539–542.

G. P. Wormser, R. J. Dattwyler, E. D. Shapiro, J. J. Halperin, A. C. Steere, M. S. Klempner, P. J. Krause, J. S. Bakken, F. Strle, G. Stanek, L. Bockenstedt, D. Fish, J. S. Dumler, and R. B. Nadelman, *Clin. Infect. Dis.*, 2006, **43**, 1089–1134.

G. D. Wright, *Adv. Drug Deliv. Rev.*, 2005, **57**, 1451–1470.

L. T. Wright and H. Schreiber, *J. Nat. Med. Assoc.*, 1949, **41**, 195–201.

Y. Wu and R. Fassihi, *Int. J. Pharmaceut.*, 2005, **290**, 1–13.

W. Yang, I. F. Moore, K. P. Koteva, D. C. Bareich, D. W. Hughes, and G. D. Wright, *J. Biol. Chem.*, 2004, **279**, 52 346–52 352.

H. Q. Yin, M. Kim, J. H. Kim, G. Kong, M. O. Lee, K. S. Kang, B. I. Yoon, H. L. Kim, and B. H. Lee, *Toxicol.*, 2006, **94**, 206–216.

B. Zechini and I. Versace, *Recent Pat. Antiinfect. Drug Discov.*, 2009, **4**, 37–50.

G. G. Zhanel, K. Homenuik, K. Nichol, A. Noreddin, L. Vercaigne, J. Embil, A. Gin, J. A. Karlowsky, and D. J. Hoban, *Drugs*, 2004, **64**, 63–88.

W. Zhang and Y. Tang, *Methods Enzymol.*, 2009, **459**, 367–393.

W. Zhang, K. Watanabe, C. C. C. Wang, and Y. Tang, *J. Biol. Chem.*, 2007, **282**, 25 717–25 725.

H. J. Zimmerman, *Hepatotoxicity: The Adverse Effects of Drugs and Other Chemicals on the Liver*, Appleton Century Crofts, New York, NY, 1978.

4.4

Chloramphenicol

Key Points

- Chloramphenicol has a broad spectrum of activity but, due to toxicity issues, its use is reserved for severe infections or for topical applications.
- Chloramphenicol inhibits protein synthesis through binding to the large ribosome subunit (50S) at the peptidyl transferase centre A site, thus preventing binding of the next charged tRNA.
- Resistance arises due to chloramphenicol acetyltransferases (CATs), which acetylate chloramphenicol so that it no longer binds to this site.

4.4.1 Discovery

Chloramphenicol was first obtained from a *Streptomyces* species isolated from the filtrate of a soil sample from a field near Caracas (Ehrlich *et al.*, 1947). For obvious reasons, the bacterium was named *Streptomyces venezuelae* and, once it had been established that this crystalline antibiotic contained covalently-linked chlorine, it was originally named chloromycetin. Chloramphenicol (Chloromycetin, Cm) was soon found to be bacteriostatic and to have a broad spectrum of activity.

4.4.2 Synthesis

Unlike some of the other antibiotics we will study in this text, chloramphenicol has a **relatively** simple structure, so that, although it was originally obtained from fermentation, once its structure had been deduced and the active diastereoisomer identified, chloramphenicol could be produced by total synthesis. As chloramphenicol contains two stereogenic (chiral) centres, it has four possible stereoisomers,[17] with only the *R,R*-diastereoisomer (shown in Figure 4.4.1) of the four having antibacterial activity. Any synthetic method

---

[17] Number of stereoisomers $= 2^n$, where $n =$ number of stereogenic centres.

Antibacterial Agents: Chemistry, Mode of Action, Mechanisms of Resistance and Clinical Applications, First Edition.
Rosaleen J. Anderson, Paul W. Groundwater, Adam Todd and Alan J. Worsley.
© 2012 John Wiley & Sons, Ltd. Published 2012 by John Wiley & Sons, Ltd.

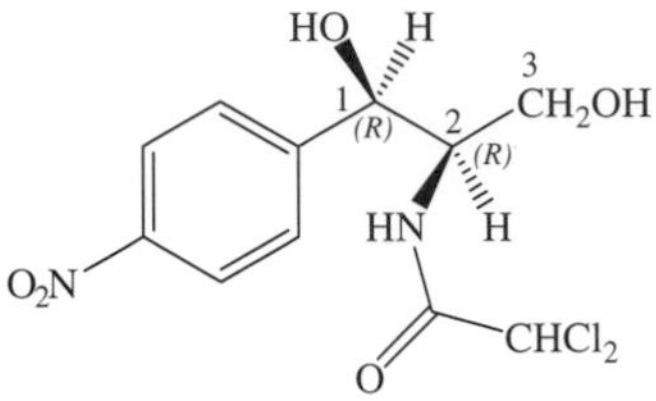

Figure 4.4.1 *Chloramphenicol*

Table 4.4.1 *Therapeutic indications for chloramphenicol*

Indication	Notes
Typhoid fever	Where use of quinolones is unsuitable
Meningitis	When patient is allergic to penicillins or cephalosporins
Epiglottitis	Given intravenously
Eye and ear infections	Topical application

thus needs to produce this isomer, preferably with as few steps as possible involving the separation of isomers, as this is both costly and wasteful (if a synthetic step produces a pair of enantiomers which need to be separated then 50% of the product will be discarded after the enantiomeric resolution[18]). The first synthesis of chloramphenicol was reported in 1949 and involved eight steps, including an enantiomeric resolution in the second-last step (a particularly wasteful process as 50% of the racemate is discarded after having been carried through the previous six steps) (Controulis *et al.*, 1949). In addition, the original work describes an explosion which occurred in one of the earlier steps, so it is probably best not to consider this route to chloramphenicol in detail and to look instead at a recent example of the many stereoselective syntheses (Scheme 4.4.1). Among the attractive features of this particular route is that it is short (only three steps and an overall yield of 23% from 4-nitrobenzaldehyde) and gives the desired, (−)-chloramphenicol, enantiomer in high purity, as the epoxide intermediate is formed in 95% enantiomeric excess (i.e. a ratio of 97.5 : 2.5 in favour of the enantiomer shown) by the Sharpless asymmetric epoxidation (K. B. Sharpless, Nobel Prize in Chemistry, 2001). The subsequent steps are all stereospecific, with the lone pair of the **imidate nitrogen** attacking the epoxide from the side opposite to the oxygen of the ring (in order to minimise the repulsion between this lone pair and those on the oxygen), resulting in an inversion of the stereochemistry at this centre, from (*S*) to (*R*) (Bhaskar *et al.*, 2004).

4.4.3 Bioavailability

When you look again at the structure of chloramphenicol, you can see that, compared to some of the other antibiotics we have encountered in this section, it is a relatively small molecule. Having two hydroxyl functional groups (which are often responsible for aiding aqueous solubility), you might think that it would be relatively water-soluble. Aromatic nitro compounds, however, are well known for their

---

[18] On a large scale, resolution (the separation of a pair of enantiomers) usually involves the formation of diastereoisomers by the reaction of a racemic mixture (1 : 1 mixture of enantiomers) with a single enantiomer of a chiral reagent, followed by the separation and conversion of the diastereoisomers back into the enantiomers.

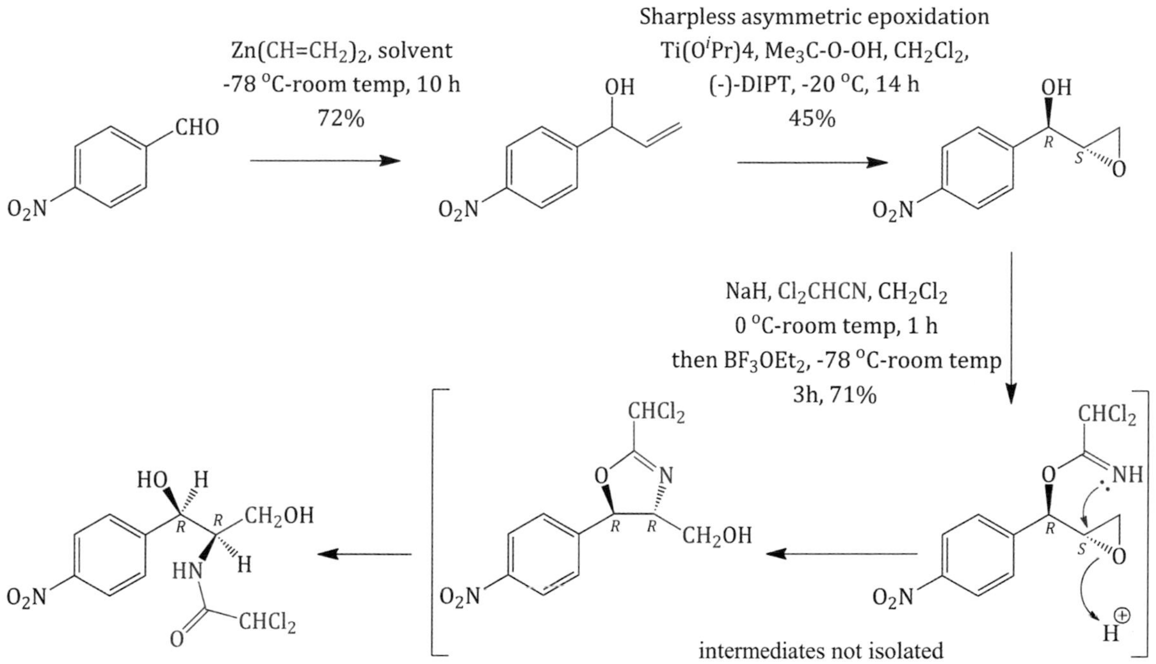

Scheme 4.4.1 *Stereoselective synthesis of chloramphenicol ((−)-DIPT = (−)-diisopropyl tartrate, $^{i}PrO_2C \cdot CHOH \cdot CHOH \cdot CO_2^{i}Pr$)*

insolubility, particularly in aqueous solutions, and chloramphenicol is no exception — it has low aqueous solubility (~2.5 g/L at 25 °C and neutral pH), with its lowest solubility occurring at pH 5.5–7.0 (i.e. in the middle of the physiologically relevant pH range). Amide NH groups are usually considered to be neutral, but that of chloramphenicol is significantly more acidic than most, due to the electronegative influence of the two chlorine atoms causing greater than usual electron flow from the amide N to the carbonyl group. However, although it is more acidic, the pK_a of 11.03[19] (Qiang and Adams, 2004) does not allow for its ready ionisation. The low solubility of chloramphenicol was sufficiently limiting to its clinical use that two prodrug forms were developed; **chloramphenicol succinate** (as the ionised sodium salt), with high aqueous solubility (suitable for IV administration, for example in an emergency situation), and **chloramphenicol palmitate**, which is essentially water-insoluble and designed for oral administration (Figure 4.4.2). A comparison of the relative bioavailability of these prodrugs showed that the orally administered palmitate has a greater and more predictable bioavailability in infants and children (Pickering *et al.*, 1980; Kauffmann *et al.*, 1981).

Chloramphenicol itself is stable, with a shelf life of several years for the crystalline form, and it retains 100% of its activity for 24 hours in solution over a pH range of 1–9.5 (Bartz, 1948); however, over a longer time period, it degrades in solution due to exposure to light (Zhou *et al.*, 2010), heat (Franje *et al.*, 2010), and different pHs (Boer and Pijnenburg, 1983; Zhou *et al.*, 2010). The succinate prodrug also suffers from instability in solution over 1 week, even when stored at −20 °C; the temperature must be decreased to −50 °C to achieve good stability over 1 month (Ambrose, 1984).

---

[19] Many chloramphenicol data sources state that it has excellent aqueous solubility and a pK_a of 5.5, but this relates to the prodrug chloramphenicol succinate.

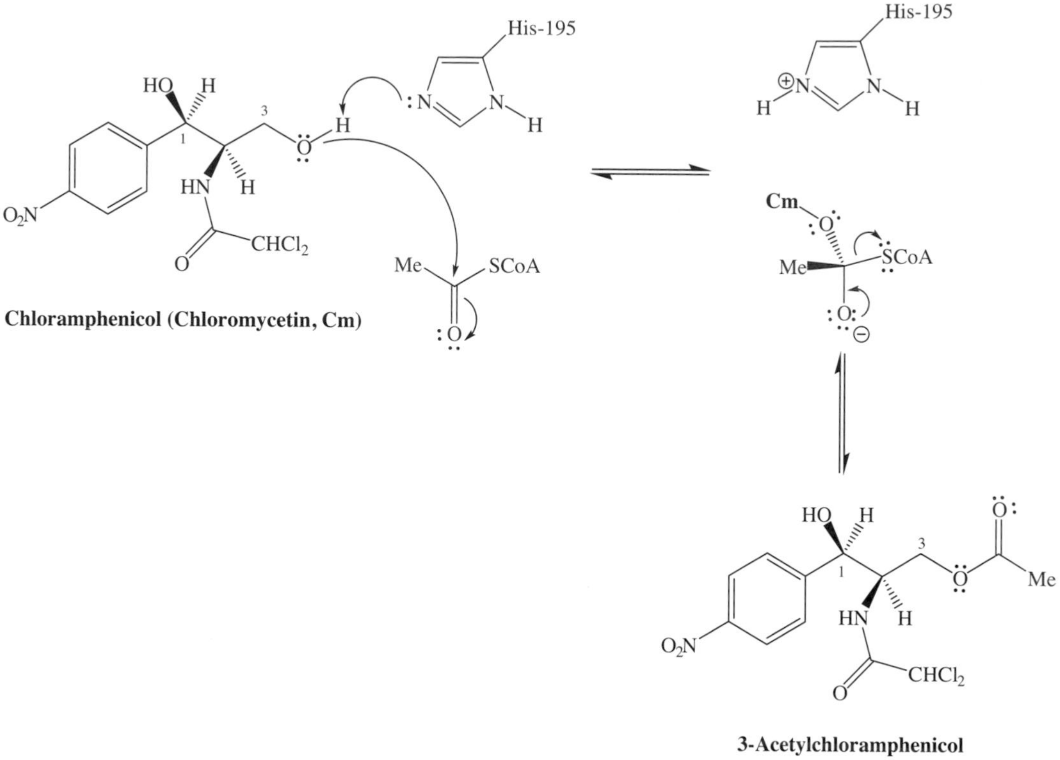

Scheme 4.4.2   *Mechanism for the acetylation of chloramphenicol*

4.4.6   Clinical applications

4.4.6.1   Spectrum of activity

Chloramphenicol has a very broad spectrum of activity; it is active against Gram positive (e.g. *Staphylococcus aureus* and *Streptococcus pneumoniae*) and Gram negative bacteria (e.g. *Neisseria meningitidis, Neisseria gonorrhoeae*, and *H. influenzae*), as well as some anaerobes (e.g. *Bacteroides fragilis*) (Cuchural *et al.*, 1988). *P. aeruginosa* is often resistant to chloramphenicol (Li *et al.*, 1994).

Despite this excellent spectrum of activity, the use of systemic chloramphenicol has been restricted to treating life-threatening infections (particularly caused by *H. influenzae*) due to the risk of developing serious

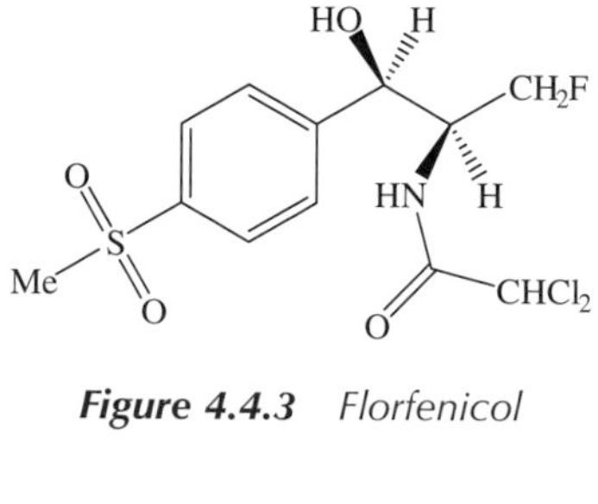

Figure 4.4.3   *Florfenicol*

haematological side effects. Chloramphenicol can also be used topically and, because there is minimal evidence to implicate topical therapy as a cause of haematological side effects, it is commonly used to treat superficial eye infections in many countries, including the UK.

4.4.6.2   Eye infections

Topical chloramphenicol is commonly used for the treatment of acute bacterial conjunctivitis and is considered the agent of choice in the UK. As the name suggests, acute bacterial conjunctivitis is an infection of the conjunctiva. It is predominantly caused by *H. influenzae, S. pneumoniae*, and *Staphylococcus* species, and is characterised by red eyes, purulent discharge, and discomfort. The condition is usually self-limiting, but topical antibiotics can be used to speed up recovery, reduce the risk of relapse, and prevent sight-threatening complications, such as keratitis (inflammation of the cornea). A Cochrane review concluded that using a broad-spectrum topical antibiotic to treat acute bacterial conjunctivitis is associated with higher clinical and microbiological remission rates than placebo (Sheikh and Hurwitz, 2006). In addition, it would appear that the benefits of topical antibiotics are more apparent if they are used at the beginning of the infection (typically days 1 to 5), rather than towards the end (Sheikh and Hurwitz, 2006).

If conjunctivitis is present in neonates, the case should urgently be referred for assessment, as neonatal conjunctivitis may quickly result in the development of a severe infection with significant complications. One possible cause of neonatal conjunctivitis is gonorrheal conjunctivitis, a condition that occurs when *N. gonorrhoeae* is passed from the mother to the baby during birth. This condition has a similar presentation to other forms of bacterial conjunctivitis, but is associated with profuse purulent discharge (Figure 4.4.4), which, if not managed appropriately, can lead to corneal ulceration and blindness. In this situation, topical chloramphenicol can be used in combination with a systemic antibiotic (such as cefotaxime) to treat the infection.

4.4.6.3   Miscellaneous

Chloramphenicol is also available as an ear-drop preparation[20] for use in otitis externa (inflammation of the external ear canal). This condition is caused by a number of factors (e.g. swimming or using ear plugs), but can also be infective in origin. If an infection is present, a topical antibacterial can be used, but chloramphenicol ear drops are not recommended, as they contain propylene glycol (as a vehicle), which can cause sensitivity problems and, in high concentrations, has been shown to cause ear damage (Morizono and Johnstone, 1975).

As described in Subsection 4.4.6.1, chloramphenicol may also be used systemically for the treatment of life-threatening infections. One situation where chloramphenicol is used systemically is in the treatment of acute bacterial meningitis. If meningitis is suspected, benzylpenicillin should be administered immediately (as empirical therapy), or if the patient is allergic to penicillin, cefotaxime can be used. In situations where a patient has a history of anaphylactic shock to penicillin or cephalosporins, chloramphenicol can be used as an alternative. In cases where the causative agent has been established (e.g. meningitis caused

---

[20] Chloramphenicol eye drops are available as a 0.5% preparation, while the ear drops are available as a 5% preparation. There have been reports of dispensing errors where a pharmacist has mistakenly dispensed ear drops instead of the eye drops. The patient often realises when the drops cause local irritation (such as stinging and burning), but perhaps more positively, the conjunctivitis clears up quicker than expected (as they are using 10 times the normal strength).

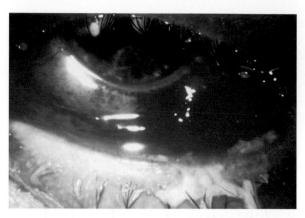

Figure 4.4.4 *A case of gonorrheal conjunctivitis that resulted in partial blindness due to the spread of N. gonorrhoeae (Image courtesy of the Public Health Image Library, ID5174. [http://phil.cdc.gov/phil/home.asp, last accessed 29th April 2011].)*

by *H. influenzae*), chloramphenicol can be used when there is a history or immediate hypersensitivity to penicillins or cephalosporins. There are, however, very few randomised controlled trials supporting these interventions due to the complex ethical issues surrounding this area (it certainly would not be ethical to compare an antibiotic intervention to a placebo in this case). A Cochrane review examining the use of preadmission antibiotics for suspected cases of meningitis identified only one randomised controlled trial comparing ceftriaxone with chloramphenicol, which concluded that both regimens were equally effective and safe in preventing morbidity and mortality associated with meningitis (Sudarsanam *et al.*, 2008).

In some circumstances, chloramphenicol is also used as a first-line agent for the treatment of meningitis. In developing countries, for example, access to cephalosporins is poor (due to the high cost), so the cheaper alternative of chloramphenicol in combination with ampicillin is recommended (World Health Organization, 1997).

Another, but perhaps less common, indication for chloramphenicol is in the treatment of acute epiglottitis – a condition characterised by inflammation of the epiglottis (the flap of tissue at the back of the throat that covers the trachea when swallowing food). As the epiglottis is in close proximity to the trachea, inflammation can restrict the supply of oxygen to the lungs, which, if not diagnosed sufficiently early, can be life-threatening. A bacterial infection is usually implicated in acute epiglottitis, most commonly *H. influenzae* type b (Hib). Thankfully, through the introduction of the Hib vaccine, epiglottitis is now considered to be a rare disease, but when treatment is required, intravenous chloramphenicol can be used to manage the infection.

Finally, chloramphenicol is occasionally used to treat typhoid fever. The beneficial effect of chloramphenicol in the treatment of typhoid fever has been known for many years,[21] but nowadays many strains of *Salmonella typhi* (the causative agent of typhoid fever) show resistance to it and, since the introduction of newer safer antibiotics, its use has become less popular. Currently, the WHO recommends chloramphenicol as

---

[21] One of the earliest studies showing the benefit of using chloramphenicol in typhoid fever was published by Patel *et al.* in the *British Medical Journal* in 1949. Six patients with moderate to severe typhoid fever were given chloramphenicol – three responded well and survived, while unfortunately the remaining three patients died. It may not seem like it now, but these results were interpreted as a significant breakthrough for the treatment of typhoid fever, with the authors stating that chloramphenicol is *'the most effective therapeutic agent available for the treatment of typhoid fever'*.

an alternative to the quinolones (see Subsection 2.1.6), providing the strain is fully susceptible (i.e. not multidrug-resistant) (World Health Organization, 2003).

4.4.7 Adverse drug reactions

The most serious adverse effect associated with chloramphenicol is aplastic anaemia. The first use of chloramphenicol in 1948 for typhoid fever was quickly followed by the first case reports of haemopoietic changes in three patients in 1950 (Volini *et al.*, 1950). The most popular explanation for the toxicity associated with chloramphenicol is that it is due to the formation of a nitroso-chloramphenicol derivative, which exhibits a variety of inhibitory actions within mitochondria, including the inhibition of polymerase activity, suggesting that mitochondrial DNA polymerase is one potential target (Abou-Khalil *et al.*, 1982; Lim *et al.*, 1984). The inhibition of protein synthesis in the bone-marrow mitochondria is dose-related and manifests itself clinically as reticulocytopenia, anaemia, leucopenia, or thrombocytopenia (which are normally reversible).

A second type of toxicity is idiosyncratic and manifests itself clinically as aplastic anaemia (when bone marrow does not produce sufficient new cells to replenish blood cells), which occurs in 1 in 24 500 to 1 in 40 800 patients (which approximates to a 13-fold increased risk of aplastic anaemia over the general population). It generally occurs several weeks or months after drug administration and has been observed with oral, parenteral, and very rarely ocular administration (Sills *et al.*, 1999).

Grey baby syndrome, often recognised with plasma levels exceeding 50 µg/dL, may manifest in neonates as nausea, vomiting, and abdominal distension, with metabolic acidosis, circulatory collapse, and impaired cardiac contractility (these symptoms are reversible with careful detoxification). The toxic effect of chloramphenicol is thought to be due to reduced glucuronidation of chloramphenicol in small infants, resulting in the accumulation of toxic chloramphenicol metabolites, which, in turn, is a result of reduced renal excretion and an immature UDP-glucuronyl-transferase enzyme system (Brunton, 2006).

Chloramphenicol has also been associated with an increased risk of childhood leukaemias (Shu *et al.*, 1987).

A reversible toxicity of chloramphenicol succinate has been observed – although this could result from the same mechanism as outlined above for the parent drug chloramphenicol after its release, other work has suggested that the prodrug may act as a competitive substrate for succinate dehydrogenase and that the chloramphenicol released may inhibit this enzyme, accounting for its toxicity (Ambekar *et al.*, 2004).

4.4.8 Drug interactions

Chloramphenicol inhibits the cytochrome P450 isoenzymes 2C19 and 3A4, resulting in elevated plasma levels of a wide variety of drugs that are metabolised by these drugs, including:

- tricyclic antidepressants (amitriptylline, clomipramine, and imipramine);
- selective serotonin reuptake inhibitors (fluoxetine, citalopram, and sertraline);
- antiepileptic drugs (phenytoin, phenobarbitone, diazepam, and primidone);
- proton pump inhibitors (omeprazole);
- azole antifungals (clotrimazole, ketoconazole, and itraconazole);
- macrolide antibiotics (clarithromycin and erythromycin);
- calcium channel blocking agents (amlodipine, diltiazem, felodipine, nifedipine, and verapamil); as well as:
- clopidrogel, gliclazide, propranolol, and tetrahydrocannabinol.

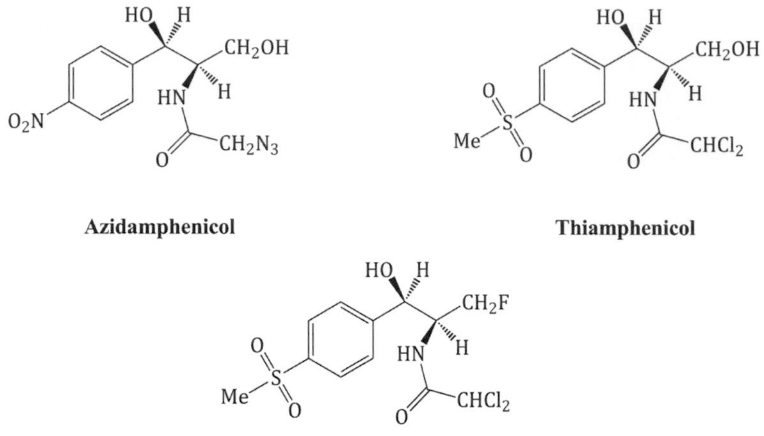

Figure 4.4.5   *Amphenicol antibiotics used around the world*

4.4.9   Recent developments

As described in the subsections above, chloramphenicol is rarely used systemically due to the risk of developing serious haematological side effects, and consequently research which focusses on the development of other agents from this class of antibiotics has halted over the years, depite the fact that several other 'amphenicol' antibiotics have made it to market around the world (Figure 4.4.5 and Table 4.4.2).

Perhaps the most significant development in recent years is the decision by the Medicines Healthcare products Regulatory Agency (MHRA) to reclassify chloramphenicol eye drops/ointment from a prescription-only medicine (POM) to a pharmacy medicine (P), so that, in the UK, chloramphenicol eye drops/ointment is available without a prescription, as long as a pharmacist supervises the sale. The authors can confirm that many pharmacy consultations during a weekend locum are related to minor eye infections, so this reclassification is a welcome addition to the list of medicines available 'over the counter'. This reclassification has, however, provoked strong debate in the medical community, with some practitioners concerned that the resulting improved availability of chloramphenicol may lead to problems with resistance. This has yet to be established, but one study has shown that, since the switch was made, the use of chloramphenicol has significantly increased (Davis *et al.*, 2009).

Table 4.4.2   *Amphenicol antibiotics licensed around the world*

Amphenicol	Registered	Indication	Reference
Azidamphenicol (also known as azidamfenicol)	Licensed in Germany	Used in the treatment of bacterial eye infections	Sachs *et al.* (2001)
Florphenicol (also known as florfenicol)	Licensed in the USA	Used in veterinary medicine	Priebe and Schwarz (2003)
Thiamphenicol	Licensed in Brazil, France, and Spain	Used to treat sexually transmitted infections	Tupasi *et al.* (1983)

References

S. Abou-Khalil, W. H. Abou-Khalil, and A. A. Yunis, *Biochem Pharmacol.*, 1982, **31**, 3823–3830.

C. S. Ambekar, J. S. Lee, B. M. Cheung, L. C. Chan, R. Liang, and C. R. Kumana, *Toxicol. In Vitro*, 2004, **18**, 441–447.

P. J. Ambrose, *Clin. Pharmacokinet.*, 1984, **9**, 222–238.

Q. R. Bartz, *J. Biol. Chem.*, 1948, **172**, 445–450.

G. Bhaskar, V. Satish Kumar, and B. Venkateswara Rao, *Tetrahedron: Asymmetry*, 2004, **15**, 1279–1283.

Y. Boer and A. Pijnenburg, *Int. J. Clin. Pharm.*, 1983, **5**, 95–101.

L. L. Brunton, 'Protein synthesis inhibitors and miscellanaeous antibacterial agents', in L. Bruton, B. Chabner, and B. Knollman, *Goodman & Gillman's The Pharmacological Basis of Therapeutics*, 11th Edn, McGraw-Hill, 2006, Chapter 46.

M. Chen, D. Howe, B. Leduc, S. Kerr, and D. A. Williams, *Xenobiot.*, 2007, **37**, 954–971.

M. Chen, B. Leduc, S. Kerr, D. Howe, and D. A. Williams, *Drug Metab. Dispos.*, 2010, **38**, 368–375.

J. Controulis, M. C. Rebstock, and H. M. Crooks, *J. Amer. Chem. Soc.*, 1949, **71**, 2463–2468.

G. J. Cuchural, Jr, F. P. Tally, N. V. Jacobus, K. Aldridge, T. Cleary, S. M. Finegold, G. Hill, P. Iannini, J. P. O'Keefe, and C. Pierson, *Antimicrob. Agents Chemother.*, 1988, **32**, 717–722.

O. Danilchanka, M. Pavlenok, and M. Niederweis, *Antimicrob. Agents Chemother.*, 2008, **52**, 3127–3134.

C. Davidovitch, A. Bashan, and A. Yonath, *Proc. Natl. Acad. Sci. USA*, 2008, **105**, 20 665–20 670.

H. Davis, D. Mant, C. Scott, D. Lasserson, and P. W. Rose, *Br. J. Gen. Pract.*, 2009, **59**, 897–900.

J. Ehrlich, Q. R. Bartz, R. M. Smith, D. A. Joslyn, and P. R. Burkholder, *Science*, 1947, **106**, 417.

C. A. Franje, S. K. Chang, C. L. Shyu, J. L. Davis, Y. W. Lee, R. J. Lee, C. C. Chang, and C. C. Chou, *J. Pharm. Biomed. Anal.*, 2010, **53**, 869–877.

T. Ghafourian, M. Barzegar-Jalali, S. Dastmalchi, T. Khavari-Khorasani, N. Hakimiha, and A. Nokhodchi, *Int. J. Pharmaceut.*, 2006, **319**, 82–97.

J. L. Hansen, P. B. Moore, and T. A. Steitz, *J. Mol. Biol.*, 2003, **330**, 1061–1075.

R. E. Kauffmann, M. C. Thirumoorthi, J. A. Buckley, M. K. Aravind, and A. S. Dajani, *J. Pediatr.*, 1981, **99**, 963–967.

X. Z. Li, D. M. Livermore, and H. Nikaido, *Antimicrob. Agents Chemother.*, 1994, **38**, 1732–1741.

L.O. Lim, W. H. Abou-Khalil, A. A. Yunis, and S. Abou-Khalil, *J. Lab. Clin. Med.*, 1984, **104**, 213–222.

T. Morizono and B. M. Johnstone, *Med. J. Aust.*, 1975, **2**, 634–638.

R. H. Mosher, D. J. Camp, K. Yang, M. P. Brown, W. V. Shaw, and L. C. Vining, *J. Biol. Chem.*, 1995, **270**, 27 000–27 006.

M. C. Nahata and D. A. Powell, *Clin. Pharmacol. Ther.*, 1981, **30**, 368–372.

M. C. Nahata and D. A. Powell, *Dev. Pharmacol. Ther.*, 1983, **6**, 23–32.

J. C. Patel, D. D. Banker, and C. J. Modi, *Brit. Med. J.*, 1949, **2**, 908–909.

L. K. Pickering, J. L. Hoecker, W. G. Kramer, S. Kohl, and T. G. Cleary, *J. Pediatr.*, 1980, **96**, 757–761.

D. A. Powell and M. C. Nahata, *Drug Intell. Clin. Pharm.*, 1982, **16**, 295–300.

S. Priebe and S. Schwarz, *Antimicrob Agents Chemother.*, 2003, **47**, 2703–2705.

Z. Qiang and C. Adams, *Water Res.*, 2004, **38**, 2874–2890.

J. E. F. Reynolds (Ed.), *Martindale: The Extra Pharmacopoeia*, 30th Edn, Pharmaceutical Press, London, 1993.

B. Sachs, S. Erdmann, T. al Masaoudi, and H. F. Merk, *Allergy*, 2001, **56**, 69–72.

F. Schlünzen, R. Zarivach, J. Harms, A. Bashan, A. Tocilj, R. Albrecht, A. Yonath, and F. Franceschi, *Nature*, 2001, **413**, 814–821.

W. V. Shaw, *Ann. Rev. Biophys. Biophys. Chem.*, 1991, **20**, 363–386.

W. V. Shaw and J. Unowsky, *J. Bacteriol.*, 1968, **95**, 1976–1978.

X. Shu, Y. Gao, M. Linet, L. Brinton, R. Gai, F. Jin, and J. Fraumeni, *Lancet*, 1987, **2**, 934–937.

M. R. Sills, D. Boenning, and T. L. Cheng, *Pediatr. Rev.*, 1999, **20**, 357–358.

A. Sheikh and B. Hurwitz, *Cochrane Database of Systematic Reviews*, 2006, Issue 2.

T. Sudarsanam, P. Rupali, P. Tharyan, O. C. Abraham, and K. Thomas, *Cochrane Database of Systematic Reviews*, 2008, Issue 1.

T. E. Tupasi, L. B. Crisologo, C. A. Torres, O. V. Calubiran, and I. de Jesus, *Br. J. Vener. Dis.*, 1983, **59**, 172–175.

I. F. Volini, I. Greenspan, L. Ehrlich, J. A. Gonner, O. Felsenfeld, and S. O. Schwartz, *JAMA*, 1950, **142**, 1333–1335.

M. W. Weber, S. R. Gatchalian, O. Ogunlesi, A. Smith, G. H. McCracken, Jr, S. Qazi, A. F. Weber, K. Olsen, and E. K. Mulholland, *Pediatr. Infect. Dis.*, 1999, **18**, 896–901.

M. Wenk, S. Vozeh, and F. Follath, *Clin. Pharmacokinet.*, 1984, **9**, 475–492.

World Health Organization, Emerging and other Communicable Diseases, Surveillance and Control. Antimicrobial and Support Therapy for Bacterial Meningitis in Children. Report of the Meeting of 18–20 June 1997, Geneva, Switzerland, 1997 (http://www.who.int/csr/resources/publications/meningitis/whoemcbac982.pdf, last accessed 9 March 2012).

World Health Organization, Communicable Disease Surveillance and Response Vaccines and Biologicals. Background Document: The Diagnosis, Treatment and Prevention of Typhoid Fever, 2003 (http://whqlibdoc.who.int/hq/2003/WHO_V&B_03.07.pdf, last accessed 9 March 2012).

A. Yonath, *Ann. Rev. Biochem.*, 2005, **74**, 649–679.

D. N. Zhou, W. Y. Huang, F. Wu, C. Q. Han, and Y. Chen, *React. Kinet. Mech. Catal.*, 2010, **100**, 45–53.

4.5

Oxazolidinones

Key Points

- Linezolid is the first in a new class of antibacterial agents active against Gram positive bacteria. It has excellent bioavailability and distribution into body tissues, making it particularly useful for complicated skin and soft tissue infections and pneumonia caused by susceptible organisms.
- Resistance is not yet a major problem due to its restricted use.
- For further information, see Leach *et al.* (2011).

The only oxazolidinone antibiotic currently approved for clinical use is linezolid (Figure 4.5.1), which is indicated for the treatment of severe infections due to Gram positive organisms, as shown in Table 4.5.1.

4.5.1 Discovery

The discovery of linezolid is an example of a careful medicinal chemistry approach to novel lead compound development, coupled with industrial competition and serendipity.

This interesting story started at E. I. duPont de Nemours & Company, during a screening programme in the 1980s, with the discovery of oxazolidinone-based molecules that displayed potent activity against Gram positive bacteria, even those known to be resistant to penicillins. DuPont announced their discovery with a series of articles between 1987 and the early 1990s (Brickner *et al.*, 2008), having patented the oxazolidinones as antibacterial agents (Gregory, 1987). The lead compounds, DuP-105 and DuP-721 (Figure 4.5.2), were comparable in potency to vancomycin (Maple *et al.*, 1989) and were found to have excellent activity against a range of streptococcal and staphylococcal strains (Leach *et al.*, 2011). Within 2 years, scientists at the Upjohn Company (now part of Pfizer Inc.) had developed their own oxazolidinone series and had compounds that matched, or even improved upon, the activity of the DuPont lead compounds. The race to find the best antibacterial oxazolidinone was on! Around this time, DuPont received results from preclinical *in vivo* trials with their lead compound (DuP-721), which identified serious toxicity issues, and they subsequently abandoned their antibacterial oxazolidinone development programme. There must have been a sharp intake

Antibacterial Agents: Chemistry, Mode of Action, Mechanisms of Resistance and Clinical Applications, First Edition.
Rosaleen J. Anderson, Paul W. Groundwater, Adam Todd and Alan J. Worsley.
© 2012 John Wiley & Sons, Ltd. Published 2012 by John Wiley & Sons, Ltd.

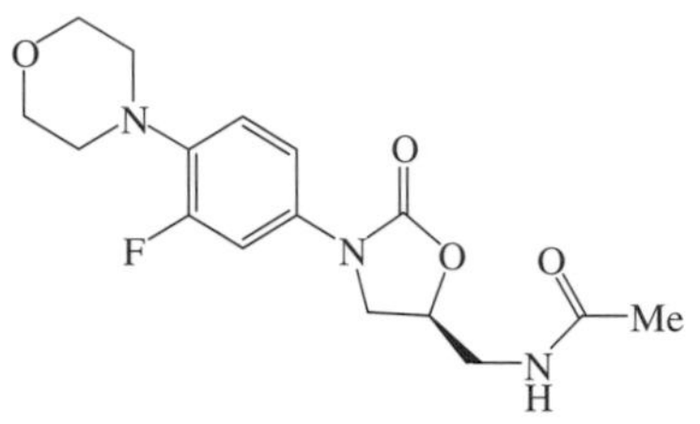

Figure 4.5.1 *Linezolid*

Table 4.5.1 *Therapeutic indications for linezolid*

Indication	Notes
Treatment of community-acquired pneumonia and nosocomial pneumonia	**Only** used when known or suspected to be caused by susceptible Gram positive bacteria
Complicated skin and soft-tissue infections	**Only** used when microbiological testing has established that the infection is known to be caused by susceptible Gram positive bacteria

of breath at Upjohn when they heard this news and they rapidly carried out their own preclinical *in vivo* screen on the racemic form of their lead compound **A**, comparing it to the racemate of DuP-721, and found that lead compound **A** was at least as efficacious as the DuPont compound, but, amazingly, produced no toxicity problems. It seems remarkable that compound **A** could have such a different *in vivo* toxicology profile, as the structural differences are very small (Figure 4.5.2), so Lady Luck had smiled on the Upjohn team.

Things did not go entirely smoothly for the Upjohn team, though, as they soon found their lead compound **A** was already included in the patent filed by DuPont. Further careful work followed, which allowed initial structure and toxicology activity relationships (discussed further in Subsection 4.5.4) to be derived, and eventually resulted in the proposal of two clinical candidates, linezolid and eperezolid (Figure 4.5.3) (Ford *et al.*, 1997; Brickner *et al.*, 2008).

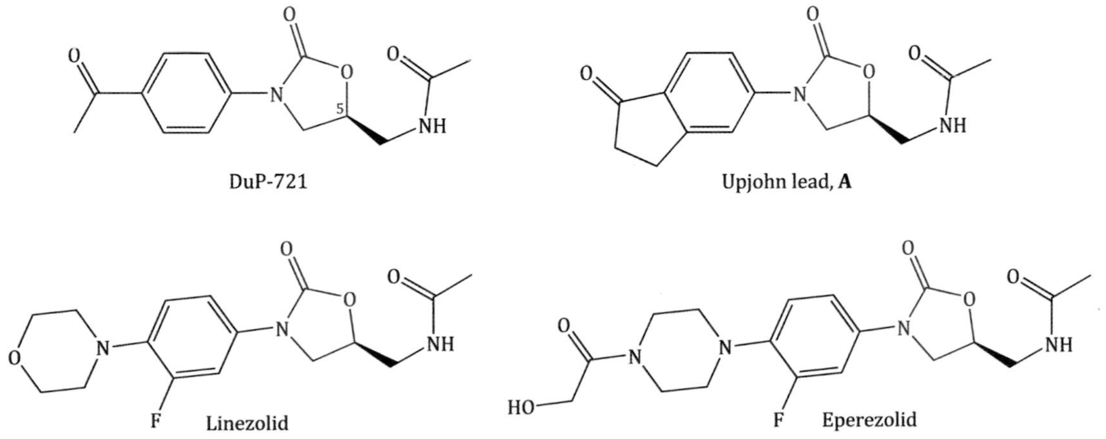

Figure 4.5.2 *Structures of oxazolidinone lead and clinical candidates (Ford et al., 1997; Brickner et al., 2008). The oxazolidinone core is highlighted in each*

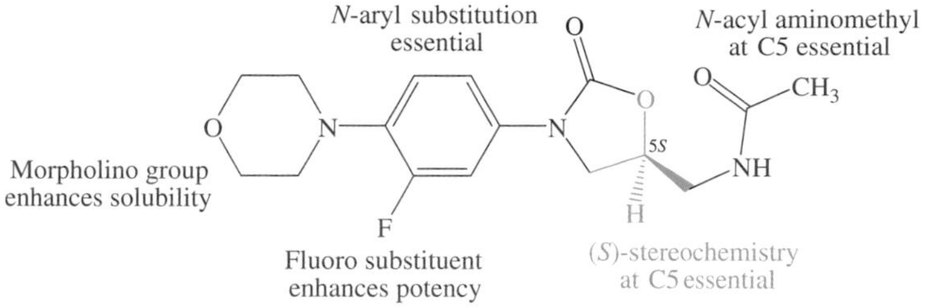

Figure 4.5.3 *Early structure–activity relationship (SAR) for the oxazolidinones, which led to the development of linezolid and eperezolid (Barbachyn and Ford, 2003; Livermore, 2003; Brickner et al., 2008)*

Both linezolid and eperezolid progressed into preclinical trials and were found to have very similar antibacterial activities. As is often seen in drug development, the choice between these similar drug candidates came down to which had the better pharmacokinetic profile.

There has also been controversy associated with the discovery of linezolid, this time with allegations of illegal marketing activities. The pharmaceutical companies Pharmacia AB and the Upjohn Company merged in 1995, and then merged with Pfizer Inc. in 2003, thus placing linezolid into Pfizer's drug portfolio. In 2009, Pfizer agreed to pay around \$2.3 bn (or £1.4 bn), 'to settle civil and criminal allegations' of having illegally promoted four unrelated drugs, including linezolid, for uses that did not have regulatory (FDA) approval. This must seem like an incredible fine, but was reported to amount to less than 3 weeks of Pfizer's sales, reflecting the huge sums of money involved in the pharmaceutical industry (http://www.nytimes.com/2009/09/03/business/03health.html, last accessed 9 March 2012). For more on the discovery of linezolid, see Brickner *et al.* (2008).

4.5.2   Synthesis

Like that of chloramphenicol, the synthesis of linezolid and its analogues is *relatively* simple.[22] The requirement for the (*S*)-stereochemistry at C5 was established early in the lead development of the oxazolidinones by DuPont and was easily solved by the use of a commercially available, enantiomerically pure, starting material.

There is more than one synthetic route to linezolid, but we will consider the initial industrial synthesis here (Scheme 4.5.1), which resulted from the optimisation of the chemistry at Upjohn and Pharmacia (Pearlman *et al.*, 1998; Brickner *et al.*, 2008). One drawback of this route is the need for expensive reagents, such as palladium on charcoal, which may have contributed to the relatively high cost of linezolid (currently estimated at about £900 for a course of 20 tablets of 600 mg each). Despite these costs, the pharmacoeconomic considerations in several countries, taking account of drug and hospitalisation costs, generally support the use of linezolid for its licensed applications.

The first three steps of the synthesis shown in Scheme 4.5.1 are routine but effective, starting with a nucleophilic aromatic substitution of the 4-fluoro group in 3,4-difluoronitrobenzene by morpholine, followed by palladium-catalysed reduction of the aromatic nitro group in **1** to give the corresponding aromatic amine **2**. Addition of a carbobenzyloxy (CBZ) group to the amine using benzylchloroformate produces the carbamate **3**, which is deprotonated by lithium *tertiary*-butoxide (a strong base) and reacted with the commercially sourced chiral reagent (*S*)-3-chloro-1,2-propanediol. The intermediate **4** is not isolated, but spontaneously cyclises,

---

[22] Planning the synthesis of a potential pharmaceutical agent, including any specified chiral centres, involves consideration of many factors, even if the target is a relatively simple molecule (Hoffmann, 2009).

CSF, with generally similar or higher concentrations in both healthy and inflamed tissues than in plasma (MacGowan, 2003; Vardakas *et al.*, 2009; Dryden, 2011). The importance of this extensive tissue penetration has been linked to the successful treatment of MRSA-related pneumonia, skin and soft tissue infections (SSTIs), Gram positive bacterial orthopedic infections, and even multiresistant *Enterococcus faecium*-related meningitis (Dryden, 2011). Sufficiently high concentrations of linezolid are reached in the aqueous humour and vitreous of the eye after oral administration to achieve antibacterial action against strains of staphylococci responsible for ocular infections (Horcajada *et al.*, 2009).

The absorption of linezolid in children appears to be similar to that in adults, whereas the distribution and plasma clearance are enhanced in infants and children (MacGowan, 2003). A pharmacokinetic model was established which showed that both the elderly and patients with lower body weights have altered distribution and clearance (Abe *et al.*, 2009). A higher plasma concentration is achieved in adult females than in males, alongside a lower total clearance and volume of distribution, whereas the serum half-life is the same in both (MacGowan, 2003; Dryden, 2011).

Linezolid is metabolised, by non-enzymatic oxidation of the morpholine ring (Scheme 4.5.2), into two major inactive metabolites, both of which are excreted in the urine, along with around 30% linezolid, accounting for >80% of the dose (MacGowan, 2003; Vardakas *et al.*, 2009).

The faecal excretion route is non-existent for linezolid, and very minor for the metabolites, and cytochrome P_{450} (CYP) enzymes do not appear to be involved in linezolid metabolism, so inducement or inhibition of liver enzymes is not expected, and there is an associated low risk of interactions with the usual CYP-dependent suspects, such as cimetidine and phenytoin (Vardakas *et al.*, 2009).

Linezolid is generally accepted to be bacteriostatic, but it shows clinical bactericidal activity against most strains of streptococci and some other bacteria (Zurenko *et al.*, 2001), perhaps explained by a linezolid-induced decrease in the amount of virulence factor released, which increases bacterial susceptibility to neutrophil phagocytosis (Gemmell and Ford, 2002; Bernardo *et al.*, 2004).

4.5.4   Mode of action and selectivity

Linezolid acts by preventing protein synthesis, but in a different manner to the agents we have studied previously in this section. Although it competes with chloramphenicol in binding assays with the 50S ribosomal subunit, it does not have the same mode of action, as it does not inhibit binding to the peptidyl transferase A site (described in Subsection 4.4.4). It has taken a number of years to identify the most likely binding site and the mode of action of linezolid and related oxazolidinones (Leach *et al.*, 2011).

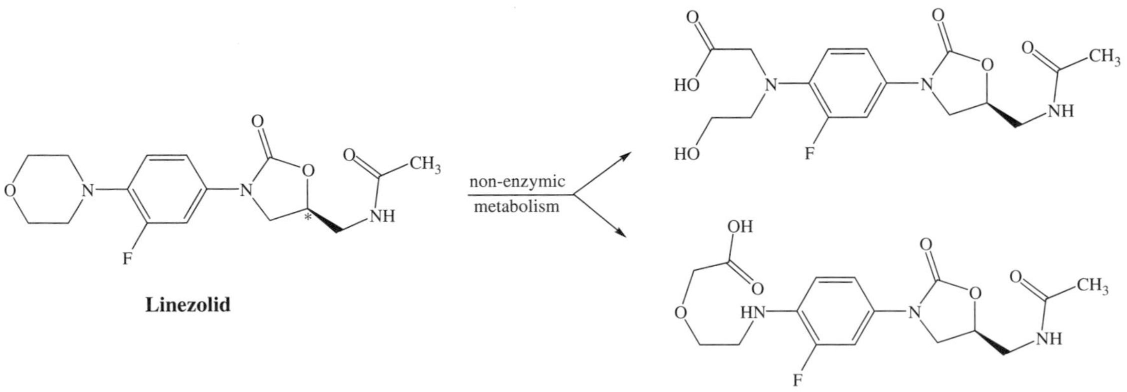

Scheme 4.5.2   *Non-enzymatic metabolism of linezolid, resulting in two major metabolites (MacGowan, 2003)*

So where does it bind and what is its mode of action? By studying the 50S ribosomal subunit of representative bacterial ribosomes (Ippolito *et al.*, 2008; Wilson *et al.*, 2008), X-ray crystallography has shown that linezolid binds to the A site, close to the catalytic centre, and disturbs the binding of the charged tRNA to this site. More specifically, linezolid binds in a pocket formed by eight RNA residues that are conserved across all bacterial 50S ribosomes: U2585, A2503, U2504, G2505, U2506, G2061, A2451, and C2452 (using the numbering system of *Escherichia coli*) (Wilson *et al.*, 2008). Although most of these residues do not need to move to accommodate linezolid, there is an induced-fit element to linezolid binding, with certain residues moving in order to better interact with linezolid. **This induced movement of residues results in alternate orientations of these bases and could be responsible for a decreased ability to bind the next charged tRNA to the A site, or linezolid binding partly in the A site may prohibit tRNA binding** (Ippolito *et al.*, 2008; Wilson *et al.*, 2008).

Selectivity for a bacterial target over a mammalian one is crucial to the clinical success of an antibacterial agent. The examples of bacterial protein synthesis inhibitors we have already studied in this section show selectivity due to a variety of reasons, including:

- Poor uptake into mammalian cells compared to bacterial cells.
- The incorporation of specific functional groups for optimum selectivity.
- Poor binding to the mammalian ribosome.

As we have seen, the aminoglycosides, macrolides, tetracyclines, and chloramphenicol are all antibiotics: they are natural molecules from a microbial source which have benefitted from Nature's design expertise to optimise the structures for their antibacterial effect. In comparison, linezolid is an antibacterial agent that is imperfect, being of human design (with little knowledge of the target or mode of action for guidance), and this is reflected in its lack of absolute selectivity for bacterial over mammalian protein synthesis. Linezolid has no effect upon mammalian cytoplasmic protein synthesis, but inhibits mitochondrial protein synthesis (Leach *et al.*, 2007), which is linked to the serious adverse effect of myelosuppression when used for an extended period (De Vriese *et al.*, 2006), causing concerns and reviews of its safety (Vinh and Rubenstein, 2009). In previous subsections, we saw how the same issue (the inhibition of mitochondrial protein synthesis) restricts the use of chloramphenicol, while some of the adverse effects of tetracyclines are due to their actions upon mitochondrial processes, and aminoglycosides also exhibit toxicity – despite Nature's expertise, perhaps it doesn't always get it right!

The huge number of linezolid analogues that have been synthesised and studied (Poce *et al.*, 2008) have fine-tuned the structural features and lipophilicity, while reducing the undesirable effects due to inhibition of human mitochondrial protein synthesis and monoamine oxidase A, enabling an updated overall structure–activity relationship (Katritzky *et al.*, 2004; Renslo *et al.*, 2006b; Prasad, 2007) and toxicity–structure relationship (Renslo, 2010) to be developed (Figure 4.5.4).

From this work, new oxazolidinone antibacterials have been, and are being, developed that retain or improve the potency, pharmacokinetics, and activity spectrum of linezolid, while reducing adverse effects. New agents in development are discussed in Subsection 4.5.9.

Further research involves developing agents that combine linezolid with another established antibacterial class, such as a quinolone (Hubschwerlen *et al.*, 2003) or nitroimidazole (Das *et al.*, 2005), resulting in at least one candidate in clinical trials (see ranbezolid in Figure 4.5.6).

4.5.5 Bacterial resistance

Gram negative bacteria are intrinsically resistant to linezolid due to efficient expulsion of the agent by bacterial efflux pumps, so we will focus here upon resistance of Gram positive bacteria to this agent. Early

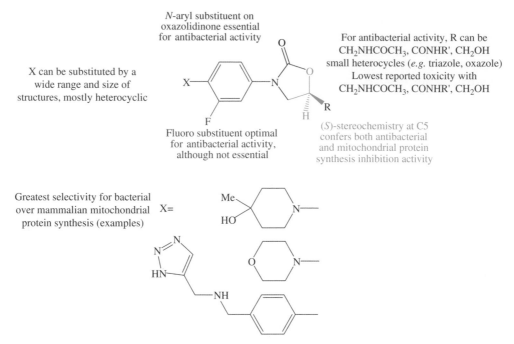

Figure 4.5.4   *Combined structure–activity and structure–toxicity relationships developed after extensive study of linezolid analogues (Renslo et al., 2006b; Prasad, 2007; Renslo, 2010)*

attempts at DuPont to promote and select Gram postitive bacteria with resistance to DuP-721 showed that resistance arose at the slow rate of about one resistant mutant per billion bacteria, giving hope that resistance to the oxazolidinones would develop only slowly (Ford *et al.*, 1997). Closer inspection identified an *in vitro* resistance frequency of $<8 \times 10^{11}$ in a *Staphyloccocus aureus* strain, suggesting a requirement for several mutation steps to occur sequentially (Zurenko *et al.*, 1996). Soon after the introduction of linezolid, however, reports of clinical resistance in *S. aureus*, *Staphylococcus epidermidis*, and vancomycin-resistant enterococci (VRE) started to appear, and these have continued to arise in many countries (Manfredi and Sabbatani, 2010).

Resistance has been found to result from modification of the ribosomal target (Treviño *et al.*, 2009), and attempts to discover the key mutations have identified a series of susceptible sites that correspond closely to the residues forming the pocket into which linezolid binds, providing additional proof of its site of action. Furthermore, most of these mutations cause cross-resistance to other antibacterial agents also known to bind to the A site, such as chloramphenicol (Wilson *et al.*, 2008). However, evidence for other resistance mechanisms has appeared, involving decreased linezolid uptake (Sierra *et al.*, 2009) and methyltransferases that modify the binding site (Morales *et al.*, 2010; LaMarre *et al.*, 2011). Some methyltransferase encoding genes we have already met, *erm* (erythromycin ribosome methylase, Subsection 4.2.5) and *tet* (tetracycline resistance, Subsection 4.3.5), are responsible for cross-resistance to several different classes of agent that inhibit bacterial protein synthesis, but they do not confer resistance to linezolid (Fines and Leclercq, 2000). The *cfr* gene encodes for a different methyltransferase, which methylates the 23S rRNA part of the 50S ribosomal subunit and causes resistance to several antibacterial classes, including linezolid, and accounts for some of the observed clinical cases of linezolid resistance (Morales *et al.*, 2010). Mutations in the indigenous methyltransferase gene *rlmN* were recently shown to confer clinical linezolid resistance

(LaMarre *et al.*, 2011), and the ease with which this gene could be mutated and inactivated causes concern for the future.

4.5.6 Clinical applications

4.5.6.1 Spectrum of activity

Linezolid has comprehensive activity against a wide range of aerobic and anaerobic Gram positive pathogens, including, perhaps most notably, meticillin-resistant *Staphylococcus aureus* (MRSA) and VRE (Noskin *et al.*, 1999). Linezolid has no clinical activity against Gram negative organisms due to the presence of efflux pumps (Ford *et al.*, 1997) (see Subsection 4.5.5 for more details) and therefore should not be used in the management of diseases caused by these pathogens. If a mixed infection is suspected (i.e. an infection caused by both Gram negative and Gram positive organisms) and linezolid is indicated, another antibacterial should also be used to provide cover against the Gram negative organism.

4.5.6.2 Complicated skin and soft-tissue infections (cSSTIs)

SSTIs are a common cause of patient morbidity and account for a significant number of hospital admissions (Edelsberg *et al.*, 2009). They can be classed as either 'uncomplicated' or 'complicated' and are caused by a variety of different pathogens – the most common being *Staphylococcus aureus*, with one study estimating that it causes around 46% of SSTIs (Rennie *et al.*, 2003).

Uncomplicated SSTIs are generally superficial infections and include impetigo, simple abscesses, and carbuncles. They are usually treated on an outpatient basis and are managed with oral/topical antibiotics, plus simple surgical incision, where appropriate. Complicated SSTIs (cSSTIs), however, involve deeper tissues, such as subcutaneous tissue or skeletal muscle, or occur in patients with other co-morbidities (e.g. diabetes or HIV) and are therefore considered to be more serious. cSSTIs may require hospitalisation, the use of IV antibiotics, and/or surgery. Examples of cSSTIs include major abscesses, a diabetic foot infection, and necrotising soft-tissue infections.

Necrotising soft-tissue infections are particularly serious and can cause progressive tissue destruction, limb loss, and possibly even death. Necrotising fasciitis, which is more commonly known to the general public as 'flesh eating disease',[24] is one example of a necrotising soft-tissue infection (Figure 4.5.5). It is managed using aggressive surgery (to remove the necrotic tissue) and antibiotics (to eradicate the infection). Initially, the condition presents with symptoms of erythema, swelling, and pain, but it can rapidly lead to severe sepsis, multi-organ failure, and death, if not promptly diagnosed and treated (Hasham *et al.*, 2005).

If a cSSTI is suspected, empirical antibiotic therapy is generally given until the causative pathogen is identified. If MRSA is responsible for the cSSTI, either linezolid or vancomycin can be used (Stevens *et al.*, 2005); both agents show good activity, but linezolid appears to be more effective than vancomycin when treating cSSTIs caused by MRSA (Weigelt *et al.*, 2005).

Linezolid should **not** be routinely used for the management of cSSTIs, or be used to treat uncomplicated SSTIs, as this may promote (or accelerate) the development of resistance. It should only be used under expert supervision, after microbiological testing has been performed in order to ensure that the infection is caused by susceptible Gram positive bacteria.

---

[24] This is a term used by the media; in reality, the bacteria do not actually eat any tissue – the damage is caused by the release of toxins.

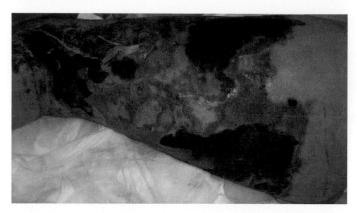

Figure 4.5.5 *The presentation of necrotising fasciitis, showing marked erythema and necrosis. In this case, the patient's blood cultures confirmed the presence of MRSA, and consequently linezolid was prescribed to manage the infection (Reprinted from P. Smuszkiewicz, I. Trojanowska and H. Tomczak, Cases J., 1, 125, 2008, Cases Network Ltd.)*

4.5.6.3 Community-acquired pneumonia and hospital-acquired pneumonia

Linezolid also plays a role in the treatment of community-acquired and hospital-acquired pneumonia. It is included in pneumonia treatment guidelines for both the UK (the British Thoracic Society) and the USA (Infectious Diseases Society of American/American Thoracic Society) (Mandell *et al.*, 2007; British Thoracic Society, 2009), but is generally withheld, alongside vancomycin, for cases specifically caused by MRSA. Studies show that linezolid is as effective as vancomycin for the treatment of MRSA infections (Beibei *et al.*, 2010; Caffrey *et al.*, 2010), but further studies are warranted to establish the long-term benefits of linezolid in comparison to other antibacterial agents.

   You may have noticed a recurrent theme of this book is that bacterial resistance can have a major impact on how well a patient responds to antibacterial therapy. Unfortunately, with antibiotics and resistance, it is often a case of 'when' and not 'if' it will happen. With new antibiotics (such as linezolid), practitioners have a responsibility to ensure that they are prescribed appropriately, and this is the reason why linezolid is kept in reserve to treat (in this case) pneumonia caused by MRSA. It would certainly not be responsible prescribing[25] if linezolid were used first-line to treat all cases of community-acquired and hospital-acquired pneumonia, as this would unnecessarily increase linezolid exposure and eventually lead to resistance.

4.5.7 Adverse drug reactions

Reports of linezolid causing rare but serious side effects, including myelosuppression, peripheral neuropathy, optic neuropathy, and lactic acidosis, have been documented. It is suggested that these effects are due to interference with mitochondrial function, and as a result several studies have investigated this effect of the oxazolidinones (Soriano *et al.*, 2005). Blood samples taken from several patients receiving linezolid had their mitochondrial activity assessed for mitochondrial respiratory complex II (succinate dehydrogenase) and complex IV (cytochrome c oxidase) (Lopez *et al.*, 2004) and were found to be normal for complex II, but below normal for complex IV. Complex II proteins are synthesised by the

---

[25] It would certainly not be sensible prescribing either – 20 tablets of linezolid cost around £890, compared to 21 capsules of amoxicillin, which cost around £1.70.

cytoplasmic ribosomes, whereas complex IV are synthesised by the mitochondrial ribosomes, so this would again suggest that mitochondrial function is altered by linezolid. Further studies have suggested that the observed lactic acidosis in several patients may be due to this mitochondrial complex IV function inhibition (De Vriese *et al.*, 2006). Inhibitory effects on mitochondrial protein synthesis have also been observed with reverse transcriptase inhibitors, which may also cause lactic acidosis and may share a similar mode of action to linezolid (Brinkman *et al.*, 1998).

Several patients who received linezolid have been reported to have experienced both peripheral and optic-nerve neuropathies (Corallo and Paull, 2002; Lee *et al.*, 2003; Bressler *et al.*, 2004). Once again, the exact mechanism for this observed toxicity is unknown, and the optical neuropathy remitted after drug withdrawal, but the peripheral neuropathies did not.

Early clinical trials before linezolid was released showed a low incidence of myelosuppression, and the mechanism for this was also unclear (Birmingham *et al.*, 2003; Moylett *et al.*, 2003).

4.5.8 Drug interactions

Linezolid acts as a weak monoamine oxidase inhibitor (MAOI) and therefore has the potential to cause serotonin syndrome when used in combination with selective serotonin reuptake inhibitors (SSRIs) (Lawrence *et al.*, 2006) or venlafaxine (Jones *et al.*, 2004). Indeed, there have been several cases of serotonin syndrome reported in the literature where patients have used both linezolid and paroxetine, sertraline, or citalopram (Lavery *et al.*, 2001; Wigen and Goetz, 2002; Bernard *et al.*, 2003).

In view of the monoamine oxidase inhibitory activity of linezolid, it should always be used with caution when given with other sympathomimetic drugs (e.g. pseudoephedrine – a product which is found in many

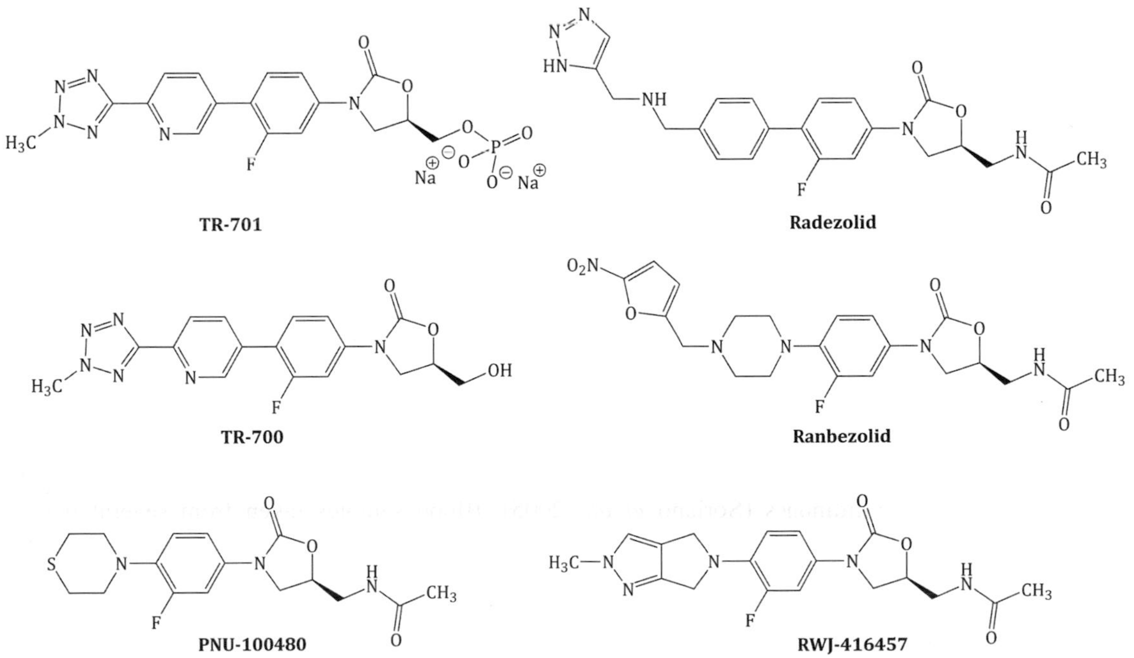

Figure 4.5.6 *Oxazolidinones in development*

Table 4.5.2   *Oxazolidinones in development*

Oxazolidinone	Comment	Reference
TR-700	Shows greater *in vitro* potency against various Gram positive bacteria when compared to linezolid	Schaadt *et al.* (2009)
TR-701 (Torezolid, a prodrug of TR-700)	A phase III study of oral TR-701 versus linezolid for the treatment of acute bacterial SSTIs is planned	ClinicalTrials.gov identifier: NCT01170221
PNU-100480	Completed phase I trials. May have a role in the treatment of TB – when used in combination with anti-TB drugs, the time taken to cure TB is shortened in a murine model	Williams *et al.* (2009)
Radezolid	Phase II trials against community-acquired pneumonia and uncomplicated SSTIs completed	Lemaire *et al.* (2010) ClinicalTrials.gov identifiers: NCT00640926; NCT00646958
Ranbezolid	Completed phase I clinical trials	Bush *et al.* (2004)
RWJ-416457	Shows greater *in vitro* activity against susceptible staphylococci and enterococci than linezolid	Livermore *et al.* (2007)
AZD5847	Structure not available. Entered phase I clinical trials to assess its safety, tolerability, and pharmacokinetics	Nuermberger *et al.* (2010)

proprietary cough and cold preparations), as there is a risk of causing severe hypertension (Hendershot *et al.*, 2001).

4.5.9   Recent developments

The development of the oxazolidinone antibiotics represents a major breakthrough in antibacterial drug design, and, indeed, linezolid is the first completely new class of antibiotic to reach the market since the discovery of nalidixic acid (a quinolone) way back in 1962. Prior to the discovery of the oxazolidinones, the development of new antibacterials generally focussed on expanding the existing classes of antibiotics (e.g. the development of third- and fourth-generation cephalosporins).

The oxazolidinones remain an attractive area of research, and there are several analogues in preclinical development (Figure 4.5.6 and Table 4.5.2). You will notice that TR-701 and TR-700 are very similar in structure; TR-700 is the active oxazolidinone, while TR-701 is an aqueous soluble prodrug form that releases the active TR-700 by the action of phosphatases *in vivo*. To date, linezolid is the only licensed oxazolidinone, but it is anticipated that many more of these agents will reach the market place in the future.

References

S. Abe, K. Chiba, B. Cirincione, T. H. Grasela, K. Ito, and T. Suwa, *J. Clin. Pharmacol.*, 2009, **49**, 1071–1078.
M. R. Barbachyn and C. W. Ford, *Angew. Chem., Int. Ed. Engl.*, 2003, **9**, 2010–2023.

L. Beibei, C. Yun, C. Mengli, B. Nan, Y. Xuhong, and W. Rui, *Int. J. Antimicrob. Agents.*, 2010, **35**, 3–12.

L. Bernard, R. Stern, D. Lew, and P. Hoffmeyer, *Clin. Infect. Dis.*, 2003, **36**, 1197.

K. Bernardo, N. Pakulat, S. Fleer, A. Schnaith, O. Utermöhlen, O. Krut, S. Müller, and M. Kronke, *Antimicrob. Agents Chemother.*, 2004, **48**, 546–555.

M. C. Birmingham, C. R. Rayner, A. K. Meagher, S. M. Flavin, D. H. Batts, and J. J. Schentag, *Clin. Infect. Dis.*, 2003, **36**, 159–168.

A. M. Bressler, S. M. Zimmer, J. L. Gilmore, and J. Somani, *Lancet Inf. Dis.*, 2004, **4**, 528–531.

S. J. Brickner, M. R. Barbachyn, D. K. Hutchinson, and P. R. Manninen, *J. Med. Chem.*, 2008, **51**, 1981–1990.

K. Brinkman, H. J. M. ter Hofstede, D. M. Burger, J. A. Smeitink, and P. P. Koopmanns, *AIDS*, 1998, **12**, 1735–1744.

British Thoracic Society, Guidelines for the Management of Community Acquired Pneumonia in Adults: Update 2009, 2009 (http://www.brit-thoracic.org.uk/Portals/0/Clinical%20Information/Pneumonia/Guidelines/CAPGuideline-full. pdf, last accessed 9 March 2012).

K. Bush, M. Macielag, and M. Weidner-Wells, *Curr. Opin. Microbiol.*, 2004, **7**, 466–476.

A. R. Caffrey, B. J. Quilliam, and K. L. LaPlante, *Antimicrob. Agents Chemother.*, 2010, **54**, 4394–4400.

ClinicalTrials.gov identifier NCT01170221, A Randomized Study of TR-701 for the Treatment of Acute Bacterial Skin and Skin Structure Infections (ABSSSI) (http://clinicaltrials.gov/ct2/show/NCT01170221?term=torezolid&rank=1, last accessed 9 March 2012).

Clinical trials.gov identifier NCT00640926, Safety and Efficacy Study of Oxazolidinone to Treat Pneumonia (http:// clinicaltrials.gov/ct2/show/NCT00640926?term=Radezolid&rank=1, last accessed 9 March 2012).

ClinicalTrials.gov identifier NCT00646958, Safety and Efficacy Study of Oxazolidinones to Treat Uncomplicated Skin Infections (http://clinicaltrials.gov/ct2/show/NCT00646958?term=Radezolid&rank=2, last accessed 9 March 2012)

C. E. Corallo and A. E. Paull, *Med. J. Aust.*, 2002, **177**, 332.

B. Das, S. Rudra, A. Yadav, A. Ray, A. V. Rao, A. S. Srinivas, A. Soni, S. Saini, S. Shukla, M. Pandya, P. Bhateja, S. Malhotra, T. Mathur, S. K. Arora, A. Rattan, and A. Mehta, *Bioorg. Med. Chem. Lett.*, 2005, **15**, 4261–4267.

A. S. De Vriese, R. van Coster, J. Smet, S. Seneca, A. Lovering, L. L. van Haute, L. J. Vanopdenbosch, J.-J. Martin, C. Ceuterick-de Groote, S. Vandecasteele, and J. R. Boelaert, *Clin. Infect. Dis.*, 2006, **42**, 1111–1117.

A. Di Paolo, P. Malacarne, E. Guidotti, R. Danesi, and M. Del Tacca, *Clin. Pharmacokinet.*, 2010, **49**, 439–447.

M. S. Dryden, *J. Antimicrob. Chemother.*, 2011, **66** (Suppl. 4), iv7–iv15.

J. Edelsberg, C. Taneja, M. Zervos, N. Haque, C. Moore, K. Reyes, J. Spalding, J. Jiang, and G. Oster, *Emerg. Infect. Dis.*, 2009, **15**, 1516–1518.

J. Einsiedel, C. Schoerner, and P. Gmeiner, *Tetrahedron*, 2003, **59**, 3403–3407.

M. Fines and R. Leclercq, *J. Antimicrob. Chemother.*, 2000, **45**, 797–802.

C. W. Ford, J. C. Hamel, D. Stapert, J. K. Moerman, D. K. Hutchinson, M. R. Barbachyn, and G. E. Zurenko, *Trends Microbiol.*, 1997, **5**, 196–200.

C. G. Gemmell and C. W. Ford, *J. Antimicrob. Chemother.*, 2002, **50**, 665–672.

W. A. Gregory, US Patent 4705799A, 10 November 1987.

G. Grunder, Y. Zysset-Aschmann, F. Vollenweider, T. Maier, S. Krähenbühl, and J. Drewe, *Antimicrob. Agents Chemother.*, 2006, **50**, 68–72.

S. Hasham, P. Matteucci, P. R. W. Stanley, and N. B. Hart, *Brit. Med. J.*, 2005, **330**, 830–833.

P. E. Hendershot, E. J. Antal, I. R. Welshman, D. H. Batts, and N. K. Hopkins, *J. Clin. Pharmacol.*, 2001, **41**, 563–572.

D. J. Herrmann, W. J. Peppard, N. A. Ledeboer, M. L. Theesfeld, J. A. Weigelt, and B. J. Buechel, *Expert Rev. Anti-Infect. Ther.*, 2008, **6**, 825–848.

R. W. Hoffmann, Elements of Synthesis Planning, Springer, 2009.

J. P. Horcajada, R. Atienza, M. Sarasa, D. Soy, A. Adán, and J. Mensa, *J. Antimicrob. Chemother.*, 2009, **63**, 550–552.

C. Hubschwerlen, J. L. Specklin, C. Sigwalt, S. Schroeder, and H. H. Locher, *Bioorg. Med. Chem.*, 2003, **11**, 2313–2319.

D. K. Hutchinson, *Curr. Topics Med. Chem.*, 2003, **3**, 1021–1042.

D. K. Hutchinson, S. J. Brickner, M. R. Barbachyn, R. B. Gammill, and M. V. Patel, World Patent WO 93/23384, 25 November 1993.

J. A. Ippolito, Z. F. Kanyo, D. Wang, F. J. Franceschi, P. B. Moore, T. A. Steitz, and E. M. Duffy, *J. Med. Chem.*, 2008, **51**, 3353–3356.

F. Islinger, P. Dehghanyar, R. Sauermann, C. Bürger, C. Kloft, M. Müller, and C. Joukhadar, *Int. J. Antimicrob. Agents*, 2006, **27**, 108–112.

S. L. Jones, E. Athan, and D. O'Brien, *J. Antimicrob. Chemother.*, 2004, **54**, 289–290.

A. R. Katritzky, D. C. Fara, and M. Karelson, *Bioorg. Med. Chem.*, 2004, **12**, 3027–3035.

J. M. LaMarre, B. P. Howden, and A. S. Mankin, *Antimicrob. Agents Chemother.*, 2011, **55**, 2989–2991.

S. Lavery, H. Ravi, W. W. McDaniel, and Y.R. Pushkin, *Psychosomatics*, 2001, **42**, 432–434.

K. R. Lawrence, M. Adra, and P. K. Gillman, *Clin. Infect. Dis.*, 2006, **42**, 1578–1583.

L. Leach, S. M. Swaney, J. R. Colca, W. G. McDonald, J. R. Blinn, L. M. Thomasco, R. C. Gadwood, D. Shinabarger, L. Xiong, and A. S. Mankin, *Mol. Cell*, 2007, **26**, 393–402.

K. L. Leach, S. J. Brickner, M. C. Noe, and P. F. Miller, *Ann. N. Y. Acad. Sci.*, 2011, **1222**, 49–54.

E. Lee, S. Burger, and J. Shah, *Clin. Infect. Dis.*, 2003, **37**, 1389–1391.

S. Lemaire, K. Kosowska-Shick, P. C. Appelbaum, G. Verween, P. M. Tulkens, and F. Van Bambeke, *Antimicrob. Agents Chemother.*, 2010, **54**, 2549–2559.

D. M. Livermore, *J. Antimicrob. Chemother.*, 2003, **51** (Suppl. S2), ii9–ii16.

D. M. Livermore, M. Warner, S. Mushtaq, S. North, and N. Woodford, *Antimicrob. Agents Chemother.*, 2007, **51**, 1112–1114.

B. B. Lohray, S. Baskaran, B. S. Rao, B. Y. Reddy, and I. N. Rao, *Tetrahedron. Lett.*, 1999, **40**, 4855–4856.

S. Lopez, O. Miro, and E. Martinez, *Antivir. Ther.*, 2004, **9**, 47–55.

A. P. MacGowan, *J. Antimicrob. Chemother.*, 2003, **51** (Suppl. S2), ii17–ii25.

L. A. Mandell, R. G. Wunderink, A. Anzueto, J. G. Bartlett, G. D. Campbell, N. C. Dean, S. F. Dowell, T. M. File, Jr, D. M. Musher, M. S. Niederman, A. Torres, and C. G. Whitney, *Clin. Infect. Dis.*, 2007, **44**, S27–S72.

R. Manfredi and S. Sabbatani, *Braz. J. Infect. Dis.*, 2010, **14**, 96–108.

P. A. Maple, J. M. Hamilton-Miller, and W. Brumfitt, *J. Antimicrob. Chemother.*, 1989, **23**, 517–525.

S. T. Marino, D. Stachurska-Buczek, D. A. Huggins, B. M. Krywult, C. S. Sheehan, T. Nguyen, N. Choi, J. G. Parsons, P. G. Griffiths, I. W. James, A. M. Bray, J. M. White, and R. S. Boyce, *Molecules*, 2004, **9**, 405–426.

T. Mathur, M. Kumar, T. K. Barman, G. R. Kumar, V. Kalia, S. Singhal, V. S. Raj, D. J. Upadhyay, B. Das and P. K. Bhatnagar, *J. Antimicrob. Chemother.*, 2011, **66**, 1087–1095.

G. Morales, J. J. Picazo, E. Baos, F. J. Candel, A. Arribi, B. Peláez, R. Andrade, M.-A. de la Torre, J. Fereres, and M. Sánchez-García, *Clin. Infect. Dis.*, 2010, **50**, 821–825.

E. H. Moylett, S. E. Pacheco, B. A. Brown-Elliott, T. R. Perry, S. Buescher, M. C. Birmingham, J. J. Schentag, J. F. Gimbel, A. Apodoca, M. A. Schwartz, R. M. Rakita, and R. J. Wallace, *Clin. Infect. Dis.*, 2003, **36**, 313–318.

W. J. Munckhof, C. Giles, and J. D. Turnidge, *J. Antimicrob. Chemother.*, 2001, **47**, 879–883.

G. A. Noskin, F. Siddiqui, V. Stosor, D. Hacek, and L. R. Peterson, *Antimicrob. Agents Chemother.*, 1999, **43**, 2059–2062.

E. L. Nuermberger, M. K. Spigelman, and W. W. Yew, *Respirology*, 2010, **15**, 764–778.

P. Smuszkiewicz, I. Trojanowska, and H. Tomczak, *Cases J.*, 2008, **1**, 125.

B. A. Pearlman, W. R. Perrault, M. R. Barbachyn, P. R. Manninen, D. S. Toops, D. J. Houser, and T. J. Fleck, US Patent 5837870, 17 November 1998.

G. Poce, G. Zappia, G. C. Porretta, B. Botta, and M. Biava, *Expert Opin. Ther. Patents*, 2008, **18**, 97–121.

J. V. N. V. Prasad, *Curr. Opin. Microbiol.*, 2007, **10**, 454–460.

R. P. Rennie, R. N. Jones, A. H. Mutnick, and the SENTRY Program Study Group (North America), *Diagn. Microbiol. Infect. Dis.*, 2003, **45**, 287–293.

A. R. Renslo, *Expert Rev. Anti Infect. Ther.*, 2010, **8**, 565–574.

A. R. Renslo, H. Gao, P. Jaishankar, R. Venkatachalam, M. Gomez, J. Blais, M. Huband, J. V. N. V. Prasad, and M. F. Gordeev, *Bioorg. Med Chem. Lett.*, 2006a, **16**, 1126–1129.

A. R. Renslo, G. W. Luehr, and M. F. Gordeev, *Bioorg. Med. Chem.*, 2006b, **14**, 4227–4240.

R. Schaadt, D. Sweeney, D. Shinabarger, and G. Zurenko, *Antimicrob. Agents Chemother.*, 2009, **53**, 3236–3239.

J. M. Sierra, M. Ortega, C. Tarragó, C. Albet, J. Vila, J. Terencio, and A. Guglietta, *J. Antimicrob. Chemother.*, 2009, **64**, 990–992.

L. Song, X. Chen, S. Zhang, H. Zhang, P. Li, G. Luo, W. Liu, W. Duan, and W. Wang, *Org. Lett.*, 2008, **10**, 5489–5492.

A. Soriano, O. Miro, and J. Mensa, *New Engl. J. Med.*, 2005, **353**, 2305–2306.

D. L. Stevens, A. L. Bisno, H. F. Chambers, E. D. Everett, P. Dellinger, E. J. Goldstein, S. L. Gorbach, J. V. Hirschmann, E. L. Kaplan, J. G. Montoya, and J. C. Wade, *Clin. Infect. Dis.*, 2005, **41**, 1373–1406.

M. S. Taylor and E. N. Jacobsen, *Proc. Natl. Acad. Sci.*, 2004, **101**, 5368–5373.

R. C. Thomas, T.-J. Poel, M. R. Barbachyn, M. F. Gordeev, G. W. Luehr, A. Renslo, U. Singh, and V. P. V. N. Josyula, US Patent US 2007/0015801, 18 January, 2007.

M. Treviño, L. Martinez-Lamas, P. A. Romero-Jung, J. M. Giráldez, J. Alvarez-Escudero, and B. J. Regueiro, *Eur. J. Clin. Microbiol. Infect. Dis.*, 2009, **28**, 527–533.

K. Z. Vardakas, I. Kioumis, and M. E. Falagas, *Curr. Drug Metab.*, 2009, **10**, 2–12.

G. E. Vargas, M. M. Afonso, and J. A. Palenzuela, *Synlett.*, 2009, 1471–1473.

J. Weigelt, K. Itani, D. Stevens, W. Lau, M. Dryden, C. Knirsch, and the Linezolid CSSTI Study Group, *Antimicrob. Agents Chemother.*, 2005, **49**, 2260–2266.

I. R. Welshman, T. A. Sisson, G. L. Jungbluth, D. J. Stalker, and N. K. Hopkins, *Biopharm. Drug Dispos.*, 2001, **22**, 91–97.

C. L. Wigen and M. B. Goetz, *Clin. Infect. Dis.*, 2002, **34**, 1651–1652.

K. N. Williams, S. J. Brickner, C. K. Stover, T. Zhu, A. Ogden, R. Tasneen, S. Tyagi, J. H. Grosset, and E. L. Nuermberger, *Am. J. Respir. Crit. Care Med.*, 2009, **180**, 371–376.

D. N. Wilson, F. Schluenzen, J. M. Harms, A. L. Starosta, S. R. Connell, and P. Fucini, *Proc. Natl. Acad. Sci.*, 2008, **105**, 13 339–13 344.

G. E. Zurenko, B. H. Yagi, R. D. Schaadt, J. W. Allison, J. O. Kilburn, S. E. Glickman, D. K. Hutchinson, M. R. Barbachyn, and S. J. Brickner, *Antimicrob. Agents Chemother.*, 1996, **40**, 839–845.

G. E. Zurenko, J. K. Gibson, D. L. Shinabarger, P. A. Aristoff, C. W. Ford, and W. G. Tarpley, *Curr. Opin. Pharmacol.*, 2001, **1**, 470–476.

Questions

(1) How many stereoisomers are possible for erythromycin?

(2) The preparation of clarithromycin from erythromycin oxime is more complicated than might be anticipated as it requires selective protection of the hydroxyl groups at 2′ and 4″ and the indirect blocking of those at 11 and 12 (by formation of an oxime at C-9), all of which are more reactive than the 6-OH. The steps required to protect these positions and introduce the methoxy group at C-6 are shown below and all involve a nucleophilic substitution. Indicate what the electrophiles **A–C** required for each of these steps are (Brunet *et al.*, 2007):

(3) Draw structures for the inactive metabolites of erythromycin produced by an esterase and a kinase (phosphorylation).

(4) Assign the stereochemistry (*R* or *S*) of the stereogenic centres in florfenicol:

(5) There are five measurable pK$_a$ values in the physiological range for tigecycline (shown below). Generate a series of acid–base equilbria and use these to indicate what form will predominate at pH 7 and to predict its lipophilicity and oral bioavailability:

(6) The major metabolic pathway for chloramphenicol (shown below) is that of glucuronidation, forming two products: the 3-*O*-glucuronide and the 1-*O*-glucuronide (minor product, from glucuronidation on the secondary alcohol). Draw structures for both these metabolites:

Glucuronic acid

5

Agents Targeting Cell-wall Synthesis

Antibacterial agents that target cell-wall synthesis include the β-lactams, glycopeptides, cycloserine, isoniazid, and daptomycin.

5.1

β-Lactam antibiotics

Key Points

- The β-lactam class of antibiotics includes penicillins, cephalosporins, carbapenems, and monobactams; many, particularly the newer examples, exhibit a broad spectrum of activity against both Gram negative and Gram positive bacteria.
- These agents inhibit the transpeptidase enzyme, also called penicillin-binding protein (PBP), which crosslinks peptidoglycan strands to provide strength and rigidity to the bacterial cell wall.
- Widespread resistance to β-lactam antibiotics has limited their clinical use, although they remain of value for the treatment of susceptible bacterial infections.

There are many β-lactam antibiotics used clinically; these include the penicillins (usually as their sodium salts) (Table 5.1.1), the cephalosporins (Table 5.1.2), the carbapenems (Table 5.1.3), and aztreonam, a monobactam (Table 5.1.4). The structures of the β-lactam antibiotics are shown in Figures 5.1.1–5.1.4 and, as the name implies, all contain the β-lactam[1] ring functionality (highlighted in red), which is key to their antibacterial activity.

5.1.1   Discovery

The discovery of penicillin represents a pivotal moment in medical history and the sequence of events which led to it illustrates the serendipity sometimes associated with the isolation and subsequent development of a therapeutic agent.

Alexander Fleming was the Medical Director of the Inoculation Department of St Mary's Hospital in London in 1928 and was interested in the cultivation of bacteria, in particular *Haemophilus influenzae*. As you

---

[1] A lactam is a cyclic amide and, in addition to the CONH, a β-lactam contains a further two carbon atoms to give a four-membered ring.

Antibacterial Agents: Chemistry, Mode of Action, Mechanisms of Resistance and Clinical Applications, First Edition.
Rosaleen J. Anderson, Paul W. Groundwater, Adam Todd and Alan J. Worsley.
© 2012 John Wiley & Sons, Ltd. Published 2012 by John Wiley & Sons, Ltd.

Table 5.1.1   *Therapeutic indications for the penicillin antibiotics*

Penicillin antibiotic	R^1 (R^2 = Na unless stated otherwise)	Indications
Benzylpenicillin	$PhCH_2$	Bacterial meningitis, endocarditis, cellulitis
Benzathine penicillin (Benzathine benzylpenicillin)	$PhCH_2$ (R^2 = $(PhCH_2NHCH_2)_2$)	Prevention of rheumatic fever Infections susceptible to prolonged low concentrations of benzylpenicillin, e.g. latent syphilis
Procaine penicillin	$PhCH_2$	Respiratory tract infections (RTIs), syphilis, cellulitis, erysipelas
(Procaine benzylpenicillin)	(R^2 = $4\text{-}H_2NC_6H_4CO_2CH_2CH_2NEt_2$)	Intramuscular injections
Phenoxymethylpenicillin	$PhOCH_2$	Cellulitis, throat infections
Ampicillin		Urinary tract infection (UTI), endocarditis
Amoxicillin		Otitis media, endocarditis, community-acquired pneumonia (CAP), UTIs, Lyme disease, *H. pylori* eradication, sinusitis, otitis media
Amoxicillin with clavulanic acid		Hospital-acquired pneumonia (HAP)
Dicloxacillin		Uncomplicated skin and soft-tissue infections (SSTIs), osteomyelitis, endocarditis
Flucloxacillin		Uncomplicated SSTIs, osteomyelitis, endocarditis
Temocillin	See Figure 5.1.1	UTIs
Piperacillin with tazobactam		HAP, community-acquired septicaemia, hospital-acquired septicaemia
Pivmecillinam	See Figure 5.1.1	UTIs
Ticarcillin with clavulanic acid	See Figure 5.1.1	HAP, community-acquired septicaemia, hospital-acquired septicaemia

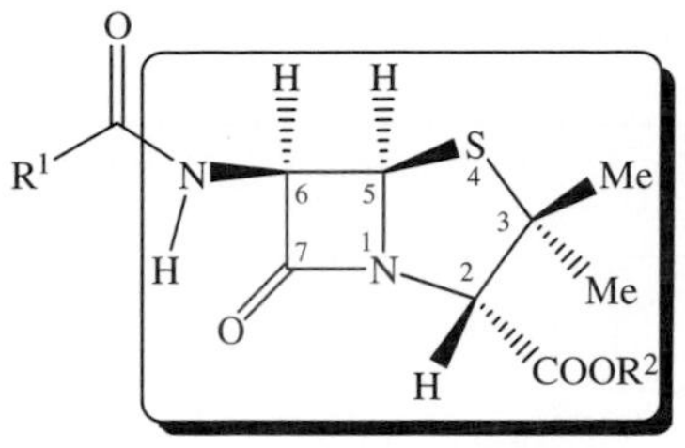

General penicillin structure

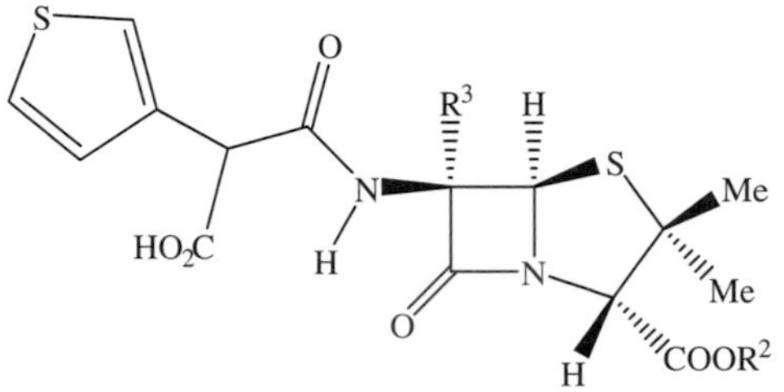

Ticarcillin (R^3 = H, R^2 = Na)
Temocillin (R^3 = OMe, R^2 = Na)

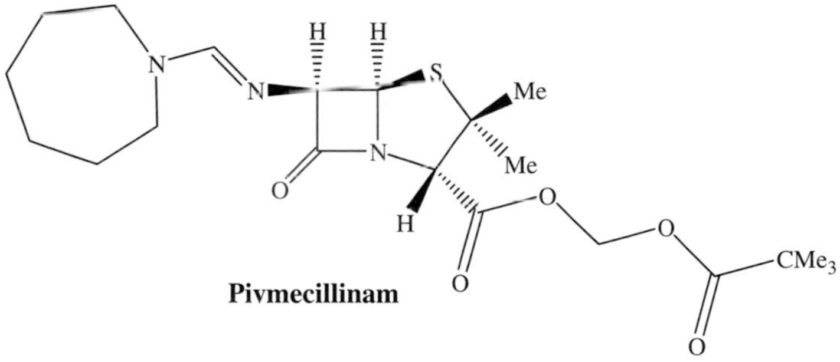

Pivmecillinam

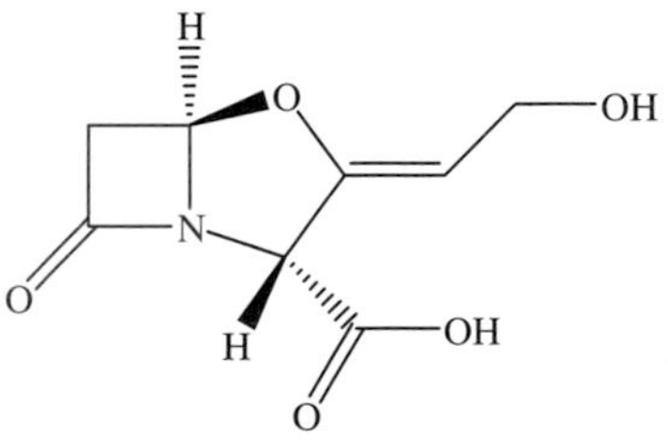

Clavulanic acid

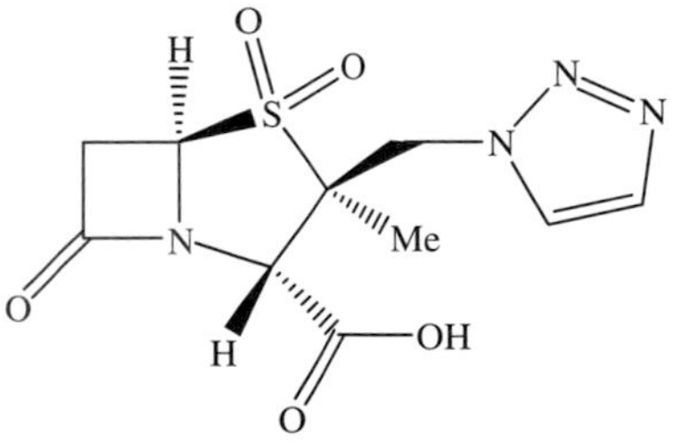

Tazobactam

Figure 5.1.1 *General structure of the penicillin antibiotics and some β-lactamase inhibitors. The β-lactam ring (shown in red in the penicillin general structure) and all other substituents within the box, with the absolute stereochemistry indicated, are essential for antibacterial activity*

Table 5.1.2   *Therapeutic indications for the cephalosporin antibiotics (colours represent generations:* first, second, third, *and* fourth*)*

Cephalosporin antibiotic	R^1	R^2	Indications
Cefalexin (Cephalexin)	Me	(phenyl with R, H, NH_2)	Acute UTI, prophylaxis of recurrent UTI
Cefalotin	CH_2–O–C(=O)–Me	(thiophene–CH_2)	Staphylococcal and streptococcal infections in patients with mild-to-moderate penicillin allergy Surgical prophylaxis UTIs due to susceptible Gram-negative bacteria
Cefazoline	CH_2–S–(thiadiazole)–Me	(tetrazole–CH_2)	As for cefalotin
Cefepime	CH_2–N$^{\oplus}$(pyrrolidine)–Me	(aminothiazole–C with =N–OMe)	*Pseudomonas aeruginosa* infections Empirical treatment of sepsis in neutropenic or otherwise immunosuppressed patients
Cefoxitin	See Figure 5.1.2	See Figure 5.1.2	Surgical prophylaxis Mixed anaerobic infections
Cefradine	Me	(cyclohexadienyl with R, H, NH_2)	UTIs
Cefadroxil	Me	(HO-phenyl with R, H, NH_2)	UTIs
Cefaclor	Cl	(phenyl with R, H, NH_2)	Uncomplicated SSTIs, acute bronchitis
Cefuroxime	CH_2–O–C(=O)–NH_2	(furan–C with =N–OMe)	CAP, HAP, community-acquired septicaemia, Lyme disease, osteomyelitis

Table 5.1.2   (*Continued*)

Cephalosporin antibiotic	R¹	R²	Indications
Cefotaxime			Salmonella, typhoid fever, epiglottitis, CAP, meningitis, HAP
Ceftazidime			HAP, hospital-acquired septicaemia
Ceftriaxone			Endocarditis, epiglottitis, meningitis, syphilis, gonorrhoea, HAP
Cefixime			Gonorrhoea
Cefpodoxime proxetil	See Figure 5.1.2	See Figure 5.1.2	UTIs, sinusitis, SSTIs

are probably already aware, several factors conspired to result in Fleming's discovery of the antibacterial activity of penicillin:

- Moulds of the relatively rare *Penicillium notatum* were being cultivated (for the development of vaccines) in the department on the floor below Fleming's laboratory and this probably resulted in the contamination of the laboratory air, and so the Petri dishes in Fleming's laboratory, by these spores.
- The contaminated Petri dishes were not incubated, which would normally promote the growth of bacteria and not fungi, and this allowed the *Penicillium* mould to grow preferentially.
- These events occurred during a particularly cool July, which also allowed the preferential growth of the *Penicillium* mould. A subsequent rise in climatic temperature at the end of the UK summer facilitated the

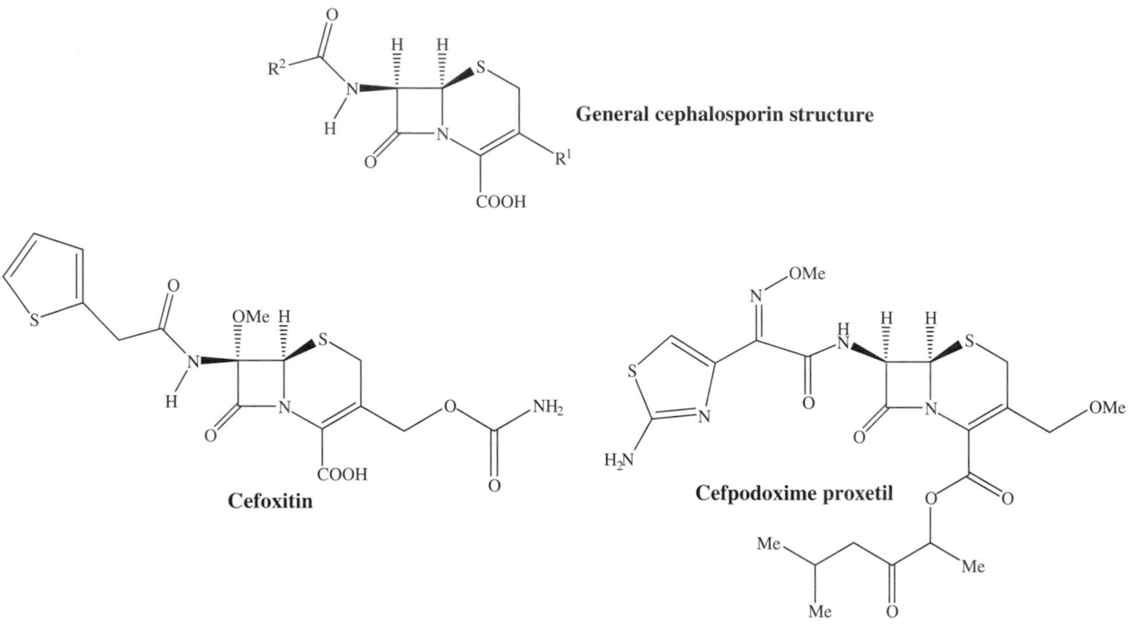

Figure 5.1.2 *Structures of the cephalosporin antibiotics, including a prodrug form*

re-establishment of *Staphylococcus* colonies, but the penicillin mould (not normally developed on such plates) was sufficiently large to demonstrate a zone of staphylococcal growth inhibition (Hare, 1970).

This sequence of events resulted in the growth of *Staphylococcus aureus* being inhibited by *Penicillium notatum*, the mould that produced significant amounts of penicillin. Further studies then demonstrated that penicillin was useful in the eradication of *Haemophilus influenzae* strains, particularly those derived from throat swabs (Fleming, 1929).

Initial attempts to isolate penicillin were performed by Raistrick and Clutterbuck, in the School of Hygiene and Tropical Medicine in London. An ether extract of the mould lacked any antibacterial activity, but Holt deduced that penicillin could be extracted into organic solvents from slightly acidic solutions, suggesting that penicillin was probably a weak acid.

In 1938, Ernst Chain and Howard Florey and their colleagues investigated the effects of penicillin on bacteria, showing that it caused cell-wall lysis. The fact that penicillin could be extracted into organic solvents demonstrated that penicillin was a small molecule and probably not an enzyme. Subsequent funding by the Rockefeller Foundation ensured that penicillin could be produced in sufficient quantities for further scientific investigation, allowing Chain to isolate approximately 100 mg of penicillin in 1940.

With modest amounts of sufficiently pure penicillin available, study of its antibiotic effects against streptococci, staphylococci, and clostridium bacteria in infected mice demonstrated its efficacy: mice injected with several small doses of penicillin survived, while untreated mice and those given a single dose did not (Chain *et al.*, 1940). Tests in human volunteers rapidly followed, confirming that oral administration destroyed the pencillin structure and antibiotic effect, while injection or infusion was effective in achieving detectable amounts of active drug systemically. In 1941, at a hospital in Oxford, England, the first patient with a serious infection, due to both staphylococci and streptococci, was recruited to the trial and the efficacy of penicillin in man was demonstrated; at this time, despite the attempts to increase the quantity of penicillin for treatment by using bedpans as larger culture vessels, production struggled to provide sufficient amounts, so renally excreted

Table 5.1.3 *Therapeutic indications for the carbapenem antibiotics*

Carbapenem	R^1	R^2	Indications
Imipenem	H	CH$_2$CH$_2$N=CHNH$_2$	Hospital-acquired septicaemia, osteomyelitis
Meropenem	Me	pyrrolidine-CONMe$_2$	Hospital-acquired septicaemia, osteomyelitis
Doripenem	Me	pyrrolidine-NHSO$_2$NH$_2$	HAP, complicated UTIs
Ertapenem	Me	pyrrolidine-CONH-C$_6$H$_4$-COOH	CAP, SSTIs

unchanged penicillin was recovered from the urine for re-use (Fletcher, 1984). Fleming, Chain, and Florey shared the Nobel Prize for Physiology or Medicine in 1945 'for the discovery of penicillin and its curative effect in various infectious diseases'; the award of the Prize so soon after the discovery of penicillin and its antibiotic effects testifies to the monumental importance of this work.

The large-scale production of penicillin continued to develop during the Second World War with the assistance of the USA, particularly the Rockefeller Foundation. Assisted by the Department of Agriculture's Northern Regional Research Laboratory in Illinois, Florey was able to substantially increase penicillin production with the use of deep fermentation tanks, the addition of corn steep liquor to the fermentation broth, and the minimisation of acid degradation by controlling pH. Commercial production of penicillin was developed sequentially by Merck, Squibb, Pfizer, Abbott Laboratories, Eli Lilly, Parke Davis, Upjohn, and Wyeth laboratories, eventually leading to the large-scale national production of penicillin.

5.1.1.1 Structure elucidation

In 1943, the British and US governments, appreciating the clinical importance of penicillin, especially in the treatment of battlefield wounds, imposed a ban on the publication of all chemical research on penicillin.

Table 5.1.4 *Therapeutic indications for the monobactam aztreonam*

Monobactam	Indications
Aztreonam	Chronic pulmonary *P. aeruginosa* infection in patients with cystic fibrosis (delivered intravenously or by inhalation), UTIs, gonorrhoea

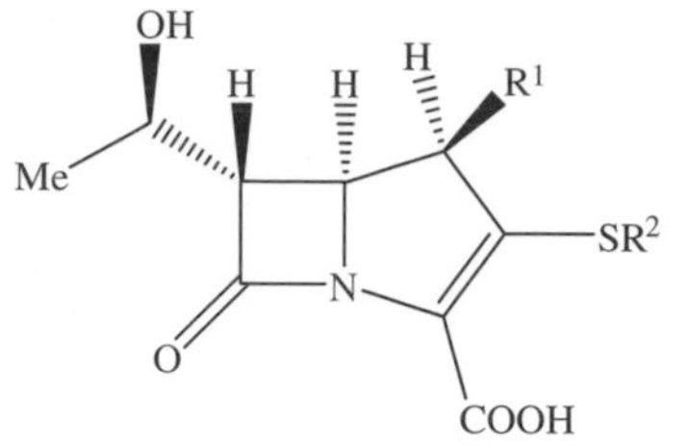

Figure 5.1.3   *General structure of the carbapenem antibiotics*

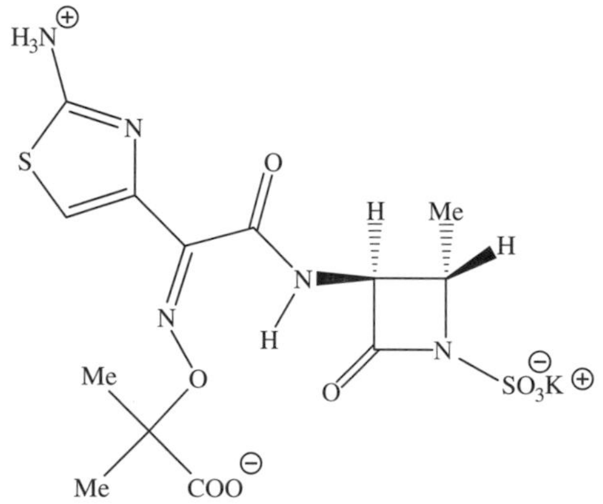

Figure 5.1.4   *Structure of the monobactam aztreonam*

Negotiations between the two governments aimed to ensure the complete and confidential exchange of information between the various research groups on both sides of the Atlantic. Much of the work performed in the early stages of penicillin discovery and development was not published at the time, but was told later through personal histories.

The initial attempts at the elucidation of the structure of penicillin were performed by Robert Robinson at the Dyson Perrins laboratory in Oxford, and by 1942 his team had combined their efforts with those of Florey. The low purity of the penicillin initially obtained (approximately 50%) severely hindered the structure determination. An X-ray crystallographic study of penicillin by Squibb Laboratories (USA) identified a structural difference from the penicillin which had been obtained in the UK, and this was later shown to be due to differences in the amide side chains (at position 6) of the two samples of penicillin. The UK sample was 2-pentenylpenicillin (latterly named penicillin F; R^1 = MeCH$_2$CH=CHCH$_2$ in the general structure in Figure 5.1.1), while the US sample was penicillin G, or benzylpenicillin. In addition to penicillins F and G, five more side-chain variants were identified in the same year, 1943. Two possible structures were proposed for penicillin: a five-membered oxazolone ring joined to a thiazolidine ring, as reported in the famous Oxford report PEN103 (Abraham *et al.*, 1943) or (correctly) a β-lactam ring fused to the thiazolidine ring (Figure 5.1.5). With the assistance of Dorothy Hodgkin[2] and X-ray crystallography, the whole structure

---

[2] Dorothy Crowfoot Hodgkin, an X-ray crystallographer at Oxford University, is particularly famous for her work on the structures of penicillin and vitamin B$_{12}$; she was awarded the Nobel Prize for Chemistry in 1964 'for her determinations by X-ray techniques of the structures of important biochemical substances'.

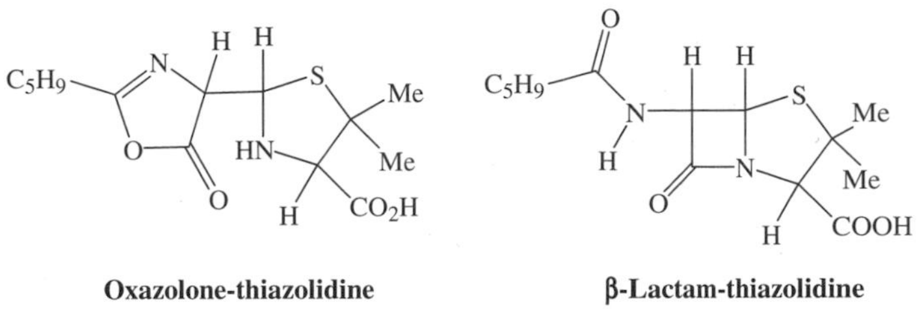

Oxazolone-thiazolidine   β-Lactam-thiazolidine

Figure 5.1.5 *Structures originally proposed for penicillin*

was eventually established in 1945 and the presence of the four-membered highly labile β-lactam ring was confirmed.

A vast number of semi-synthetic penicillin antibiotics have been made (outlined in Subsection 5.1.2), which show a range of activities and efficacies; Table 5.1.1 shows the small number still remaining in clinical use. These agents have been limited by resistance to their action as inhibitors of cell-wall synthesis; we will look at this problem in more detail in Subsection 5.1.5.

5.1.1.2 Cephalosporins

Encouraged by the discovery and success of penicillin and other antibiotics discovered from microbial sources, Guiseppe Brotzu, who was Professor of Hygiene at the University of Cagliari in Italy, isolated *Cephalosporium acremonium*[3] from Sardinian sewers in 1945 and showed that this mould inhibited the growth of typhoid bacilli on agar plates. Brotzu had limited resources to continue his studies, but instead sent a culture of the mould, along with a copy of his notes, to Oxford, where Chain and Florey continued the work with the isolation of two antibiotic agents, one of which was identified as a new type of penicillin, while the other was a different kind of agent, a 'cephalosporin'. A few years later, Edward Abraham and Guy Newton, in the same Oxford laboratory, discovered another cephalosporin from *C. acremonium*, naming it cephalosporin C. By this time, resistance to penicillin had begun to limit its clinical use in the treatment of staphylococcal infections, but cephalosporin C showed a broad spectrum of antibiotic activity and was effective in protecting mice from penicillin-resistant staphylococcal infections, raising interest in this new class of antibiotics (Abraham and Newton, 1956). The structure of the cephalosporins was again confirmed by Dorothy Hodgkin using X-ray crystallography and was found to include the same β-lactam ring, in this case fused to a dihydrothiazine ring (Hodgkin and Maslen, 1961). A huge number of semi-synthetic cephalosporins have been made, each with their own spectrum of activity; referring back to Table 5.1.2, you can see that their names start with 'ceph' or 'cef' and they can be classified into 'generations':

- **First generation** Vary in their spectrum of activity, but are all inactivated by β-lactamase, a class of enzyme found in resistant bacteria and discussed in depth in Subsection 5.1.5.
- **Second generation** More stable to β-lactamases and have greater activity across Gram negative bacteria. Some can penetrate into the CNS, so are of value in treating meningitis.
- **Third generation** Greater β-lactamase stability, a wider spectrum of activity, the ability to cross the meninges, and may be used for surgical prophylaxis.
- **Fourth generation** Resistant to a wide range of β-lactamases, excellent activity against Gram positive and Gram negative bacteria, including *Pseudomonas aeruginosa* and bacteria resistant to other classes of antibacterial agent.

---

[3] This mould was later renamed *Acremonium chrysogenum*.

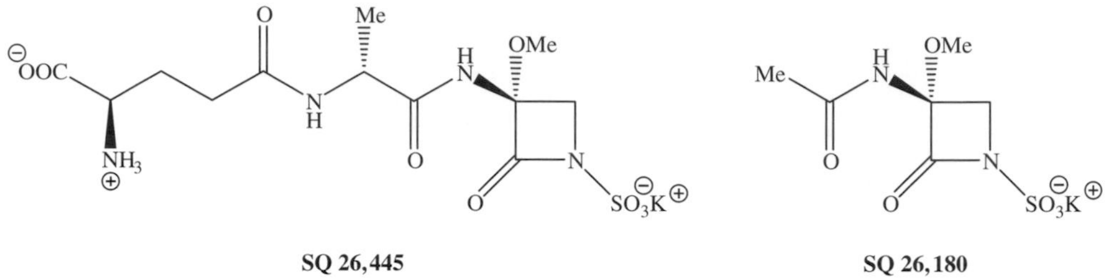

SQ 26,445   SQ 26,180

Figure 5.1.6   *Monobactams: SQ 26 445 (from a strain of* Gluconobacter*) and SQ 26 180 (from* Chromobacterium violaceum*) (Sykes et al., 1981)*

5.1.1.3   Carbapenems

The carbapenem thienamycin was discovered much later by the Merck laboratories, with its initial isolation from *Streptomyces cattleya* in 1976 (Kahan *et al.*, 1979). As you will have spotted from Figures 5.1.1–5.1.3, the major difference between the penicillins, cephalosporins, and carbapenems is the position of the sulfur atom and the unsaturation between C2 and C3. Although other naturally occurring carbapenems have been discovered, thienamycin is the most potent, with high antibiotic activity over a broad spectrum of aerobic and anerobic bacteria, both Gram positive and Gram negative, including pencillin-resistant species. It is, however, unstable to a wide range of chemical conditions, contributing to low yields from *S. cattleya* fermentation, despite considerable work to optimise the conditions. Successful synthetic carbapenems, with wide antibacterial activity, have since been developed and are discussed in Subsection 5.1.2.

5.1.1.4   Monobactams

As we have already seen, the discovery of the penicillins sparked enormous interest in the discovery of other antibiotics. In 1978, to facilitate the testing of large numbers of compounds, the Squibb Institute for Medical Research developed a high-throughput screening programme, which involved a strain of *Bacillus lichenformis* engineered to be very sensitive to β-lactam antibiotics, to search for new β-lactam agents. Three years later, over 1 million bacterial isolates had been screened and seven monocyclic β-lactams, given the family name **monobactams**, had been isolated and identified (Figure 5.1.6) (Sykes *et al.*, 1981). These natural monobactams, including SQ 26 445 and SQ 26 180, have only limited antibacterial properties, but structure–activity studies, requiring the synthesis of many chemically modified monbactams, led to aztreonam: the only monobactam used clinically.

5.1.2   Synthesis

5.1.2.1   Penicillins

The presence of a very labile ring and three stereogenic centres (with only the diastereoisomer shown in Figure 5.1.1 having biological activity) means that the penicillins present a significant synthetic challenge. The first total synthesis to give a penicillin (in this case penicillin V) in any significant quantity was achieved by Sheehan and Henerey-Logan in 1959 (having started this work in 1948!) (Scheme 5.1.1) (Sheehan and Henerey-Logan, 1959). Some notable features of this synthesis are:

- The formation of the thiazolidine ring, from the coupling of D-penicillamine and the malonaldehyde half-ester, gave only two of the four possible diastereoisomers (these are epimers at the 1′-position).

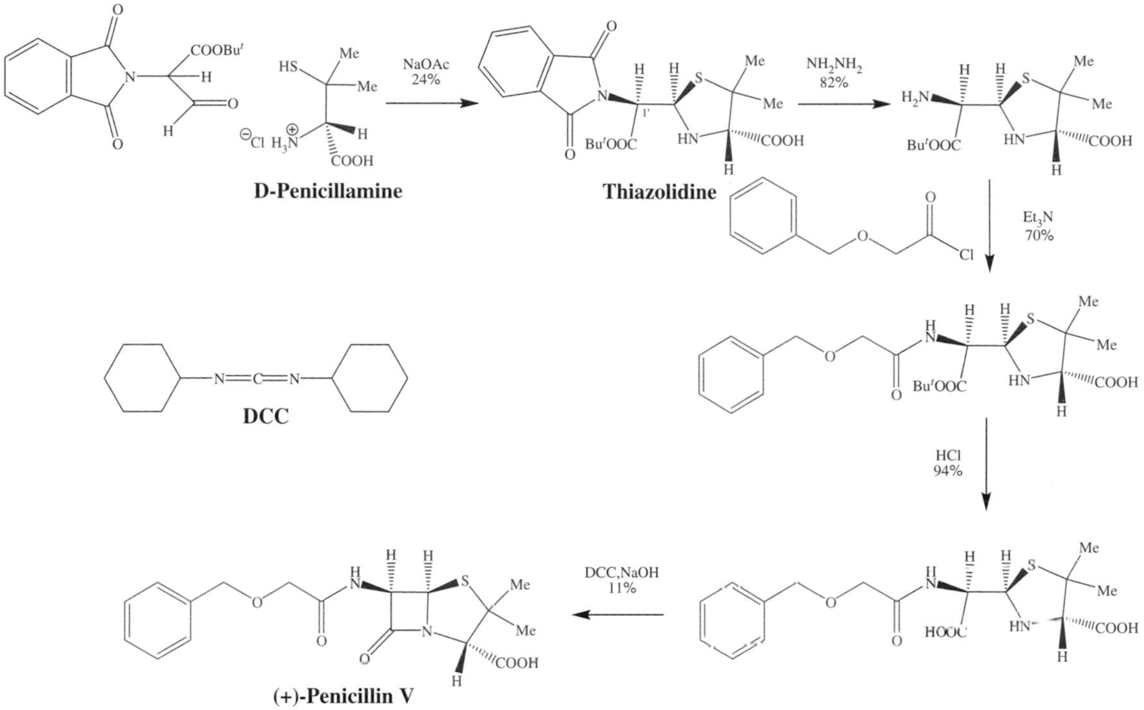

Scheme 5.1.1 *Synthesis of penicillin V (Sheehan and Henerey-Logan, 1959)*

- Deprotection of the amino group, followed by amide bond formation and hydrolysis of the *tertiary*-butyl ester, gave the precursor to the fused ring system.
- Ring closure to the β-lactam was accomplished with dicyclohexylcarbodiimide (DCC), a reagent commonly used in amide bond formation.

Such total syntheses of penicillins were never going to be able to provide sufficient quantities for clinical use (the overall yield for the steps shown in Scheme 5.1.1 is 1.4%), so a combination of fermentation and organic synthesis was quickly employed to produce a wide range of penicillins. The β-lactam ring is essential for the antibacterial activity of these agents, while structure–activity relationships have shown that any variations in the groups attached to the fused β-lactam-thiazolidine ring structure or the stereochemistry at the 3-, 5-, or 6-positions (highlighted in the box in Figure 5.1.1), leads to a decrease in antibacterial activity. As a result, as you can see from Table 5.1.1, the penicillins usually differ only in the nature of the amide side chain (R^1) at position 6. A useful synthetic precursor is, therefore, 6-aminopenicillanic acid, from which a whole range of **semi-synthetic** penicillins are available by acylation of the amino group (Scheme 5.1.2). An important finding in the early days of penicillin production was that this key intermediate can be obtained by fermentation in the absence of a monosubstituted carboxylic acid (R^1COOH) or through the enzyme-catalysed hydrolysis of the 6-amide group of benzylpenicillin (penicillin G) by an acylase (amide hydrolase) (Scheme 5.1.2). The relatively straightforward synthetic procedure of acylation (addition of R^1CO) was then used to obtain a range of penicillins.

It's reasonable to ask here why such a range of penicillins is required, as surely their very similar structures will mean that their properties, including their antibacterial spectrum, as well as any bacterial resistance to

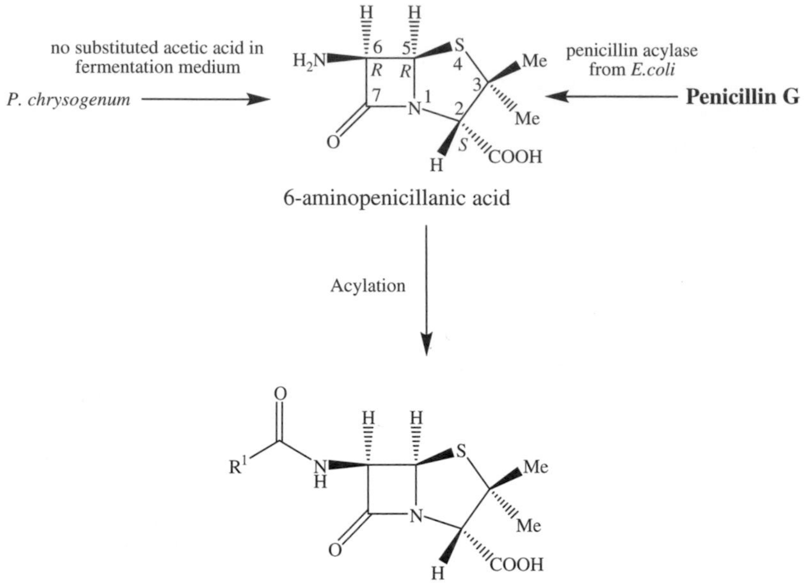

Scheme 5.1.2   *Semi-synthetic production of penicillins*

them, will be almost identical. This is not actually the case and there are a number of reasons why the availability of a range of penicillins is desirable, including:

- **Having a number of possible routes of administration** Penicillins such as dicloxacillin, flucloxacillin, phenoxymethylpenicillin, and amoxicillin have increased acid stability, which makes them most suited for oral administration (we will look at the effect of acid, such as that encountered in the stomach, when we look at bioavailability in Subsection 5.1.3). Other penicillins, such as the highly aqueous soluble piperacillin and ticarcillin, have been developed for injection alongside β-lactamase inhibitors to allow rapid treatment of life-threatening infections, such as septicaemia and peritonitis.
- **Enhancing bioavailability and decreasing side effects** Prodrug esters, such as pivmecillinam, can be employed to enhance the absorption of drugs which have poor oral bioavailability due to their zwitterionic (doubly charged) nature, which results in them being poorly absorbed from the stomach. We will also consider this further in Subsection 5.1.3.
- **Having agents with different antibacterial spectra** Variation of the C-6 amide R^1 group can be used to obtain penicillins with a different antibacterial spectrum; for example, piperacillin has an extended spectrum of activity (including activity against Gram positive and negative organisms, aerobic and anaerobic), while ticarcillin is active only against Gram negative organisms and can be used in the treatment of infections due to *P. aeruginosa*.

5.1.2.2   Cephalosporins

What really makes the β-lactams exceptional as antibacterial agents (aside from the history of their discovery) is the number of them which are in clinical use. Like the penicillins, there are a number of cephalosporins in clinical use, each of which has its own spectrum of antibacterial activity and physicochemical properties. Once again, a total synthesis would not enable sufficient quantities of the individual antibiotics to be prepared, so these agents are also semi-synthetic and can be obtained from 7-aminocephalosporanic acid (7-ACA). In this case, however, the amino precursor to the cephalosporins cannot be hydrolysed by an acylase, or obtained via fermentation, and so chemical degradation of the amide side chain of the fermentation product, cephalosporin C, is employed (Scheme 5.1.3) (Campos Muñiz *et al.*, 2007).

This is not quite as easy as it might seem, as the chemical hydrolysis of the side-chain secondary amide must be accomplished in the presence of a highly strained cyclic tertiary amide and a chemically labile ester group. Harsh methods, such as those involving acid or base reflux, cannot be used and the hydrolysis therefore has to take advantage of a unique property of the secondary amide. The chlorination of amides with inorganic chlorides, such as PCl_3, PCl_5, or $SOCl_2$, gives rise to a very reactive class of compounds known as imidoyl chlorides, in which a reactive $Cl-C=N$ group replaces the amide bond. Formation of an imidoyl chloride is

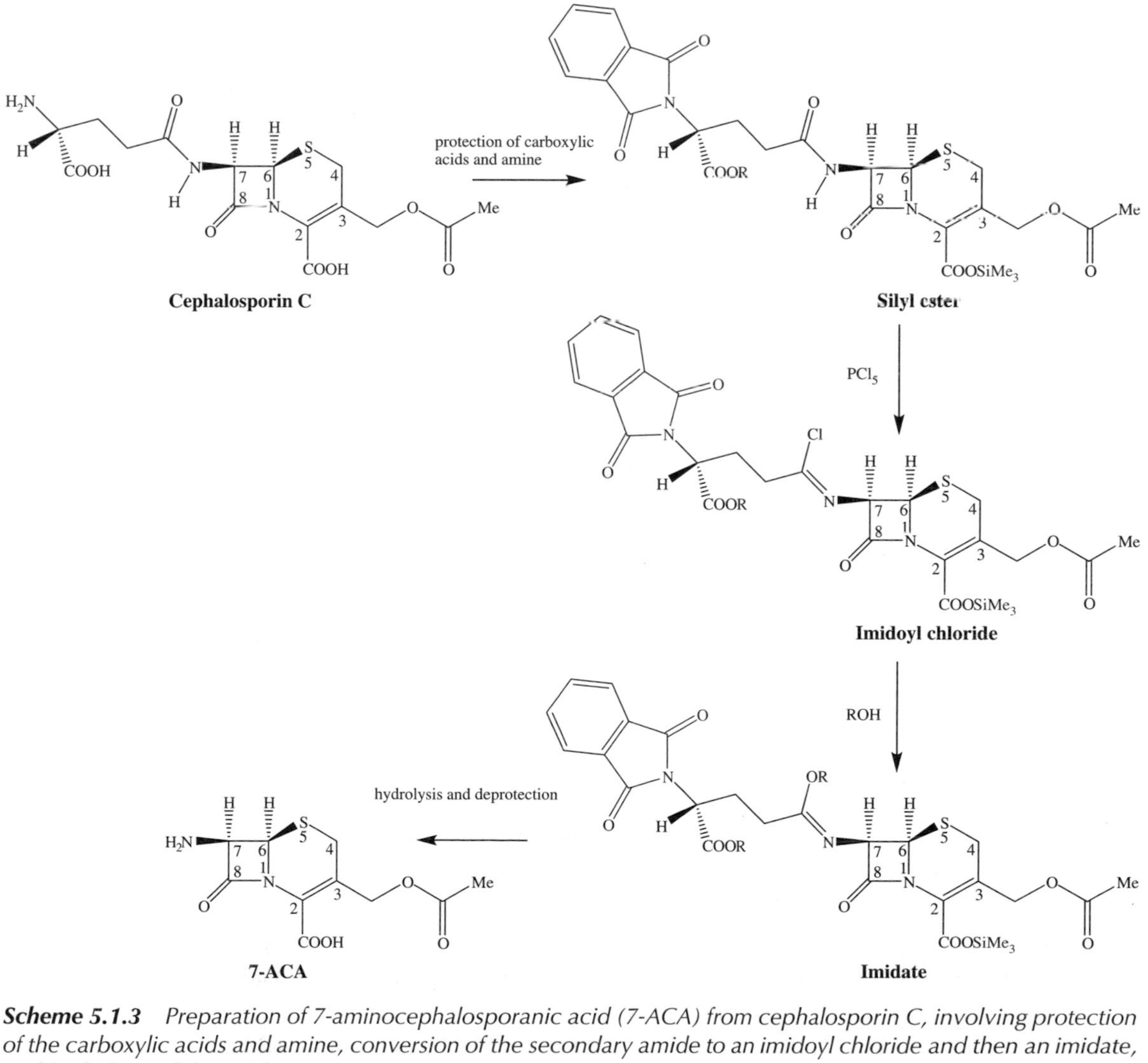

Scheme 5.1.3 *Preparation of 7-aminocephalosporanic acid (7-ACA) from cephalosporin C, involving protection of the carboxylic acids and amine, conversion of the secondary amide to an imidoyl chloride and then an imidate, and hydrolysis of the imidate*

possible for the side-chain secondary amide, but not for the β-lactam, and this is the basis for the chemoselectivity[4] of this process. The imidoyl chloride is then reacted with an alcohol to form an imidate, which is readily hydrolysed to an amine (in this case, 7-ACA) and an ester (Scheme 5.1.3). In a similar manner to 6-APA, 7-ACA can then be acylated to give the first- to fourth-generation cephalosporins. The possibility of using enzymes to carry out this conversion has also been investigated (Parmar *et al.*, 1998).

5.1.2.3   Carbapenems

As you can see from Table 5.1.3, there are only four carbapenems, but, like the penicillins and cephalosporins, each has an individual spectrum of activity and properties, making it distinct from the rest of the group. The carbapenems have remained active against a wide range of β-lactam-resistant bacteria and, except for ertapenem, are active against the notoriously treatment-resistant *P. aeruginosa*, resulting in an important position for these members of the antibiotic armoury (although recent developments are threatening this position: read on to Subsection 5.1.5). We met the natural carbapenem thienamycin in Subsection 5.1.1.3; although highly potent and broad-spectrum, its instability prevents its clinical use. One major limiting problem is due to ring-opening of the β-lactam ring of thienamycin by the nucleophilic side-chain amino group of another thienamycin molecule. Except for thienamycin itself, there is no naturally available intermediate for the semi-synthesis of carbapenems, and as a consequence, despite the structural complexity and the functional group sensitivity, the production of thienamycin analogues has been pursued via synthetic chemistry, perhaps explaining why there are so few of these β-lactam antibiotics available. Imipenem, the first synthetic clinical carbapenem, was prepared from thienamycin (refer to Table 5.1.3 and Figure 5.1.3 for the structures); it seems remarkable that the minor structural change between thienamycin and imipenem should make such a difference, yet imipenem is significantly more stable with many clinical applications. The basicity of the amidine group, which is protonated at physiological pH, results in the loss of the side chain's nucleophilic properties and greater stability of the β-lactam ring (Kesado *et al.*, 1980). Unfortunately, imipenem was found to cause nephrotoxicity and must be co-administered with an inhibitor (you can read more about this in Subsection 5.1.6); in the later carbapenems – meropenem, ertapenem, and doripenem – a methyl group at C1 essentially eliminates this problem. Most syntheses of carbapenems rely on the intramolecular condensation of monocyclic β-lactam **1**, in which the α-keto ester group reacts with the β-lactam nitrogen atom to give the key intermediate **2** (Scheme 5.1.4).

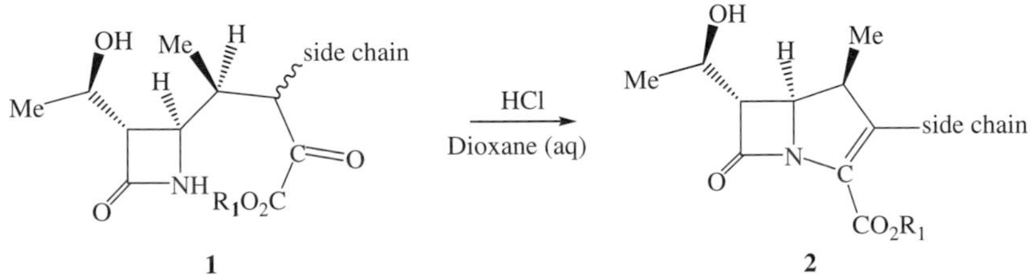

Scheme 5.1.4   *Key intramolecular condensation step in the synthesis of carbapenems (Prashad et al., 1998)*

---

[4] Chemoselectivity is the preferential reaction of a reagent with one of a number of functional groups.

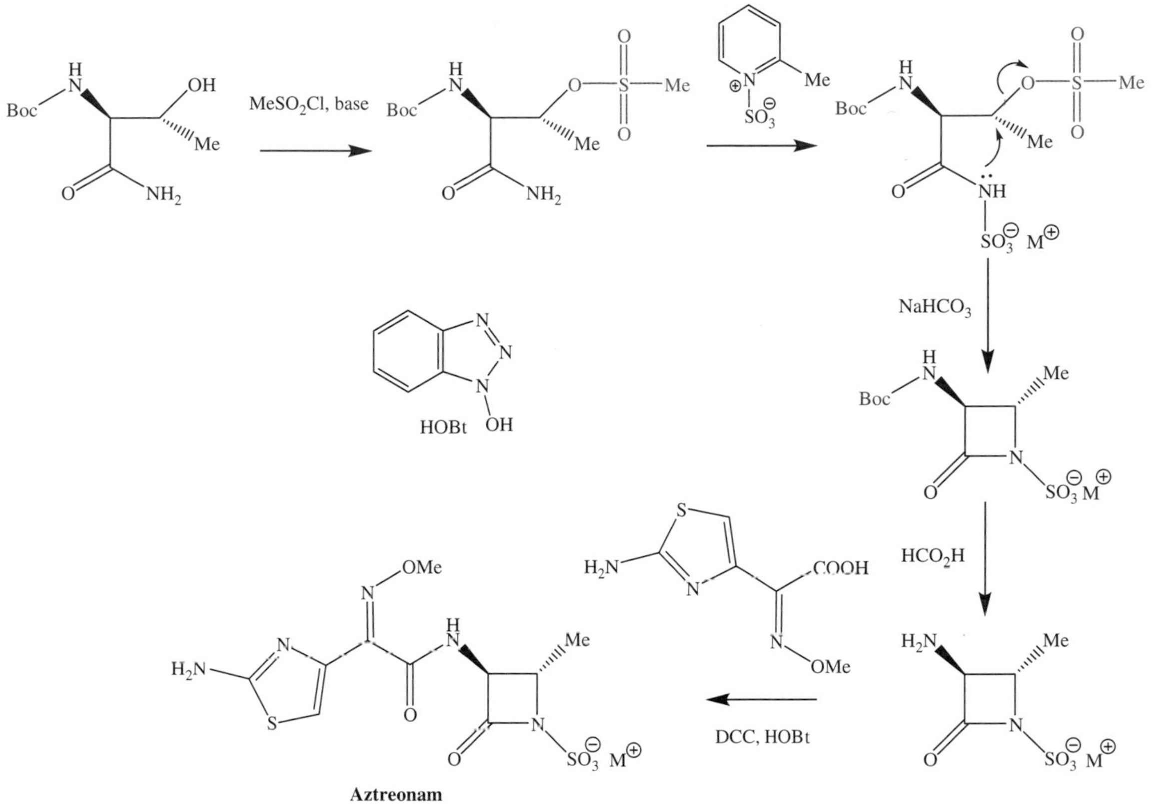

Scheme 5.1.5   *Squibb synthesis of aztreonam (Cimarusti et al., 1983)*

5.1.2.4   Monobactams

No intermediates are available for the semi-synthesis of aztreonam, and total synthesis is the only option. Despite appearing to be relatively simple, aztreonam has an unusual functional group (an acyl sulfamate) and two stereogenic centres, and so even a relatively simple synthetic route, such as that by the Squibb group, has five steps (Scheme 5.1.5) (Cimarusti *et al.*, 1983). You'll notice that this route again requires a protecting group and a mesylate group (MeSO$_3$), which acts as the leaving group in the cyclisation to the azetidinone. The final step involves fairly routine amide-bond-forming conditions, with DCC (see Scheme 5.1.1) and 1-hydroxybenzotriazole (HOBt).

5.1.3   Bioavailability

The β-lactam antibiotics are time-dependent agents, which means that their clinical success is influenced predominantly by the length of time for which the drug concentration is maintained above the minimum inhibitory concentration (MIC) at the site of infection (T > MIC) (MacGowan, 2011). The penicillins and cephalosporins have a limited post-antibiotic effect (PAE) on Gram negative bacteria, but a longer effect of 1.2–7.1 hours on Gram positive bacteria and *H. influenzae* (MacGregor and Graziani, 1997). The carbapenems, by contrast, exert a PAE of several hours against both Gram positive and Gram negative bacteria

(Mouton *et al.*, 2000). Whatever the β-lactam agent and infecting organism, though, the T > MIC parameter is the major consideration and it is believed that maintaining the antibiotic concentration above the MIC for >60% of the dosing interval is optimal for therapeutic success in the clinic (Turnidge and Paterson, 2007).

The carboxylic acid group that is integral to the β-lactam antibiotics allows their formulation as a salt, most often as the sodium salt, suitable for infusion or injection, which is a positive feature of these agents. In fact, some β-lactam antibiotics (e.g. piperacillin, ticarcillin) have been developed specifically for IV and intramuscular (IM) administration. Oral administration, however, offers a wider flexibility of application, is usually more economically favourable, and is generally more acceptable to patients. Oral administration of many of this class of agents (especially the penicillins, e.g. benzylpenicillin and meticillin) is limited by their acid-sensitivity – they are inactivated in the stomach by acidic hydrolysis, which opens up the β-lactam ring and renders them inactive. The extent of inactivation is sufficient to reduce the bioavailability significantly, and other penicillins were developed in order to overcome this problem and facilitate oral use. Consider the functional groups present in even a simple example β-lactam: benzylpenicillin contains carboxylic acid, thioether, and amide groups, none of which are renowned for being particularly sensitive to acid, so what is it about this molecule that makes it prone to acid hydrolysis? Well, the relatively unusual four-membered lactam ring should grab your attention, as it this which introduces a high degree of strain to the molecule:

- First, you will remember that the normal bond angle for an sp^2 hydridised carbon (as found in a carbonyl group) is around 120°, yet in the β-lactam ring it is close to 90°; even the sp^3 hydridised carbon atoms in this ring have a reduced bond angle (from 109° to 90°). We can conclude that this ring **amide carbonyl** suffers from **angle strain**.
- Second, as a four-membered ring has little flexibility, the substituents on each ring atom are partly eclipsed,[5] which adds **torsional strain** to this system.
- Third, in linear molecules, amide carbonyls are stabilised by resonance with the N lone pair (which is part of the reason for the usual stability of amide bonds[6]), but this cannot happen effectively in this cyclic amide as the fused thiazolidine ring restricts the possible conformations of this molecule, such that it adopts a wide V-shape and the N is not able to delocalise its electrons into the carbonyl π-orbital.[7] Flattening the bicyclic structure to allow resonance to occur would form an impossibly strained system (Figure 5.1.7). Additional **steric strain** caused by the bicyclic system leads to increased chemical instability of the carbonyl group.
- One more structural aspect adds to the lack of stability to acid: the side-chain amide group attached to the β–lactam ring. This group can attack the ring amide carbonyl C, causing opening of the four-membered ring

---

[5] 'Eclipsed' is the term used to indicate that the bonds on a carbon atom lie directly behind those of an adjacent atom, leading to the maximum repulsion between the electron clouds of these bonds. The more energetically favourable conformation, in which the bonds lie inbeteen those of the adjacent atom, is said to be staggered.

eclipsed   staggered

[6] The amide bond is often called the peptide bond – this is the same strong, stable bond that joins amino acids together and provides a stable backbone with restricted conformational freedom, which helps to define the structure and function of peptides and proteins.

[7] Remember that delocalised systems have to be planar to allow overlap of the π- and/or p-orbitals involved.

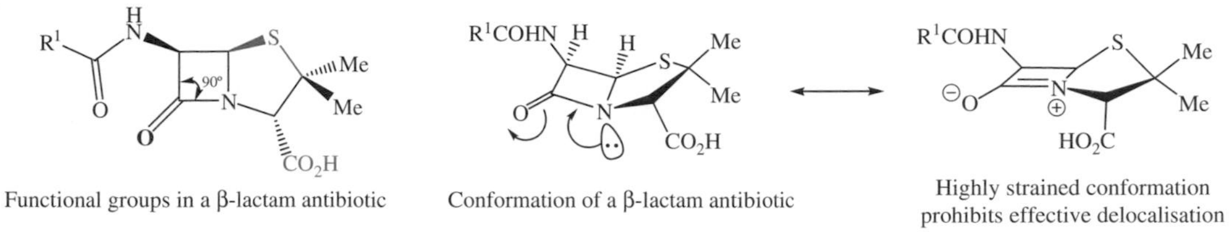

Figure 5.1.7 *Structural features of β-lactam antibiotics responsible for their inherent reactivity*

and the eventual formation of two products: penillic acids and penicillenic acids (Scheme 5.1.6). The reaction is promoted under acidic conditions and is the major reason for the decomposition of some penicillins in the stomach after oral administration.

The first three of these structural features are inherent to all penicillins, so cannot be changed without altering the potency (usually with a drastic reduction in or total loss of activity), and are of lesser importance to acid stability anyway. The contribution from the final, more major, problem can be reduced by forming penicillin analogues with an electron-withdrawing group on the amide side chain at C-6; the attack by this group on the ring carbonyl is then suppressed and these β-lactams are sufficiently stable to acidic conditions for oral administration. Phenoxymethylpenicillin (penicillin V) (Figure 5.1.8) has a more acid-stable structure than benzylpenicillin – you can see that they only differ by an O atom as part of the R^1 side chain, which exerts enough electron-withdrawing effect on the side-chain amide carbonyl of penicillin V to inhibit the cyclisation on to the ring carbonyl group (Scheme 5.1.6). This increases the half-life for stability in acid from about 30 minutes for benzylpenicillin to more than 5 hours for phenoxymethylpenicillin (Barza and Weinstein, 1976), thus allowing more than enough time for it to be absorbed from the gastrointestinal (GI) tract. Although penicillin V has a lower potency than pencillin G, its greater bioavailability after oral administration allowed it to largely replace the penicillin G injections in clinical use in the late 1950s (Hart *et al.*, 1956). Similarly, steric bulk in the side chain suppresses cyclisation; an example is amoxicillin, and this penicillin is also suitable for oral administration.

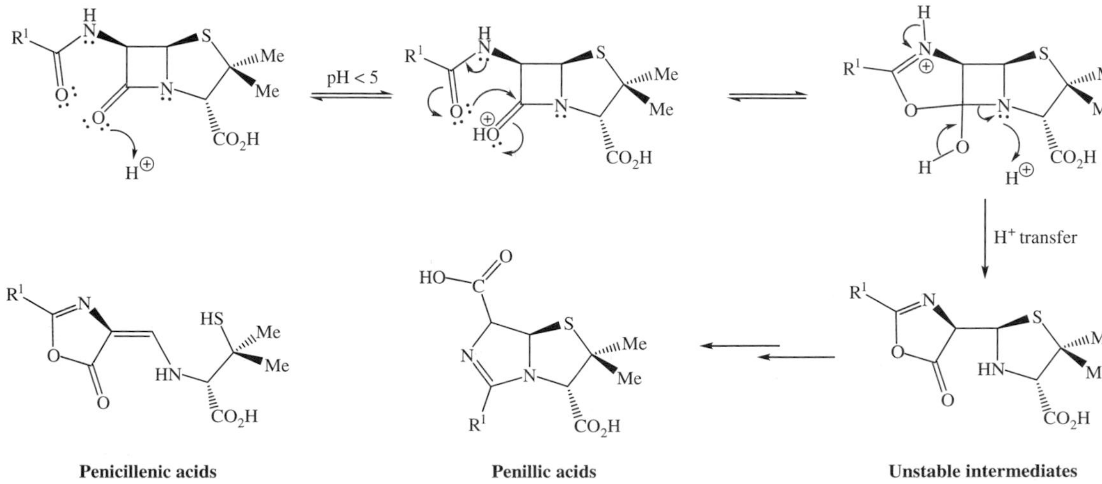

Scheme 5.1.6 *Acid-catalysed degradation of β-lactam penicillins to penillic and penicillenic acids*

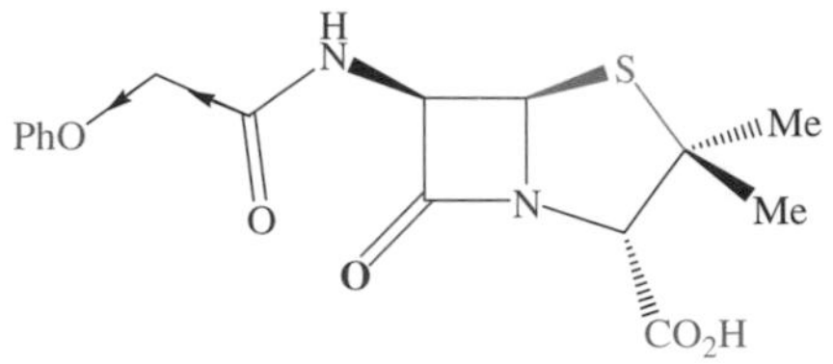

The electron withdrawing effect of the ether O atom on the side-chain amide carbonyl is enough to inhibit attack on the ring amide carbonyl

Phenoxymethylpenicillin (Penicillin V)

Figure 5.1.8   *Electron-withdrawing substituents on the 6-amino group of penicillin inhibit attack of the linear amide carbonyl group on the ring amide carbonyl carbon and enhance acid stability*

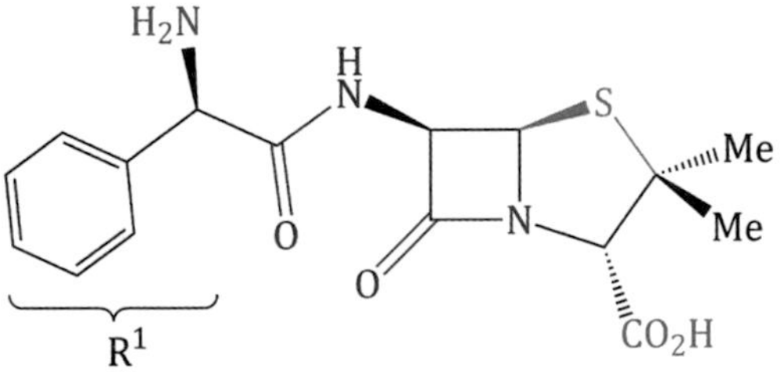

Figure 5.1.9   *Ampicillin*

Another barrier to oral bioavailability is that of low lipophilicity seen in particularly polar molecules with a low logP value – these molecules are usually poorly absorbed across biological membranes and suffer from low bioavailability. As β-lactam antibiotics remaining in the GI tract can kill commensal bacteria[8] and cause diarrhoea, it is obviously something to avoid. Most β-lactam antibiotics have only one ionisable group, the carboxylic acid. However, some have more than one group that can ionise at the various pH values seen in the GI tract, and these suffer from particularly poor absorption across the gastric mucosa.

Consider the structure of ampicillin in Figure 5.1.9: how many ionisable groups are there in this molecule, where are they, and at what pH values are they ionised? Is there a pH value at which ampicillin is un-ionised?

We have already considered the functional groups that are found on all penicillins: the ring amide, side-chain amide, thioether, and carboxylic acid groups. The only extra group in ampicillin is an amine on the R^1 side chain. Of these groups, you will have spotted that only the carboxylic acid (pK_a 2.5) and amine (pK_a 7.2) are ionisable at the pH range found in the GI tract. If we construct acid–base equilibria for ampicillin (as we did for the aminoglycosides) at the approximate pH values it would encounter after oral administration, we can work out its charge in each environment (Scheme 5.1.7).

As the equilibria show, ampicillin is ionised and highly polar at every location in the GI tract, particularly in the duodenum, where it exists as a zwitterion; not surprisingly, it has low oral bioavailability. Yet ampicillin, and other β-lactam antibiotics that have similarly ionisable groups, have very useful antibacterial activity and so are important members of this class. One solution to this problem has been the use of prodrugs.[9] Of the many

---

[8] Commensal bacteria are the normal flora found in animal systems that exist in symbiosis with their host; for example, various species of staphylococci are commonly found on the skin, in the armpits and nostrils, while *Escherichia coli* and *Bacteroides* species are often found in the GI tract. Other species, such as streptococci, are found in the mouth. While they remain in these usual locations, there is little harm to host or bacteria, but infection can occur if they are able to pass into the blood and become systemic.

[9] A prodrug is an inactive form of a drug that has no pharmacological action itself, but releases the active parent drug *in vivo*, usually by enzymic action. Many prodrugs rely on hydrolysis of an ester group by an esterase, as these are widely distributed in the body and usually release the parent drug within 15 minutes.

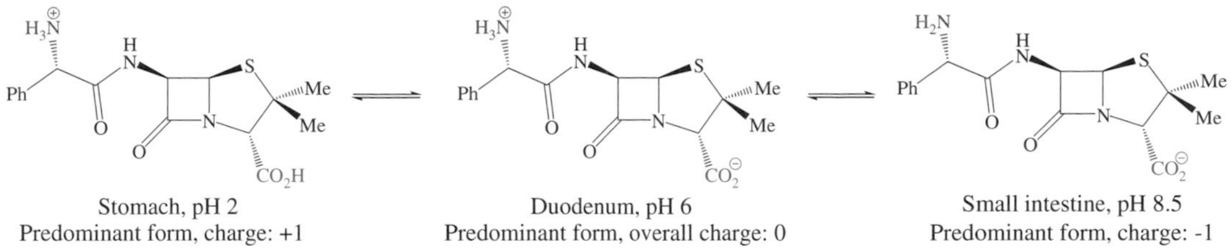

Scheme 5.1.7 *Acid–base equilibria for ampicillin*

possible prodrug groups for carboxylic acids (Rautio *et al.*, 2008), esters are most commonly used to overcome the physicochemical problems presented by carboxylate-containing pharmaceutical products (Beaumont *et al.*, 2003), as the almost ubiquitous distribution of esterases in the body provides a reliable and rapid method for the release of the active parent drug; ester prodrugs are hydrolysed mostly by serum esterases after absorption from the GI tract. Usually, a simple ester, like an ethyl group, is satisfactory, but in this case the ethyl ester of ampicillin, and of other β-lactams, is not a good esterase substrate; although it is often claimed that the size and shape of the penicillin provides steric hindrance and prevents the ester group from accessing the active site of the esterase for efficient hydrolysis, it has been shown that the simple esterified penicillins are rapidly degraded by attack at the β-lactam ring (Nielsen and Bundegaard, 1988). A double-ester approach was developed to overcome this problem and has been successfully applied to a wide range of drugs, including the β-lactam antibiotics. Examples can be found in Table 5.1.1: pivmecillinam and cefpodoxime proxetil, along with others found in clinical use, such as pivampicillin and cefuroxime axetil. Taking pivmecillinam as an example, we can see that the carboxylate of mecillinam has been derivatised by an extended double ester, called an α-acyloxyalkylester. the first ester group is formed by the derivatisation of mecillinam's carboxylate group, while the second ester group is removed from the penicillin by several atoms' distance, which facilitates access to the esterase and enables hydrolysis of the second ester group to form an unstable intermediate (Scheme 5.1.8).

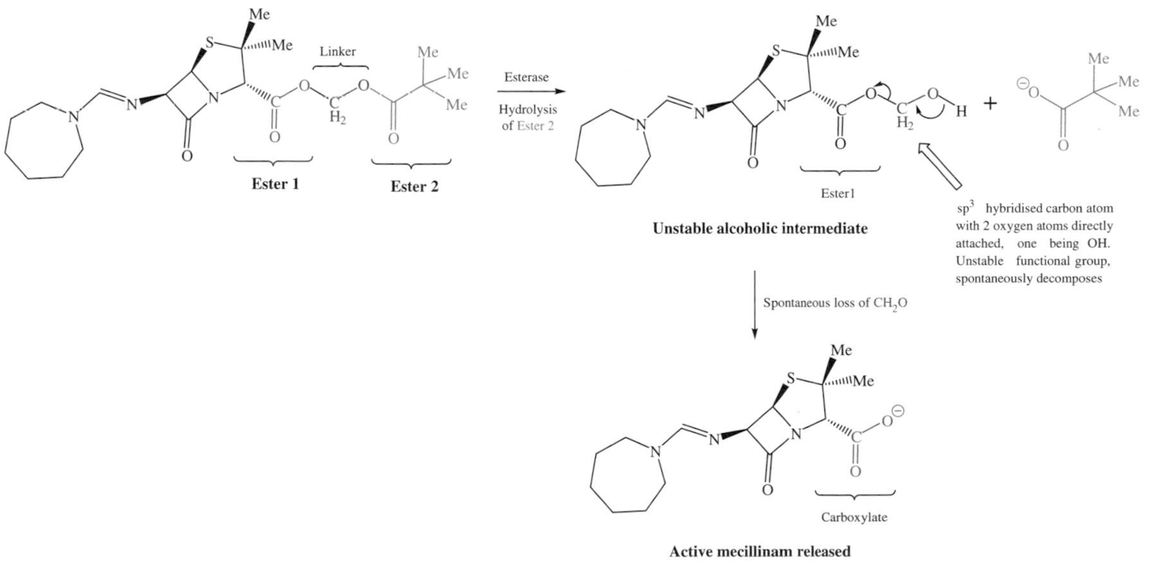

Scheme 5.1.8 *Esterase hydrolysis of pivmecillinam, an α-acyloxyalkyl ester (double ester), through an unstable intermediate that spontaneously decomposes to release the active antibiotic*

Ester 2 is readily hydrolysed by the esterase to an unstable alcoholic intermediate and trimethylacetate. The sp^3 hybridised methylene linker group has both an ether and an alcohol group attached and decomposes spontaneously to release formaldehyde (CH_2O) and the free active antibiotic mecillinam. Other α-acyloxyalkyl esters break down by the same esterase-driven process (in cases when the methylene linker is substituted with an alkyl group, it releases a substituted aldehyde (RCHO) rather than methanal).

Benzathine penicillin and procaine penicillin are not prodrugs, but are salts of benzylpenicillin for intramuscular injection; the first part of each (as named) is an amine base and forms the cation part of the salt. These salts are sparingly soluble in aqueous solution: 25 mg/mL for benzathine penicillin and up to 4.5 mg/mL for procaine penicillin. Clearly, they are not for IV use, but when injected deep into muscle tissue they slowly dissolve and release the antibiotic over time, allowing prolonged serum levels of penicillin to be achieved. In procaine penicillin, the base is also a local anaesthetic, with obvious benefits for an injected form. Of these two prodrugs, only benzathine penicillin is available in the UK, but procaine penicillin is used in several countries worldwide; however, both are narrow-spectrum antibiotics and are sensitive to enzymes that inactivate penicillins.

The bioavailability of the un-derivatised acid-sensitive penicillins, for example ampicillin, is reduced by their oral administration with food, which may be due in part to increased levels of gastric acid when there is food in the stomach, since those penicillins esterified on the carboxylate group are less affected by food (Winstanley and Orme, 1989). It is a similar story with the cephalosporins, for which, interestingly, the carboxylate ester prodrug forms seem to be better absorbed when administered with food, possibly because the increased levels of GI esterases promote ester hydrolysis to release the parent drug, which can also be absorbed (Wise, 1990).

Once absorbed, the pharmacokinetics of the various β-lactam-based antibiotics varies according to the side chain (R^1 in Figures 5.1.1–5.1.5) and the base (counter-ion) used to make the carboxylate salt. Have a look at the structures of the cephalosporins in clinical use (Table 5.1.2) and decide which will be more lipophilic and likely to be well-absorbed and distributed into a variety of tissues, remembering that some are also acid-sensitive. You can match them to their percentage bioavailability in Table 5.1.5 and won't be surprised to find that the agents with higher lipophilicity tend to penetrate into tissue better; in fact, penicillins in general are widely distributed throughout tissues and body fluids after absorption. There is a clear correlation between serum half-life and tissue penetration: the longer the agents are in the body, the more likely they are to distribute into the tissues. The lipophilicity also helps these agents to avoid renal excretion, as they are easier to re-absorb by the renal tubules, which means they are excreted more slowly (Wise, 1990).

The high degree of hydrophilicity of certain β-lactam antibiotics limits their access to the cerebrospinal fluid (CSF), which relies upon the opening of tight junctions between cells. This paracellular entry route is restricted in healthy subjects (in whom less than 1% of the serum concentration is reached in the CSF) but is increased by inflammation, so that therapeutic levels in the CSF are readily achieved in cases of meningitis (about 5–10% of the serum levels), enabling their use in emergency situations. A specific active transport system for penicillin and ceftriaxone exists in cerebral capillaries, but is not highly active and conveys little of these agents from the blood into the CSF, the major route still being paracellular (Lutsar *et al.*, 1998). Beyond exploiting the leaky meninges, the penicillins also benefit from reduced active transport out of the CSF during infection, which promotes their accumulation during meningeal infection (Kadurina *et al.*, 2003).

Excretion of the β-lactam antibiotics is largely through renal mechanisms; penicillin G is actively excreted by the renal proximal tubules, at a rate which exceeds filtration and can reach up to 1.8 g/hour in adults (Pallasch, 1988; Kadurina *et al.*, 2003). When protein binding is particularly high and renal excretion is particularly low, a higher half-life results. For example, ceftriaxone and ertapenem have high protein binding, correspondingly low renal excretion, and much increased half-life values (Table 5.1.5).

Due to their relative safety, when susceptibility is established, β-lactam antibiotics are often used to treat infections in neonates. As usual, as a group, the pharmacokinetic parameters differ in these patients from those observed in healthy adults. During the first month of life, about 75% of the total bodyweight is water, leading to

Table 5.1.5   *Selected pharmacokinetic data for β-lactam antibiotics after oral administration (data adapted from Price, 1970; Barza and Weinstein, 1976; Meyers et al., 1983; Drusano et al., 1984; Wise, 1990; Kinzig et al., 1992; Nathwani and Wood, 1993; Barradell and Bryson, 1994; Mattie, 1994; Rains et al., 1995; Perry and Brogden, 1996; Brogden and Spencer, 1997; MacGregor and Graziani, 1997; Mouton et al., 2000; Majumdar et al., 2002; Matthews and Lancaster, 2009.)*

β-lactam	Dose used for data (mg)	Oral bioavailability (%)	Peak serum level,[a] C_{max} (mg/L)	Serum half-life, $t_{1/2}$ (h)	Protein binding (%)
Penicillins					
Amoxicillin	500	90	14	1.03	17–20
Amoxicillin/clavulanic acid	500	60	3.3/5.2	1	18/25
Ampicillin	500	30–55	5–15	0.7–1.4	20
Dicloxacillin	500	35–76	10–18	0.3–0.9	95–97
Flucloxacillin	500	>80	11–15	0.7–1.3	95
Mecillinam[b]	400	75–80	5.3	1.0–1.4	5–10
Penicillin G (benzylpenicillin)	500	15–30	1.5–2.7	0.5	75–89
Penicillin V (phenoxymethyl-penicillin)	500	60–73	3–5	0.5	75–89
Piperacillin	2000 (IV)	*	28–30	0.9	16
Temocillin	1000 (IV)	*	172	4.3–5.4	85
Ticarcillin	1000 (IV)	*	250	1.2	50–60
Cephalosporins					
Cefaclor	500	52–95	15–25	0.6–1.0	20–25
Cefadroxil	500	>88	9.5	1.6	18–20
Cefalexin (Cephalexin)	500	90–100	15–20	0.6–1.3	10–19
Cefepime	2000 (IV)	*	57.5	1.3–2.3	16–19
Cefixime	400	30–50	3.7–4.8	3.1–3.8	48–69
Cefotaxime	1000 (IV)	*	81–102	0.82–1.43	25–40
Cefpodoxime[c]	200	29–53	2.1–4.5	2.2–2.8	18–30
Cefradine (Cephradine)	500	90–95	10–20	0.6–1.3	10–19
Ceftazidime	1000 (IV)	*	59–83	0.1–0.6	10–17
Ceftriaxone	1000 (IV)	*	110–130	6–9	90
Cefuroxime[d]	500	30–52	4.4–7.9	1–2	33–50
Carbapenems					
Doripenem	500 (IV)	*	20.2–31.4	1	5.4–15.2
Ertapenem	1000 (IV)	*	30–34	3.8–4.4	>90
Imipenem[e]	1000 (IV)	*	52.1–66.9	0.9–1.1	9
Meropenem	1000 (IV)	*	39–58	0.82–1.1	9
Monobactam					
Aztreonam	1000 (IV)	*	125	1.7–2.0	56

[a] C_{max} values for IV/IM-administered antibiotics relate to maximum plasma concentration after infusion; plasma levels were generally two- to fourfold lower after IM doses.

[b] After oral administration of the prodrug pivmecillinam; half-life of prodrug in whole blood <10 minutes at pH 7.4 and 37 °C (Roholt *et al.*, 1975).

[c] After oral administration of the prodrug cefpodoxime proxetil.

[d] After oral administration of the prodrug cefuroxime axetil.

[e] Co-administered with cilastatin, which inhibits degradation of imipenem to toxic products by dehydropeptidase-1 in the kidneys (the nephrotoxicity of some carbapenems is discussed in Subsection 5.1.5).

*Oral bioavailability of IV/IM-administered agents not relevant; overall bioavailability generally 100% for these administration routes.

an increased volume of distribution for aqueous soluble drugs, such as penicillins, while the lower concentrations of proteins in the circulation alter the ratio of unbound cephalosporins. Renal excretion in neonates is also significantly changed – both glomerular filtration and tubular secretion are decreased, leading to reduced total body clearance and renal clearance. As a result, the maximum serum concentration can be significantly higher, requiring continuous monitoring and dose adjustment (Paap and Nahata, 1990).

5.1.3.1   Bacterial uptake of β-lactams

As you will see in Subsection 5.1.4, β-lactam antibiotics act upon PBP (penicillin-binding protein), which is located on the inner bacterial plasma membrane, so, unlike some other antibacterial agents, the β-lactams do not need to enter the cytoplasm of the bacterial cell to reach their target, reducing the physical barriers they need to cross. The simple membrane of Gram positive bacteria is freely permeable to small molecules, even those as hydrophilic as β-lactams, causing these bacteria to be more generally susceptible to antibiotics (Lambert, 2002). However, the more complex cell membrane of Gram negative bacteria presents a chemotherapeutic challenge, especially for *P. aeruginosa*, which has one of the least permeable bacterial membranes. How can hydrophilic molecules, even relatively low-molecular-weight ones like β-lactams, access the PBPs in the periplasmic space?

To reach this target, the antibiotics must cross the outer membrane, either by diffusion or by passing through a porin (outer membrane proteins, OMPs). We have already looked at the polarity of the β-lactams in this section; some of these molecules are zwitterionic and all (except some of the prodrugs) have at least one charge at physiological pH, conferring poor lipophilicity for passive diffusion across a complex glycopeptide cell wall. You will not be surprised then to learn then that the main route for uptake of β-lactams into Gram negative bacterial cells is via porins (Hancock and Bell, 1988; Denyer and Maillard, 2002). When we consider resistance to β-lactam antibiotics later, you will see that the reliance of these agents upon porins for entry into Gram negative bacteria makes them vulnerable to resistance through alterations to the porins, thereby indirectly reducing their ability to access the PBP target. In fact, susceptibility to β-lactam antibiotics is a balance between porin-facilitated entry and clearance; the latter of these processes can involve hydrolysis or removal of the β-lactam from the bacterial cell, both of which we will meet when we consider resistance.

5.1.4   Mode of action and selectivity

Like the other agents which target the bacterial cell wall, the penicillins represent the ideal scenario in terms of selectivity, since they interfere with a structural component of the prokaryotic cell wall that is not present in mammalian cells. You would probably anticipate that an agent targeting the mechanical strength of the bacterial cell wall (have a look again at Subsection 1.1.2.3 for a reminder of the structure of the cell wall of bacteria), which protects the cell from the influx of water due to osmosis, would be selectively toxic, but would you expect it to be bacteriostatic or bactericidal? If the agent targets the growing cell wall then you might think that it would be bacteriostatic, since mature cells already have a developed cell wall, which was formed during cell division. You might expect that such a structure would not be affected by agents which target the cell wall since the key crosslinks responsible for its strength are already in place.

The β-lactams are, however, bactericidal, affecting both developing and mature cells. Bacterial cell walls are not static structures once formed and there is a balance between the synthesis of peptidoglycan[10] (catalysed by the PBPs) and its hydrolysis (catalysed by murein hydrolases). Murein hydrolases play a role in a number of processes, including the regulation of cell growth, the turnover of peptidoglycan during growth, the separation

---

[10] Peptidoglycan is also referred to as murein.

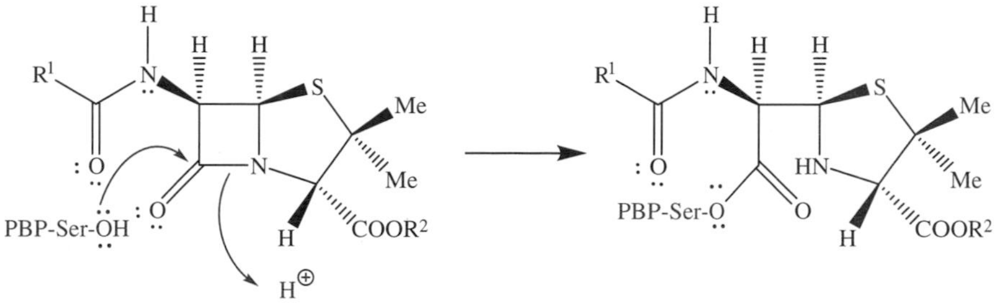

Scheme 5.1.9 *Interaction of a penicillin with a penicillin-binding protein (PBP)*

of daughter cells during cell division, and autolysis (Vollmer *et al.*, 2008). If peptidoglycan synthesis is inhibited (for example, by the β-lactams) while murein hydrolase activity is not, the net result will be the degradation of the peptidoglycan layer, resulting in a weakening of the cell wall and, ultimately, cell death.

As we saw in Subsection 1.1.3.4, the PBPs[11] play a number of roles in the formation of the crosslinks which give bacterial cell walls their strength. It is these enzymes that are the targets of the penicillins (as you have no doubt guessed from their name), which bind to the active site and thus inhibit the formation of the peptidoglycan-strengthening crosslinks. The range of enzymatic processes which are catalysed by the PBPs includes **D-alanine carboxypeptidase** (the removal of D-ala from the peptidoglycan precursor), **peptido-glycan transpeptidase** (in which one peptide bond is broken and replaced by a peptide bond to a different amino acid), and **peptidoglycan endopeptidase** (in which peptide bonds between non-terminal amino acids, i.e. within the peptide chain, are broken).

If you have another look at Figure 1.1.15 (in Section 1) you will see that in the formation of Gram positive peptidoglycan, a serine hydroxyl at the active site of the PBP attacks the terminal D-ala residue of the pentapeptide side chain to give an acyl-enzyme intermediate, which is subsequently attacked by the pentaglycine side chain. The β-lactams interfere with this D-alanine carboxypeptidase-catalysed step, being sufficiently similar in structure to the D-Ala-D-Ala dipeptide unit at the carboxy terminus of the pentapeptide strand to bind at the active site of the PBP and to acylate this key serine residue (e.g. serine 70 in *E. coli*), once again forming an ester. In this case, however, the attack by the pentaglycine side chain is presumably blocked by the thiazolidine portion (shown in blue in Scheme 5.1.9) of the penicillin-derived intermediate, which remains covalently linked to the key serine residue at the active site. This blocks the first step in crosslink formation and the overall transpeptidation process is thus inhibited.

All of the β-lactams act in the same way, blocking peptidoglycan crosslink formation and so reducing the strength of the bacterial cell wall, their selectivity being due to the fact that mammalian cells do not have a cell wall.

5.1.5 Bacterial resistance

As was stated earlier, the β-lactam ring is essential for activity so there are no prizes for guessing one of the ways in which bacteria can develop resistance. What might be surprising is how quickly resistance to the β-lactams arose. In fact, an enzyme capable of hydrolysing β-lactams antibiotics was first identified in 1940, even before the first penicillin was used clinically (Abraham and Chain, 1940). Just a few years later, the first penicillin-destroying enzyme was found in *S. aureus*, one of the early clinical targets for these agents

---

[11] Bacteria usually express a number of PBPs, with different molecular weights, PBP1 being high-molecular-weight, PBP2 lower-molecular-weight, and so on. These enzymes are usually capable of catalysing more than one of the processes listed.

(Kirby, 1944). When considering bacterial resistance for the β-lactam antibiotics, we will look at the most common, and clinically important, mechanisms of resistance, and also at the ways in which these mechanisms can be transferred from one bacterium to another, resulting in the spread of resistance.

As we have seen with other antibacterials, there are several different mechanisms that cause resistance. The four mechanisms known to cause resistance to β-lactams in a wide range of bacteria, and of particular clinical significance, are variations on those we have already met:

- **Inactivation of the β-lactam antibiotic by β-lactamase enzymes.** As this is the most important mechanism of resistance for the β-lactam antibiotics (due to the widespread occurrence of β-lactamase enzymes in clinically relevant bacteria) we will look at this more closely below.
- **Modification of the target protein.** This reduces its affinity for the β-lactam antibiotic. Mutations to the active site of PBP, through the horizontal transfer of the PBP2a or PBP2b gene, result in reduced binding of the β-lactam, and cell-wall synthesis can be maintained in the presence of penicillins and cephalosporins.
- **Decreased uptake of antibiotic into the bacterial cells.** As we have already seen, β-lactam antibiotics act upon PBP, located in the inner plasma membrane. In order to access their target in Gram negative bacteria, the antibiotics cross the outer membrane either by passive diffusion or by using the porin channels. As we saw earlier, β-lactams are mostly polar molecules and have poor lipophilicity for passive diffusion across a complex glycopeptide cell wall, so uptake via the porins is the main route of entry and mutations and modifications to the porins can thus reduce, or even prevent, the entry of β-lactams. Low-functioning porins can result from a point mutation or an insertion in the corresponding porin gene – these are often associated with β-lactamase expression.
- **Increased efflux of antibacterial agents.** This efflux, including that of β-lactam antibiotics, by up-regulated efflux pumps gives rise to multidrug resistance. It can be linked to intrinsic or acquired resistance mechanisms. We have already looked at the effects of efflux pumps in Section 4, when we considered the tetracycline antibiotics.

For more on bacterial resistance to the β-lactams, see Drawz and Bonomo (2010).

5.1.5.1   β-Lactamases

Resistance to β-lactam antibiotics through the production of β-lactamase enzymes has had a huge detrimental effect upon the use of these agents for clinical applications. The activity of penicillins and cephalosporins upon a wide range of Gram positive and Gram negative bacteria, the low incidence of adverse effects, and the relative ease of semi-synthesis of a vast number of analogues with varying activity and specificity profiles, made them first-line for many infections. As β-lactamase-conferred resistance has grown, the use of the original penicillins and cephalosporins has been considerably curtailed. Second-, third-, and even fourth-generation β-lactam antibiotics were developed, but the rapid evolution of new β-lactamases with an extended spectrum of hydrolytic activity (**extended-spectrum β-lactamases** (ESBLs)) has almost matched the pace of new β-lactam development (Paterson and Bonomo, 2005). In fact, the number of distinct β-lactamases grew from around 250 known in 2000 to about 650 in 2009, mostly due to the explosion in the number of ESBLs during this time (Bush and Jacoby, 2010).

The different β-lactamases can be grouped into four distinct classes and there are two different systems for classifying them:

- A structural classification that categorises the β-lactamases as class A, B, C, or D based on their amino acid sequences (Ambler, 1980), which has been revised to accommodate the huge number of new β-lactamases being identified (Hall and Barlow, 2005; Pfeifer *et al*., 2010).

- A functional system that uses the structures of the β-lactam substrates and inhibitors to classify the enzymes into groups 1, 2, 3, and 4, which has also been updated recently (Bush, 1989; Bush *et al.*, 1995; Bush and Jacoby, 2010).

The vast majority of β-lactamases are serine β-lactamases – as the name suggests, these enzymes use a serine residue in the active site to attack the β-lactam; we will look at the mechanism of this reaction shortly. In the Bush system (β-lactam structural classification), the serine β-lactamases are sub-divided into groups 1 and 2, whereas the Ambler system (a.a. sequence classification) subdivides them into three classes, A, C, and D, in recognition of their distinguishing activities and structural features. There is a smaller, but increasingly significant, group of β-lactamases with a zinc ion in the active site that facilitates the hydrolysis of β-lactams: these are the metallo-β-lactamases (class B or group 3). Lastly, there are a few β-lactamases that have not been classified into any of the previous classes or groups; these are pulled together into group 4 (and do not have an Ambler classification). A systematic description of the various β-lactamase enzymes would be a sizeable review in its own right, so we will restrict ourselves here to a summary table of some examples of clinically relevant β-lactamases, with reference to both systems (Table 5.1.6). The Bush classification system includes other subgroups not included in this table – these subgroups further refine the serine β-lactamases in particular, according to the specific group of β-lactam substrates or inhibitors.

You might wonder why the cephalosporinases and oxacillinases are separated into different classes in the Ambler system, when they are serine β-lactamases. The first β-lactamases studied were penicillinases (group 2a) – they hydrolysed penicillins, such as penicillin V and penicillin G, but were inactive against cephalosporins and oxacillins (penicillinase-resistant semisynthetic penicillin derivatives, such as cloxacillin, dicloxacillin, and flucloxacillin). When the β-lactamases evolved to hydrolyse cephalosporins and oxacillins, it was logical to classify the enzymes that could hydrolyse these distinct groups of β-lactam antibiotics separately.

There is some evidence that serine-β-lactamases evolved from the β-lactam target PBP (Suvorov *et al.*, 2007), and the details on their mechanism of action lend support to this theory. As the name suggests, β-lactamases hydrolyse the β-lactam ring, with the first step, resulting in a covalent enzyme-lactam intermediate **A**, being identical to the way in which β-lactams exert their antibacterial action in reacting with PBP (see Subsection 5.1.4). In Scheme 5.1.10, the upper sequence resembles the reaction of PBP (represented as Enz-Ser-OH) with a the carbonyl of the β-lactam, while the lower sequence shows the part unique to β-lactamases: their ability to use water to hydrolyse the covalent enzyme-lactam intermediate **3** to penicilloic acid, releasing the β-lactamase enzyme to inactivate further β-lactam molecules (Scheme 5.1.10).

The rapid spread of β-lactamases across the world and between different bacteria is a major source of concern, with the worst-case scenario being a return to the pre-antibacterial age, when infections went untreated due to lack of available agents. As with other resistance determinants, β-lactamases can be spread either vertically (through chromosomal transmission) or horizontally (via plasmids). The horizontal transfer of broad-spectrum resistance-conferring β-lactamases is of particular concern, as until now bacteria expressing high resistance to antibacterials have been located mostly in hospitals (where they are the cause of life-threatening infections). Now, however, highly active β-lactamases from clinical bacteria are being transferred to normal gut flora, such as *E. coli*, and carried out into community settings, leading to their greater occurrence in soil and water and increased opportunities for contamination and infection.

Carbapenemases deserve a special mention here as they present worldwide concerns: despite their name, they are promiscuous in their β-lactamase action and confer resistance to almost all clinical β-lactam antibiotics. Until about 20 years ago, there were only a few β-lactamases that could hydrolyse carbapenems, which are stable to the action of most β-lactamases due to the hydroxyl-containing side-chain (refer to Figure 5.1.3 for the structure) that sterically blocks enzymic attack on the β-lactam carbonyl group, and we were confident that serious infections could be overcome; these early carbapenemases were metallo-

Table 5.1.6 Classification of selected clinically relevant β-lactamases (Bush and Jacoby, 2010; Drawz and Bonomo, 2010; Pfeifer et al., 2010)

Type of enzyme	Ambler class	Bush group	Clinically relevant bacteria	Typical β-lactam substrates	Inhibited by CA or TZB[a]	Example enzymes
Serine β-lactamases	A: penicillinases	2a	Enterobactericeae and nonfermenting bacteria	Penicillins (excluding oxacillins)	+	PC1
		2b	Klebsiella pneumoniae	Penicillins, narrow-range (early) cephalosporins	+	TEM-1, TEM-2, SHV-1
		2be[b]		Extended-spectrum cephalosporins and monobactams	+	CTX-M, SHV-2, TEM-3
		2e		Extended-spectrum cephalosporins	+	CepA
	carbapenemases	2f		Penicillins, cephalosporins, carbapenems	±	KPC-2, IMI-1
	D: oxacillinases	2d	Enterobactericeae; Acinetobacter baumannii	Penicillins, including oxacillins	±	OXA-1, OXA-10
	C: cephalosporinases	1	Enterobacter spp., Citrobacter spp.	Cephalosporins	−	AmpC, CMY-2, FOX-1
Metallo-β-lactamases	B: metallo-β-lactamases (carbapenemases)	3a	Enterobactericeae and nonfermenting bacteria	Broad range of β-lactam substrates, including carbapenems	−	IMP-1, VIM-1, VIM-2, NDM-1
		3b		Preferentially carbapenems	−	CphA, Sfh-1

[a] Common β-lactamase inhibitors: CA, clavulanic acid; TZB, tazobactam. +, strong inhibition; ±, weak or variable inhibition; −, no inhibition. See later in this section for a discussion of the mode of action of β-lactamase inhibitors.

[b] The notation '2be' indicates that this is a subgroup of 2b with an extended spectrum.

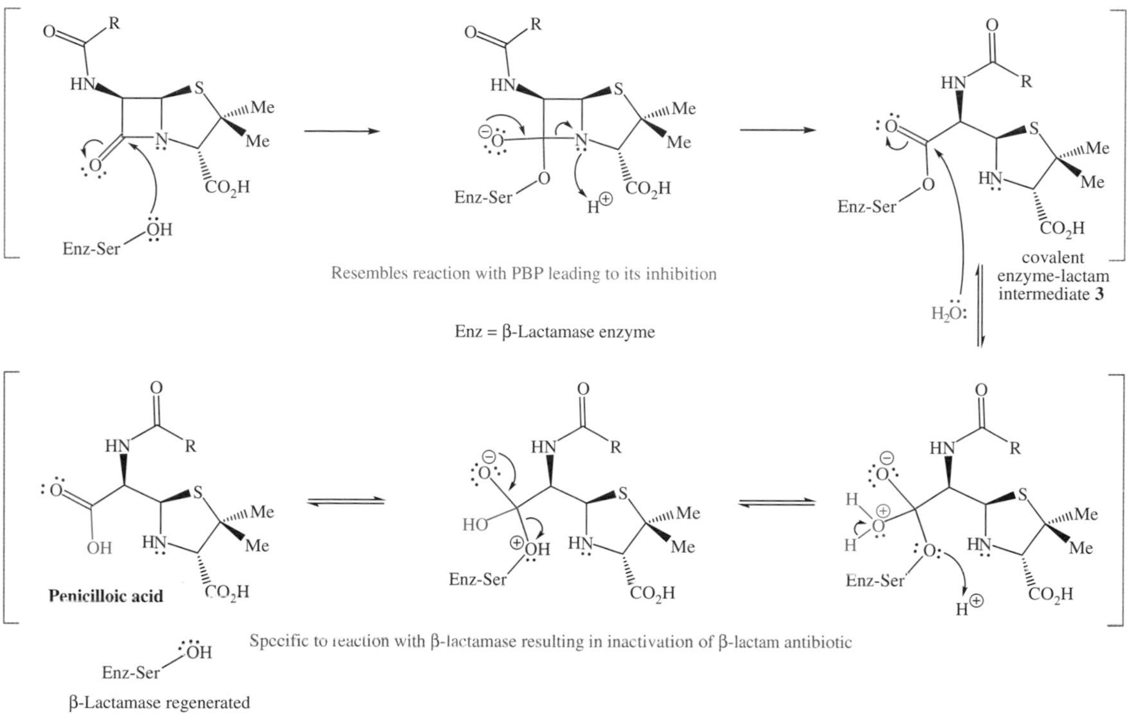

Resembles reaction with PBP leading to its inhibition

Enz = β-Lactamase enzyme

covalent enzyme-lactam intermediate **3**

Specific to reaction with β-lactamase resulting in inactivation of β-lactam antibiotic

β-Lactamase regenerated

Scheme 5.1.10 *Inactivation of β-lactam antibiotics by β-lactamase enzymes*

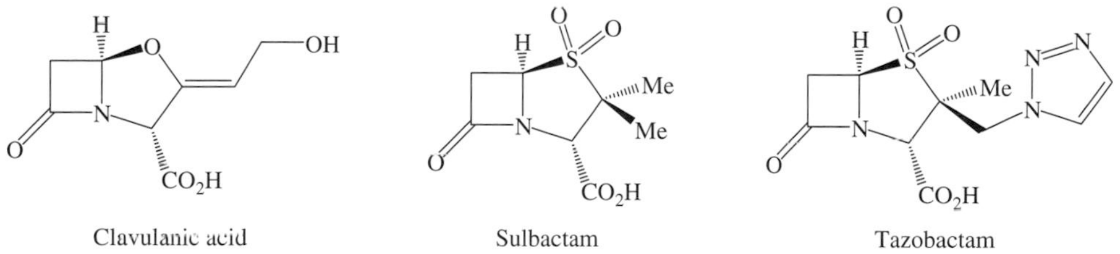

Clavulanic acid   Sulbactam   Tazobactam

Figure 5.1.10 *β-Lactamase inhibitors*

β-lactamases that arose in Gram positive bacteria, were species-specific, and were limited to a relatively slow chromosomal (vertical) transmission. Recently, however, several plasmid-borne carbapenemases have arisen in Gram negative bacteria, such as KPC-1 from *Klebsiella pneumoniae*, OXA-23 from *Acinetobacter baumannii*, and IMP-1 from *P. aeruginosa*, all clinically relevant infecting species (Queenan and Bush, 2007; Pfeifer *et al.*, 2010). This was a particularly worrying discovery, as carbapenems had continued to be active against most clinically relevant bacteria (especially Gram negative bacterial infections) and the plasmid vector facilitates more rapid horizontal transmission, not only only within a particular bacterial species, but now between genera. The ESBL β-lactamase KPC-1 was first isolated and characterised from a carbapenem-resistant strain of *Klebsiella pneumoniae* in a hospital (KPC = *Klebsiella pneumoniae* carbapenemase) (Yigit *et al.*, 2001). *K. pneumoniae* is an example of a Gram negative *Enterobacter* species and has been linked to hospital-acquired infections, such as pneumonia, with a high mortality rate in immunocompromised patients. Since then, KPC enzymes have become a worldwide problem and confer resistance to all β-lactams. The ready transmission of the plasmid carrying the bla_{KPC-1} gene[12] has ensured its rapid spread, even being transferred to other members of the *Enterobacteriaceae* genus, as well as to *P. aeruginosa* (Sidjabat *et al.*, 2009; Richter *et al.*, 2011). As we learned in Subsection 1.1.7, the broad-spectrum β-lactamase NDM-1 (New Delhi metallo-β-lactamase-1) emerged as a cause of clinical resistance in 2009 (Yong *et al.*, 2009) and has quickly become a serious worldwide threat to health (Walsh, 2010; Walsh and Toleman, 2011). It is able to hydrolyse and inactivate all β-lactam antibiotics except aztreonam, but, as it is common for NDM-1 to be co-expressed with CMY and CTX-M enzymes (one of which hydrolyses monobactams; see Table 5.1.6 for a reminder), any β-lactam antibiotic used for treatment can be deactivated. Although NDM-1 was discovered initially in *K. pneumoniae*, the plasmid has rapidly spread to other Gram negative bacteria, carrying not only bla_{NDM-1}, but also a host of genes responsible for resistance to other classes of drugs, making NDM-1-expressing bacteria almost untouchable by antibacterial agents (Liang *et al.*, 2011).

It might feel like a very gloomy picture of the future is being presented here. The current position does look a bit bleak, but there is hope. . . First, there are several β-lactamase inhibitors available, and some of these are co-administered with certain β-lactams to improve the clinical outcome; we'll discuss this in some detail later. Second, there are new β-lactamase inhibitors in development, with the distinct possibility that inhibitors with improved efficacy, even against KPC β-lactamases, may be available in the not-too-distant future, as discussed in more detail in Subsection 5.1.9.

The rationale for β-lactamase inhibitors follows the argument that inhibition of the enzyme that hydrolyses the β-lactam antibiotics would prevent this inactivation and so allow the antibiotic to act upon bacterial PBP in the usual way, restoring bacterial susceptibility to β-lactam action. Examples of successful β-lactamase inhibitors include clavulanic acid, sulbactam, and tazobactam (Figure 5.1.10).

Note that all the inhibitors in Figure 5.1.10 incorporate a β-lactam ring, which is the basis of their inhibitory activity. Clavulanic acid was originally isolated from *Streptomyces clavuligerus* and, although it has weak antibacterial activity, its suicide inhibition of the β-lactamases has proved greatly beneficial in the protection of β-lactams. The mode of action of these agents is similar to the way in which β-lactams react with PBP and to the first step of the reaction of serine-based β-lactamases with penicillins: once again, a serine residue in the active site attacks the reactive β-lactam carbonyl group to form a covalent enzyme-inhibitor complex **4**. Ring opening of the oxazolidine produces imine **5**; at this stage, the β-lactamase can be regenerated by deacylation, presumably assisted by a water molecule. Alternatively, the imine can tautomerise to a *cis*-enamine **6**, which in turn can isomerise into the corresponding *trans*-enamine **7**. Both of the enamines then undergo further reactions, leading to irreversible inhibition of the enzyme (Scheme 5.1.11) (Padayatti *et al.*, 2005). This mechanism is general for β-lactamase inhibitors with a β-lactam structure – the so-called β-lactam-β-lactamase inhibitors (Pagan-Rodriguez *et al.*, 2004; Padayatti *et al.*, 2005). Of course, these inhibitors are not active against metallo-β-lactamases, which do not use serine in their active site.

In the crystal structure of β-lactamase with clavulanic acid, a decarboxylated form of the imine can be seen covalently bound to serine-70 (Figure 5.1.11).

For more on β-lactamase-conferred resistance, see Hawkey and Jones (2009) and Bush (2010).

5.1.6 Clinical applications

5.1.6.1 Spectrum of activity

The β-lactam antibiotics are active against a wide range of bacteria and, due to the structural variations present in the different core stuctures, there is considerable diversity in the spectrum of activity. It is important,

---

[12] The genes that code for β-lactamases are given the notation *bla* (for **β**–**la**ctamase), while the exact enzyme is indicated by a subscript, e. g. bla_{CTX-M} is the gene that codes for the CTX-M β-lactamase enzyme.

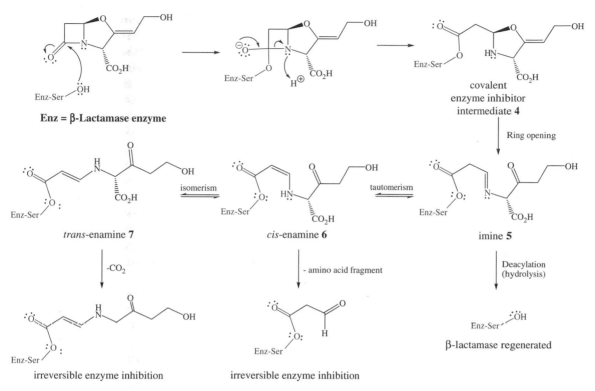

Scheme 5.1.11   *Reaction of clavulanic acid with active site serine residue (Padayatti et al., 2005)*

Figure 5.1.11   *Crystal structure of SHV-1 covalently bound to clavulanic acid; serine-70 covalently attached to a decarboxylated imine product is shown in yellow (PDB 2A49) (Padayatti et al., 2005)*

therefore, to know the spectrum of activity for each antibiotic in order to ensure that when a β-lactam is indicated, the most suitable one is prescribed (it is not appropriate to simply prescribe 'a β-lactam antibiotic'). From the **penicillins**:

- Benzylpenicillin shows activity against Gram positive and Gram negative cocci.
- Phenoxymethylpenicillin has a similar spectrum to benzylpenicillin, but has reduced activity (Nathwani and Wood, 1993; Wright, 1999).
- Both ampicillin and amoxicillin are considered to have broad antibacterial spectra, are active against certain Gram positive (e.g. *Streptococcus pneumoniae*) and Gram negative (e.g. *E. coli*) bacteria, but are, unfortunately, inactivated by penicillinases.
- Flucloxacillin and temocillin, however, are penicillinase-stable and are therefore used to treat infections caused by penicillinase-producing staphylococci and Gram negative bacteria (with the exception of *Pseudomonas aeruginosa* and Acinetobacter species), respectively (Sutherland *et al.*, 1970; Livermore and Tulkens, 2009).
- Pivmecillinam, an orally active prodrug of mecillinam, also has activity against Gram negative bacteria, including *E. coli* and other *Enterobacteriaceae*; it is not active against *P. aeruginosa* or enterococci.

As you may have gathered, the production of β-lactamases by certain bacteria has significantly affected the activity of some penicillins. To overcome this problem, several penicillins have been formulated in combination with a β-lactamase inhibitor. Examples of these combinations include:

- Amoxicillin with clavulanic acid (this combination is often referred to as co-amoxiclav), which is active against bacteria that are resistant to amoxicillin through the production of β-lactamases (e.g. *S. aureus* and *E. coli*).
- Piperacillin with tazobactam and ticarcillin with clavulanic acid, which both show activity against Gram positive and Gram negative bacteria, including *P. aeruginosa* (Fuchs *et al.*, 1984; Livermore and Tulkens, 2009).

The **cephalosporins** are broad-spectrum antibiotics with varying activity against Gram positive and Gram negative bacteria. Generally speaking, the first-generation cephalosporins have greatest activity against aerobic Gram positive cocci, including meticillin-susceptible *Staphylococcus aureus* (MSSA), whereas the later-generation agents show greater activity towards Gram negative bacteria (with some showing activity against *P. aeruginosa*) (Marshall and Blair, 1999). One important thing to note about the cephalosporins is that many of these agents are listed in the literature as being used to treat acute UTIs (see Table 5.1.2). This may have been true many years ago, but nowadays, in clinical practice, where there are other therapeutic options, the use of the cephalosporins as first-line agents to treat acute UTIs is discouraged (despite their having good activity against many urinary pathogens) due to an increased risk of developing *Clostridium difficile* infections (especially in elderly patients) (Gerding, 2004).

The **carbapenems** are also broad-spectrum agents which show activity against many pathogenic Gram positive and Gram negative bacteria and so are, clinically, very useful agents. Imipenem, meropenem, and doripenem are active against *P. aeruginosa*, while ertapenem is not (so should obviously not be used when this organism is implicated) (Zhanel *et al.*, 2007).

Aztreonam, a **monobactam**, has activity against Gram negative bacteria (including *P. aeruginosa*, *Neisseria meningitidis*, and *Neisseria gonorrhoeae*), but is not active against Gram positive bacteria and should not, therefore, be given (as monotherapy, at least) when these organisms are implicated (Hellinger and Brewer, 1999).

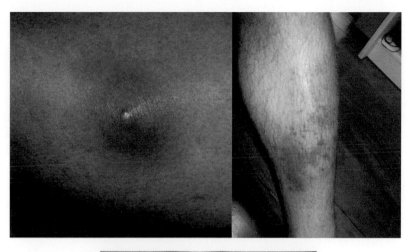

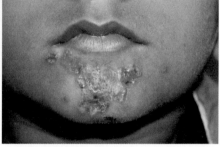

Figure 5.1.12 *Clinical presentation of a furuncle (top left) (Reprinted from 'Furuncle', Dermatology Atlas, Online, [http://www.edoctoronline.com/dermatology-atlas.asp?c=4&id=24862&m=3] 2011.), cellulitis (top right) (Reprinted from 'Cellulitis', Wikipedia, Online, [http://en.wikipedia.org/wiki/Cellulitis].), and impetigo (bottom)*

5.1.6.2 Uncomplicated skin and soft-tissue infections

In Section 4.5.6, we discussed the use of linezolid in the treatment of complicated skin and soft-tissue infections (cSSTIs). The β-lactams (the penicillins, in particular) also have a role in the treatment of skin and soft-tissue infections but, in contrast to linezolid, can be used to treat uncomplicated SSTIs. To recap, complicated infections tend to involve deeper tissues or skeletal muscle, while uncomplicated infections tend to be superficial and are considered less serious than complicated infections. Common pathogens implicated in uncomplicated SSTIs include *Staphylococcus aureus* and *Streptococcus pyogenes*. Examples of uncomplicated SSTIs include impetigo, furuncles[13] (more commonly known as boils), and mild cellulitis. Cellulitis is an acute infection of the dermis and subcutaneous tissue; it usually presents with a spreading area of redness, and inflammation – the skin surface in cellulitis has even been described as looking like orange peel (a condition called 'peau d'orange'). Impetigo is a highly contagious infection of the superficial layers of the skin and is characterised by the presence of golden/yellow crusty scabs, while a furuncle is an infection of a hair follicle (Figure 5.1.12).

As these conditions are commonly caused by *S. aureus*, flucloxacillin is often used first-line to manage these infections (remember that flucloxacillin is stable to many *S. aureus* β-lactamases, so is a sensible choice). In the case of impetigo, if the condition is confined to small areas, topical antibiotics (such as fusidic acid) are

---

[13] You may have also heard the term 'carbuncle', which is used to describe the condition when multiple hair follicles become infected. The lesions tend to be broad and painful, and pus will usually drain through multiple sites.

recommended (Koning *et al.*, 2003), but if the condition is widespread, systemic treatment with flucloxacillin is recommended.

In the case of a furuncle, antibiotics are warranted if the patient has other co-morbidities (such as immunosuppression or diabetes), the lesion is a carbuncle, or the patient is in pain or suffering from a fever. Flucloxacillin is again recommended as first-line therapy, but specialist advice should be sought if meticillin-resistant *Staphylococcus aureus* (MRSA) is suspected or confirmed.

For the treatment of mild cellulitis, flucloxacillin is also recommended as first-line therapy, and if the cellulitis is severe, flucloxacillin can be used in combination with benzylpenicillin, although one set of guidelines has suggested using benzylpenicillin with ciprofloxacin and clindamycin, instead of flucloxacillin (CREST, 2005).

5.1.6.3   Osteomyelitis

Osteomyelitis is a condition characterised by inflammation of the bone (or bone marrow) due to the presence of microorganisms and is managed, in part, by the β-lactam antibiotics. Symptoms of osteomyelitis include fever, erythema, and pain and tenderness over an area of bone. The causative organism, as is the case for SSTIs, is most commonly *S. aureus*, although other microorganisms, such as *P. aeruginosa* and *Candida albicans*, have also been implicated (Lew and Waldvogel, 2004). You might think that bones are usually resistant to infections, and you would be correct, but there are a numbers of ways in which a bone can become infected: for example, infection can spread from a contiguous infection (such as in cellulitis), from trauma to the bone (such as a bone fracture or a surgical operation), or when there is vascular insufficiency to the bone (which is sometimes observed in diabetic patients). Osteomyelitis is categorised as acute, sub-acute, or chronic, depending on how long the bone infection takes to develop; the acute condition is said to occur within 2 weeks after the initial infection, the sub-acute within 1 or 2 months, and the chronic in 2 months or more (Carek *et al.*, 2001).

The inflammation caused by the infection (through the release of inflammatory factors) can cause ischaemia, which may lead to necrosis of the bone (as the bone is not getting a sufficient blood supply) and, ultimately, the formation of sequestra (dead bone). The ischaemia also poses a problem for the pharmacological treatment of osteomyelitis, as antibiotics, which travel to the site of infection via the bloodstream, are not able to reach the site of infection and therefore have a limited effect – this problem (and the resulting sequestra) is typically observed in chronic osteomyelitis (when surgery is often required).

For patients with acute osteomyelitis, antibiotics can be used successfully to manage the condition, but for patients with chronic osteomyelitis a combined approach using surgery (to remove the dead bone) and antibiotics is often needed. As osteomyelitis is a serious condition (and can have severe complications, on rare occasions requiring amputation), it is important to isolate the causative organism, which is often done by taking a bone biopsy from the infected bone, in order to choose the most appropriate antibiotic therapy.

If MSSA is identified as the causative organism then either flucloxacillin or a second-generation cephalosporin (e.g. cefuroxime) is recommended. If *P. aeruginosa* is identified, the treatment choice is based on sensitivity analysis of the isolated strain; possible options include cefepime, meropenem, imipenem, or piperacillin in combination with tazobactam. If, however, the infection is 'mixed' (i.e. contains aerobic and anaerobic bacteria), co-amoxiclav is generally the treatment of choice (Lew and Waldvogel, 2004).

Large doses of antibiotics are also required to treat osteomyelitis. To give you an indication of how large, we will use flucloxacillin as an example: for the treatment of uncomplicated SSTIs (as discussed above), a typical daily dose of flucloxacillin is 2 g; in osteomyelitis, however, the daily dose of flucloxacillin can be up to 8 g.

Generally speaking, IV antibiotics should be given for 4–6 weeks, and a PICC[14] line is often used for this purpose. In some cases the patient will have to stay in hospital for the full course of antibiotics, but on occasion the antibiotics can also be administered on an outpatient basis.

---

[14] PICC stands for 'peripherally inserted central catheter' and is used as a form of intravenous access to deliver antibiotics over a prolonged period of time without subjecting the patient to frequent injections.

5.1.6.4   Pneumonia

Pneumonia is another condition in which the β-lactam antibiotics are commonly used. We have previously discussed the management of community-acquired pneumonia (CAP) in Section 4, when we outlined the use of the macrolide antibiotics (and the role of amoxicillin). In addition to this, the β-lactams also have a role in the management of hospital-acquired pneumonia (known as HAP) and ventilator-associated pneumonia (known as VAP). HAP is classed as a pneumonia that occurs more than 48 hours after hospital admission, while VAP is defined as a pneumonia developing 48 hours after initiation of mechanical ventilation. Both conditions can be further subclassified as 'early onset' and 'late onset', with the former occurring within 4 days of hospital admission (or mechanical ventilation in the case of VAP) and the latter occurring after 5 or more days. Early-onset HAP or VAP is most frequently associated with drug-susceptible community-like pathogens (e.g. *S. pneumoniae* or *H. influenzae*), whereas the late-onset conditions tend to be associated with multidrug-resistant bacteria (e.g. MRSA or *P. aeruginosa*) and therefore lead to poorer patient outcomes (Lynch, 2001; Chastre and Fagon, 2002).

HAP is thought to affect around 0.5–1% of patients (American Thoracic Society, 1995), whereas the incidence of VAP is significantly higher and is thought to complicate around 24% of patients receiving mechanical ventilation (Torres *et al.*, 1990), numbers that should give you an idea of how much of a problem HAP and VAP are to health care providers across the world, given the large number of patients admitted to hospital every day. Both HAP and VAP thus remain significant causes of patient morbidity and mortality.

In Subsection 4.2.6.3, we introduced you to the CURB65 score, which is used to assess the severity of CAP. There is also a scoring system that can be used to aid the diagnosis of HAP/VAP. Generally speaking, both HAP and VAP can be difficult to diagnose, which, in part, is due to the fact that there is no universally accepted gold standard – this can unfortunately lead to mis-diagnosis (and hence a delay in the appropriate treatment). The scoring system used is the clinical pulmonary infection score (often abbreviated as CPIS) and examines the following six criteria: temperature, blood leukocytes, oxygenation, pulmonary radiography, tracheal secretions, and culture of tracheal aspirate (Chandler and Hunter, 2009; File, 2010).

Once a patient has been diagnosed with HAP/VAP, it is important to ensure that the treatment is based on the best available evidence, and fortunately there have been a number of international guidelines produced for support. For example, in the UK, the British Society for Antimicrobial Chemotherapy has produced a set of evidence-based guidelines (Masterton *et al.*, 2008), while in the USA a joint document has been produced by the Infectious Diseases Society of America and the American Thoracic Society (American Thoracic Society, 2004). When deciding on which antibiotic to prescribe, it is important to establish whether the patient has recently received any other antibiotics, as this can influence the choice of treatment. For example, if a patient developed early-onset HAP or VAP and they **have not recently received other antibiotics**, then empirical use of either co-amoxiclav or cefuroxime is recommended (Masterton *et al.*, 2008). For patients with early-onset HAP or VAP who **have recently received antibiotics**, piperacillin/tazobactam or a third-generation cephalosporin (such as cefotaxime or ceftriaxone) is an appropriate choice. In cases of **late-onset HAP or VAP**, patients should receive antibacterial treatment with activity against *P. aeruginosa* (as this is the most common pathogen in this patient group), so piperacillin/tazobactam is often used in this case (other options include ciprofloxacin, meropenem, and ceftazidime) (Masterton *et al.*, 2008).

Now we know which antibiotic to prescribe, you may wonder how long the patient should receive it for (i.e. what is the duration of treatment?). This is a good question and can in part be answered by a randomised double blind trial that looked at giving antibiotics for either 8 or 15 days for confirmed VAP, in which Chastre and colleagues found that patients treated for 8 days had neither excess mortality nor more recurrent infections than those treated for 15 days (Chastre *et al.*, 2003). This trial also found that the emergence of multidrug-resistant pathogens was lower in those who received the shorter, 8-day course of antibiotics, and as a result it is recommended that, for patients who respond to empirical antibiotic therapy, the routine course of treatment should not exceed 8 days.

5.1.6.5   Miscellaneous

Tables 5.1.1–5.1.4 contain a very brief summary of the therapeutic indications for the β-lactam antibiotics. We cannot discuss every possible indication, and we have already discussed some of the disease states for which several of the β-lactam antibiotics are used (and made reference to the particular β-lactam indicated in each subsection), which included the eradication of *H. pylori* (Subsection 2.3.6.5), gonorrhoea (Subsection 2.1.6.4), Lyme disease (Subsection 4.3.6.5), infective endocarditis (Subsection 4.1.6.2), chronic *P. aeruginosa* infection in patients with cystic fibrosis, meningitis (Subsection 4.1.6.4), and prophylaxis of rheumatic fever (Subsection 3.1.6.5).

5.1.7   Adverse drug reactions

The adverse reactions associated with the β-lactams have been described extensively in the medical literature and range from anaphylaxis, urticaria, Stevens–Johnson syndrome, and acute exanthematic pustulosis to drug hypersensitivity syndrome. These reactions can be broadly grouped into two categories: those occurring immediately (primarily concerned with IgE-mediated responses) and those with a delayed reaction (mediated by T cells).

The adverse reactions most often associated with the β-lactams have become known as classic penicillin hypersensitivity and fall within the immediate (primarily concerned with IgE-mediated responses) category.

5.1.7.1   Mechanism of hypersensitivity

As we have seen, the β-lactam ring structure can be cleaved *in vivo*, resulting in serum binding and covalent adduct formation with cell-membrane proteins. These products are referred to as hapten carrier conjugates, which are processed by antigen-presenting cells to eventually be presented to T cells. The T cells react with B cells and thus produce immunoglobulin E (IgE). This β-lactam interaction, as with most other hypersensitivity reactions, provokes an immunological cascade, with the subsequent release of inflammatory mediators.

The first steps in hapten formation by the β-lactams are illustrated in Scheme 5.1.12. As you can see, this process is initiated by the formation of penicillenic and penicilloic acids, the former of which we encountered in Scheme 5.1.6, in Subsection 5.1.3. These inactive penicillin derivatives then react with a range of amino acid side chains from proteins, such as lysine and cysteine, to form adducts.

This process occurs *in vitro* as well as *in vivo* and is facilitated by low pH and also the presence of cations, especially copper, zinc, and iron. The penicillenic acid reacts irreversibly with lysine ε-amino groups to form penicillinoyl-amine haptenic[15] groups (known as the 'major determinant' because this form is most commonly observed; it is usually formed with human serum albumen lysine groups) (Levine and Ovary, 1961).

Penicillenic acid can also react with other amino acid side chains to form other haptens (e.g. penicillenic acid-cysteine mixed disulfide), which are known as 'minor determinants' because they are formed less frequently than the major determinant (Svensson *et al.*, 2001). The major determinant and minor determinants can be applied to the skin to establish whether a patient is going to be hypersensitive to a

---

[15] A hapten is a small molecule that illicits an immune response when joined to a protein molecule.

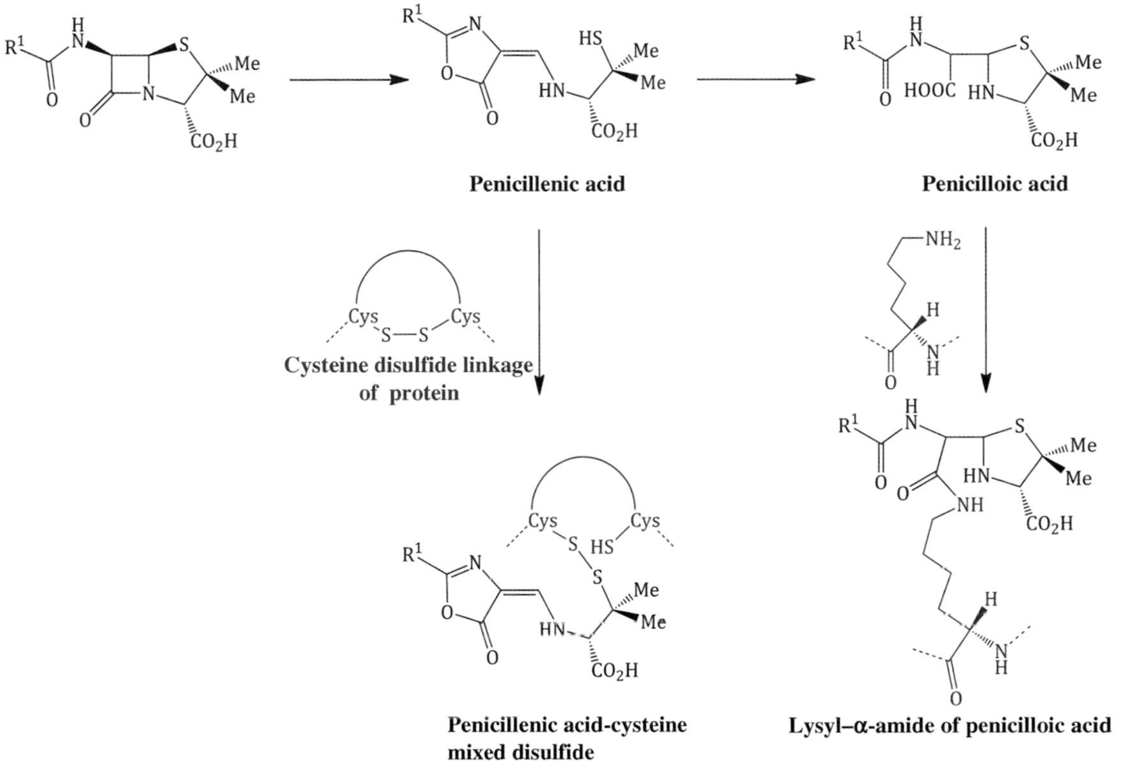

Scheme 5.1.12 *Possible hapten formation by the β-lactam antibiotics*

penicillin. If a wheal-type skin reaction with erythema is produced (a positive test), there is a good chance the patient will be hypersensitive to penicillin (and, because of the structural similarity between agents, there is also a chance that the patient will also be hypersensitive to other β-lactams, such as cephalosporins, penems, and monobactams).

Around 10% of patients who are allergic to penicillins are also allergic to cephalosporins. It has been suggested that the lower rates of hypersensitivity to cephalosporins in comparison to the penicillins are best explained by the reactivity of the two systems. Indeed, the high degree of torsion experienced in the penicillin nucleus (the thiazolidine and β-lactam ring) results in greater reactivity when compared to the less-strained heterocycles found in the cephalosporin nucleus. Cross-reactivity between cephalosporins and penicillins has, however, been observed, particularly between those that share identical/similar side chains. Among patients with a selective response to amoxicillin, 38% were shown to develop cross-reactivity with cefadroxil, a cephalosporin with a side chain identical to that of amoxicillin (Miranda *et al.*, 1996).

5.1.7.2 Clinical features of β-lactam hypersensitivity

Symptoms associated with β-lactam hypersensitivity usually occur within 1 hour of IV drug administration (or around 6 hours after oral administration) and usually follow a general pattern, commencing with generalised erythema and urticarial rash, with extension to the whole body (Figure 5.1.13); in severe cases, angioedema,

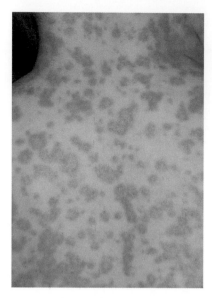

Figure 5.1.13   *Urticarial rash observed in penicillin hypersensitivity*

tachycardia, hypotension, bronchospasm, and laryngeal oedema can occur. If the rash occurs more than 3 days after the administration of the penicillin, it is unlikely to be a true hypersensitivity reaction and is considered less serious (as the risk of anaphylaxis is very low).

Patient hypersensitivity to β-lactams is well documented – its diagnosis has been described by the European Academy of Allergy and Clinical Immunology and involves skin testing, *in vitro* testing, and drug provocation tests (Torres *et al.*, 2003). Patients who experience an immediate hypersensitivity response to β-lactam-containing compounds (i.e. within 1 hour of IV administration) should be recommended to avoid them in the future.

The adverse effect of nephrotoxicity of the carbapenem imipenem has already been mentioned in Section 5.1.2.3; this carbapenem is degraded to toxic products in the renal tubules by an enzyme dehydropeptidase-1 (DHP-1) and must be co-administered with cilastatin, a DHP-1 inhibitor. This toxicity was tuned out in the later carbapenems, so that meropenem, ertapenem, and doripenem do not require a DHP-1 inhibitor to be co-administered.

5.1.8   Drug interactions

There are relatively few drug interactions reported with the β-lactam group of antibiotics that manifest into serious clinical events. One such interaction that has been reported is between the penems and valproic acid, leading to significantly increased clearance of valproate and possibly resulting in seizures (Fudio *et al.*, 2006; Spriet *et al.*, 2007; Tobin *et al.*, 2009). The rapidity of plasma concentration reduction of valproate with concomitant use of drugs (such as the carbapenems) cannot be explained with enzyme induction, which normally takes several days to occur, while interference with plasma protein binding is also unlikely. Animal studies have suggested that the combined use of penems with valproic acid most likely inhibits the hydrolytic enzyme involved in the hydrolysis of valproic acid glucuronide to valproic acid, resulting in decreased valproic acid plasma levels (Nakajima *et al.*, 2004).

5.1.8.1   β-Lactams and combined oral contraceptives

Opinion varies on a possible interaction between broad-spectrum β-lactam antibiotics and the combined oral hormonal contraceptives. The generally accepted pharmacological interaction is that those antibiotics exhibiting 'broad-spectrum' activity – penicillins, cephalosporins, and so on – reduce the 'enterohepatic shunt' of oral combined contraceptives. In other words, the enzymatic activity of the gut flora (bacteria) assists in the deconjugation of oestrogen metabolites (removal of sugars attached to oestrogen), releasing ethinylestradiol, which facilitates its reabsorption from the duodenum/jejunum, allowing it to exert its pharmacological effect via the bloodflow to the ovaries. Some oestrogen passes to the liver, where again it is conjugated (with simple sugars) and then re-enters the GI tract via the bile. This forms a cycle known as the enterohepatic shunt and effectively allows minute doses of oestrogen to exert a pharmacological action via 'recycling' (Manwaring, 1943). It is commonly believed that broad-spectrum antibiotics may temporarily eradicate the gut flora responsible for the 'deconjugation', effectively reducing the amount of absorbed ethinylestradiol and theoretically diminishing the overall contraceptive effect of such products.

An additional consideration is antibiotic-induced vomiting and/or diarrhoea, whereby the overall absorption of orally administered ethinylestradiol may be reduced. A review of the literature has concluded that only a small group of women may be susceptible to oral hormonal contraceptive failure, most likely those with significant enterohepatic shunt mechanism (Back *et al.*, 1980; Back and Orme, 1990; Shenfield, 1993). There have been a few isolated case reports describing contraceptive failure in woman taking the combined oral contraceptive pill in combination with a penicillin (e.g. ampicillin) (Dossetor, 1975) or a cephalosporin (e.g. cefalexin) (DeSano and Hurley, 1982), while other studies have not shown a statistically significant increase in the number of pregnancies observed with broad-spectrum antibiotic use (Helms *et al.*, 1997).

There is clearly a lot of information to take in about this drug interaction and it is important that we give the patient the correct clinical information, as the personal and ethical consequences of an unwanted pregnancy are very serious. At present, the clinical evidence does not support an interaction between a combined oral contraceptive and a broad-spectrum antibiotic, and consequently the advice on how to manage this interaction has changed. Originally, the advice given while taking a broad-spectrum antibiotic and a combined oral contraceptive was to use alternative methods of contraception (e.g. a barrier method) during antibiotic treatment and for 7 days afterwards. Nowadays, however, it is recommended that **no** additional contraceptive precautions are required while taking this combination of drugs (Faculty of Sexual & Reproductive Healthcare, 2011).

5.1.8.2   Methotrexate and penicillins

There have been several case reports linking methotrexate toxicity with the concurrent use of penicillins, including amoxicillin (Ronchera *et al.*, 1993), flucloxacillin (Mayall *et al.*, 1991), piperacillin (Zarychanski *et al.*, 2006), and phenoxymethylpenicillin (Nierenberg and Mamelok, 1983). It is thought that the penicillins compete with methotrexate in the kidney tubules for excretion, which results in less methotrexate being excreted (and more being retained in the body). This drug interaction is clinically significant, and because methotrexate has a narrow therapeutic index, there have been a number of fatalities reported in the literature. Although this drug interaction is not commonly observed, given the seriousness of the interaction, the combination of a penicillin and methotrexate should always be used with caution.

5.1.8.3   Warfarin and β-lactams

There have been several reports showing a potential drug interaction when warfarin is used concurrently with β-lactam antibiotics; some β-lactams appear to increase anticoagulation, while others decrease it. For example, a case report on the use of warfarin and dicloxacillin showed that the prothrombin time, (*S*)-, and

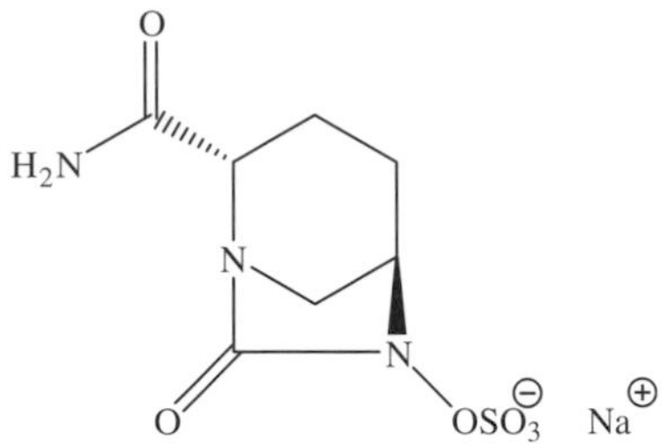

Figure 5.1.14   *The structure of the non-β-lactam-β-lactamase inhibitor NXL104*

(*R*)-warfarin concentrations fell by 17, 25, and 20%, respectively, after the initiation of dicloxacillin (and thus decreasing anticoagulation) (Mailloux *et al.*, 1996), while a similar case has also been reported with flucloxacillin (which, looking at the structural similarity between the two compounds, is not so surprising) (Garg and Mohammed, 2009). It has been suggested that the use of dicloxacillin or flucloxacillin may increase the metabolism of warfarin – with one article reporting that flucloxacillin appears to induce expression of CYP 3A4 (Huwyler *et al.*, 2006), which, in theory, could increase the metabolism of (*R*)-warfarin (and thus decrease the anticoagulant effect).

In contrast, however, increased anticoagulant effects have been reported with other penicillins (e.g. amoxicillin and benzylpenicillin), and occasionally some cephalosporins (e.g. cefazolin), when used in combination with warfarin (Baxter, 2006). In addition, the manufacturers of the penems and aztreonam also advise caution when these agents are used in combination with warfarin, due to the potential of an increased risk of bleeding. It would, therefore, be prudent to closely monitor the INR of patients taking warfarin in combination with a β-lactam antibiotic and advise patients to report any signs of bruising or bleeding (which may be a sign of warfarin toxicity).

5.1.9   Recent developments

Since the discovery of benzylpenicillin by Fleming in 1928, a lot has happened in terms of antibiotic resistance, with one significant example being the rise of the β-lactamase enzymes (Section 5.1.5). As we have already discussed, one strategy employed to overcome this problem was the development of the β-lactamase inhibitors, and indeed this approach has, over the years, proved fruitful. Once more, however, the β-lactamases threaten to hinder the progress made against the management of bacterial infections by rendering many clinically used β-lactams useless. As we have seen, the KPC enzymes have high hydrolytic activity against many clinically used β-lactams (including the carbapenems) and, rather worryingly, are not inhibited by the standard β-lactam-β-lactamase inhibitors (such as clavulanic acid, tazobactam, and sulbactam) (Papp-Wallace *et al.*, 2010). One compound, however, that may offer some hope is NXL104, a non-β-lactam-β-lactamase inhibitor[16] that shows inhibitory activity against KPC β-lactamases (Figure 5.1.14) (Stachyra *et al.*, 2009). This compound, used in combination with ceftazidime, has shown to be very effective in two murine infection models (one of which was septicaemia and the other a thigh infection) due to highly resistant KPC-producing *Klebsiella pneumoniae* isolates – which would have otherwise been resistant to ceftazidime alone (Endimiani *et al.*, 2011). As a result of this, and other encouraging data, a phase II clinical trial is being

---

[16] This sounds complicated, but all it means is that NXL104 acts as a β-lactamase inhibitor but does not itself contain a β-lactam ring.

undertaken, evaluating the use of NXL104 in combination with ceftazidime for the management of complicated UTIs (Bassetti *et al.*, 2011).

References

E. P. Abraham and E. Chain, *Nature*, 1940, **146**, 837.

E. P. Abraham and G. G. F. Newton, *Biochem. J.*, 1956, **63**, 628–634.

E. P. Abraham, W. Baker, E. Chain, and R. Robinson.Further Studies on the Degradation of Penicillin. VII. British Ministry of Supply Penicillin Production Committee Report No PEN103, 22 October 1943.

American Thoracic Society, *Am. J. Respir. Crit. Care Med.*, 1995, **153**, 1711–1725.

American Thoracic Society, Guidelines for the Management of Adults with Hospital-acquired, Ventilator-associated, and Healthcare-associated Pneumonia, 2004 (http://ajrccm.atsjournals.org/cgi/reprint/171/4/388, last accessed 13 March 2012).

R. P. Ambler, *Phil. Trans. R. Soc. Lond. B*, 1980, **289**, 321–331.

D. J. Back and M. L. Orme, *Clin. Pharmacokinet.*, 1990, **18**, 472–484.

D. J. Back, A. M. Breckenridge, F. E. Crawford, K. J. Cross, M. L. Orme, A. Percival, and P. H. Rowe, *J. Steroid Biochem.*, 1980, **13**, 95–100.

L. B. Barradell and H. M. Bryson, *Drugs*, 1994, **47**, 471–505.

M. Barza and L. Weinstein, *Clin. Pharmacokinet.*, 1976, **1**, 297–308.

M. Bassetti, F. Ginocchio, and M. Mikulska, *Crit. Care*, 2011, **15**, 215.

K. Baxter (Ed.), *Stockley's Drug Interactions*, 7th Edn, Pharmaceutical Press, London, 2006, pp. 274–304.

K. Beaumont, R. Webster, I. Gardner, and K. Dack, *Curr. Drug Metab.*, 2003, **4**, 461–485.

R. N. Brogden and C. M. Spencer, *Drugs*, 1997, **53**, 483–510.

K. Bush, *Antimicrob. Agents Chemother.*, 1989, **33**, 259–263.

K. Bush, *Crit. Care*, 2010, **14**, 224.

K. Bush and G. A. Jacoby, *Antimicrob. Agents Chemother.*, 2010, **54**, 969–976.

K. Bush, G. A. Jacoby, and A. A. Medeiros, *Antimicrob. Agents Chemother.*, 1995, **39**, 1211–1233.

C. Campos Muñiz, T. E. Cuadra Zelaya, G. Rodriguez Esquivel, and F. J. Fernández, *Rev. Latinoam. Microbiol.*, 2007, **49**, 88–98.

P. J. Carek, L. M. Dickerson, and J. L. Sack, 2001, *Am. Fam. Physician*, **63**, 2413–2421.

E. B. Chain, H. W. Florey, and A. D. Gardner, *Lancet*, 1940, **2**, 226–228.

B. Chandler and J. Hunter, *J. Intensive Care Soc.*, 2009, **10**, 29–33.

J. Chastre and J. Y. Fagon, *Am. J. Respir. Crit. Care Med.*, 2002, **165**, 867–903.

J. Chastre, M. Wolff, J. Y. Fagon, S. Chevret, F. Thomas, D. Wermert, E. Clementi, J. Gonzalez, D. Jusserand, P. Asfar, D. Perrin, F. Fieux, and S. Aubas for the PneumA Trial Group, *JAMA*, 2003, **290**, 2588–2598.

C. M. Cimarusti, D. P. Bonner, H. Breuer, H. W. Chang, A. W. Fritz, D. M. Floyd, T. P. Kissick, W. H. Koster, D. Kronenthal, F. Massa, R. H. Mueller, J. Pluscec, W. A. Slusarchyk, R. B. Sykes, M. Taylor, and E. R. Weaver, *Tetrahedron*, 1983, **39**, 2577–2589.

CREST, Guidelines for the Management of Cellulitis, 2005 (http://www.acutemed.co.uk/docs/Cellulitis%20guidelines,%20CREST,%2005.pdf, last accessed 13 March 2012).

S. P. Denyer and J.-Y. Maillard, *J. Appl. Microbiol. Symp. Suppl.*, 2002, **92** (Suppl. 1), 35S–45S.

E. A. DeSano, Jr., and S. C. Hurley, *Fertil. Steril.*, 1982, **37**, 853–854.

J. Dossetor, *Brit. Med. J.*, 1975, **4**, 467–468.

S. M. Drawz and R. A. Bonomo, *Clin. Microbiol. Rev.*, 2010, **23**, 160–201.

G. L. Drusano, H. C. Standiford, C. Bustamante, A. Forrest, G. Rivera, J. Leslie, B. Tatem, D. Delaportas, R. R. MacGregor, and S. C. Schimpff, *Antimicrob. Agents Chemother.*, 1984, **26**, 715–721.

A. Endimiani, K. M. Hujer, A. M. Hujer, M. E. Pulse, W. J. Weiss, and R. A. Bonomo, *Antimicrob. Agents Chemother.*, 2011, **55**, 82–85.

Faculty of Sexual & Reproductive Healthcare, Drug Interactions with Hormonal Contraceptives, 2011 (http://www.fsrh.org/pdfs/CEUGuidanceDrugInteractionsHormonal.pdf, last accessed 13 March 2012).

T. M. File, Jr, *Clin. Infect. Dis.*, 2010, **51**, S42–S47.

A. Fleming, *Br. J. Exp. Pathol.*, 1929, **10**, 226–236.

C. Fletcher, *Brit. Med. J.*, 1984, **289**, 1721–1723.

P. C. Fuchs, A. L. Barry, C. Thornsberry, and R. N. Jones, *Antimicrob. Agents Chemother.*, 1984, **25**, 392–394.

S. Fudio, A. Carcas. E. Pinana, and R. Ortega, *J. Clin. Pharm. Ther.*, 2006, **31**, 393–396.

A. Garg and M. Mohammed, *Ann. Pharmacother.*, 2009, **4397**, 1374–1375.

B. G. Hall and M. Barlow, *J. Antimicrob. Chemother.*, 2005, **55**, 1050–1051.

R. E. W. Hancock and A. Bell, *Eur. J. Clin. Microbiol. Infect. Dis.*, 1988, **7**, 713–720.

R. Hare, *The Birth of Penicillin*, Allen and Unwin, London, 1970.

F. D. Hart, D. Burley, R. Maley, and G. Brown, *Brit. Med. J.*, 1956, **1**(Medical Memoranda), 496–497.

W. C. Hellinger and N. S. Brewer, *Mayo Clin. Proc.*, 1999, **74**, 420–434.

S. E. Helms, D. L. Bredle, and J. Zajic, *J. Am. Acad. Dermatol.*, 1997, **365**, 705–710.

D. C. Hodgkin and E. N. Maslen, *Biochem. J.*, 1961, **79**, 393–402.

J. Huwyler, M. B. Wright, H. Gutmann, and J. Drewe, *Curr. Drug Metab.*, 2006, **7**, 119–126.

M. Kadurina, G. Bocheva, and S. Tonev, *Clin. Dermatol.*, 2003, **21**, 12–23.

J. S. Kahan, F. M. Kahan, and R. Goegelman, *J. Antibiot.*, 1979, **32**, 1–12.

T. Kesado, T. Hashizume, and Y. Asahi, *Antimicrob. Agents. Chemother.*, 1980, **17**, 912–917.

M. Kinzig, F. Sorgel, B. Brismas, and C. E. Nord, *Antimicrob. Agents Chemother.*, 1992, **36**, 1997–2004.

W. M. Kirby, *Science*, 1944, **99**, 452–453.

S. Koning, A. P. Verhagen, L. W. A. van Suijlekom-Smit, A. D. Morris, C. Butler, and J. C. van der Wouden, *Cochrane Database of Systematic Reviews*, 2003, Issue 2.

P. A. Lambert, *J. Appl. Microbiol. Symp. Suppl.*, 2002, **92**(Suppl. 1), 46S–54S.

B. B. Levine and Z. Ovary, *J. Exp. Med.*, 1961, **114**, 875.

D. P. Lew and F. A. Waldvogel, *Lancet*, 2004, **364**, 369–379.

Z. J. Liang, L. Li, Y. Wang, L. Chen, X. Kong, Y. Hong, L. Lan, M. Zheng, C. Guang-Yang, H. Liu, X. Shen, C. Luo, K. K. Li, K. Chen, and H. Jiang, *PLoS One*, 2011, **6**, e23606.

D. M. Livermore and P. M. Tulkens, *J. Antimicrob. Chemother.*, 2009, **63**, 243–245.

I. Lutsar, G. H. McCracken, Jr, and I. R. Friedland, *Clin. Infect. Dis.*, 1998, **27**, 1117–1129.

J. Lynch, *Chest*, 2001, **119**, 373S–384S.

A. MacGowan, *Curr. Opin. Pharmacol.*, 2011, **11**, 1–7.

R. R. MacGregor and A. L. Graziani, *Clin. Infect. Dis.*, 1997, **24**, 457–467.

A. T. Mailloux, B. E. Gidal, and C. A. Sorkness, *Ann. Pharmacother.*, 1996, **30**, 1402–1407.

W. H. Manwaring, *Cal. West Med.*, 1943, **59**, 257.

W. F. Marshall and J. E. Blair, *Mayo Clin. Proc.*, 1999, **74**, 187–195.

R. G. Masterton, A. Galloway, G. French, M. Street, J. Armstrong, E. Brown, J. Cleverley, P. Dilworth, C. Fry, A. D. Gascoigne, A. Knox, D. Nathwani, R. Spencer, and M. Wilcox, *J. Antimicrob. Chemother.*, 2008, **62**, 5–34.

S. J. Matthews and J. W. Lancaster, *Clin. Therapeut.*, 2009, **31**, 42–63.

H. Mattie, *Clin. Pharmacokinet.*, 1994, **26**, 99–106.

B. Mayall, G. Poggi, and J. D. Parkin, *Med. J. Aust.*, 1991, **155**, 480–484.

B. R. Meyers, E. S. Srulevitch, J. Jacobson, and S. Z. Hirschman, *Antimicrob. Agents Chemother.*, 1983, **24**, 812–814.

A. Miranda, M. Blanca, J. M. Vega, F. Moreno, M. J. Carmona, and J. J. Gracia, *J. Allergy Clin. Immunol.*, 1996, **98**, 671–677.

J. W. Mouton, D. J. Touw, A. M. Horrevorts, and A. A. T. M. M. Vinks, *Clin. Pharmacokinet.*, 2000, **39**, 185–201.

Y. Nakajima, M. Mizobuchi, M. Nakamura, H. Takagi, H. Inagaki, G. Kominami, M. Koike, and T. Yamaguchi, *Drug Metab. Dispos.*, 2004, **32**, 1383–1391.

D. Nathwani and M. J. Wood, *Drugs*, 1993, **45**, 866–894.

N. M. Nielsen and H. Bundegaard, *J. Pharm. Pharmacol.*, 1988, **40**, 506–509.

D. W. Nierenberg and R. D. Mamelok, *Arch. Dermatol.*, 1983, **119**, 449–450.

C. M. Paap and M. C. Nahata, *Clin. Pharmacokinet.*, 1990, **19**, 280–318.

P. S. Padayatti, M. S. Helfand, M. A. Totir, M. P. Carey, P. R. Carey, R. A. Bonomo, and F. van den Akker, *J. Biol. Chem.*, 2005, **280**, 34 900–34 907.

D. Pagan-Rodriguez, X. Zhou, R. Simmons, C. R. Bethel, A. M. Hujer, M. S. Halfand, Z. Jin, B. Guo, V. E. Anderson, L. M. Ng, and R. A. Bonomo, *J. Biol. Chem.*, 2004, **279**, 19 494–19 501.

T. J. Pallasch, *Anaesth. Prog.*, 1988, **35**, 133–146.

K. M. Papp-Wallace, C. R. Bethel, A. M. Distler, C. Kasuboski, M. Taracila, and R. A. Bonomo, *Antimicrob. Agents Chemother.*, 2010, **54**, 890–897.

D. L. Paterson and R. A. Bonomo, *Clin. Microbiol. Rev.*, 2005, **18**, 657–686.

C. M. Perry and R. N. Brogden, *Drugs*, 1996, **52**, 125–158.

Y. Pfeifer, A. Cullik, and W. Witte, *Int. J. Med. Microbiol.*, 2010, **300**, 371–379.

A. S. Prashad, N. Vlahos, P. Fabio, and G. B. Feigelson, *Tet. Letters*, 1998, **39**, 7035–7038.

K. E. Price, *Adv. Appl. Microbiol.*, 1970, **11**, 17–75.

C. P. Rains, H. M. Bryson, and D. H. Peters, *Drugs*, 1995, **49**, 577–617.

J. Rautio, H. Kumpulainen, T. Heimbach, R. Oliyai, D. Oh, T. Järvinen, and J. Savolainen, *Nature Rev. Drug Discov.*, 2008, **7**, 255–270.

S. N. Richter, I. Frasson, C. Bergo, S. Parisi, A. Cavallaro, and G. Palù, *J. Clin. Microbiol.*, 2011. **49**, 2040–2042.

K. Roholt, B. Nielsen, and E. Kristensen, *Chemother.*, 1975, **21**, 146–166.

C. L. Ronchera, T. Hernández, J. E. Peris, F. Torres, L. Granero, N. V. Jiménez, and J. M. Plá, *Ther. Drug Monit.*, 1993, **15**, 375–379.

J. C. Sheehan and K. R. Henerey-Logan, *J. Am. Chem. Soc.*, 1959, **81**, 3089–3094.

C. M. Shenfield, *Drug Safety*, 1993, **9**, 21–37.

H. E. Sidjabat, F. P. Silveira, B. A. Potoski, K. M. Abu-Elmagd, J. M. Adams-Haduch, D. L. Paterson, and Y. Doi, *Clin. Infect. Dis.*, 2009, **49**, 1736 1738.

I. Spriet, J. Govens, W. Messerrman, L. Wilmer, and W. Van Paesschen, *Ann. Pharmacother.*, 2007, **41**, 1130–1136.

T. Stachyra, P. Levasseur, M. C. Péchereau, A. M. Girard, M. Claudon, C. Miossec, and M. T. Black, *J. Antimicrob. Chemother.*, 2009, **64**, 326–329.

R. Sutherland, E. A. P. Croydon, and G. N. Rolinson, *Brit. Med. J.*, 1970, **4**, 455–460.

M. Suvorov, S. B. Vakulenko, and S. Mobashery, *Antimicrob. Agents Chemother.*, 2007, **51**, 2937–2942.

C. K. Svensson, E. W. Cowen, and A. A. Gaspari, *Pharmacol. Rev.*, 2001, **53**, 357–379.

J. K. Tobin, L. K. Golightly, S. D. Kick, and M. A. Jones, *Drug Metab. Drug Interact.*, 2009, **24**, 153–182.

A. Torres, R. Aznar, J. M. Gatell, P. Jiménez, J. González, A. Ferrer, R. Celis, and R. Rodriguez-Roisin, *Am. Rev. Respir. Dis.*, 1990, **142**, 523–528.

M. J. Torres, M. Blanca, A. de Weck, J. Fernandez, P. Demoly, A. Romano, and W. Abereer, ENDA & EAACI Interest Group on Drug Hypersensitivity, *Allergy*, 2003, **58**, 854–863.

J. Turnidge and D. L. Paterson, *Clin. Microbiol. Rev.*, 2007, **20**, 391–408.

T. R. Walsh, *Int. J. Antimicrob. Agents*, 2010, **36**, S8–14.

T. R. Walsh and M. A. Toleman, *Exp. Rev. Anti Infect. Ther.*, 2011, **9**, 137–141.

P. A. Winstanley and M. L'E. Orme, *Br. J. Clin. Pharmac.*, 1989, **28**, 621–628.

R. Wise, *J. Antimicrob. Chemother.*, 1990, **26** (Suppl. E), 13–20.

A. J. Wright, *Mayo Clin. Proc.*, 1999, **74**, 290–307.

H. Yigit, A. M. Queenan, G. J. Anderson, A. Domenech-Sanchez, J. W. Biddle, C. D. Steward, S. Alberti, K. Bush, and F. C. Tenover, *Antimicrob. Agents Chemother.*, 2001, **45**, 1151–1161.

D. Yong, M. A. Toleman, C. G. Giske, H. S. Cho, K. Sundman, K. Lee, and T. R. Walsh, *Antimicrob. Agents Chemother.*, 2009, **53**, 5046–5054.

R. Zarychanski, K. Wlodarczyk, R. Ariano, and E. Bow, *J. Antimicrob. Ther.*, 2006, **58**, 228–230.

G. G. Zhanel, R. Wiebe, L. Dilay, K. Thomson, E, Rubinstein, D. J. Hoban, A. M. Noreddin, and J. A. Karlowsky, *Drugs*, 2007, **67**, 1027–1052.

5.2

Glycopeptide antibiotics

Key Points

- Vancomycin is used for the treatment of meticillin-resistant *Staphylococcus aureus* (MRSA) infections.
- The glycopeptides bind to D-ala-D-ala and form a 'protective layer', thus preventing some of the steps involved in cell wall assembly.
- Resistance to vancomycin can involve the production of an alternative ligase, which produces D-ala-D-lac. D-ala-D-lac performs the same role in cell wall assembly as D-ala-D-ala but has greatly reduced binding affinity for the glycopeptides.

The glycopeptide antibiotics, teicoplanin (a mixture of nine components: A2-1–A2-5 and RS-1–RS-4) and vancomycin (Figure 5.2.1), are used in the prophylaxis and treatment of severe infections caused by Gram positive bacteria which are penicillin-resistant (or in individuals who are penicillin-intolerant). The therapeutic indications for these antibiotics are given in Table 5.2.1.

5.2.1 Discovery

Vancomycin was discovered when, in 1952, a missionary in Borneo sent a soil sample to a friend at the pharmaceutical giant Eli Lilly and Co. It was soon found that *Streptomyces orientalis* (also referred to as *Amycolatopsis orientalis*), an organism present in the sample, produced a compound which had activity against most Gram positive (including penicillin-resistant staphylococci), and some anaerobic, organisms. Originally referred to as 05865, this compound is now known as vancomycin, with its name being derived from the word 'vanquish' (a strong claim, which could easily have been the kiss of death for the drug) (Levine, 2006).

Like vancomycin, the teicoplanins were discovered from a soil isolate, this time from an organism called *Actinoplanes teichomyceticus* (Parenti *et al.*, 1978).

Antibacterial Agents: Chemistry, Mode of Action, Mechanisms of Resistance and Clinical Applications, First Edition.
Rosaleen J. Anderson, Paul W. Groundwater, Adam Todd and Alan J. Worsley.
© 2012 John Wiley & Sons, Ltd. Published 2012 by John Wiley & Sons, Ltd.

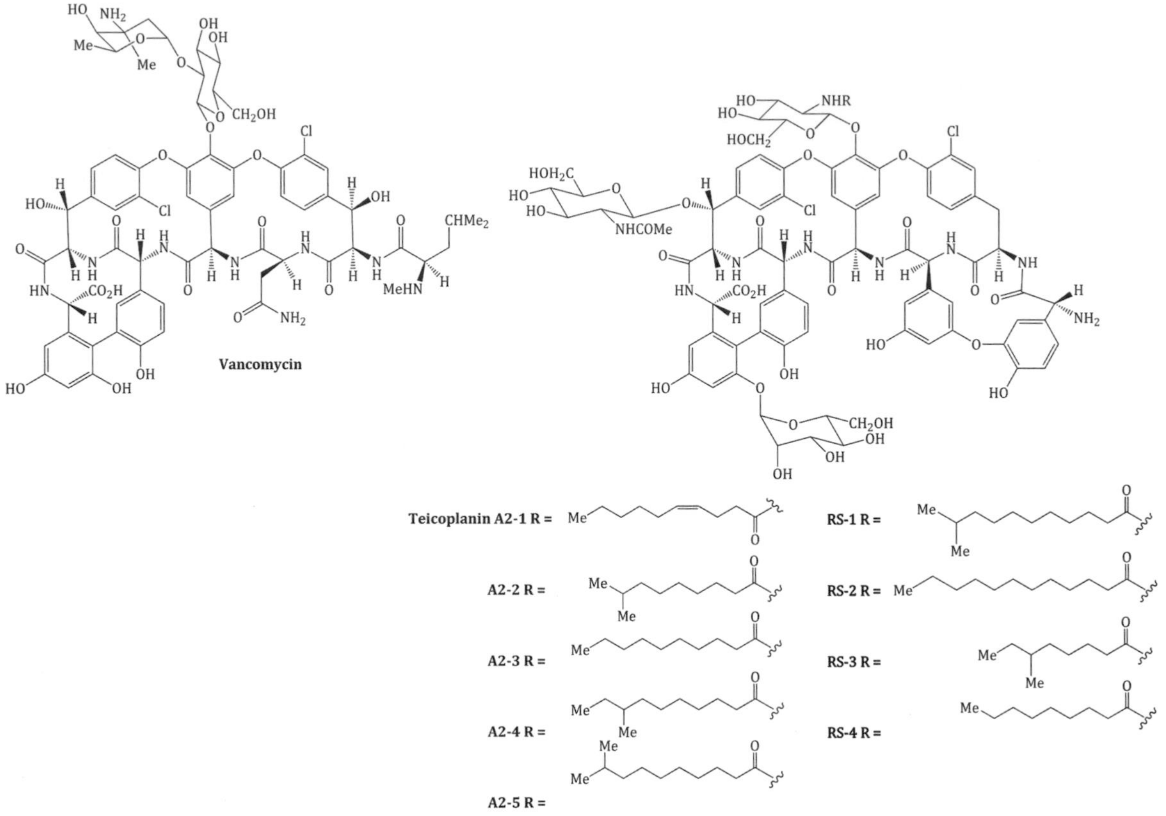

Figure 5.2.1   *Glycopeptide antibiotics*

Table 5.2.1   *Therapeutic indications for the glycopeptide antibiotics*

Glycopeptide antibiotic	Indications
Teicoplanin	Serious Gram positive infections, e.g. septic arthritis or MRSA
Vancomycin	Endocarditis MRSA, *Clostridium difficile*-associated diarrhoea

5.2.2   Synthesis

The structural complexity of these antibiotics, with their cyclic rings incorporating unusual amino acids, unnatural sugar units, atropisomerism,[17] and numerous stereogenic centres (nine in the vancomycin aglycone[18] and eight in that of teicoplanin), meant that the elucidation of their structure was a slow process (Barna and Williams, 1984). This structural complexity also means that total synthesis of these antibiotics, while fascinating from an academic viewpoint, will, once again, never be able to produce sufficient quantities

---

[17] Atropisomerism is a form of chirality which arises due to restricted rotation in a molecule (in this case, in each of the cyclic tripeptide units) and results in non-superimposable stereoisomers.
[18] In natural product chemistry, the portion of a molecule which does not include the sugar units present in the natural product is known as the aglycone.

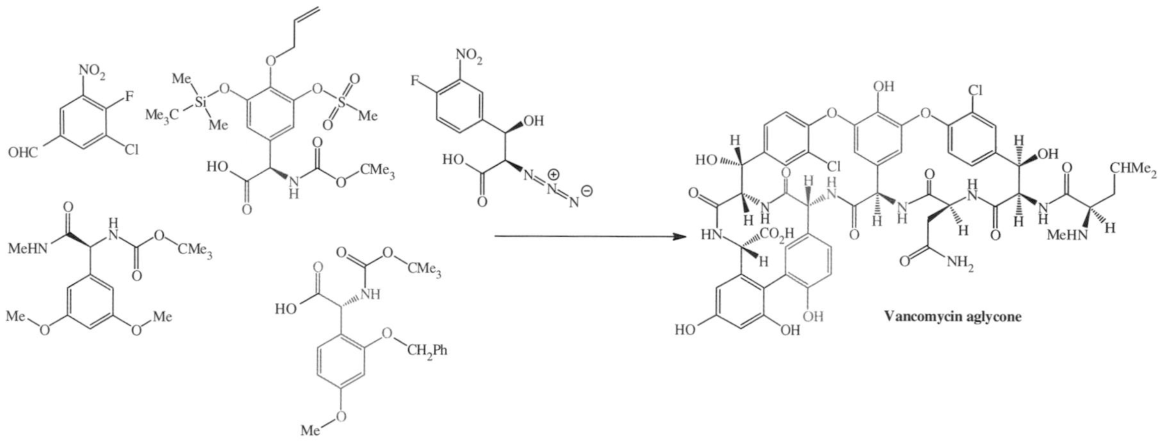

Scheme 5.2.1 *Building blocks used in the Evans synthesis of the vancomycin aglycone (Evans et al., 1998)*

for clinical use. The challenge associated with the total synthesis of the glycopeptides antibiotics has again attracted some of the world's leading synthetic chemists, and the first reported synthesis of the aglycone portion, with the longest linear synthetic sequence consisting of 40 steps, is outlined in Scheme 5.2.1 (Evans *et al.*, 1998).

The availability of the vancomycin aglycone soon led to the total synthesis of vancomycin and the possibility of the incorporation of alternative disaccharide units (Thompson *et al.*, 1999). As mentioned above, these total syntheses were never going to be sufficient for the production of sufficient quantities of the drug, and vancomycin is produced on an industrial scale by the fermentation of *A. orientalis* (Padma *et al.*, 2002).

Teicoplanin has also been prepared by a total synthesis (Boger *et al.*, 2000), but is produced on an industrial scale by a submerged culture of *A. teichomyceticus*. Rather unusually, this microorganism is sensitive to the antibiotic it produces (a bit of a design fault, to say the least) (Jung *et al.*, 2009) and so innovative methods have been employed to enable large-scale production, including mutagenesis so that the bacterial cells become more tolerant to teicoplanin (Jung *et al.*, 2008) and, once it is formed, trapping the antibiotic on an absorbent resin (Lee *et al.*, 2003).

5.2.3 Bioavailability

As a result of its multiple anionic and cationic charged groups and hydrophilicity, vancomycin is poorly absorbed from the gastrointestinal (GI) tract and has a bioavailability of less than 2% (Anderson *et al.*, 2001). In addition, the plasma half-life of vancomycin is 3–13 hours (Felmingham, 1993), so it is usually administered IV, with doses of 500 mg every 6 hours (or 1 g every 12 hours). For twice-daily dosing, the trough level should be 15–20 mg/L as this minimises the chances of resistance developing (Rybak *et al.*, 2009). No relationship appears to exist between the peak concentrations, trough concentrations, and pharmacodynamic parameters (such as time above the MIC or AUC/MIC) and the eradication of the microorganism eradication or patient outcome (Rybak, 2006). Oral vancomycin is used only for the treatment of *Clostridium difficile*-associated diarrhoea as, being poorly absorbed, the orally administered vancomycin remains in the gut and so acts very effectively on the *C. difficile* present.

Vancomycin is eliminated by the kidneys (with 80–90% recovered unchanged in urine within 24 hours of a single dose), so the dose should be altered in renally impaired patients, using the patient's creatinine clearance rate to determine the appropriate dose and schedule.

Vancomycin penetrates into most body spaces, with values for the apparent volume of distribution of 0.4–1 L/Kg (Rybak *et al.*, 2009).

Teicoplanin is also excreted by the kidney but is 90% bound to serum albumin (protein binding for vancomycin is less than 50%) and, as it has a significantly longer half-life, only requires once-daily administration (Felmingham, 1993; Yu *et al.*, 1995).

For the same reasons of ionisation, hydrophilic nature, and large size that confer poor oral bioavailability, vancomycin and teicoplanin do not traverse the Gram negative bacterial membrane, which is presumably why they are inactive against these species. As their site of action is the bacterial cell wall (as explained in Subsection 5.2.4) and the cell wall of Gram positive bacteria is less complex and so easier to breach, the glycopeptide antibiotics are able to access their site of action sufficiently well to exert their bacteriostatic effect. For more on the bioavailability of the glycopeptides, see Rybak *et al.* (2009).

5.2.4   Mode of action and selectivity

The glycopeptides interfere with cell-wall peptidoglycan synthesis as a result of their binding affinity for the D-ala-D-ala amino acid sequence present in the transpeptidation precursors. As you will remember from the introductory section, the biosynthesis of peptidoglycan involves three main stages: i) synthesis of the cell-wall precursors, ii) transportation of the precursors across the cell membrane by a carrier lipid (undecaprenyl in this case), and iii) cell-wall synthesis, involving coupling of the precursors into polymeric chains and the crosslinking of these chains (Scheme 5.2.2). As eukaryotic cells do not have a cell wall, the glycopeptides are selective bactericidal agents.

The glycopeptides contain hydrogen-bond donor and acceptor groups and the aglycone portion forms five hydrogen bonds to the terminal D-ala-D-ala sequence of the precursor pentapeptide chain (three of these hydrogen bonds are shown in Figure 5.2.2) (Barna and Williams, 1984).

By binding to the D-ala-D-ala sequence, the glycopeptides inhibit a number of processes which are involved in the assembly of the cell wall, but they do not do this by directly acting on the enzymes responsible; instead they form a 'protective layer' preventing these steps, with the D-ala-D-ala unit being held within a deep cleft of the antibiotic. One of the earliest steps in cell-wall synthesis involves the transglycosylase-catalysed transfer of a disaccharide peptide from its lipid carrier to the growing peptidoglycan chain. Despite being bound to a site of this disaccharide-peptide unit that is remote from the disaccharide portion, these antibiotics inhibit this process, possibly through a steric clash with some region of the transglycosylase enzyme lying close in space to the terminal D-ala-D-ala of the peptide chain (Scheme 5.2.3a) (Reynolds, 1989).

A later step in the formation of the cell wall is the penicillin-binding protein-catalysed removal of the terminal D-ala residue (by D-alanine carboxypeptidase). This process involves the formation of an acyl-enzyme intermediate, which is subsequently attacked by a glycine amino group to give a new peptide bond, thus crosslinking the peptidoglycan chains (catalysed by peptidoglycan transpeptidase, also called PBP) (Scheme 5.2.3b). As the glycopeptide antibiotics bind to the D-ala-D-ala sequence, they once again prevent this process from taking place (Reynolds, 1989).

In summary, as a result of vancomycin's binding to the D-ala-D-ala sequence at the end of the pentapeptide chain of the peptidoglycan precursor, transglycosylase and transpeptidase activity are inhibited and cell-wall formation is disrupted.

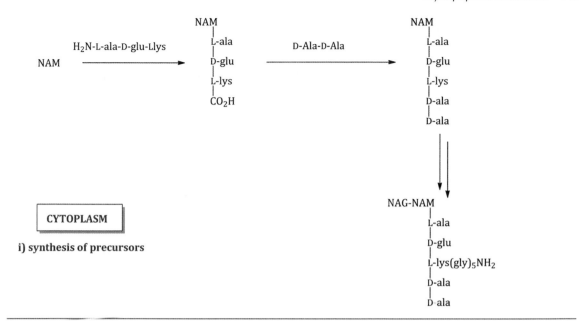

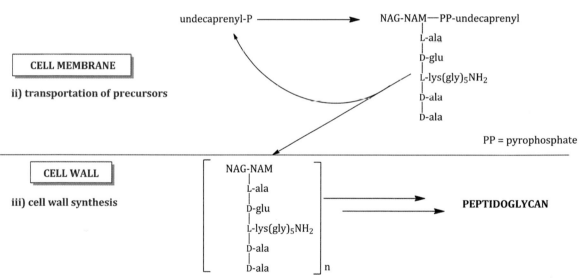

Scheme 5.2.2 *Peptidoglycan biosynthesis*

5.2.5 Bacterial resistance

Now that we know that the glycopeptide antibiotics act by specifically binding to the D-ala-D-ala dipeptide sequence, it should come as no surprise that bacterial resistance to these agents involves the modification of this sequence through alterations in the ligase activity (the enzyme which joins the two amino acids to produce the dipeptide).

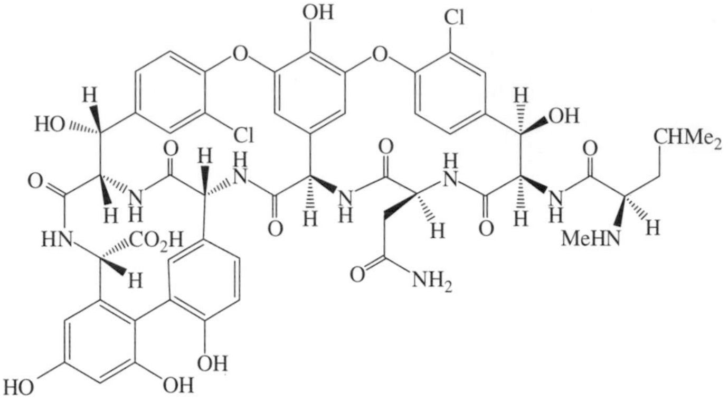

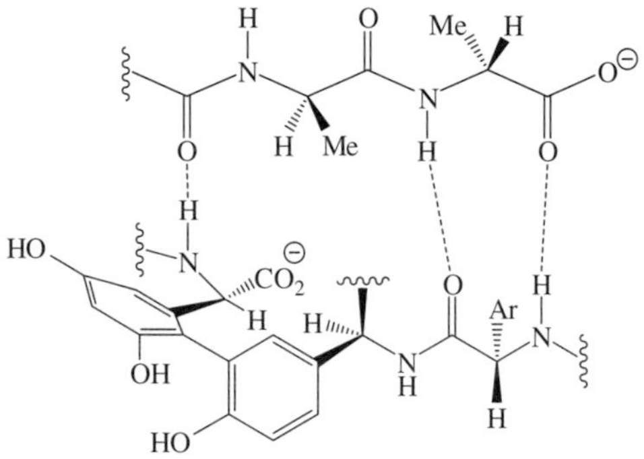

***Figure* 5.2.2**   *Some of the hydrogen bonds (red) formed between vancomycin and* D-ala-D-ala. *The portion of vancomycin corresponds to the region highlighted in blue in the full structure of the aglycone*

There are five types of glycopeptide resistance in vancomycin-resistant enterococci (VRE), classified as VanA, B, C, D, and E. The first four of these are fairly well understood, and share some degree of similarity, so we will concentrate on the VanA phenotype, in which resistance is mediated by the transposon containing the nine genes of the *vanA* gene cluster. Four of these genes are responsible for the molecular processes giving rise to vancomycin resistance; *vanH* produces VanH, a dehydrogenase which reduces pyruvate to D-lactate (D-lac), for subsequent incorporation into D-ala-D-lac by VanA (the product of the *vanA* gene), a D-ala-D-lactate ligase which synthesises an ester (D-Ala-D-lac) rather than an amide (D-ala-D-ala) (Table 5.2.2). The D-ala-D-lactate sequence replaces D-ala-D-ala in the pentapeptide chain (the resulting chain is now not, of course, a pentapeptide). Crucially, D-ala-D-lac has 1000-fold reduction in affinity for vancomycin (being able to form fewer hydrogen bonds) but can still be attacked by glyNH$_2$ and so act as precursor for crosslink formation (Scheme 5.2.4) (Malathum and Murray, 1999). Such altered ligases are also produced by vancomycin-producing microorganisms (Kuzin *et al.*, 2000).

The generation of such a vancomycin-resistant peptidoglycan precursor would be insufficient for high-level resistance as long as the normal ligase was producing the D-ala-D-ala sequence (to which vancomycin could bind) and this was still being incorporated into the pentapeptide precursor. In order to overcome this, *vanX* produces VanX, a D,D-dipeptidase which cleaves the D-ala-D-ala unit, while *vanY* produces VanY, a D,D-carboxypeptidase which cleaves the terminal D-ala from the pentapeptide (Scheme 5.2.5) (Malathum and Murray, 1999), so that there is no remaining D-ala-D-ala to be used in peptidoglycan synthesis.

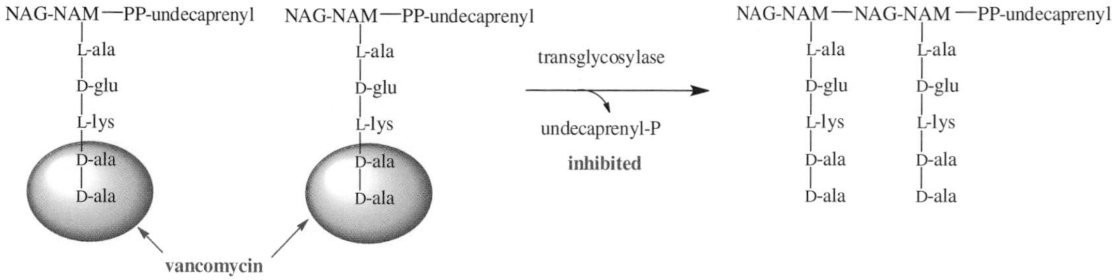

a Transglycosylase inhibition

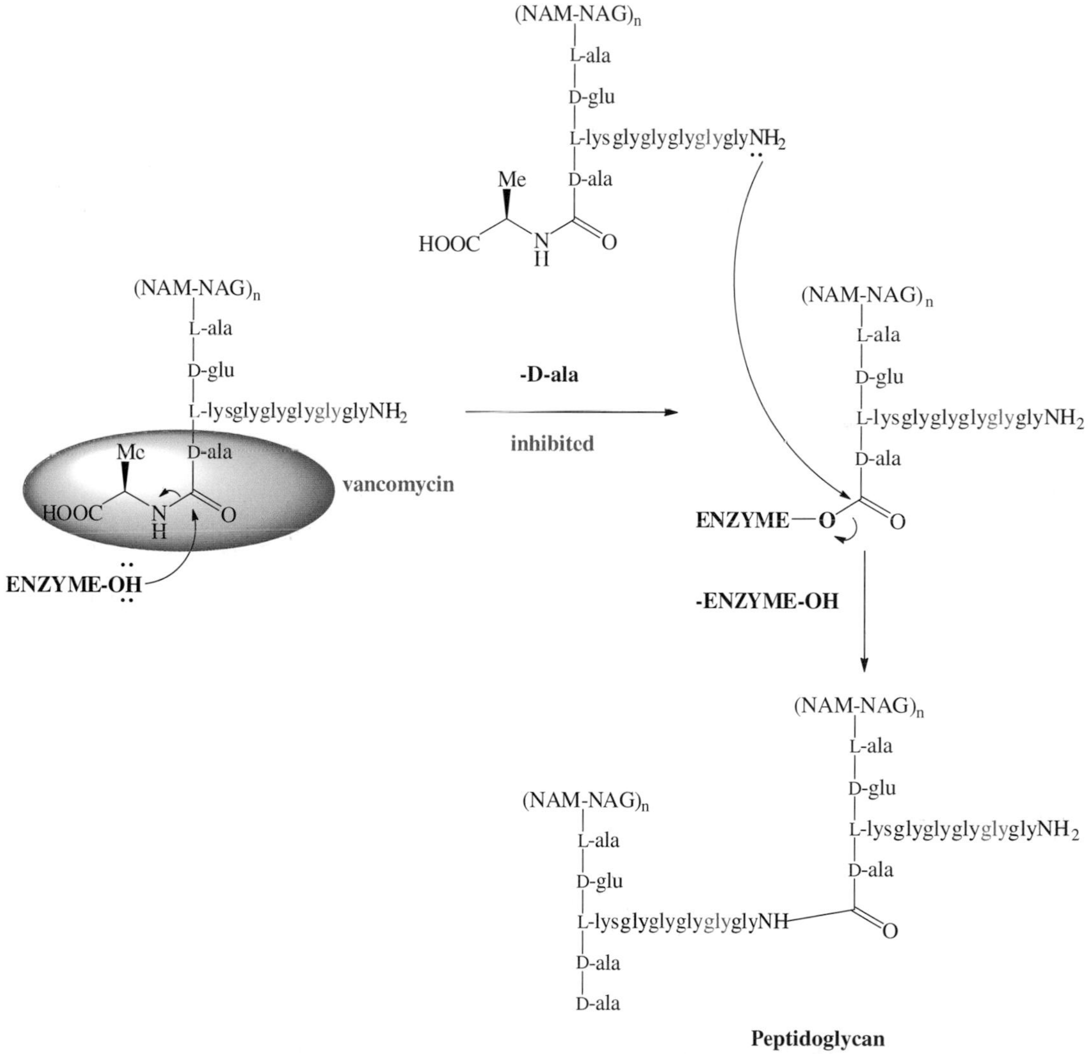

b Transpeptidase inhibition

Scheme 5.2.3 *Schematic representation of vancomycin inhibition of (a) transglycosylase and (b) peptidoglycan transpeptidase activity*

Table 5.2.2   *VRE VanA resistance phenotype genes and their enzyme product/function*

Gene	Enzyme product and function
vanS	VanS (membrane-associated sensor kinase,[a] which recognises the presence, or an effect, of vancomycin)
vanR	VanR (cytoplasmic response regulator which receives a signal from VanS then activates its own promoter and those of VanH, VanA, and VanX)
vanH	VanH (dehydrogenase responsible for reduction of pyruvate to D-lactate)
vanA	VanA (D-ala-D-lac ligase which utilises the D-lactate product of VanH)
vanX	VanX (a D,D-dipeptidase which hydrolyses D-ala-D-ala to individual D-ala units, thus reducing the cellular quantities of the dipeptide unit and preventing its incorporation into the pentapeptide chain)
vanY	VanY (a D-carboxypeptidase which removes the terminal D-ala from the pentapeptide chain to give a tetrapeptide)
vanZ	Function unknown, confers low-level resistance to teicoplanin only
ORF1	Transposase which binds to the end of transposon *Tn1546* and catalyses its movement to another part of the genome
ORF2	Resolvase which is responsible for the recombination of transposon *Tn1546* and the bacterial genome

[a] Kinases catalyse the phosphorylation of a substrate.

VRE were isolated in the late 1980s in the UK and have now been detected worldwide; the majority being *Enterococcus faecium* and resistant to multiple antibiotics. The transferable nature of vancomycin resistance in the laboratory, and the appearance of acquired vancomycin-resistance genes in other organisms, is a growing health care concern. For example, the first *Staphylococcus aureus* strains with reduced vancomycin susceptibility were isolated in 1997, and there are various organisms with reduced susceptibility, including

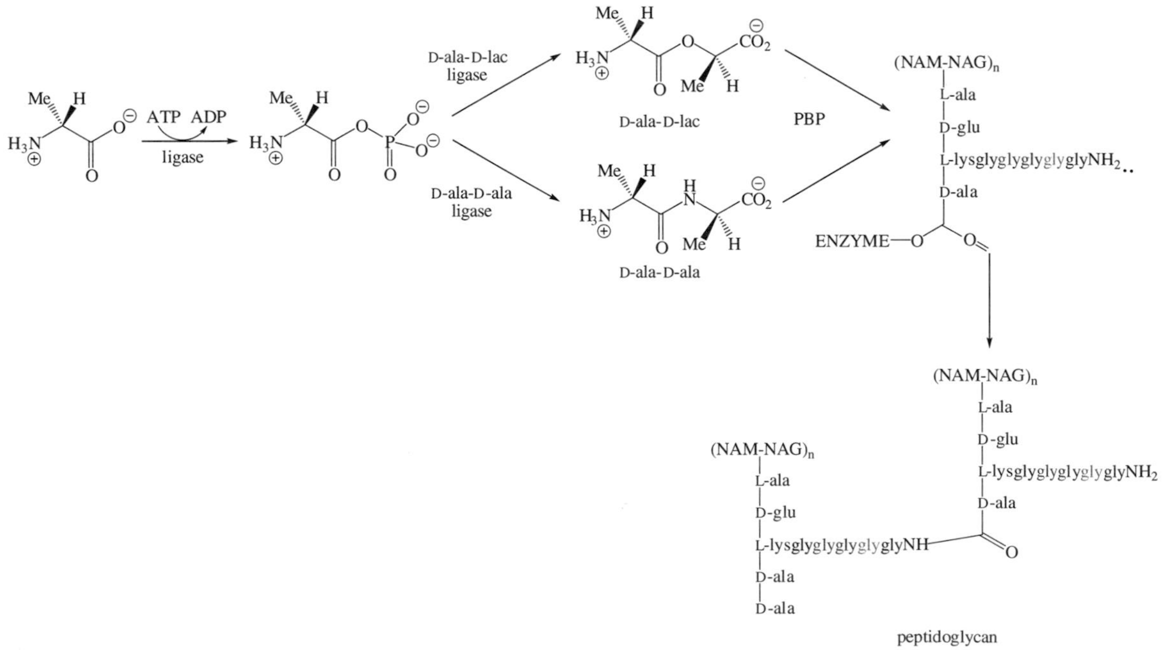

Scheme 5.2.4   *Peptidoglycan precursors incorporating D-ala or D-lac*

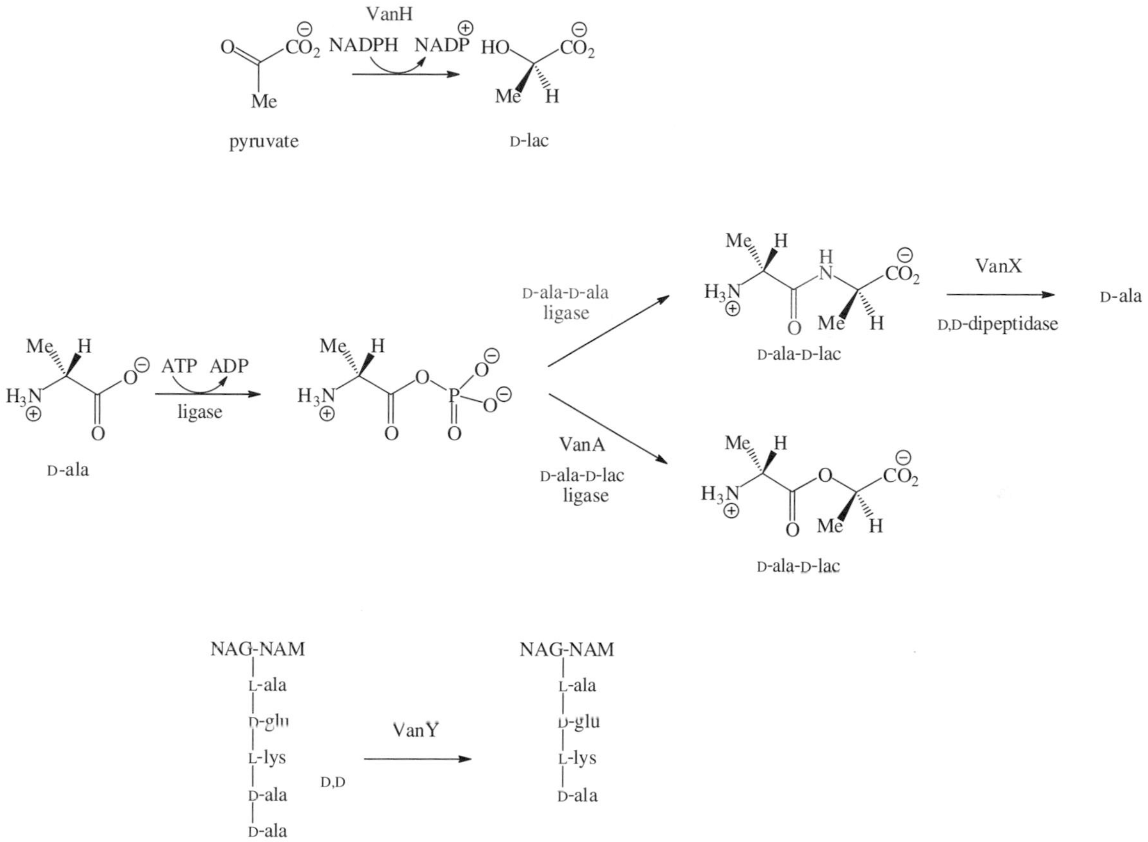

Scheme 5.2.5   *Vancomycin resistance in VanA-type isolates (Malathum and Murray, 1999)*

vancomycin-resistant *Staphylococcus aureus* (VRSA) (for which the vancomycin MIC is greater than 16 mg/L) and vancomyin-intermediate *Staphylococcus aureus* (VISA) (vancomycin MIC 4–8 mg/L), in which resistance is at least partly due to a thickened cell wall, resulting in decreased vancomycin penetration and decreased peptidoglycan crosslinking (Rong and Leonard, 2010). The *vanA* gene cluster has now found its way via a plasmid into meticillin-resistant *Staphylococcus aureus* (MRSA), leading to vancomycin resistance (Weigel *et al.*, 2003). Vancomycin-resistant MRSA has been identified in a number of locations – leaving worryingly few treatment options for infections caused by this pathogenic organism.

5.2.6 Clinical applications

5.2.6.1 Spectrum of activity

The glycopeptides vancomycin and teicoplanin show activity against a range of aerobic and anaerobic Gram positive bacteria (many of which show resistance to other antibiotics), including, perhaps most notably, staphylococci (e.g. MRSA), enterococci (e.g. *E. faecalis*), and clostridia species (e.g. *C. difficile*). However, there are also some Gram positive bacteria that are inherently resistant to the glycopeptides, including *Leuconostoc*, *Pediococcus*, and *Lactobacillus* species (Swenson *et al.*, 1990) – although thankfully these organisms are seldom implicated in causing disease. Furthermore, due to the large size of the glycopeptides and their ionisation and hydrophilicity, they are unable to penetrate the outer membrane of Gram negative bacteria and are thus considered to be inactive against these organisms.

5.2.6.2   *Clostridium difficile* infection

We have already discussed *Clostridium difficile* infection (CDI) in Section 2, when we outlined the clinical uses of the 5-nitroimidazole antibiotics (see Subsection 2.3.6). Vancomycin can also be used to treat CDI, but is generally reserved for occasions when a patient has a 'severe' infection, so it is important to establish the severity of the infection in order to ensure that the patient receives the most appropriate treatment.

Generally, with mild CDI patients pass more than three stools per day, which are categorised as type 5 (soft blobs passed easily) to 7 (watery, with no solid pieces) on the Bristol Stool Chart. In cases of moderate CDI, again the patient will pass more than three stools per day of type 5–7 (as for minor CDI), but will also have a raised white cell count. When the CDI is classed as minor or moderate, treatment with metronidazole is usually sufficient as trial data suggests that the efficacy of metronidazole and vancomycin is similar in this patient group (Zar *et al.*, 2007).

For severe CDI, however, the patient may present with an elevated temperature, a raised white cell count, and a rising serum creatinine level. Interestingly, in this situation, the number of stools passed per day may be a less reliable indicator of the severity of infection. In this case, oral vancomycin should be used, as trial data suggests that it is superior to metronidazole in this patient group. Indeed, one clinical trial showed that among patients with severe *Clostridium difficile*-associated diarrhoea (CDAD), treatment with metronidazole resulted in a cure rate of 76%, while vancomycin resulted in a statistically significant higher cure rate of 97% (Zar *et al.*, 2007). Generally, a dose of 125 mg of vancomycin four times daily is used for the treatment of severe CDI, but if the infection is considered to be life-threatening, doses as high as 500 mg four times daily have been indicated.

5.2.6.3   Serious Gram positive infections resistant to other antibiotics

The glycopeptides vancomycin and teicoplanin are also used parenterally to manage serious Gram positive infections resistant to other antibiotics. They should only be used when a resistant organism, such as MRSA, is either strongly suspected or has been identified. It is important that guidelines are followed when prescribing the glycopeptides (or indeed, any other antibiotic), to minimise the threat of bacterial resistance.

When MRSA is either suspected or has been identified, the glycopeptides can be used in the treatment of endocarditis, community acquired pneumonia (CAP), hospital acquired pneumonia (HAP), septicaemia, osteomyelitis, septic arthritis, and complicated skin and soft tissue infection (cSSTIs). There appears to be more clinical experience with vancomycin, but as teicoplanin has a similar spectrum of activity, it is, for some conditions, considered an appropriate alternative to vancomycin. In addition, if a patient has an allergy to the recommended first-line treatment for one of the above conditions, it may be appropriate, in some circumstances, to use a glycopeptide antibiotic (e.g. a patient with a penicillin allergy who has endocarditis), although the clinical guidelines should always be consulted before any prescribing decisions are made.

5.2.7   Adverse drug reactions

As mentioned above, vancomycin and teicoplanin have been used extensively in the treatment of Gram positive infections. It is commonly understood that vancomycin may cause a number of side effects when specific serum concentrations are achieved. In comparison, teicoplanin has a longer half life and may be given as an IV bolus dose or via IM injection, with a lower incidence of nephrotoxicity and ototoxicity (Wood, 1996).

The most common adverse events observed with vancomycin administration are skin disorders, which we shall now look at in further detail.

5.2.7.1 Red man syndrome (vancomycin)

Vancomycin is responsible for the cause of several hypersensitivity reactions, which may range from localised skin reactions to cardiovascular collapse, the most common of which is red man syndrome, a rate-dependent infusion reaction and so not a true allergic reaction. Red man syndrome is characterised by erythema, flushing, and pruritis (generally affecting the neck, face, and upper body) (Figure 5.2.3). Pain, hypotension, and dyspnoea may also occur (Symons *et al.*, 1985; Hepner and Castells, 2003). These anaphylactoid-type reactions are thought to occur via a direct non-immunoglobulin E-mediated release of mediators, particularly histamine, from mast cells or from complement activation. Red man syndrome is rarely life-threatening, but severe cardiovascular toxicity, should it occur, may be fatal (Mayhew and Deutsch, 1985). Administration of diphenhydramine has been shown to reduce the incidence of red man syndrome (Wallace *et al.*, 1991), while a comparison of 1- and 2-hour IV-infusion of vancomycin showed that slower infusion rates resulted in a lower incidence of histamine release and red man syndrome (Healy *et al.*, 1990; Kiazawa *et al.*, 2006).

Isolated cases of vancomycin-induced linear IgA bullous disease have been reported (Neughebauer *et al.*, 2002) – a rare skin-blistering disorder which, unlike red man syndrome, does not appear to be dose-dependent (i.e. infusion rate-dependent), is reversible upon drug removal, and recurs upon resumption of

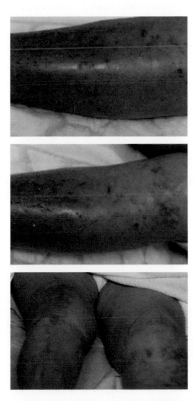

Figure 5.2.3 Patient with vancomycin-related red man syndrome (Reprinted from M. Nallasivan, F. Maher and K. Murthy, BMJ Case Rep, 2009, available from: http://casereports.bmj.com/content/2009/bcr.03.2009.1705 [Last accessed 11.3.2012] with permission of BMJ.)

treatment with the drug. Linear IgA bullous dermatosis is an autoimmune blistering disease which occurs due to the release of IgA at the dermis–epidermis junction. Severe skin reactions are not experienced with teicoplanin (Davey and Williams, 1991), but hypersensitivity syndrome has been observed and is estimated to be comparable to that of the β-lactams (Perrett and McBride, 2004).

5.2.7.2   Thrombocytopenia

Vancomycin has been implicated in thrombocytopenia, albeit rarely, with limited evidence that this reaction is induced by immunological mechanisms (Von Dryglaski *et al.*, 2007). Studies have detected platelet reactive antibodies (IgG and IgM class) with vancomycin-associated thrombocytopenia, which can be removed upon drug withdrawal (Christie *et al.*, 1990; Marraffa *et al.*, 2003).

5.2.7.3   Nephrotoxicity

In one study of 527 patients (in which half were treated with teicoplanin and half with vancomycin), nephrotoxicity was observed in both the teicoplanin (1.4% patients) and vancomycin (0.8% of patients) groups (Menichetti *et al.*, 1994). The exact mechanism of nephrotoxicity for vancomycin and teicoplanin is not fully understood, but is thought to occur via oxidative stress mechanisms (Yoshihiro *et al.*, 2002), and, in particular, the stimulation of oxygen consumption and elevated cellular adenosine triphosphate concentrations. These results probably indicate that vancomycin is most likely involved in oxidative phosphorylation, increasing oxygen free radicals, with subsequent proximal cell damage (King and Smith, 2004). Early cases of vancomycin nephrotoxicity were attributed to poor manufacturing processes (Levine, 2006; Farber and Moellering, 1983), while the increased doses of vancomycin which are recommended in light of MRSA infections have resulted in an increased incidence of nephrotoxicity.

5.2.8   Drug interactions

The risk of nephrotoxicity and ototoxicity may be enhanced with vancomycin administered concomitantly with other drugs that have similar adverse effects, such as the aminoglycosides (gentamicin, tobramycin, etc.) (Rybak *et al.*, 1983; Bertino *et al.*, 1993; Pauly *et al.*, 1990). There are also data to suggest that the use of furosemide, dobutamine, or dopamine with vancomycin may reduce the serum levels of vancomycin (Pea *et al.*, 2000).

5.2.9   Recent developments

In each section so far, we have taken the opportunity to inform you of any compounds that are in early stages of development, or any that are undergoing clinical trials. The glycopeptide antibiotics certainly have products in development, and indeed one compound, telavancin (Figure 5.2.4), a derivative of vancomycin, has recently been approved by the FDA for the treatment of SSTIs (Corey *et al.*, 2009). One recent study has shown that telavancin has superior bactericidal activity against heterogeneously vancomycin-intermediate *S. aureus* (hVISA) than vancomycin or linezolid (Leonard *et al.*, 2011).

While research developing new antibiotic derivatives (or in some cases, new classes of antibiotics) is very encouraging, there is, unfortunately, research being undertaken examining the mechanisms of antibiotic resistance, which is less positive. In Subsection 5.2.5, we outlined the mechanisms associated with glycopeptide resistance, and a new resistance gene cluster from *E. faecium* has recently been disclosed (Xu *et al.*, 2010). The new cluster type, called *vanM*, produces a protein, VanM, which, like VanA, is thought to

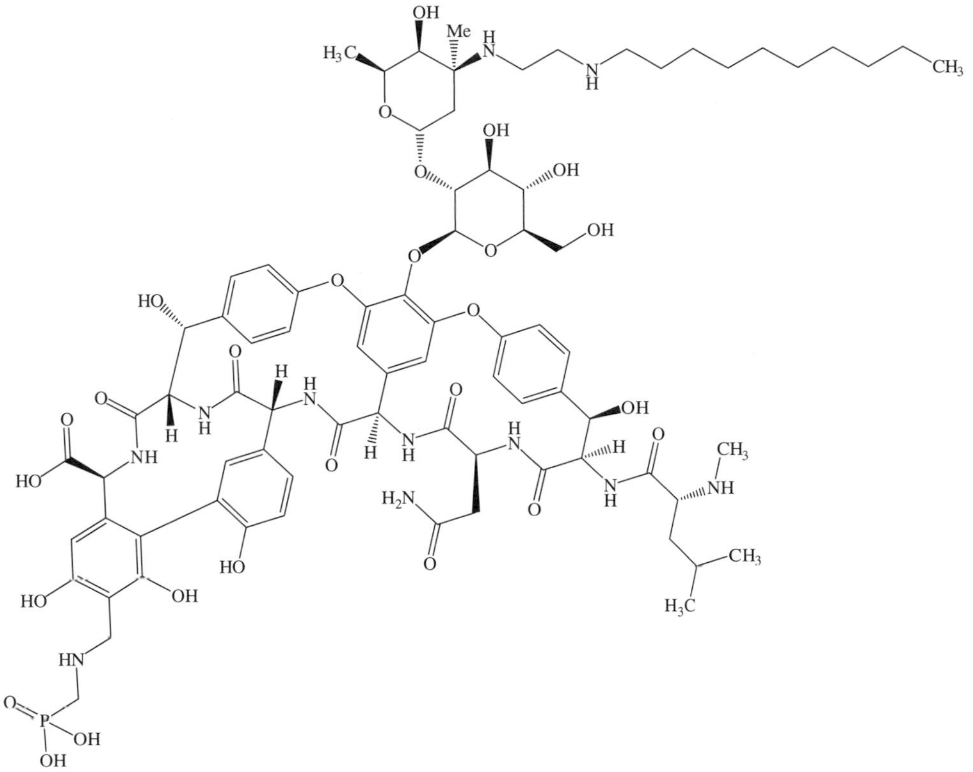

Figure 5.2.4 *Telavancin, a new glycopeptide for the treatment of cSSTIs*

act as a D-ala-D-lac ligase, resulting in the synthesis of peptidoglycan precursors with low affinities for the glycopeptide antibiotics. This new gene cluster was isolated from hospital patients in China and serves as a further reminder that research into developing new antibiotics is paramount in order to continue the progress made against the management of infectious diseases.

References

K. E. Anderson, L. A. Eliot, B. R. Stevenson, and J. A. Rogers, *Pharmaceut. Res.*, 2001, **18**, 316–322.

J. C. J. Barna and D. H. Williams, *Ann. Rev. Microbiol.*, 1984, **38**, 339–357.

D. L. Boger, S. H. Kim, S. Miyazaki, H. Strittmatter, J.-H. Weng, Y. Mori, O. Rogel, S. L. Castle, and J. J. McAtee, *J. Amer. Chem. Soc.*, 2000, **122**, 7416–7417.

D. J. Christie, N. van Buren, S. S. Lennon, and J. L. Putnam, *Blood*, 1990, **75**, 518–523.

G. R. Corey, M. E. Stryjewski, W. Weyenberg, U. Yasothan and P. Kirkpatrick, *Nat. Rev. Drug. Discov.*, 2009, **8**, 929–930.

P. G. Davey and A. H. Williams, *J. Antimicrob. Chemother.*, 1991, **27** (Suppl. B), 69–73.

D. A. Evans, M. R. Wood, B. W. Trotter, T. I. Richardson, J. C. Barrow, and J. L. Katz, *Angew. Chem. Intn. Edn. Engl.*, 1998, **37**, 2700–2704.

B. F. Farber and R. C. Moellering, Jr, *Antimicrob. Agents Chemother.*, 1983, **23**, 138–141.

D. Felmingham, *J. Antimicrob. Chemother.*, 1993, **32**, 663–666.

D. P. Healy, J. V. Sahai, S. H. Fuller, and R. E. Polk, *Antimicrob. Agents Chemother.*, 1990, **34**, 550–554.

D. L. Hepner and M. C. Castells, *Anesth. Analg.*, 2003, **97**, 1381–1395.

H.-M. Jung, S.-Y. Kim, P. Prahbu, H.-J. Moon, I.-W. Kim, and J.-K. Lee, *Appl. Microbiol. Biotechnol.*, 2008, **80**, 21–27.

H.-M. Jung, M. Jeya, S.-Y. Kim, H.-J. Moon, R. K. Singh, Y.-W. Zhang, and J.-K. Lee, *Appl. Microbiol. Biotechnol.*, 2009, **84**, 417–428.

T. Kiazawa, Y. Ota, N. Kada, Y. Morisawa, A. Yoshida, K. Koike, and S. Kimura, *Intern. Med.*, 2006, **45**, 317–321.

D. W. King and M. A. Smith, *Toxicol. In Vitro.*, 2004, **18**, 797–803.

A. P. Kuzin, T. Sun, J. Jorczak-Baillass, V. L. Healy, C. T. Walsh, and J. R. Knox, *Structure with Folding & Design*, 2000, **8**, 463–470.

J.-C. Lee, H.-R. Park, D.-J. Park, K. H. Son, K.-H. Yoon, Y.-B. Kim, and C.-J. Kim, *Biotechnol. Lett.*, 2003, **25**, 537–540.

S. N. Leonard, Y. G. Szeto, M. Zolotarev, I. V. Grigoryan, *Int. J. Antimicrob. Agents*, 2011, **37**, 558–561.

D. P. Levine, *Clin. Infect. Dis.*, 2006, **42**, S5–12.

K. Malathum and B. E. Murray, *Drug Res. Updates,1;* 1999, **2**, 224–243.

J. Marraffa, R. Guharoy, D. Duggan, F. Rose, and S. Nazeer, *Pharmacotherapy*, 2003, **23**, 1195–1198.

J. F. Mayhew and S. Deutsch, *Can. Anaesth. Soc. J.,1;* 1985, **32**, 65.

F. Menichetti, P. Martino, G. Bucaneve, G. Gentile, D. D'Antonio, V. Liso, P. Ricci, A.M. Nosari, M. Buelli, and M. Carotento, *Antimicrob. Agents Chemother.*, 1994, **38**, 2041–2046.

M. Nallasivan, F. Maher, and K. Murthy, *BMJ Case Rep.*, 2009 (http://casereports.bmj.com/content/2009/bcr.03.2009.1705, last accessed 11 March 2012).

B. I. Neughebauer, G. Negron, S. Pelton, R. W. Plunkett, E. H. Beutner, and R. Magnussen, *Am. J. Med. Soc.*, 2002, **323**, 273–278.

P. N. Padma, A. B. Rao, J. S. Yadav, and G. Reddy, *Appl. Biochem. Biotechnol.*, 2002, **102–103**, 395–405.

F. Parenti, G. Berreta, M. Berti, and V. Arioli, *J. Antibiot.*, 1978, **31**, 276–283.

D. J. Pauly, D. M. Musa, M. R. Lestico, M. J. Lindstrom, and C. M. Hetsko, *Pharmacotherapy*, 1990, **10**, 378–382.

C. M. Perrett and S. R. McBride, *Brit. Med. J.*, 2004, **328**, 1292.

P. E. Reynolds, *Eur. J. Clin. Microbiol. Infect. Dis.*, 1989, **8**, 943–950.

S. L. Rong and S. N. Leonard, *Ann. Pharmacother.*, 2010, **44**, 844–850.

M. J. Rybak, *Clin. Infect. Dis.*, 2006, **42**, S35–39.

M. Rybak, B. Lomaestro, J. C. Rotschafer, R. Moellering, W. Craig, M. Billeter, J. R. Dalovisio, and D. P. Levine, *Amer. J. Health-Syst. Pharm.*, 2009, **66**, 82–98.

J. M. Swenson, R. R. Facklam, and C. Thornsberry, *Antimicrob. Agents Chemother.*, 1990, **34**, 543–549.

N. L. Symons, A. F. Hobbes, and H. K. Leaver, *Can. Anaesth. Soc. J.*, 1985, **32**, 178–181.

C. Thompson, M. Ge, and D. Kahne, *J. Amer. Chem. Soc.*, 1999, **121**, 1237–1244.

A. Von Dryglaski, B. R. Curtis, D. W. Bougie, J. G. McFarland, S. Ahl, I. Limbu, K. R. Baker, and R. H. Aster, *N. Eng. J. Med.*, 2007, **356**, 904–910.

M. R. Wallace, J. Y. R Mascola, and E. C. Oldfield, *J. Infect. Dis.*, 1991, **164**, 1180–1185.

L. M. Weigel, D. B. Clewell, S. R. Gill, N. C. Clark, L. K. McDougal, S. E. Flannagan, J. F. Kolonay, J. Shetty, G. E. Kilgore, and F. C. Tenover, *Science*, 2003, **302**, 1569–1571.

M. J. Wood, *J. Antimicrob. Chemother.*, 1996, **37**, 209–222.

X. Xu, D. Lin, G. Yan, X. Ye, S. Wu, Y. Guo, D. Zhu, F. Hu, Y. Zhang, F. Wang, G. A. Jacoby and M. Wang, *Antimicrob. Agents Chemother.*, 2010, **54**, 4643–4647.

N. Yoshihiro, T. Shigekazu, M. Yukiko, H. Kazuhiro, T. Hiromu, I. Masayasu, O. Shigeru, and K. Hirokai, *Redox Report*, 2002, **7**, 317–319.

D. K. Yu, E. Nordbrock, S. J. Hutcheson, E. W. Lewis, W. Sullivan, V. O. Bhargava, and S. J. Weir, *J. Pharmacokin. Biopharmaceut.*, 1995, **23**, 25–39.

F. A. Zar, S. R. Bakkanagari, K. M. Moorthi, and M. B. Davis, *Clin. Infect. Dis.*, 2007, **45**, 302–307.

5.3

Cycloserine

Key Points

- D-Cycloserine (DCS) is a natural antibiotic with good oral bioavailability and activity against both Gram positive and Gram negative bacteria.
- D-Cycloserine interferes with two enzymes in cell-wall biosynthesis: L-alanine racemase, which forms D-alanine from L-alanine, and D-ala-D-ala-ligase, which forms the D-ala-D-ala dipeptide that is required for peptidoglycan crosslinking.
- DCS is a reserve drug for the treatment of multiple-drug-resistant tuberculosis (MDR-TB).

D-Cycloserine (DCS) (Figure 5.3.1) is an orally available antibiotic, which is used as part of a combination therapy regimen in a second-line treatment for *Mycobacterium tuberculosis*.

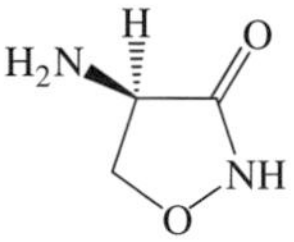

Figure 5.3.1 *D-cycloserine (DCS)*

5.3.1 Discovery

DCS, which was originally referred to as oxamycin or seromycin, or by its IUPAC name (D-4-amino-3-isoxazolidone) (Kuehl *et al.*, 1955), is a metabolite produced by *Streptomyces garyphalus*, *orchidaceus*, and *lavendulae*. It was originally isolated from an extract from *Streptomyces lavendulae* and was found to be a bactericidal agent with a broad spectrum of antibacterial activity.

Studies in *S. garyphalus* have shown that the biosynthesis of DCS involves the nucleophilic attack of hydroxyurea on *O*-acetylserine (Scheme 5.3.1) (Svensson and Gatenbeck, 1982).

Antibacterial Agents: Chemistry, Mode of Action, Mechanisms of Resistance and Clinical Applications, First Edition.
Rosaleen J. Anderson, Paul W. Groundwater, Adam Todd and Alan J. Worsley.
© 2012 John Wiley & Sons, Ltd. Published 2012 by John Wiley & Sons, Ltd.

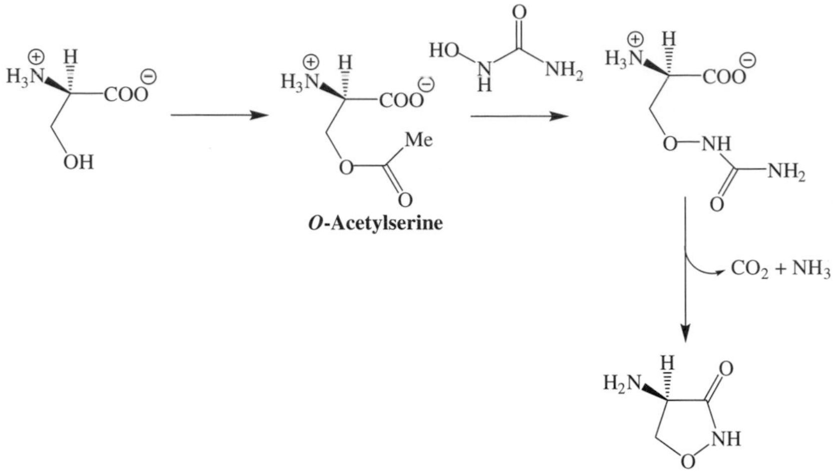

Scheme 5.3.1   *Biosynthesis of DCS*

5.3.2   Synthesis

As it is a small-molecule antibiotic, DCS can be obtained by either fermentation (Shull *et al.*, 1956; Harned, 1957) or a relatively straightforward synthesis (Scheme 5.3.2) (Plattner *et al.*, 1957).

The original synthesis of DCS led to a racemic mixture of the cycloserine enantiomers, which had to be resolved using tartaric acid (Stammer *et al.*, 1955). The first stereospecific DCS synthesis used as the starting material an ester of D-serine (the unnatural stereoisomer) and gave DCS in an overall yield of approx 8% (Scheme 5.3.2). The simple chemical transformations involved in this synthesis include:

- Side-chain chlorination of the serine ester (an example of a nucleophilic substitution), followed by ester hydrolysis (Shiraiwa *et al.*, 1996) to give the β-chloro-D-alanine **1**.
- Anhydride **2** formation by the reaction of the α-amino acid **1** with phosgene ($COCl_2$).
- Formation of a hydroxamic acid **3**, from the reaction of the anhydride **2** with hydroxylamine.
- Cyclisation to give DCS, with the stereochemistry at the 4-position being derived from that of the α-carbon atom of the D-serine starting material (Plattner *et al.*, 1957).

5.3.3   Bioavailability

Due to its high water solubility and low molecular weight, DCS has an apparent volume of distribution which is similar to that of total body water (Zhu *et al.*, 2001). Following a single oral 500 mg dose, the peak plasma concentration (C_{max}) was 12–30 µg/mL, with a corresponding T_{max} (time to peak plasma concentration) of 0.25–2.5 hours. The half-life can vary widely, between 4 and 30 hours, with a mean value of 10 hours. Typical MICs for DCS are approximately 10 µg/mL (Berning and Peloquin, 1999). Steady-state concentrations after a 500 mg dose of DCS every 12 hours would produce C_{max} : MIC ratios of about 2.4 and 24-hour concentration : MIC ratios of approximately 1, so that serum concentrations would equal, or exceed, the MIC for most of the dosing interval. A 500 mg dose of DCS every 12 hours would thus be a suitable treatment regimen, with initial

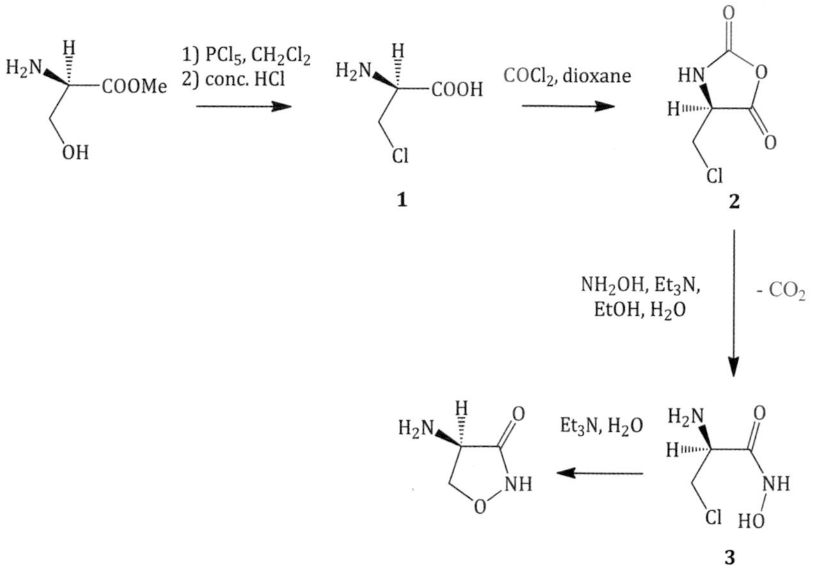

Scheme 5.3.2   *Synthesis of DCS (Plattner et al., 1957)*

low dosages of DCS (250 mg every 12–24 hours) gradually being increased (over several days) to this level. The pharmacokinetics of DCS are minimally affected by orange juice and antacids, whereas a high-fat meal delays absorption (Zhu *et al.*, 2001).

DCS is primarily renally excreted, so that dosage reduction is required for patients with reduced kidney function (Peloquin, 1991).

Due to its structural resemblance to D-alanine, DCS is recognised by the D-alanine proton motive-force-dependent transport system and is taken into bacterial cells by this active transport system that supplies the bacterium with amino acids from its environment. This gives DCS an enormous advantage and activity over a broad spectrum of bacteria, both Gram positive and Gram negative (Chopra, 1988).

5.3.4   Mode of action and selectivity

Peptidoglycan (or murein) is composed of linear glycan chains crosslinked by peptide units, and it is the nature of these peptide crosslinks which is important here, as DCS targets two enzymes that are crucial to the synthesis of the D-alanine-D-alanine (D-ala-D-ala) dipeptide unit subsequently incorporated into a pentapeptide chain. D-Ala-D-ala is required for the synthesis of the bacterial cell wall, but D-alanine is the unnatural enantiomer of alanine and, in a rare example of the destruction of chirality in nature, bacteria produce D-ala from the natural enantiomer, L-ala. Alanine racemase (ALR) catalyses this process, producing a racemic alanine mixture (50:50) starting from either alanine enantiomer, using pyridoxal 5′-phosphate (PLP) as the co-factor to cause racemisation via an imine (Schiff's base), thereby increasing the acidity of the α-hydrogen of the alanine (Scheme 5.3.3) (Watanabe *et al.*, 2002). Initially the PLP is bound to the enzyme through imine formation with a lysine residue (lysine-39 in *Bacillus stearothermophilus*) of one of the monomer units of the ALR (to give a PLP-lys complex). Trans-amination, by reaction of L-ala with the PLP-lysine, gives a new imine

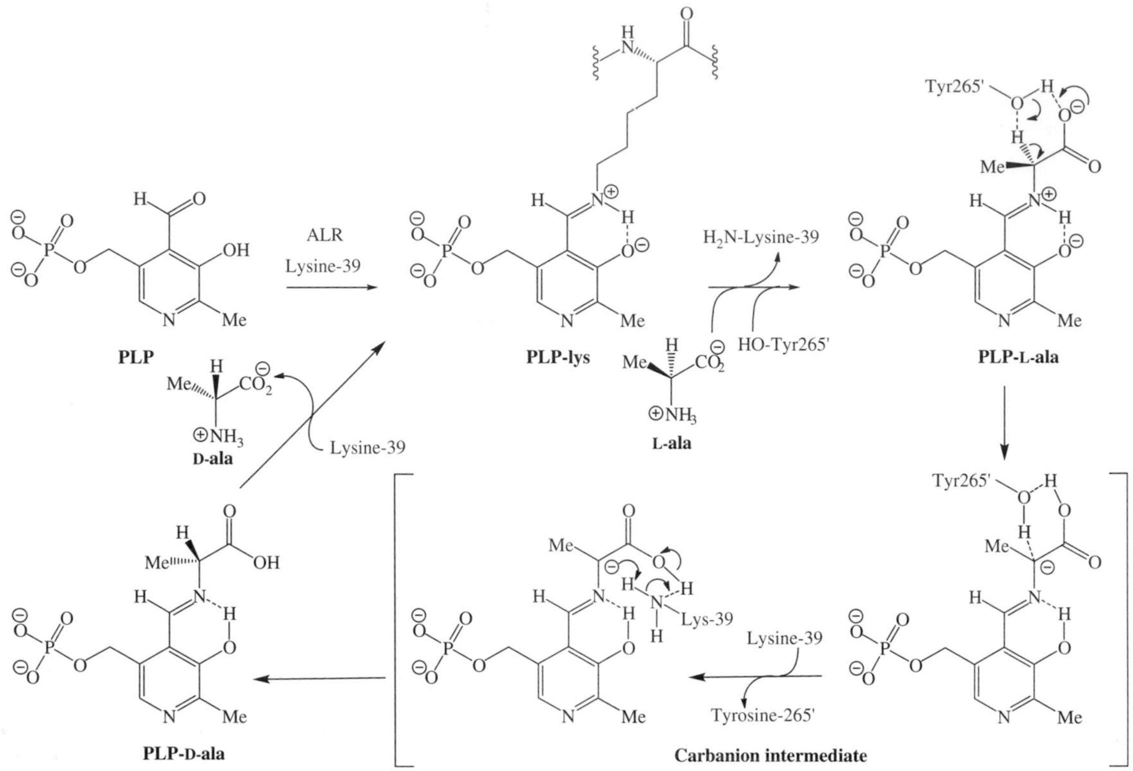

Scheme 5.3.3   *Mechanism of* Bacillus stearothermophilus *ALR-catalysed alanine racemisation (only reprotonation from above is shown, for clarity)*

(PLP-L-ala) at the active site, in which the acidity of the α-hydrogen of the alanine is increased, as the positively charged N of the protonated imine is electron-withdrawing (as is the carboxyl group), and because deprotonation gives a resonance-stabilised intermediate. Deprotonation of this PLP-ala complex, by a tyrosine residue (Tyr265' from the other half of the ALR dimer in *Bacillus stearothermophilus*) gives a planar carbanion intermediate (i.e. the portion of the intermediate highlighted in blue is flat, enabling the stabilisation of the intermediate through extensive delocalisation of the negative charge), which can be reprotonated from either face (i.e. from above or below), giving the PLP Schiff's base complex of either of the pair of alanine enantiomers. A further trans-amination with the active-site lysine-39 is the final step in the catalytic cycle and regenerates the PLP-lys complex ready for reaction with another L-ala molecule (Scheme 5.3.3).

By inhibiting this enzyme (Lambert and Neuhaus, 1972), through the formation of an adduct with PLP which binds to the active site and/or removes PLP from the system via the formation of a stable aromatic adduct (3-hydroxyisoxazolepyridoxamine phosphate (3-hydroxyisoxazole-PMP)) (Scheme 5.3.4 and Figure 5.3.2) (Fenn *et al.*, 2003; Noda *et al.*, 2004a), DCS prevents the formation of the D-ala required for crosslink formation. DCS is obviously similar enough in structure to be mistaken for alanine in the first step in this mechanism, and the formation of the PLP-DCS adduct leads to a reduction in the PLP available for the catalysis of the alanine racemisation.

DCS is also a competitive inhibitor of D-ala-D-ala ligase (Ddl), the enzyme responsible for the coupling of the two D-ala residues to give the D-ala-D-ala dipeptide (Strominger *et al.*, 1960; Noda *et al.*, 2004b). This

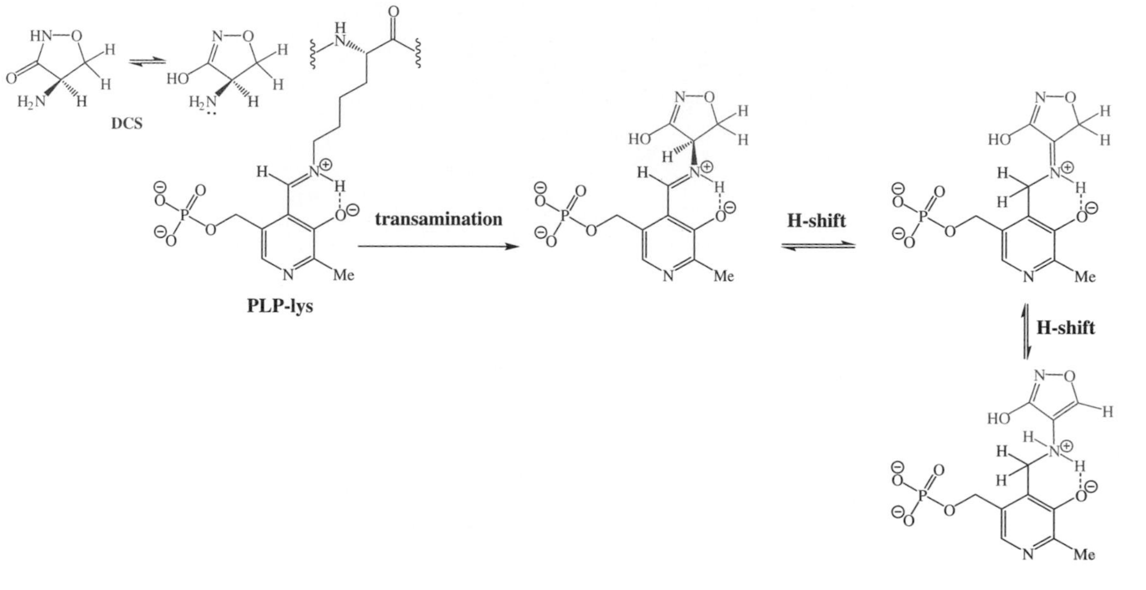

Scheme 5.3.4 *Proposed DCS-PLP adducts (Fenn et al., 2003; Noda et al., 2004a)*

inhibition of the Ddl active site appears to be due to binding to the site for the *N*-terminal ᴅ-alanine residue (i.e. that which is phosphorylated in Step 1, shown in red) (Scheme 5.3.5) (Zawadzke *et al.*, 1991).

5.3.5 Bacterial resistance

Resistance to DCS in *E. coli* has been shown to arise due to mutations in the *cycA* gene, which codes for a permease involved in DCS uptake, resulting in a decrease in the cellular DCS levels (Fehér *et al.*, 2006). In addition, in *Mycobacterium smegmatis* the over-expression of both the *alr* (which codes for the ALR) and the *ddl* (which codes for the ᴅ-ala-ᴅ-ala ligase) genes leads to resistance, with the level of resistance produced by over-expression of the *ddl* gene being lower than that produced by over-expression of the *alr* gene (Caceres *et al.*, 1997; Feng and Barletta, 2003). The over-expression of these genes results in increased levels of these enzymes, such that there are insufficient levels of DCS for their inhibition.

5.3.6 Clinical applications

5.3.6.1 Spectrum of activity

DCS shows activity against Gram negative (e.g. *Escherichia coli*) and Gram positive bacteria (e.g. *Staphylococcus aureus*), and, perhaps most notably, mycobacteria (e.g. *M. tuberculosis*). Despite this broad spectrum of antibacterial activity, the clinical applications of DCS are limited due to the serious adverse effects associated with its use.

5.3.6.2 Tuberculosis

DCS is classed as a group 4 agent (alongside *p*-aminosalicylic acid, ethionamide, protionamide, and terizidone) by the World Health Organization and is therefore used as a second-line or reserve drug for

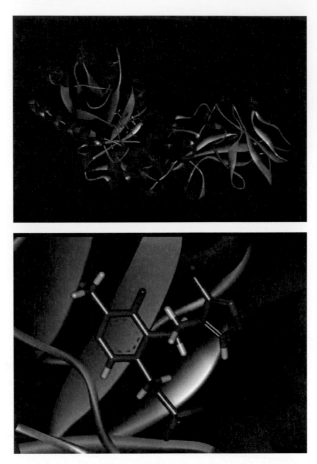

Figure 5.3.2   *Crystal structure of the PLP-D-cycloserine adduct bound to alanine racemase (ALR) (PDB 1VFS) (Noda et al., 2004b)*

the management of MDR-TB (World Health Organization, 2010). In most cases, when an anti-TB regimen requires the addition of a group 4 agent and the cost of treatment is a barrier, ethionamide (or protionamide) is used. If the cost of treatment is not an issue then *p*-aminosalicylic acid is usually the group 4 agent added to the regimen. If an anti-TB regimen requires the addition of a second group 4 agent, then DCS should be used.

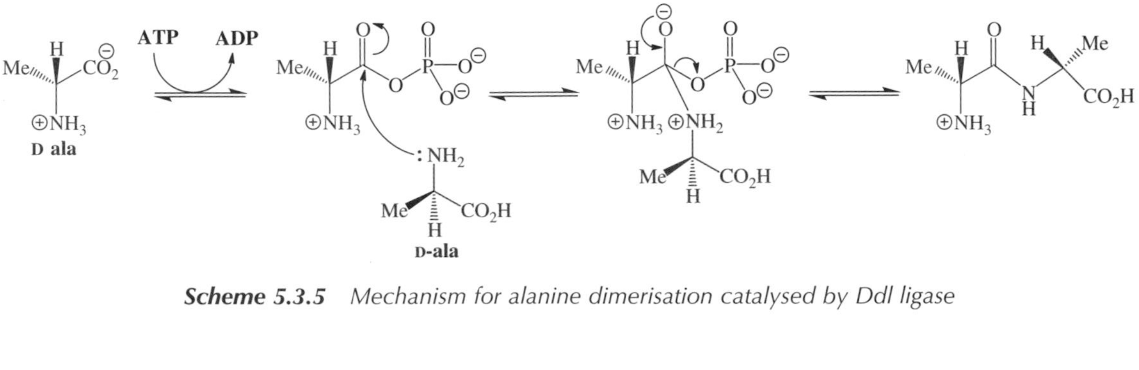

Scheme 5.3.5   *Mechanism for alanine dimerisation catalysed by Ddl ligase*

As is the case for all medicines, there are always risks and benefits associated with treatment. DCS offers the advantage of having a high gastric tolerance (compared to the other group 4 agents) and it does not have the problem of cross-resistance, which commonly occurs with other drugs used to treat TB (rifampicin and rifabutin, for example) (Caminero *et al.*, 2010). Unfortunately, however, due to the pharmacology of DCS (in particular the partial-agonist activity at NMDA receptors – as discussed in Subsection 5.3.7), it can cause adverse psychiatric reactions (including suicidal tendencies), which limits the clinical usefulness of this drug.

5.3.7 Adverse drug reactions

DCS acts as a co-agonist at the glycine binding site of the NR1 subunit of the NMDA receptor, present in the CNS (Hood *et al.*, 1989; McBain *et al.*, 1989). DCS alone is unable to activate NMDA receptors and it requires glutamate binding of the receptor in order to exert its effect (Olive *et al.*, 2012).

DCS can cause a variety of neurotoxic reactions and these appear to be dose-related (generally seen in doses above 500 mg daily) (Kwon *et al.*, 2008). These neurotoxic reactions range from seizures to myoclonic jerks, encephalopathy, and psychoses (Fujita *et al.*, 2008). DCS has been known to induce seizures, especially in patients with existing epilepsy, and in these cases the drug should be discontinued (Seaworth, 2002; Blumberg *et al.*, 2003).

Several skin reactions have been reported, ranging from a lichenoid skin eruption to an allergic manifestation (Shim *et al.*, 1995).

5.3.8 Drug interactions

The maximal absorption of DCS and terizidone in combination can be reduced by the effects of acidic beverages, such as orange juice, but this reduction is not considered to be clinically significant. Co-administration with a fatty meal may, however, reduce absorption of DCS by 16–31% (Zhu *et al.*, 2001).

5.3.9 Recent developments

In previous sections we have mainly concentrated on the derivatisation of existing classes of antibiotics. However, there are not many research groups (or indeed big pharmaceutical companies) looking to develop DCS derivatives, as there are many other, and perhaps more attractive, agents to study. One stream of research that is gaining considerable interest is not related to the antibacterial effects of DCS, but, rather strangely, to the activity at NMDA receptors (and ironically, this is one of the reasons why DCS is seldom used as an antibacterial agent). The partial-agonist activity at NMDA receptors may give DCS a role in the treatment of negative symptoms in schizophrenia (Goff *et al.*, 2008) (although the evidence from a recent meta-analysis was not encouraging (Tsai and Lin, 2010)) and dementia associated with Parkinson's disease (Ho *et al.*, 2011). In addition, the use of DCS also appears to enhance cognitive behavioural therapy in panic and obsessive–compulsive disorder (OCD) (Wilhelm *et al.*, 2008; Otto *et al.*, 2010). In the future, therefore, the clinical applications of DCS may lie away from the treatment of tuberculosis and towards the management of various dementia and anxiety-related disorders.

References

S. E. Berning and C. A. Peloquin, 'Antimycobacterial agents: cycloserine', in V. L. Yu, T. C. Merigan, Jr, and S. L. Barriere (Eds), *Antimicrobial Therapy and Vaccines*, Williams and Wilkins, Baltimore, MD, 1999, 638–642.

H. M. Blumberg, W. J. Burman, R. E. Chaisson, C. L. Daley, S. C. Etkind, and L. N. Friedman, *Am. J. Respir. Crit. Care Med.*, 2003, **167**, 603–662.

N. E. Caceres, N. B. Harris, J. F. Wellehan, V. Kapur, and R. G. Barletta, *J. Bacteriol.*, 1997, **179**, 5046–5055.

J. A. Caminero, G. Sotgiu, A. Zumla, and G. B. Migliori, *Lancet Infect. Dis.*, 2010, **10**, 621–629.

I. Chopra, *Parasitol.*, 1988, **96** (Suppl.), S25–S44.

T. Fehér, B. Cseh, K. Umenhoffer, I. Karcagi, and G. Pósfai, *Mutat. Res.*, 2006, **595**, 184–190.

Z. Feng and R. G. Barletta, *Antimicrob. Agents and Chemother.*, 2003, **47**, 283–291.

T. D. Fenn, G. F. Stamper, A. A. Morollo, and D. Ringe, *Biochem.*, 2003, **42**, 5775–5783.

J. Fujita, K. Sunada, H. Hayashi, K. Hayashihara, and T. Saito, *Kekkaku*, 2008, **83**, 21–25.

D. C. Goff, C. Cather, J. D. Gottlieb, A. E. Evins, J. Walsh, L. Raeke, M. W. Otto, D. Schoenfeld, and M. F. Green, *Schizophr. Res.*, 2008, **106**, 320–327.

R. L. Harned,US Patent PATN 2789983, 23 April 1957.

Y. J. Ho, S. C. Ho, C. R. Pawlak, and K. Y. Yeh, *Behav. Brain Res.*, 2011, **219**, 280–290.

W. F. Hood, R. P. Compton, and J. B. Monahan, *Neurosci. Lett.*, 1989, **98**, 91–95.

F. A. Kuehl, R. P. Buhs, I. Putter, R. Ormond, J. E. Lyons, L. Chaiet, F. J. Wolf, N. R. Trenner, R. L. Peck, E. Howe, B. D. Hunnewell, G. Downing, E. Newstead, and K. Folkers, *J. Amer. Chem. Soc.*, 1955, **77**, 2344–2345.

H. M. Kwon, H. K. Kim, J. Cho, Y. H. Hong, and H. Nam, *Eur. J. Neurol.*, 2008, **15**, e60–e61.

M. P. Lambert and F. C. Neuhaus, *J. Bacteriol.*, 1972, **110**, 1978–1987.

C. J. McBain, N. W. Kleckner, S. Wyrick, and R. Dingledine, *Mol. Pharmacol.*, 1989, **36**, 556–565.

M. Noda, Y. Kawahara, A. Ichikawa, Y. Matoba, H. Matsuo, D.G. Lee, T. Kumagai, and M. Sugiyama, *J. Biol. Chem.*, 2004a, **279**, 46 143–46 152.

M. Noda, Y. Matoba, T. Kumagai, and M. Sugiyama, *J. Biol. Chem.*, 2004b, **279**, 46 153–46 161.

M. F. Olive, R. M. Cleva, P. W. Kalivas, and R. J. Malcolm, *Pharmacol. Biochem. Behav.*, 2012, **100**, 801–810.

M. W. Otto, D. F. Tolin, N. M. Simon, G. D. Pearlson, S. Basden, S. A. Meunier, S. G. Hofmann, K. Eisenmenger, J. H. Krystal, and M. H. Pollack, *Biol. Psychiatry*, 2010, **67**, 365–370.

C. A. Peloquin, 'Antituberculosis drugs: pharmacokinetics', in L. B. Heifets (Ed.), *Drug Susceptibilities in the Chemotherapy of Mycobacterial Infections*, CRC Press , Boca Raton, FL, 1991, pp. 59–88.

P. A. Plattner, A. Boller, H. Frick, A. Fürst, B. Hegedus, H. Kirchensteiner, S. Majnoni, R. Schläpfer, and H. Spiegelberg, *Helv. Chim. Acta*, 1957, **160**, 1531–1552.

B. J. Seaworth, *Infect. Dis. Clin. North Am.*, 2002, **16**, 73–105.

J. H. Shim, T. Y. Kim, H. O. Kin, and C. W. Kim, *Dermatol.*, 1995, **191**, 142–144.

T. Shiraiwa, H. Miyazaki, M. Ohkubo, A. Ohta, A. Yoshioka, T. Yamane, and H. Kurokawa, *Chirality*, 1996, **8**, 197–200.

G. M. Shull, J. B. Routein, and A. C. Finlay,US Patent 2773878, 11 December 1956.

C. H. Stammer, A. N. Wilson, F. W. Holly, and K. Folkers, *J. Amer. Chem. Soc.*, 1955, **77**, 2346–2347.

J. L. Strominger, E. Ito, and R. H. Threnn, *J. Amer. Chem. Soc.*, 1960, **82**, 998–999.

M.-L. Svensson and S. Gatenbeck, *Arch. Microbiol.*, 1982, **131**, 129–131.

G. E. Tsai and P. Y. Lin, *Curr. Pharm. Des.*, 2010, **16**, 522–537.

A. Watanabe, T. Yoshimura, B. Mikami, H. Hayashi, H. Kagamiyama, and N. Esaki, *J. Biol. Chem.*, 2002, **277**, 19 166–19 172.

S. Wilhelm, U. Buhlmann, D. F. Tolin, S. A. Meunier, G. D. Pearlson, H. E. Reese, P. Cannistraro, M. A. Jenik, and S. L. Rauch, *Am. J. Psychiatry*, 2008, **165**, 335–341.

World Health Organization, Treatment of Tuberculosis Guidelines, 4th Edn, 2010 (http://whqlibdoc.who.int/publications/2010/9789241547833_eng.pdf, last accessed 11 March 2012).

M. Zhu, D. E. Nix, R. D. Adam, J. M. Childs, and C. A. Peloquin, *Pharmacotherapy*, 2001, **28**, 891–897.

5.4

Isoniazid

Key Points

- Isoniazid inhibits the synthesis of mycolic acids, which are the building blocks for the mycobacterial cell wall.
- Isoniazid is used in combination with other agents to treat active tuberculosis infection; it is included in both the initial phase and continuation phase of therapy.
- Isoniazid is also used as monotherapy to prevent the development of active tuberculosis from latent infection.

Isoniazid (INH) (Figure 5.4.1) is a key component of the combination therapy used to treat tuberculosis (TB) and latent TB.

5.4.1 Discovery

As is the case with a number of drugs, most notably azidothymidine (AZT), a nucleoside reverse transcriptase inhibitor (NRTI) used in the treatment of HIV infections, isoniazid was first synthesised many years before the discovery of the antimycobacterial activity which led to its use in the treatment of TB. Isoniazid (**isoni**cotinic **a**cid hydra**zid**e, INH) was originally synthesised in 1912 as part of a purely organic chemistry investigation (Meyer and Mally, 1912) and it was not until 1952 that its potential in the treatment of the dreaded TB (or consumption) was discovered (Wang, 2005).

Interestingly, the discovery of isoniazid was partly due to a Nobel Prize winner we have already encountered in Section 3.1, Gerhard Domagk, who had been screening compounds against *Mycobacterium tuberculosis* (the organism responsible for TB) in the drug-discovery programme which resulted in the sulfonamides. A sulfonamide, sulfathiazole, had been shown to have weak activity against tubercule bacilli and a series of analogues was prepared, with the greatest activity found for compounds which were precursors of the thiazole ring, the thiosemicarbazones, and this eventually led to thiacetazone (Tb 1/698), later named conteben (Figure 5.4.2).

Antibacterial Agents: Chemistry, Mode of Action, Mechanisms of Resistance and Clinical Applications, First Edition.
Rosaleen J. Anderson, Paul W. Groundwater, Adam Todd and Alan J. Worsley.
© 2012 John Wiley & Sons, Ltd. Published 2012 by John Wiley & Sons, Ltd.

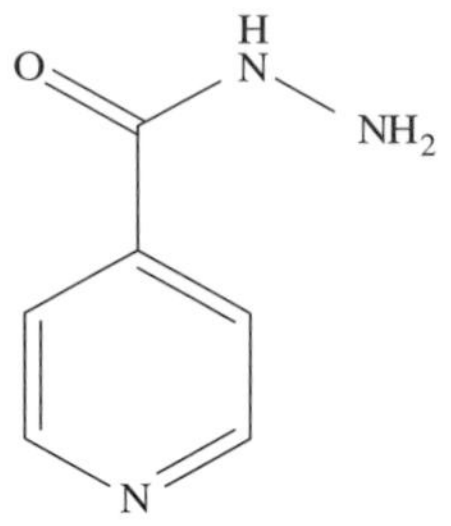

Figure 5.4.1 *Isoniazid (INH)*

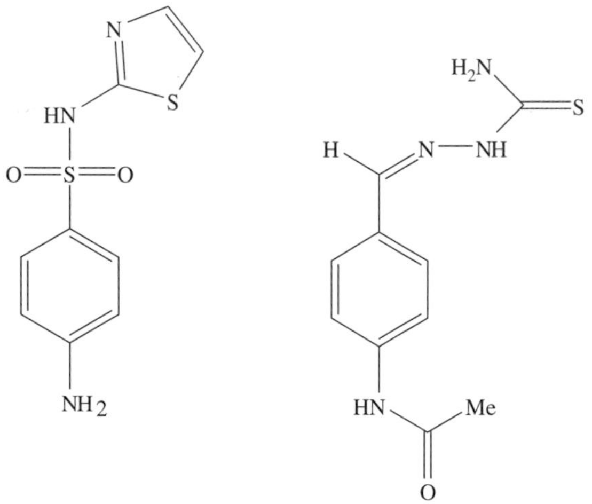

Sulfathiazole   **Conteben**

Figure 5.4.2 *Sulfathiazole and conteben (thiacetazone), showing the similarity of the* **thiazolylamine** *and* **thiosemicarbazone** *portions*

News of the trials of conteben in Germany in the late 1940s was soon picked up in the rest of the world and H. Corwin Hinshaw (Stanford University) and Walsh McDermott (Cornell Medical School) visited Europe to observe the effects of this drug on TB patients. The Americans returned home with a sample of conteben, and an amazing series of coincidences resulted in four independent laboratories working on isoniazid at the same time – at Roche and Squibb in the USA, Wellcome in the UK, and Domagk in Germany, all of whom thought that the intellectual property belonged to them! Having been synthesised 40 years previously, isoniazid was actually free of any patent restrictions and production of isoniazid (brand name, Pycazide) began in the UK in 1951 at Herts Pharmaceuticals Ltd. The worldwide interest in the 1950s in nicotinamide/isoniazid analogues for the treatment of TB led to other agents, including pyrazinamide and ethionamide, both of which are still used in some combination-treatment regimens (Greenwood, 2008).

5.4.2   Synthesis

As mentioned earlier, isoniazid was first synthesised in 1912, from isonicotinic acid ethyl ester and hydrazine (Meyer and Mally, 1912), and recent synthetic studies have concentrated on the use of more readily available starting materials such a 4-cyanopyridine (Scheme 5.4.1) (Sycheva *et al.*, 1971).

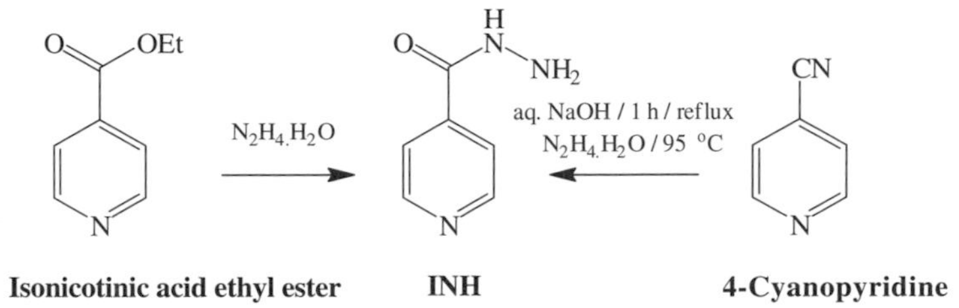

Scheme 5.4.1 *Synthesis of isoniazid (INH)*

5.4.3 Bioavailability

Isoniazid has excellent bioavailability after oral administration as a single agent, at around 100% absorption, with peak blood concentrations achieved after 0.9–1 hours, peak mean serum concentrations of 3.6 μg/mL, and a half-life of 1.9 hours (Peters, 1960; Ellard and Gammon, 1976). As isoniazid is used as one component of combination therapy for TB, the pharmacokinetic parameters have also been investigated as part of a combination therapy; fortunately, they do not differ significantly from the single-agent parameters (Zwolska *et al.*, 2002).

The absorption of isoniazid is affected by the presence of food (Melander *et al.*, 1976) and by aluminum hydroxide (Hurwitz and Schlozman, 1974), with the result that it should not be taken with food or less than 1 hour before antacid formulations. TB is a complicating factor for many patients with AIDS, some of whom are already receiving didanosine, which incorporates antacid into the formulation to reduce stomach acidity and minimise the acid-catalysed hydrolysis of didanosine; fortunately, the small amount of antacid in the didanosine formulation has little effect upon the absorption of isoniazid, allowing its concurrent administration (Gallicano *et al.*, 1994).

The bioavailability of isoniazid, like that of the sulfonamides, is affected by the rate at which it is metabolised by the human cytochrome enzyme, *N*-acetyltransferase 2 (NAT2) (Hughes, 1954; Evans *et al.*, 1960). A patient with rapid acetylation of isoniazid metabolises and excretes it at a faster rate than is appropriate for the antibacterial activity, with the result that sub-therapeutic doses are achieved, with associated poor therapeutic outcomes. The situation is little better for those with slow acetylation status, as there is increased opportunity for high serum concentrations to be achieved, due to decreased metabolism and excretion of the isoniazid, and adverse events are more common for these patients; you will discover this in more detail in Subsection 5.4.7. As acetylator status is inherited and can be determined before treatment by genotype analysis, the personalisation of isoniazid treatment according to NAT2 status has been considered, and this would help to avoid adverse events caused by this problem (Evans, 1989; Spielberg, 1996; Butcher *et al.*, 2002). When genotype analysis becomes less expensive and more routine, this may be a development for the future (Kinzig-Schippers *et al.*, 2005).

5.4.4 Mode of action and selectivity

As mentioned in Section 1, mycobacteria are aerobic organisms, characterised by thicker cell walls than Gram positive and negative organisms. The mycobacterial cell wall is hydrophobic/waxy in nature, being rich in very-long-chain fatty acids, known as mycolic acids, and is resistant to agents which interfere with cell wall synthesis, such as β-lactams. Isoniazid is actually a prodrug and is converted *in vivo* to the active form by the mycobacterial enzyme catalase-peroxidase (KatG) (Zhou *et al.*, 2006). Catalase-peroxidases protect bacteria from the cellular damage caused by reactive oxygen species and KatG is responsible for the high resistance of *M. tuberculosis* to hydrogen peroxide (Manca *et al.*, 1999).

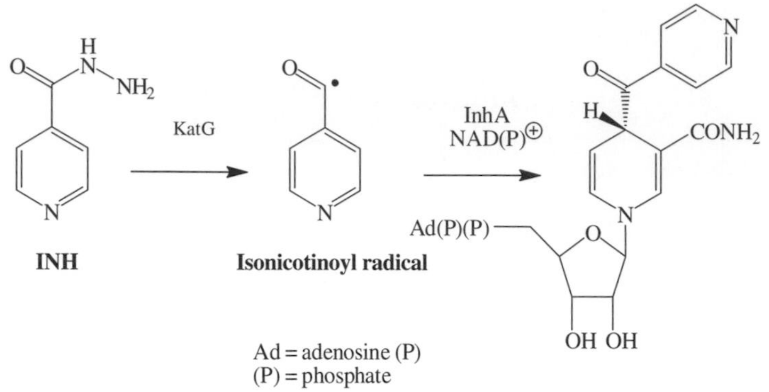

Scheme 5.4.2   *INH-NAD(P) adduct formation*

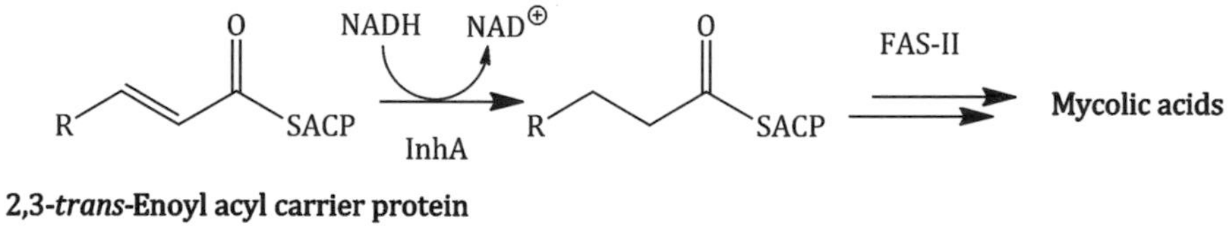

Scheme 5.4.3   *InhA reduction of enoylACP*

KatG is believed to activate isoniazid to the highly reactive isonicotinoyl radical, which complexes with the coenzymes NAD^+ or $NADP^+$ (bound in the active site of InhA) to give an INH-NAD(P) adduct (Scheme 5.4.2), in which the isonicotinoyl group has replaced the 4S-hydrogen involved in the hydride transfer reduction of enoyl-ACP substrates (Rozwarski *et al.*, 1998; Cade *et al.*, 2010).

This adduct inhibits InhA, an NADH-dependent enoyl-acyl carrier protein reductase which catalyses the reduction of 2,3-*trans*-enoylACP through competition with the NADH enzyme cofactor (Scheme 5.4.3 and Figure 5.4.3) (Argyrou *et al.*, 2007). InhA preferentially reduces long-chain fatty acid precursors to give products that are utilised by the mycobacterial type II fatty acid synthase system (FAS-II) to produce the very-long-chained α-branched mycolic acids, which are key components of the mycobacterial cell wall (Rozwarski *et al.*, 1998).

Isoniazid thus indirectly blocks the formation of the mycobacterial cell wall, through the formation of an adduct with NAD(P), which inhibits enoyl ACP reductase, thus accounting for its bacteriostatic/bactericidal activity (Kaur *et al.*, 2009). Other mechanisms of action are still being investigated for INH, including the generation of nitric oxide (NO) radicals from INH and these radicals' enhancement of the antimycobacterial effect of isoniazid (Timmins *et al.*, 2004).

5.4.5   Bacterial resistance

As might be expected, *M. tuberculosis* resistance to isoniazid may be the result of a number of mechanisms, including:

- Mutations/deletions in the *katG* gene regulatory region, resulting in a decrease in INH-NAD(P) adduct formation. This effect is not solely due to a loss of catalase-peroxidase activity, and in fact the major mutation found in clinical isolates (serine 315 to threonine) (González *et al.*, 1999) leads to a conformational change in the substrate access channel of the enzyme, resulting in decreased isoniazid binding (Cade *et al.*, 2010).

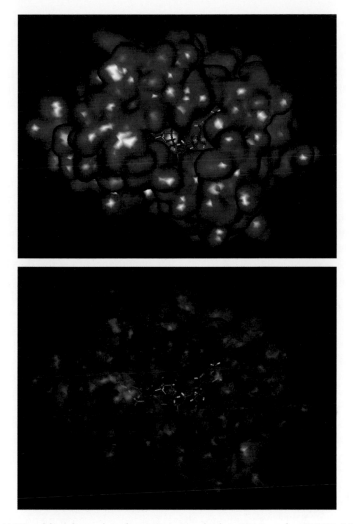

Figure 5.4.3 *INH-NAD adduct bound to the active site of InhA (PDB code: 2NV6) (Vilchèze et al., 2006)*

- Over-expression of the *InhA* gene, so that there is insufficient INH-NADP complex for inhibition, results in resistance in *M. tuberculosis* strains harbouring *InhA* plasmids (Larsen *et al.*, 2002), while 25% of the mutations in the *InhA* gene found in clinical isolates are in the promoter or structural regions (with consequences for NADH and INH-NADP binding) (Telenti *et al.*, 1997).
- Increased arylamine *N*-acetyltransferase (NAT) activity, as a result of increased expression of the *NAT* gene (*N*-acetylation reduces the activity of INH) (Payton *et al.*, 1999).
- For more on the bacterial resistance of isoniazid, see Dye and Espinal (2001).

5.4.6   Clinical applications

5.4.6.1   Spectrum of activity

As isoniazid targets the formation of the mycobacterial cell wall, it will not be a surprise to learn that it has activity against several types of mycobacteria – most notably against *M. tuberculosis*, making it a key component in the chemotherapeutic management of TB (Zhang *et al.*, 1996). Isoniazid also shows activity against other members of the *M. tuberculosis* complex (MTC),[19] but it lacks activity against *Mycobacterium leprae* (one of the causative agents of leprosy) and the *Mycobacterium avium* complex (MAC), and consequently should not be used to treat diseases caused by these pathogens (Inderlied *et al.*, 1993; Eiglmeier *et al.*, 1997). Interestingly, it is thought that the *katG* gene (the gene which codes for the catalase-peroxidase that activates isoniazid) of *M. leprae* has been inactivated, which helps explain the apparent lack of activity of isoniazid against *M. leprae* (and thus leprosy) (Eiglmeier *et al.*, 1997).

5.4.6.2   Treatment of TB

Isoniazid, as discussed in Section 2.2.6.1, is used in combination with other antibacterial agents to treat active TB. Indeed, because of its high activity towards *M. tuberculosis* and its relatively cheap cost, it is recommended that isoniazid be used in the **initial phase** (alongside rifampicin, pyrazinamide, and ethambutol) and the **continuation phase** (alongside rifampicin) of TB treatment, unless there is a specific contraindication or bacterial resistance is suspected. Isoniazid should never be used as a monotherapy to treat active TB (although see Subsection 5.4.6.3), as this would encourage the development of resistance, which is already becoming a significant clinical problem.

   TB (or consumption, as it was known – because patients were apparently consumed from within) is mostly caused by *M. tuberculosis* infection of pulmonary alveoli and has been responsible for many millions of deaths. The symptoms of TB are a chronic cough (with blood-tinged sputum), fever, and weight loss. If untreated, TB kills more than half of those affected. The earliest treatments for TB involved 'fresh country air', so many patients were sent to TB sanatoria. The dreaded 'consumption' has been responsible for the deaths of sufferers from every social class (including many notable figures from history, e.g. King Louis XIII of France in 1643, King Edward VI of England in 1553, General Simón Bolívar (the liberator of much of South America) in 1830, Erwin Schrödinger (the 1933 Nobel Laureate for Physics) in 1961, at least three members of the Brontë family of writers, Vivien Leigh (Scarlett O'Hara in *Gone with the Wind*) in 1967, and Frédéric Chopin (composer) in 1849, with countless others having contracted TB but survived (including Bishop Desmond Tutu (after whom a TB centre in South Africa has been named) and Tom Jones (whose first single was titled 'Chills and Fever')). Today it is estimated that one-third of the world's population has been exposed to TB (with 90% of the exposed population having latent (asymptomatic) TB) and that there is almost one new infection per second (8.8 million new cases in 2010), with 1.4 million deaths in 2010, the majority of which were in Africa (World Health Organization, 2011). As mentioned above, resistance is a growing problem; in 2000, 3% of TB strains were multidrug resistant (MDR-TB), while in a study performed between 2000 and 2004, 20% of TB isolates were MDR and 10% were XDR (extensively drug resistant) (Centers for Disease Control and Prevention, 2006). In 2010, an estimated 650,000 cases were MDR-TB (World Health Organisation, 2011).

5.4.6.3   Prevention of TB

Isoniazid can also be used as a preventative therapy in susceptible patients who have had close contact with active TB, or those who have become tuberculin positive.[20] In contrast to its use in the treatment of active TB,

---

[19] The MTC is a group of mycobacteria that cause human and/or animal TB. The MTC includes species such as *M. tuberculosis*, *Mycobacterium africanum* (a cause of human TB in sub-Saharan Africa), *Mycobacterium bovis* (a major cause of TB in cattle), and *Mycobacterium microti* (a cause of TB in rodents).

[20] The tuberculin skin test is a screening tool used to identify the presence of *M. tuberculosis*: tuberculin purified protein derivative (PPD), derived from *M. tuberculosis*, is injected into a patient's skin and a strong skin reaction after 48–72 hours indicates an immune response, which suggests the presence of *M. tuberculosis*.

isoniazid can be used as monotherapy to prevent the development of active TB, and is usually taken once daily for six months.

A Cochrane review concluded that isoniazid is effective at preventing active TB in diverse at-risk patients, showing that isoniazid prophylaxis prevents the development of active TB in approximately 60% of individuals in various at-risk groups. In addition, among contacts with a positive tuberculin skin test, for every 35 patients treated with isoniazid for six months, one case of active TB will be prevented (Smieja *et al.*, 1999). As with all drug treatments, the benefits versus risks should always be considered, however, as around 1 in every 200 patients given isoniazid prophylaxis will also develop isoniazid-induced hepatitis (Smieja *et al.*, 1999).

One group of patients who are at increased risk of contracting TB are those infected with HIV/AIDS. Latent TB infections occur when a patient is infected with *M. tuberculosis* but no clinical symptoms are observed, and as we said in Subsection 5.4.6.2, it has been estimated that one-third of the world's population has been exposed to TB and that 90% of those exposed have latent TB. In healthy individuals with a strong immune system, the latent infection is often controlled and the lifetime risk of developing active TB is low, at around 10%. However, in cases when the immune system is weakened, such as in HIV/AIDS patients, the risk of developing active disease from latent infection increases to around 30% (De Cock, 1995). Thankfully, if isoniazid is given to patients co-infected with latent TB (especially with a positive tuberculin skin test) and HIV, the risk of developing active TB is reduced (Akolo *et al.*, 2010). More research is needed to determine the optimal duration of preventative therapy but, given the large numbers of patients co-infected with latent TB and HIV, the prophylactic use of isoniazid may have a significant impact on HIV-associated morbidity and mortality.

5.4.7   Adverse drug reactions

A common adverse effect associated with isoniazid is hepatic toxicity, which is associated with approximately 20% of patients receiving the drug for TB, with severe hepatotoxicity being shown in 1% of exposed patients (Scharer and Smith, 1969; Garibaldi *et al.*, 1972).

Another serious adverse effect associated with the use of isoniazid is systemic lupus erythematosus, which occurs in approximately 1% of patients (Rubin, 2005).

As we saw in Subsection 5.4.3, isoniazid is primarily metabolised into acetylisoniazid (acetylated by *N*-acetyltransferase, NAT2). Individual subjects are either fast or slow acetylators (this is genetically determined); with chronic use, slow acetylation may lead to higher blood levels with increased toxicity. Acetylisoniazid is then further hydrolysed to isonicotinic acid and acetylhydrazine, which can be renally excreted. Isonicotinic acid is conjugated with glycine. Acetylhydrazine may be further metabolised to diacetylhydrazine and it is then suggested that it is converted by CYP 2E1 into hydrazine, leading to hepatotoxicity (Yard and McKennis, 1955; Ellard and Gammon, 1976; Sarich *et al.*, 1996). Other side effects associated with isoniazid are skin rash, sideroblastic anaemia, and peripheral neuropathy, which is exhibited by those patients who are slow acetylators and/or when high doses of isoniazid are used (7.5–9.5 mg/kg) daily (Devadetta *et al.*, 1960). The degree of neuropathy also depends on the overall nutritional state of the patient and, as a result, poorly nourished patients lacking in pyridoxine (vitamin B6, an important component in protein metabolism) are often at a higher risk (Rudman and Williams, 1983). Simple supplementation of pyridoxine in large amounts may contribute to peripheral neuropathy (Schaumberg *et al.*, 1983), so lower doses of pyridoxine (100–200 mg daily) should be used in patients with isoniazid-induced peripheral neuropathy (Nisar *et al.*, 1990).

Sideroblastic anaemia, in which the body has plentiful iron reserves but cannot incorporate them into haemoglobin, has been reported to have been caused by the long-term use of isoniazid (Sharp *et al.*, 1990; Yoshimoto *et al.*, 1992). The reasons for the poor haem incorporation into erythrocytes in sideroblastic anaemia remains elusive but isoniazid does have an effect upon *delta*-aminolevulinic acid synthase (ALAS), which acts upon proporphyrinogen II and subsequent haem formation (Bottomley and Muller-Eberhard, 1988). Isoniazid-induced sideroblastic anaemia is treated by withdrawing the drug and administering pyridoxine (the precursor of pyridoxal phosphate, which is a co-factor associated with ALAS function).

5.4.8   Drug interactions

Isoniazid is metabolised in two phases (described above), in which it acts as both an inhibitor and then an inducer of the isoenzyme CYP 2E1 (Self *et al.*, 1999). In addition, isoniazid is a weak monoamine oxidase inhibitor and has the potential to inhibit histaminase, thus causing many drug–food interactions (Hauser and Baier, 1982).

5.4.8.1   Phenytoin

There are several reports of isoniazid interacting with phenytoin, resulting in an increased serum concentration of phenytoin and, in some cases, patient toxicity (Miller *et al.*, 1979). Isoniazid is seldom used alone and is often combined with rifampicin, which is known to induce the metabolism of phenytoin, and as such the combination of rifampicin with isoniazid in a typical TB treatment regimen will reduce the plasma levels of phenytoin (the induction effect of rifampicin upon the metabolism of phenytoin far outweighs the inhibition by isoniazid), so all patients receiving this combination of drugs need to be carefully monitored (Kay *et al.*, 1985).

5.4.8.2   Carbamazepine

In a similar manner, carbamazepine toxicity may be experienced if given concurrently with isoniazid (Block, 1982; Fleenor *et al.*, 1991).

5.4.8.3   Paracetamol (acetaminophen)

Isoniazid and paracetamol are both suggested to be metabolised via CYP 2E1. Isoniazid inhibits CYP 2E1 and so raises the levels of *N*-acetyl-*para*-benzoquinoneimine, which is the toxic metabolite of paracetamol, thus explaining the reports of toxicity experienced when both drugs are administered to some patients (Moulding *et al.*, 1991; Crippin, 1993; Chien *et al.*, 1997). The effect of slow and fast acetylators has also been investigated. Slow acetylators of isoniazid were given both paracetamol and isoniazid, and reduced levels of urinary oxidative metabolites were collected, whereas increased levels were found with rapid acetylators of isoniazid (Chien *et al.*, 1997).

5.4.8.4   Benzodiazepines

Isoniazid treatment causes hepatic impairment of the *N*-demethylation of diazepam, resulting in a prolonged half-life of this benzodiazepine (Ochs *et al.*, 1981). Isoniazid also increased the half-life of triazolam, but not oxazepam, as the latter is metabolised by glucuronidation only, thus avoiding the interaction with isoniazid (Ochs *et al.*, 1983).

5.4.8.5   Vitamin D

Vitamin D is hydroxylated, in both the liver and the kidneys, by CYP 27B1, and a study has shown that the administration of isoniazid reduced the serum concentration of calcium, 25-hydroxyvitamin D, and $1\alpha,25$-dihydroxyvitamin D (Brodie *et al.*, 1981).

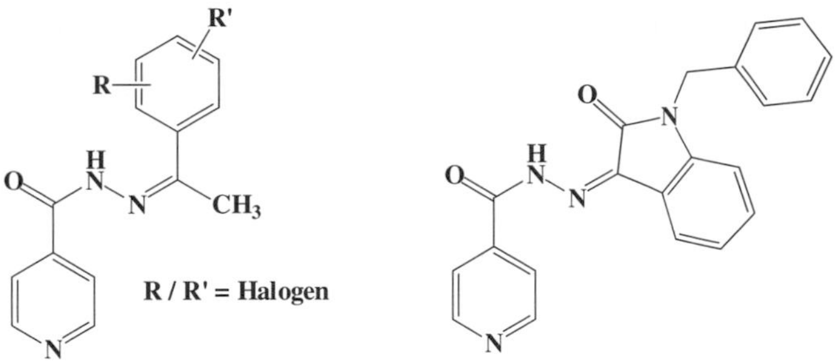

Figure 5.4.4 *Halogenated isoniazid analogues (left) and an example of an isatin-isoniazid derivative (left). The structure of the isoniazid scaffold is shown in red*

5.4.8.6 Foods

As mentioned previously, isoniazid is a known monoamine oxidase inhibitor and so results in the reduced clearance/metabolism of foods high in monoamines, such as tyramine (Hauser and Baier, 1982). High plasma levels of tyramine (found in strong cheeses, etc.) may cause flushing, palpitations, and headache, with sometimes an increase in systolic blood pressure, so these foods should be avoided in patients taking isoniazid.

As isoniazid is also a histaminase inhibitor, foods high in histamine (e.g. tuna, shellfish, cured pork, parmesan cheese) should also be avoided, due to the reduced metabolism of histamine (Baciewicz and Self, 1985).

5.4.9 Recent developments

Surprisingly, given the advent of MDR-TB (when TB is resistant to isoniazid and rifampicin), research into developing isoniazid derivatives is not a hot area at present. Many pharmaceutical companies are focussing on developing novel agents for the treatment of TB rather than expanding an existing class – especially one that has been around since the 1950s. There have been several attempts to prepare isoniazid derivatives, with one group preparing halogenated analogues (Vigorita *et al.*, 1994) and, more recently, another group preparing a set of isatin (indolin-2,3-dione) derivatives (Figure 5.4.4) (Aboul-Fadl *et al.*, 2003). When tested *in vitro* against *M. tuberculosis*, the halogenated analogues had inferior activity to isoniazid, but several isatin derivatives showed activity against an isoniazid-resistant strain of *M. tuberculosis*. Further work has shown that the isatin derivatives only have minor effects on mouse hepatic enzymatic activity and are thus likely to be free of major CYP drug–drug interactions (Franklin *et al.*, 2010), although it remains to be seen what role these compounds might have in the actual management of TB.

References

C. Akolo, I. Adetifa, S. Shepperd, and J. Volmink, *Cochrane Database of Systematic Reviews*, 2010, Issue 1.

A. Argyrou, M. W. Vetting, and J. S. Blanchard, *J. Amer. Chem. Soc.*, 2007, **129**, 9582–9583.

A. M. Baciewicz and T. H. Self, *South Med. J.*, 1985, **78**, 714–718.

S. H. Block, *Pediatrics*, 1982, **69**, 494–495.

S. S. Bottomley and U. Muller-Eberhard, *Semin. Hematol.*, 1988, **25**, 282–302.

M. J. Brodie, A. R. Boobis, C. J. Hillyard, G. Abeyasekera, I. MacIntyre, and B. K. Park, *Clin. Pharmacol. Ther.*, 1981, **30**, 363–367.

N. J. Butcher, S. Boukouvala, E. Sim, and R. F. Minchin, *Pharmacogenom. J.*, 2002, **2**, 30–42.

C. E. Cade, A. C. Diouhy, K. F. Medzihradszky, S. P. Salas-Castillo, and R. A. Ghiladi, *Protein Sci.*, 2010, **19**, 458–474.

Centers for Disease Control and Prevention, *Morb., Mort. Weekly Report*, 2006, **55**, 301.

J. Y. Chien, R. M. Peter, C. M. Nolan, C. Wartell, J. T. Slattery, S. D. Nelson, R. L. Carithers, Jr, and K. E. Thummel, *Clin. Pharmacol. Ther.*, 1997, **61**, 24–34.

J. Crippin, *Am. J. Gastroenterol.*, 1993, **88**, 590–592.

K. M. De Cock, A. Grantand, and J. D. Porter, *Lancet*, 1995, **345**, 833–836.

S. Devadetta, P. R. J. Gangadharam, R. H. Andrews, W. Fox. C. V. Ramakrishnan, J. B. Selkon, and S. Velu, *Bull. World Health Organ.*, 1960, **23**, 587–598.

K. Eiglmeier, H. Fsihi, B. Heym, and S. T. Cole, *FEMS Microbiol. Lett.*, 1997, **149**, 273–278.

G. A. Ellard and P. T. Gammon, *J. Pharmacokinet. Biopharm.*, 1976, **4**, 83–113.

D. A. P. Evans, *Pharmacol. Ther.*, 1989, **42**, 157–234.

D. A. P. Evans, K. A. Manley, and V. A. McKusick, *Brit. Med. J.*, 1960, **2**, 485–491.

M. E. Fleenor, J. W. Harden, and G. Curtis, *Chest*, 1991, **99**, 1554.

K. Gallicano, J. Sahai, G. Zaror-Behrens, and A. Pakuts, *Antimicrob. Agents Chemother.*, 1994, **38**, 894–897.

R. A. Garibaldi, R. E. Drusin, S. H. Ferebee, and M. B. Gregg, *Am. Rev. Respir. Dis.*, 1972, **106**, 357–365.

N. Gonzalez, M. J. Torres, J. Aznar, and J. C. Palomares, *Tubercule and Lung Disease*, 1999, **79**, 187–190.

D. Greenwood, *Antimicrobial Drugs. Chronicle of a Twentieth Century Medical Triumph*, Oxford University Press, USA, 2008.

M. J. Hauser and H. Baier, *Drug. Intell. Clin. Pharm.*, 1982, **16**, 617–618.

H. B. Hughes, *Am. Rev. Tuberculosis*, 1954, **70**, 266–273.

A. Hurwitz and D. L. Schlozman, *Am. Rev. Respir. Dis.*, 1974, **109**, 41–47.

C. B. Inderlied, C. A. Kemper, and L. E. Bermudez, *Clin. Microbiol. Rev.*, 1993, **6**, 266–310.

P. Kaur, S. Agarwal, and S. Datta, *PLoS ONE*, 2009, **4**, e5923.

L. Kay, J. P. Kampmann, T. L. Svendsen, B. Vergman, J. E. Hansen, L. Skovsted, and M. Kristensen, *Brit. J. Clin. Pharmacol.*, 1985, **20**, 323–326.

M. Kinzig-Schippers, D. Tomalik-Scharte, A. Jetter, B. Scheidel, V. Jakob, M. Rodamer, I. Cascorbi, O. Doroshyenko, F. Sörgel, and U. Fuhr, *Antimicrob. Agents Chemother.*, 2005, **49**, 1733–1738.

M. H. Larsen, C. Vilchèze, L. Kremer, G. S. Besra, L. Parsons, M. Salfinger, L. Heifets, M. H. Hazbon, D. Alland, J. C. Sacchettini, and W. R. Jacobs, *Mol. Microb.*, 2002, **46**, 453–466.

C. Manca, S. Paul, C. E. Barry, V. H. Freedman, and G. Kaplan, *Infect. Immun.*, 1999, **67**, 74–79.

A. Melander, K. Danielson, A. Hanson, L. Jansson, J.C. Rerup, B. Scherstén, T. Thulin, and E. Wåhlin, *Acta Med. Scand.*, 1976, **200**, 93–97.

H. Meyer and J. Mally, *Monatsh. Chemie*, 1912, **33**, 393–414.

R. R. Miller, J. Porter, and D. J. Greenblatt, *Chest*, 1979, **75**, 356–358.

T. Moulding, A. Redeker, and G. Kamel, *Ann. Intern. Med.*, 1991, **114**, 431.

F. J. Murray, *Am. Rev. Respir. Dis.*, 1962, **86**, 729–732.

M. Nisar, S. W. Watkin, R. C. Bucknall, and R. A. L. Agnew, *Thorax*, 1990, **45**, 419–420.

H. R. Ochs, D. J. Greenblatt, G. M. Roberts, and H. J. Dengler, *Clin. Pharmacol. Ther.*, 1981, **29**, 671–678.

H. R. Ochs, D. J. Greenblatt, and M. Knuchel, *Brit. J. Clin. Pharmacol.*, 1983, **16**, 743–746.

M Payton, R. Auty, R. Delgoda, M. Everett, and E. Sim, *J. Bacteriol.*, 1999, **181**, 1343–1347.

J. H. Peters, *Am. Rev. Resp. Dis.*, 1960, **81**, 485–497.

D. A. Rozwarski, G. A. Grant, D. H. R. Barton, W. R. Jacobs, and J. C. Sacchettini, *Science*, 1998, **279**, 98–102.

R. L. Rubin, *Toxicol.*, 2005, **209**, 135–147.

D. Rudman and P. J. Williams, *N. Engl. J. Med.*, 1983, **309**, 488–489.

T. C. Sarich, M. Youssefi, T. Zhou, S. P. Adams, R. A. Wall, and J. M. Wright, *Arch. Toxicol.*, 1996, **70**, 835–840.

L. Scharer and J. P. Smith, *Ann. Int. Med.*, 1969, **71**, 1113–1120.

H. Schaumberg, J. Kaplan, and A. Windebank, *N. Engl. J. Med.*, 1983, **309**, 445–448.

T. H. Self, C. R. Chrisman, A. M. Baciewicz, and M. S. Bronze, *Am. J. Med. Sci.*, 1999, **317**, 304–311.

R. A. Sharp, J. G. Lowe, and R. N. Johnston, *Brit. J. Clin. Pract.*, 1990, **44**, 706–707.

M. Smieja, C. Marchetti, D. Cook, and F. M. Smaill, *Cochrane Database of Systematic Reviews*, 1999, Issue 1.

S. P. Spielberg, *J. Pharmacokinet. Biopharmaceut.*, 1996, **24**, 509–519.

T. P. Sycheva, T. N. Pavlova, and M. N. Shchulina, *Khimiko-Farmatsevticheskii Zhurnal (Pharm. Chem. J.)* 1971, **6**, 696–698.

A. Telenti, N. Honore, C. Bernasconi, J. March, A. Ortega, B. Heym, H. E. Takiff, and S. T. Cole, *J. Clin. Microbiol.*, 1997, **35**, 719–723.

G. S. Timmins, S. Master, F. Rusnak, and V. Deretic, *Antimicrob. Agents Chemother.*, 2004, **48**, 3006–3009.

C. Vilchèze, F. Wang, M. Arai, M. H. Hazbón, R. Colangeli, L. Kremer, T. R. Weisbrod, D. Alland, J. C. Sacchettini, and W. R. Jacobs, Jr, *Nat. Med.*, 2006, **12**, 1027–1029.

L. Wang, *Chem. & Eng. News*, 2005, **83**, 8325 (http://pubs.acs.org/cen/coverstory/83/8325/8325isoniazid.html, last accessed 11 March 2012).

World Health Organisation, Tuberculosis Factsheet 2011 (www.who.int/tb/publications/2011/factsheet_tb_2011.pdf).

A. S. Yard and H. McKennis, Jr, *J. Pharmacol. Exp. Ther.*, 1955, **114**, 391–397.

S. Yoshimoto, M. Takeuchi, and A. Tada, *Rinsho Ketsueki*, 1992, **33**, 986–990.

Y. Zhang, S. Dhandayuthapani, and V. Deretic, *Proc. Natl. Acad. Sci. USA.*, 1996, **93**, 13 212–13 216.

X. Zhou, H. Yu, S. Yu, F. Wang, J. C. Sacchettini, and R. S. Magliozzo, *Biochem*, 2006, **45**, 4131–4140.

Z. Zwolska, E. Augustynowicz-Kopeć, and H. Niemirowska-Mikulska, *Acta Pol. Pharm.*, 2002, **59**, 448–452.

5.5

Daptomycin

Key Points

- Daptomycin is approved for the treatment of MRSA-linked complicated skin and skin-structure infections (cSSSIs) and bacteraemia, including right-sided endocarditis.
- The mode of action is not well understood, but involves disruption of the cell wall of Gram positive bacteria after insertion of daptomycin, facilitated by its lipophilic 'tail'.
- Daptomycin was only recently introduced and resistance is not yet common; to avoid the development of resistance, it should be used rarely and only for the indications above.

Cubicin (daptomycin for injection) was approved by the FDA in 2003 for the treatment of complicated SSSIs caused by MRSA, and in 2006 the FDA approved a new indication: the treatment of *S. aureus* bacteraemia due to MRSA, including right-sided endocarditis.

5.5.1 Discovery

As mentioned in Section 1, daptomycin is one of only two truly novel antibacterial agents launched in the last 30 years. Daptomycin is a lipopeptide and was originally isolated by researchers at Eli Lilly and Co. from a strain of *Streptomyces roseosporus* in a soil sample from Mount Ararat[21] in Turkey. As you can see from the structure in Figure 5.5.1, daptomycin contains 13 amino acids (not all of which are proteinogenic), 10 of which form a **depsipeptide**[22] ring, with the other three attached to this ring through a threonine residue. This depsipeptide forms the core of the family of lipopeptides (called A21978C) isolated from this organism. The 13-amino-acid core is common to all members of the A21978C complex and it is only the lipid portion which differs, with the fatty acid side chain being ***n*-decanoyl** (C_{10}) in daptomycin (which was originally called

---

[21] Are you wondering where you've heard the name Mount Ararat before? According to the Book of Genesis in the Bible, it is the final resting place of Noah's Ark.

[22] A depsipeptide is a peptide in which at least one amide (CONH) bond has been replaced by an **ester link**.

Antibacterial Agents: Chemistry, Mode of Action, Mechanisms of Resistance and Clinical Applications, First Edition.
Rosaleen J. Anderson, Paul W. Groundwater, Adam Todd and Alan J. Worsley.
© 2012 John Wiley & Sons, Ltd. Published 2012 by John Wiley & Sons, Ltd.

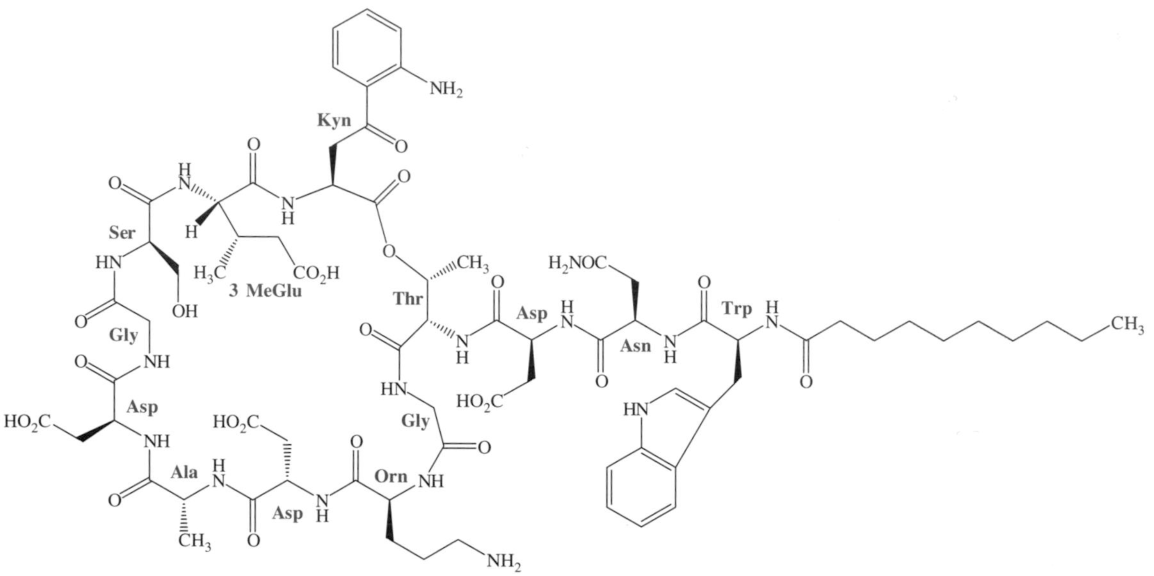

Figure 5.5.1   *Daptomycin*

LY 146032), a minor component of the mixture. Once the antibacterial activity of these lipopeptides had been determined, daptomycin was the agent chosen for clinical development as this fatty acid side chain was associated with the highest *in vivo* efficacy and lowest toxicity in animals (Eisenstein *et al.*, 2010). As you might imagine, all the possible fatty acid derivatives of the core peptide (as well as some containing other functional groups) were prepared and evaluated for their antibacterial activity, with only a few being selected for further evaluation. Amazingly, even a one-carbon difference in the fatty acid chain was found to lead to a dramatic increase in toxicity, as when mice were dosed with 1 g/kg (IV) of the *n*-undecanoyl (C_{11}) derivative, 100% were killed, compared to 20% for the *n*-decanoyl (C_{10}) analogue (daptomycin).

For more on the discovery of daptomycin, see Tally and DeBruin (2000), Baltz *et al.* (2005), and Eisenstein *et al.* (2010).

5.5.2   Synthesis

As is the case for the other very complex antibiotics discussed throughout this book, the total synthesis of daptomycin is not a viable option for the manufacture of sufficient quantities of this drug for clinical use. Daptomycin can be produced by one of two methods, both of which have some similarity to the synthesis of the penicillins (which we covered in Section 5.1). Eli Lilly, the company that first isolated daptomycin, developed a method which converted the A21978C complex to the core peptide (using a deacylase enzyme isolated from *Actinoplanes utahensis* NRRL 12052), which could then be re-acylated with different lipid side chains (Scheme 5.5.1) (DeBono *et al.*, 1988). As shown in Scheme 5.5.1, this process is not as straightforward as you might expect and involves a couple more steps than you might have anticipated. The de-acylated intermediate is re-acylated with an active ester (in this case, the **2,3,5-trichlorophenyl** ester of **decanoic acid**) to give daptomycin.

As it involves four steps, some of which give yields in the 50–60% range, the de-acylation/re-acylation method was not deemed to be cost-effective for full-scale manufacture of daptomycin, so an alternative

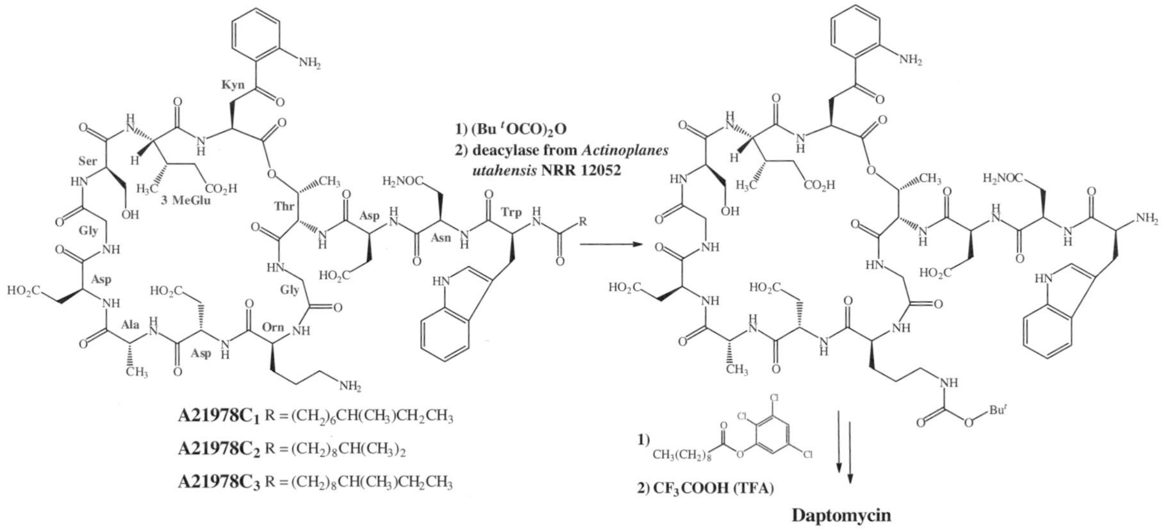

Scheme 5.5.1   *De-acylation of the A21978C complex, followed by re-acylation*

method was developed, in which decanoic acid was fed into the fermentation cultures of *S. roseosporus*, thus producing daptomycin directly (Huber *et al.*, 1988).

5.5.3   Bioavailability

Daptomycin is administered intravenously and has an apparent volume of distribution of 0.1 L/kg, which may be affected by a high degree of plasma protein binding (90–95%). It has an elimination half-life of approximately 9 hours and, as it is eliminated unchanged by renal excretion (approx. 54%), the dose interval should be increased to 48 hours in patients with impaired renal function. In a study conducted in 24 healthy male and female volunteers treated with daptomycin for 7–14 days, it showed linear pharmacokinetics and, at 4 mg/kg, had a mean C_{max} of 54.6 µg/mL and an AUC of 494 mg/hour/mL in patients treated for 7 days (Dvorchik *et al.*, 2003).

As daptomycin interacts with, and inserts into, the lipid cell wall of Gram positive bacteria, it does not have to cross the entire membrane or enter the cell, only requiring access through the outer components of the membrane. Gram positive bacteria have a more readily traversed outer membrane, while the Gram negative bacteria are resistant due to their more complex membrane structure.

For more on the bioavailability of daptomycin, see Jeu and Fung (2004).

5.5.4   Mode of action and selectivity

You are probably thinking that, as daptomycin was discovered relatively recently and approved for use only within the last 10 years, its mode of action will have been the subject of extensive research investigations and so will be fully understood. Surprisingly, especially for a drug which is in clinical use, the mode of action of daptomycin is by no means completely understood, and so we can only mention here what the evidence so far suggests.

What is clear about the daptomycin mode of action is that calcium ions are essential for its rapid bactericidal activity against Gram positive bacteria (Steenbergen *et al.*, 2005), which is diminished in the presence of alternative divalent ions. It was initially suggested that daptomycin interferes with lipoteichoic acid (LTA) synthesis, as it inhibits biosynthesis of LTA in *Enterococcus hirae* before that of other macromolecules (which gives it kinetic specificity) (Canepari *et al.*, 1990; Boaretti *et al.*, 1993). In this organism, daptomycin also exhibits dose-dependent effects, inhibiting only LTA synthesis at doses near the MIC, but multiple pathways at higher doses. As we saw in Section 1 (Figure 1.1.7), LTA is a cell-surface macromolecule which is anchored to the cytoplasmic membrane and extends into the peptidoglycan layer of the cell wall. The dose and kinetic specificity exhibited by daptomycin in *E. hirae* were not mirrored in *Staphylococcus aureus*, suggesting that a different mode of action is operating, and this is consistent with the fact that LTAs show marked structural diversity among Gram positive bacteria, meaning that it is unlikely that they would function as a single target for daptomycin.

The mode of action of daptomycin in *S. aureus* or *Enterococcus faecalis* is currently believed to involve the insertion of daptomycin into the lipid bilayer (Figure 5.5.2) (Laganas *et al.*, 2003; Silverman *et al.*, 2003). This insertion is facilitated by the lipid tail of the antibiotic, which initially promotes weak hydrophobic interactions with the lipids of the phospholipid bilayer. What happens before this insertion into the lipid bilayer, and how daptomycin inserting into the membrane leads to cell death, is not yet entirely clear. One study has shown that daptomycin experiences two structural transitions on contact with a cell membrane (Jung *et al.*, 2004). First, a structural transition occurs with the simple interaction of daptomycin and calcium itself, followed by an interaction with calcium and the phosphatidyl glycerol. These interactions promote mild disturbances in the lipid membrane and cause content leakage. This allows interaction with the cytoplasmic membrane and alters permeability with enhanced daptomycin-membrane adhesion. It is believed that daptomycin oligomerisation[23] (which is promoted by binding to Ca^{2+}) creates a large pore in the membrane, which allows potassium efflux, membrane depolarisation, and eventually cell death (Laganas *et al.*, 2003; Silverman *et al.*, 2003). This mode of action is consistent with the fact that daptomycin has reduced *in vivo* activity against

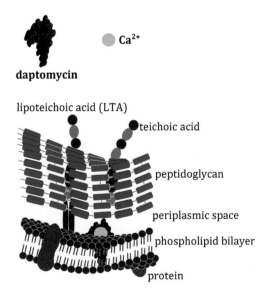

Figure 5.5.2   *Mode of action of daptomycin (**decanoyl chain**) (Robbel and Marahiel, 2010)*

---

[23] Oligomerisation is the process in which a small number of monomer units combine to form a polymer.

Streptococcus pneumoniae, despite having potent bactericidal activity against this organism *in vitro*. In the lungs, daptomycin is thought to interact with pulmonary surfactant (a component of epithelial lining fluid which consists primarily of the phospholipid dipalmitoylphosphatidylcholine) in preference to the phospholipid bilayer of these bacteria (Silverman *et al.*, 2005).

5.5.5   Bacterial resistance

As daptomycin has not been in clinical use for long, it should come as no surprise that resistance is not yet an issue. There are a few examples of resistance, both *in vivo* and *in vitro*, so daptomycin should be used sparingly and only for those diseases for which it has been approved.

In the laboratory, sub-lethal concentrations of daptomycin have been shown to result in single point mutations, and the accumulation of these mutations over time correlates with a steady increase in the daptomycin MIC (Friedman *et al.*, 2006; Mishra *et al.*, 2009; Gould, 2011). These point mutations occurred in three different proteins: MprF (a lysylphosphatidylglycerol synthetase), YycG (a histidine kinase), and RpoB and RpoC (the β and β′ subunits of RNA polymerase). Interestingly, and perhaps significantly, sequence analysis of clinical isolates, in which increases in the daptomycin MIC emerged after treatment, showed point mutations in the *mprF* and *yycG* genes. As daptomycin inserts into the phospholipid bilayer as part of its mode of action, it is not surprising that these isolates contain a mutation in a gene for an enzyme which has an influence on the nature of the phospholipid content (MprF catalyses the addition of lysine to membrane phosphatidylglycerol). The *yycG* gene is believed to be involved in cell permeability and numerous genes are regulated by YycG, the enzyme product which is responsible for the phosphorylation of histidine.

There have been a number of cases of *S. aureus* which are not susceptible to daptomycin, but these are often associated with vancomycin-unresponsive strains (indeed, one patient developed a daptomycin-resistant MRSA infection despite being daptomycin-naïve) (van Hal *et al.*, 2011). As mentioned in Subsection 5.2.5, vancomycin-resistant (VRSA) and vancomycin-intermediate *S. aureus* (VISA) have thickened cell walls, and it is proposed that daptomycin resistance in such isolates is due to the inability of daptomycin to diffuse through these thicker cell walls to its site of action at the lipid membrane. Daptomycin-resistant strains of vancomycin-resistant enterococci (VRE) have also been reported (Kamboj *et al.*, 2011).

Thankfully, clinical daptomycin resistance is still rare and, as there are established daptomycin surveillance programmes around the world, which monitor the *in vitro* activity of this agent (Sader *et al.*, 2005a, 2005b), it is to be hoped that, through careful stewardship of the use of this antibiotic, this will remain the case.

For more on the bacterial resistance of daptomycin, see Boucher and Sakoulas (2007).

5.5.6   Clinical applications

As mentioned above, daptomycin is only active against Gram positive bacteria. Its spectrum of activity does, however, include Gram positive organisms that tend to be resistant to other classes of drugs (e.g. MRSA and VRE) (Bell *et al.*, 2010). In cases where a 'mixed infection' is suspected, another agent should be added (in addition to daptomycin) to ensure sufficient coverage against Gram negative bacteria.

Daptomycin is used in the treatment of both cSSTIs and right-sided infective endocarditis – we have met both of these conditions before when we discussed the use of linezolid (for cSSTIs, Subsection 4.5.6) and the aminoglycosides (for infective endocarditis, Subsection 4.1.6).

When used for infective endocarditis, daptomycin is licensed for the treatment of right-sided endocarditis due to *S. aureus*. You may wonder why it is only licensed for the treatment of right-sided infective endocarditis and not left-sided infective endocarditis. If we look at the summary of product characteristics[24] for Cubicin (daptomycin for injection), this may help to answer the question: Novartis, the marketing authorisation holder of Cubicin, cites a clinical trial showing that daptomycin is effective in the treatment of right-sided infective endocarditis due to *S. aureus* (to be more precise, it is not inferior to standard therapy) (Fowler *et al.*, 2006). In this trial, the number of patients with right-sided infective endocarditis was surprisingly small (19 patients in the daptomycin arm and 16 patients in the comparator arm), but the number of patients with left-sided endocarditis was even smaller (nine patients in each arm). The results for right-sided infective endocarditis show that eight patients (42.1%) were successfully treated in the daptomycin arm, while seven patients (43.8%) were successfully treated in the comparator arm. In the case of left-sided endocarditis, however, only one patient (11.1%) was treated successfully with daptomycin, compared to two patients (22.2%) with standard therapy. Given the relatively small patient numbers in this trial, and the apparent lack of successful treatment outcomes in left-sided infective endocarditis, Novartis states that the efficacy of Cubicin has not been demonstrated in patients with left-sided infective endocarditis due to *S. aureus* and it should therefore only be used in patients with right-sided infective endocarditis due to *S. aureus*.

In contrast to right-sided infective endocarditis, the licensing application of daptomycin for the treatment of cSSTIs was based on a pooled analysis of two phase III clinical trials (DAP-SST-98-01 and DAP-SST-99-01) and therefore had larger patient numbers. Data from these studies show that daptomycin is not inferior to vancomycin when used in combination with a penicillinase-resistant penicillin. Among 902 clinically evaluable patients with cSSTIs, successful treatment was reported in 83.4% of patients who used daptomycin and 84.2% of patients who used vancomycin and a penicillinase-resistant penicillin (Arbeit *et al.*, 2004). It remains to be seen what role daptomycin has in the management of cSSTIs but, like linezolid and vancomycin, it can be used to manage cSSTIs caused by MRSA (Stevens *et al.*, 2005).

As we have already said, daptomycin (along with linezolid) is one of only two truly novel antibacterial agents launched in the last 30 years and clinicians have a responsibility to be vigilant and only use it when specifically indicated in order to minimise the possibility of resistance developing. For the two indications described, therefore, the prescribing of daptomycin should always be based on antibacterial susceptibility and, where appropriate, advice from a specialist clinician.

5.5.7   Adverse drug reactions

A daptomycin patient registry was established by the manufacturer and shows that 18% of patients (out of a total of 1073) have experienced adverse events (DePestel *et al.*, 2010). The most common adverse event observed was an elevation in creatinine phosphokinase (CPK), which occurred in 2.1% of patients. Other adverse events listed include anaemia, eosinophilia, sinus tachycardia, abdominal pain, increased blood pressure, anorexia, and dehydration. Nausea, vomiting, and diarrhoea have been observed in 1.0% of patients and constitute the second largest observed adverse event. Some adverse events led to a discontinuation of treatment, with the most common of these being elevated CPK, rash, rhabdomyolysis, and renal insufficiency. Another study demonstrated elevated CPK levels (10 times normal level) in patients receiving high-dose

---

[24] The summaries of product characteristics (or SPCs) for all UK-licensed medicines are available free of charge from the website www.medicines.org.uk, which is well worth a visit.

daptomycin (8 mg/kg mean dose) (Figueroa *et al.*, 2008), while a further study demonstrated leucopenia (low white blood cells) and amblyopia ('a lazy eye') in Taiwansese patients, which resolved spontaneously (Liang *et al.*, 2009).

Animal models have demonstrated that musculoskeletal toxicity is related to dosing frequency, with an increased frequency causing marked muscle tissue changes, rather than a single dose (Oleson *et al.*, 2000). It has been suggested that longer periods between doses allow for tissue regeneration, while the mechanism of action may contribute to myoblast/myofibril toxicity (Kostrominova *et al.*, 2010). In addition, daptomycin has been shown to act as a surfactant and may accumulate in alveolar epithelium tissue, which may explain the pulmonary adverse effects (Jung *et al.*, 2004; Lal and Assimacopoulos, 2010; Cobb *et al.*, 2007). Two cases of life-threatening eosinophilic pneumonia have been reported, along with 36 possible cases, causing the FDA to issue a warning to clinicians.

5.5.8 Drug interactions

There are several cases of healthy subjects taking simvastatin (40 mg daily) concomitantly with daptomycin (4 mg/kg) who did not experience an increase in adverse effects when compared to placebo (Cubist Pharmaceuticals Inc., 2008), but five such patients experienced raised CPK levels (an indicator of rhabdomyolysis) and one patient who had previously taken simvastatin developed rhabdomyolysis after receiving daptomycin (Echeverria *et al.*, 2005). It is reported to be difficult to distinguish between rhabdomyolysis that may have occurred in patients receiving statins alone and that in patients receiving a combination of daptomycin and statin, and the manufacturers recommend that the use of statins be arrested while patients receive daptomycin

The use of NSAIDs (both Cox I and Cox II inhibitors) may reduce the renal excretion of daptomycin, and in doing so may increase renal toxicity (Novartis, 2009), and renal function should be monitored if both agents are used.

Daptomycin does not appear to affect the INR of patients who received a single 25 mg dose of warfarin (Cubist Pharmaceuticals Inc., 2008). Daptomycin may cause a concentration-dependant false prolongation of prothrombin time (Cubist Pharmaceuticals Inc., 2008); the clinical significance of this is still uncertain, but the manufacturers suggest careful monitoring of INR.

5.5.9 Recent developments

As daptomycin is a large, complex molecule, it is more difficult to derivatise compared to smaller, less-complicated molecules, such as linezolid, so there are very few daptomycin analogues in development. Gu *et al.* examined three daptomycin-related compounds with varying acyl tail lengths purified from the fermentation broth of a recombinant strain of *S. roseosporus*. The analogues exhibited potent activity against *S. aureus* and were equivalent or, in some cases, superior to daptomycin (Gu *et al.*, 2007).

Another recent example of work relating to daptomycin involved the generation of hybrid molecules of daptomycin or the structurally related lipopeptide, A54145 (Nguyen *et al.*, 2010). This work relates to the use of daptomycin in CAP, which is questionable, despite the fact that it has good *in vitro* activity against *S. pneumoniae*. As was mentioned earlier, the poor *in vivo* efficacy may be due to lung surfactant (a mixture of phospholipids and proteins that facilitate the exchange of oxygen) sequestering daptomycin and preventing it from exhibiting its effect. The mean inhibitory concentration for daptomycin against *S. aureus* in the absence of surfactant was 0.5 μg/mL, but in the presence of surfactant, the mean inhibitory concentration increased to 64 μg/mL (i.e. in the presence of surfactant, daptomycin is 128 times less active). In the presence of surfactant, several of the compounds described showed lower mean inhibitory concentrations (i.e. were more active) against *S. aureus* when compared to daptomycin. Unfortunately, however, the same compounds had higher mean inhibitory concentrations (i.e. were less active) in the absence of surfactant when compared to daptomycin (meaning that they are not ideal candidates for further development). Hopefully this research will identify a compound that has superior activity to daptomycin and also retains its activity in the presence of surfactant, meaning that, in future, an antibiotic from this class could have a role in the treatment of CAP.

References

R. D. Arbeit, D. Maki, F. P. Tally, E. Campanaro, and B. I. Eisenstein, *Clin. Infect. Dis.*, 2004, **38**, 1673–1681.

R. H. Baltz, V. Miao, and S. K. Wrigley, *Nat. Prod. Rep.*, 2005, **22**, 717–741.

J. M. Bell, J. D. Turnidge, H. S. Sader, and R. N. Jones. *Pathology*, 2010, **42**, 470–473.

M. Boaretti, P. Canepari, M. M. Lleó, and G. Satta, *J. Antimicrob. Chemother.*, 1993, **31**, 227–235.

H. W. Boucher and G. Sakoulas, *Clin. Infect. Dis.*, 2007, **45**, 601–608.

P. Canepari, M. Boaretti, M. M. Lleó, and G. Satta, *Antimicrob. Agents Chemother.*, 1990, **34**, 1220–1226.

E. Cobb, R. C. Kimborough, K. M. Nugent, and M. P. Phy, *Ann. Pharmacother.*, 2007, **41**, 696–701.

Cubist Pharmaceuticals, Inc., US Prescribing Information, Cubicin (Daptomycin), August 2008.

M. DeBono, B. J. Abbott, R. M. Molloy, D. S. Fukuda, A. H. Hunt, V. M. Daupert, F. T. Counter, J. L. Ott, C. B. Carrell, L. C. Howard, L. V. Boeck, and R. L. Hamill, *J. Antibiot.*, 1988, **41**, 1093–1105.

D. D. DePestel, E. Hershberger, K. C. Lamp, and P. N. Malani, *Am. J. Geriatr. Pharmacol.*, 2010, **8**, 551–561.

B. H. Dvorchik, D. Brazier, M. E. DeBruin, and R. D. Arbeit, *Antimicrob. Agents Chemother.*, 2003, **47**, 1318–1323.

B. I. Eisenstein, F. B. Oleson, and R. H. Baltz, *Clin. Infect. Dis.*, 2010, **50**, S10–15.

E. Echeverria, P. Datta, J. Cadena, and J. S. Lewis, *J. Antimicrob. Chemother.*, 2005, **55**, 599–600.

D. A. Figueroa, E. Mangini, M. Amodio-Groton, B. Vardianos., A. Melchert, C. Fana, W. Wehbeh, C. M. Urban and S. Segal-Maurer, *Clin. Infect. Dis.*, 2008, **49**, 177–180.

V. G. Fowler, Jr, H. W. Boucher, G. R. Corey, E. Abrutyn, A. W. Karchmer, M. E. Rupp, D. P. Levine, H. F. Chambers, F. P. Tally, G. A. Vigliani, C. H. Cabell, A. S. Link, I. DeMeyer, S. G. Filler, M. Zervos, P. Cook, J. Parsonnet, J. M. Bernstein, C. S. Price, G. N. Forrest, G. Fätkenheuer, M. Gareca, S. J. Rehm, H. R. Brodt, A. Tice, and S. E. Cosgrove, S. aureus Endocarditis and Bacteremia Study Group, *N. Engl. J. Med.*, 2006, **355**, 653–665.

L. Friedman, J. D. Adler, and J. A. *Silverman, Antimicrob. Agents Chemother.*, 2006, **50**, 2137–2145.

I. M. Gould, *J. Antimicrob. Chemother*, 2011, **66S4**, iv17–21.

J. Q. Gu, K. T. Nguyen, C. Gandhi, V. Rajgarhia, R. H. Baltz, P. Brian, and M. Chu, *J. Nat. Prod.*, 2007, **70**, 233–240.

F. M. Huber, R. L. Pieper, and A. J. Tietz, *J. Biotechnol.*, 1988, **7**, 283–292.

L. Jeu and H. B. Fung, *Clin. Therap.*, 2004, **26**, 1728–1757.

D. Jung, A. Rozek, M. Okon, and R. E. Hancock, *Chem. Biol.*, 2004, **11**, 949–957.

M. Kamboj, N. Cohen, K. Gilhuley, N. E. Babady, S. K. Seo, and K. A. Sepkowitz, *Infect. Control Hosp. Epidem.*, 2011, **32**, 391–394.

T. Y. Kostrominova, S.Coleman, F. B. Oleson, J. A. Faulkner, and L. M. Larkin, *In Vitro Cell. Dev. Biol.*, 2010, **46**, 613–618.

V. Laganas, J. Alder, and J. A. Silverman, *Antimicrob. Agents Chemother.*, 2003, **47**, 2682–2684.

Y. Lal and A. P. Assimacopoulos, *Clin. Inf. Dis.*, 2010, **50**, 737–740.

S. H. Liang, W. H. Sheng, Y. T. Huang, F. L. Wu, and S. C. Chang, *Chemother.*, 2009, **55**, 91–96.

N. Mishra, S.-J. Yang, A. Sawa, A. Rubio, C. C. Nast, M. R. Yeaman, and A. S. Bayer, *Antimicrob. Agents Chemother.*, 2009, **53**, 2312–2318.

K. T. Nguyen, X. W. He, D. C. Alexander, C. Li, J. Q. Gu, C. Mascio, A. Van Praagh, L. Mortin, M. Chu, J. A. Silverman, P. Brian, and R. H. Baltz, *Antimicrob. Agents Chemother.*, 2010, **54**, 1404–1413.

Novartis, UK Summary of Product Characteristics, Cubicin (Daptomycin), 2009.

F. B. Oleson, C. L. Berman, and J. B. Kirkpatrick, *Antimicrob. Agents Chemother.*, 2000, **44**, 2948–2953.

L. Robbel and M. A. Marahiel, *J. Biol. Chem.*, 2010, **285**, 27 501–27 508.

H. S. Sader, T. R. Fritsche, and R. N. Jones, *Diagn. Microbiol. Infect. Dis.*, 2005a, **53**, 329–332.

H. S. Sader, T. R. Fritsche, J. M. Streit, and R. N. Jones, *J. Chemother.*, 2005b, **17**, 477–483.

J. A. Silverman, N. G. Perlmutter, and H. M. Shapiro, *Antimicrob. Agents Chemother.*, 2003, **47**, 2538–2544.

J. A. Silverman, L. I. Mortin, A. D. G. VanPraagh, T. Li, and J. Alder, *J. Infect. Dis.*, 2005, **191**, 2149–2152.

J. N. Steenbergen, J. Adler, G. M. Thorne, and F. P. Tally, *J. Antimicrob. Chemother.*, 2005, **55**, 283–288.

D. L. Stevens, A. L. Bisno, H. F. Chambers, E. D. Everett, P. Dellinger, E. J. Goldstein, S. L. Gorbach, J. V. Hirschmann, E. L. Kaplan, J. G. Montoya, and J. C. Wade, *Clin. Infect. Dis.*, 2005, **41**, 1373–1406.

F. P. Tally and M. F. DeBruin, *J. Antimicrob. Chemother.*, 2000, **46**, 523–526.

S. J. van Hal, D. L. Paterson, and I. B. Gosbell, *Eur. J. Clin. Microbiol. Infect. Dis.*, 2011, **30**, 603–610.

Questions

(1) Although the stereochemical configuration at the α-carbon atom of serine does not change during the synthesis of DCS (Scheme 5.3.2), the stereochemical (*R*) or (*S*) descriptor does. Assign the stereochemistry at this centre in each of the compounds shown in Scheme 5.3.2 and explain why the descriptor changes.

(2) The stability of the carbanion intermediate provides a strong driving force for the deprotonation of the PLP-L-ala complex in the mechanism for ALR (Scheme 5.3.3). Draw resonance forms for this intermediate.

(3) Draw the product of the D-ala-D-ser ligase which is produced in VanC- and VanE-type resistance in VRE.

(4) Why should isoniazid not be used in the management of leprosy?

(5) Have a look again at the structure of daptomycin (Figure 5.5.1) and list any amino acids which are non-proteinogenic or have an unusual configuration.

(6) In Scheme 5.5.1, what is the function of the Boc (Bu^{t}OCO) group, which is added in the first step by the reaction of A21978C with di-*tert*-butyl dicarbonate (Bu^{t}OCO)$_2$O)?

(7) Suggest reagents for the conversion of thienamycin to imipenem.

Thienamycin   Imipenem

Index

Antibacterial Agents: Chemistry, Mode of Action, Mechanisms of Resistance and Clinical Applications, First Edition.
Rosaleen J. Anderson, Paul W. Groundwater, Adam Todd and Alan J. Worsley.
© 2012 John Wiley & Sons, Ltd. Published 2012 by John Wiley & Sons, Ltd.